Lecture Notes in Mathematics 1800

Editors:
J.-M. Morel, Cachan
F. Takens, Groningen
B. Teissier, Paris

T0202964

Springer
Berlin
Heidelberg
New York
Hong Kong
London
Milan
Paris
Tokyo

Ofer Gabber   Lorenzo Ramero

Almost Ring Theory

Springer

Ofer Gabber
IHES
Le Bois-Marie
35, route de Chartres
91440 Bures-sur-Yvette, France
e-mail: gabber@ihes.fr

Lorenzo Ramero
Institut de Mathématiques
Université de Bordeaux I
351, cours de la Libération
33405 Talence, France
e-mail: ramero@math.u-bordeaux.fr

Cataloging-in-Publication Data applied for
Bibliographic information published by Die Deutsche Bibliothek

Die Deutsche Bibliothek lists this publication in the Deutsche Nationalbibliografie;
detailed bibliographic data is available in the Internet at http://dnb.ddb.de

---

The cover figure entitled "Prescano" is reproduced by kind permission of
Michel Mendès France

Mathematics Subject Classification (2000):
13D10, 13B40, 12J20, 14G22, 18D10, 13D03

ISSN 0075-8434
ISBN 3-540-40594-1 Springer-Verlag Berlin Heidelberg New York

This work is subject to copyright. All rights are reserved, whether the whole or part of the material is concerned, specifically the rights of translation, reprinting, reuse of illustrations, recitation, broadcasting, reproduction on microfilm or in any other way, and storage in data banks. Duplication of this publication or parts thereof is permitted only under the provisions of the German Copyright Law of September 9, 1965, in its current version, and permission for use must always be obtained from Springer-Verlag. Violations are liable for prosecution under the German Copyright Law.

Springer-Verlag Berlin Heidelberg New York a member of BertelsmannSpringer
Science + Business Media GmbH

http://www.springer.de

© Springer-Verlag Berlin Heidelberg 2003
Printed in Germany

The use of general descriptive names, registered names, trademarks, etc. in this publication does not imply, even in the absence of a specific statement, that such names are exempt from the relevant protective laws and regulations and therefore free for general use.

Typesetting: Camera-ready TeX output by the author

SPIN: 10896419   41/3142/du-543210 - Printed on acid-free paper

CONTENTS

1. INTRODUCTION

1.1. Motivations and a little history. Almost mathematics made its official debut in Faltings' fundamental article [33], the first of a series of works on the subject of p-adic Hodge theory, culminating with [34]. Although almost ring theory is developed here as an independent branch of mathematics, stretching somewhere in between commutative algebra and category theory, the original applications to p-adic Hodge theory still provide the main motivation and largely drive the evolution of the subject.

Indeed, one of the chief aims of our monograph is to supply adequate foundations for [34], and to pave the way to further extensions of Faltings' methods (especially, of his deep "almost purity" theorem), that we plan to present in a future work. For these reasons, it is fitting to begin this introduction with some background, leading up to a review of the results of [33]. (Besides, we suspect that all but the most dedicated expert of p-adic Hodge theory will require some inducement before deciding to plunge into close to 300 pages of foundational arcana.)

The starting point of p-adic Hodge theory can be located in Tate's paper [74] on p-divisible groups. An important example of p-divisible group is the p-primary torsion subgroup A_{p^∞} of an abelian scheme A defined over the valuation ring K^+ of a complete discretely valued field K of characteristic zero. We assume that the residue field κ of K is perfect of characteristic $p > 0$; also, let π be a uniformizer of K^+, K^{a} the algebraic closure of K and denote by C the completion of K^{a}; the Galois group $G := \mathrm{Gal}(K^{\mathrm{a}}/K)$ acts linearly on the étale cohomology of A, and actually A_{p^∞} and the Galois module $H^1_{\text{ét}}(A_{K^{\mathrm{a}}}, \mathbb{Z}_p)$ determine each other. G also acts semilinearly on C, whence a natural *continuous semilinear* action of G on the tensor product of Galois modules

$$H^\bullet_{\text{ét}}(A_{K^{\mathrm{a}}}, C) := H^\bullet_{\text{ét}}(A_{K^{\mathrm{a}}}, \mathbb{Z}_p) \otimes_{\mathbb{Z}_p} C.$$

At first sight, it would seem that, in replacing a linear Galois representation by a semilinear one, we are trading a simpler object by a more complicated one. In fact, the opposite holds: as a consequence of his general study of p-divisible groups, Tate showed that for every $i \leq 2\dim(A)$ there exists a natural equivariant isomorphism

$$(1.1.1) \qquad H^i_{\text{ét}}(A_{K^{\mathrm{a}}}, C) \simeq \bigoplus_{j+k=i} H^j(A, \Omega^k_A) \otimes_K C(-k)$$

where, for every integer $j \in \mathbb{Z}$, we define $C(j) := \mathbb{Q}_p(j) \otimes_{\mathbb{Q}_p} C$, and $\mathbb{Q}_p(j)$ is the j-th tensor power of the one-dimensional p-adic representation $\mathbb{Q}_p(1)$ on which G acts as the p-primary cyclotomic character. Tate conjectured that an equivariant decomposition such as (1.1.1) should exist for any smooth projective variety defined over K. To put things in perspective, let us turn to consider the archimedean counterpart of (1.1.1): if X is a smooth, proper complex algebraic variety, one can combine deRham's theorem with Grothendieck's theorem on algebraic deRham cohomology [42], to deduce a natural isomorphism

$$(1.1.2) \qquad\qquad H^{\bullet}(X^{\mathrm{an}}, \mathbb{Z}) \otimes_{\mathbb{Z}} \mathbb{C} \simeq H^{\bullet}_{\mathrm{dR}}(X)$$

between the singular and deRham cohomologies. The two sides of (1.1.2) contribute complementary information on X; namely, singular cohomology supplies an integral structure for $H^{\bullet}(X^{\mathrm{an}}, \mathbb{R})$ (the lattice of periods) and deRham cohomology gives the Hodge filtration: neither of these two structures is reducible to the other.

The above conjecture of Tate is rather startling because it implies that in the non-archimedean case, étale cohomology and deRham cohomology should not be complementary: rather, étale cohomology, viewed as a Galois module, would already detect, if not quite the Hodge filtration, at least its associated graded subquotients, each of them clearly recognizable by the different weight (or Tate twist) with which it appears in $H^{\bullet}_{\text{ét}}(A_{K^{\mathrm{a}}}, C)$ (now this graded Galois module is known as the Hodge-Tate cohomology and often denoted $H_{\mathrm{HT}}(-)$).

On the other hand, working around the same time as Tate, Grothendieck realized that the deRham cohomology of an abelian scheme carries more structure than it would appear at first sight: using his crystalline Dieudonné theory he showed that $H^1_{\mathrm{dR}}(A)$ comes with a canonical K_0-structure (where K_0 is the field of fractions of the ring $W(\kappa)$ of Witt vectors of κ), namely the K_0-vector space $M \otimes_{W(\kappa)} K_0$ where M is the Dieudonné module of the special fibre of A_{p^∞} (see [43]). Furthermore, this K_0-vector space is endowed with an automorphism ϕ which is semilinear, *i.e.* compatible with the Frobenius automorphism of K_0. Grothendieck even proved that A_{p^∞} is determined up to isogeny by $H^1_{\mathrm{dR}}(A)$ together with its Hodge filtration, K_0-structure and automorphism ϕ. Taking into account the above theorem of Tate, he was then led to ask the question of describing an algebraic procedure that would allow to pass directly from $H^1_{\mathrm{dR}}(A)$ to $H^1_{\text{ét}}(A_{K^{\mathrm{a}}}, \mathbb{Q}_p)$ without the intermediary of the p-divisible group A_{p^∞}; he also expected that such a procedure should exist for the cohomology in arbitrary degree (he baptized[1] this as the problem of the "mysterious functor").

The question in degree one was finally solved by Fontaine, several years later ([37]); he actually constructed a functor in the opposite direction, *i.e.* from the category of p-adic Galois representations to the category of filtered K-modules with additional structure as above. The construction of Fontaine hinges on a remarkable ring (actually a whole hierarchy of increasingly complex rings), endowed with both a Galois action and a filtration (and eventually, additional subtler structures). The simplest of such rings is the graded ring $B_{\mathrm{HT}} := \oplus_{i \in \mathbb{Z}} C(i)$, with its obvious multiplication; with its help, Tate's decomposition can be rewritten as an isomorphism of graded K-vector spaces:

---

[1]The name entered the folklore, even though Grothendieck apparently only ever used it orally, and we could find no trace of it in his writings

$$(H^i_{\text{ét}}(A_{K^a}, \mathbb{Q}_p) \otimes_{\mathbb{Q}_p} B_{\text{HT}})^G \simeq \bigoplus_{j+k=i} H^j(A, \Omega^k_A).$$

Fontaine proved that his functor solves Grothendieck's problem for the H^1, and proposed a precise conjecture (the C_{cris} conjecture) in arbitrary degree, for schemes X that are proper and smooth over K and have good reduction over K^+.

1.2. **The method of almost étale extensions.** The C_{cris} conjecture is now completely proved, as well as some later extensions (*e.g.* to schemes with not necessarily good reduction, or not necessarily proper). There are actually at least two very different methods that both have led to a proof: one – due to Fontaine, Kato, Messing and crowned by Tsuji's work [75] – relies on so-called syntomic cohomology and a delicate study of vanishing cycles; the other, due to Faltings, is based on his theory of almost étale extensions (for the case of varieties of good reduction, Niziol has found yet another method, that uses a comparison theorem from étale cohomology to K-theory as a go-between: see [63]).

We won't say anything about the first and the third approaches, but we wish to give a rough overview of the method of almost étale extensions, which was first presented in [33], where Faltings used it to prove the sought comparison with Hodge-Tate cohomology; in subsequent papers the method has been refined and amplified, and its latest incarnation is contained in [34]. However, many important ideas are already found in [33], so it is on the latter that we will focus in this introduction.

For simplicity we will assume that our varieties have good reduction over K^+, hence we let X be a smooth, connected and projective K^+-scheme. The idea is to construct an intermediate cohomology $\mathcal{H}(X)$, with values in C-vector spaces, receiving maps from both étale and Hodge-Tate cohomology, and prove that the resulting natural transformations

$$H_{\text{ét}}(X_{K^a}, \mathbb{Z}_p) \otimes_{\mathbb{Z}_p} C \to \mathcal{H}(X) \quad \text{and} \quad H_{\text{HT}}(X_K) \to \mathcal{H}(X)$$

are isomorphisms of functors. In order to motivate the definition of $\mathcal{H}(X)$, it is instructive to consider first the case of a point, *i.e.* $X = \operatorname{Spec} K^+$. In this case étale cohomology reduces to Galois cohomology, and the calculation of the latter was the main technical result in [74].

Tate's calculation can be explained as follows. The valuation v of K extends uniquely to any algebraic extension, and we want to normalize the value group in such a way that $v(p) = 1$ in every such extension. Let E be a finite Galois extension of K, with Galois group G_E. Typically, one is given a discrete $E^+[G_E]$-module M (such that the Γ-action on M is *semilinear*, that is, compatible with the G_E-action on E^+), and is interested in studying the (modified) Tate cohomology $\widehat{H}^i := \widehat{H}^i(G_E, M)$ (for $i \in \mathbb{Z}$). (Recall that $\widehat{H}^i$ agrees with Galois cohomology $R^i\Gamma^{G_E} M$ for $i > 0$, with Galois homology for $i < -1$, and for $i = 0$ it equals $M^{G_E}/\operatorname{Tr}_{E/K}(M)$, the G_E-invariants divided by the image of the trace map).

In such a situation, the scalar multiplication map $E^+ \otimes_{\mathbb{Z}} M \to M$ induces natural cup product pairings

$$\widehat{H}^i(G_E, E^+) \otimes_{\mathbb{Z}} \widehat{H}^j \to \widehat{H}^{i+j}.$$

Especially, the action of $(E^+)^{G_E} = K^+$ on $\widehat{H}^i$ factors through $K^+/\mathrm{Tr}_{E/K}(E^+)$; in other words, the image of E^+ under the trace map annihilates the modified Tate cohomology.

If now the extension E is *tamely ramified* over K, then $\mathrm{Tr}_{E/K}(E^+) = K^+$, so the $\widehat{H}^i$ vanish for all $i \in \mathbb{Z}$. Even sharper results can be achieved when the extension is *unramified*. Indeed, in such case E^+ is a G_E-torsor for the étale topology of K^+, hence, some basic descent theory tells us that the natural map

$$E^+ \otimes_{K^+} R\Gamma^{G_E} M \to M[0]$$

is an isomorphism in the derived category of the category of $E^+[G_E]$-modules (where we have denoted by $M[0]$ the complex consisting of M placed in degree zero).

In Tate's paper [74] there occurs a variant of the above situation : instead of the finite extension E one considers the algebraic closure K^a of K, so that $G_{K^a} = G$ is the absolute Galois group of K, and the discrete G-module M is replaced by the *topological* module $C(\chi)$, obtained by "twisting" the natural G-action on C via a continuous character $\chi : G \to K^\times$. Then the relevant $H^\bullet$ is the *continuous* Galois cohomology $H^\bullet_{\mathrm{cont}}(G, C(\chi))$, which is defined in general as the homology of a complex of continuous cochains. Under the present assumptions, H^i can be computed by the formula:

$$H^i_{\mathrm{cont}}(G, C(\chi)) := (\varprojlim_n H^i(G, K^{a+}(\chi) \otimes_{\mathbb{Z}} \mathbb{Z}/p^n\mathbb{Z})) \otimes_{\mathbb{Z}} \mathbb{Q}.$$

Let now K_∞ be a totally ramified Galois extension with Galois group H isomorphic to $\mathbb{Z}_p$. Tate realized that, for cohomological purposes, the extension K_∞ plays the role of a maximal totally ramified Galois extension of K. More precisely, let L be any finite extension of K, and set $L_n := L \cdot K_n$, where K_n is the subfield of K_∞ fixed by $H^{p^n} \simeq p^n \cdot \mathbb{Z}_p$. The extension $K_n \subset L_n$ is unramified if and only if the different ideal $\mathscr{D}_n := \mathscr{D}_{L_n^+/K_n^+}$ equals L_n^+. In case this fails, the valuation $v(\delta_n)$ of a generator δ_n of $\mathscr{D}_n$ will be a strictly positive rational number, giving a quantitative measure for the ramification. With this notation, [74, §3.2, Prop.9] reads

$$(1.2.1) \qquad \lim_{n\to\infty} v(\delta_n) = 0$$

(indeed, $v(\delta_n)$ approaches zero about as fast as p^{-n}). In this sense, one can say that the extension $K_\infty \subset L_\infty := L \cdot K_\infty$ is *almost unramified*. One immediate consequence is that the maximal ideal $\mathfrak{m}$ of K_∞^+ is contained in $\mathrm{Tr}_{L_\infty/K_\infty}(L_\infty^+)$. If, additionally, L is a Galois extension of K, we can consider the subgroup

$$G_\infty := \mathrm{Gal}(L_\infty/K_\infty) \subset \mathrm{Gal}(L/K)$$

and the foregoing implies that $\mathfrak{m}$ annihilates $H^i(G_\infty, M)$, for every $i > 0$, and every $L_\infty^+[G_\infty]$-module M. More precisely, the homology of the cone of the natural morphism

$$(1.2.2) \qquad L_\infty^+ \otimes_{K_\infty^+} R\Gamma^{G_\infty} M \to M[0]$$

is annihilated by $\mathfrak{m}$ in all degrees, *i.e.* it is *almost zero*. Equivalently, one says that the maps on homology induced by (1.2.2) are *almost isomorphisms* in all degrees.

A first generalization of (1.2.1) can be found in the work [38] by Fresnel and Matignon; one interesting aspect of this work is that it does away with any consideration of local class field theory (which was used to get the main estimates in [74]); instead, Fresnel and Matignon write a general extension L as a tower of monogenic subextensions, whose structure is sufficiently well understood to allow a direct and very explicit analysis. The main tool in [38] is a notion of different ideal $\mathcal{D}_{E^+/K^+}$ for a possibly infinite algebraic field extension $K \subset E$; then the extension K_∞ considered in [74] is replaced by any extension E of K such that $\mathcal{D}_{E^+/K^+} = (0)$, and (1.2.1) is generalized by the claim that $\mathcal{D}_{F^+/E^+} = F^+$, for every finite extension $E \subset F$.

In some sense, the arguments of [38] anticipate those used by Faltings in the first few paragraphs of [33]. There we find, first of all, a further extension of (1.2.1): the residue field of K is now not necessarily perfect, instead one assumes only that it admits a finite p-basis; then the relevant K_∞ is an extension whose residue field is perfect, and whose value group is p-divisible. The assertion (1.2.1) under such assumptions represents the one-dimensional case of the almost purity theorem. In order to state and prove the higher dimensional case, Faltings invents the method of "almost étale extensions", and indeed sketches in a few pages a whole program of "almost commutative algebra", with the aim of transposing to the almost context as much as possible of the classical theory. So, for instance, if A is a given K_∞^+-algebra, and M is an A-module, one says that M is *almost flat* if, for every A-module N, the natural map of complexes

$$M \overset{\mathbf{L}}{\otimes}_A N \to M \otimes_A N[0]$$

induces almost isomorphisms on homology in all degrees. Similarly, M is *almost projective* if the same holds for the map of complexes

$$\mathrm{Hom}_A(M, N)[0] \to R\mathrm{Hom}_A(M, N).$$

Then, according to [33], a map $A \to B$ of K_∞^+-algebras is called *almost étale* if B is almost projective as an A-module and as a $B \otimes_A B$-module (moreover, B is required to be *almost finitely generated* : the discussion of finiteness conditions in almost ring theory is a rather subtle business, and we dedicate the better part of section 2.3 to its clarification).

With this new language, the almost purity theorem should be better described as an almost version of Abhyankar's lemma, valid for morphisms $A \to B$ of K^+-algebras that are étale in characteristic zero and possibly wildly ramified on the locus of positive characteristic. The actual statement goes as follows. Suppose that A admits global étale coordinates, that is, there exists an étale map $K^+[T_1^{\pm 1}, ..., T_d^{\pm 1}] \to A$ (following Faltings, one calls *small* such an algebra); whereas in the tamely ramified case a finite ramified base change $K^+ \to K^+[\pi^{1/n}]$ (with $(p, n) = 1$) suffices to kill all ramification, the infinite extension

$$A \to A_\infty := A[T_1^{\pm 1/p^\infty}, ..., T_d^{\pm 1/p^\infty}] \otimes_{K^+} K_\infty^+$$

is required in the wildly ramified case, to kill *almost all* ramification, which means that the normalization B_∞ of $A_\infty \otimes_A B$ is almost étale over A_∞.

Faltings has proposed two distinct strategies for the proof of his theorem : the first one, presented in [33], consists in adapting Grothendieck's proof of Zariski-Nagata's purity[2]; a more recent one ([34]) uses the action of Frobenius on some local cohomology modules, and is actually valid under more general assumptions (one does not require the existence of étale coordinates, but only a weaker semi-stable reduction hypothesis on the special fibre).

As a corollary, one deduces cohomological vanishings generalizing the foregoing : indeed, suppose that the extension of fraction fields $\mathrm{Frac}(A) \subset \mathrm{Frac}(B)$ is Galois with group G_B; then, granting almost purity, B_∞ is an "almost G_B-torsor" over A_∞, therefore, for any $B_\infty[G_B]$-module M, the natural map of complexes

$$B_\infty \otimes_{A_\infty} R\Gamma^{G_B} M \to M[0]$$

induces almost isomorphisms on homology.

We are now ready to return to the construction of $\mathscr{H}(X)$. Let A be a K^+-algebra that is small in the above sense; let d be the relative dimension of A over K^+. We denote by $\overline{A}$ the integral closure of A in a maximal algebraic extension of the field of fractions of A which is unramified over $A[p^{-1}]$; also let $\overline{A}^\wedge$ be the π-adic completion of $\overline{A}$: this is an algebra over the π-adic completion C^+ of $K^{\mathrm{a}+}$. The fundamental group

$$\Delta(A) := \pi_1(\mathrm{Spec}\, A \otimes_{K^+} K^{\mathrm{a}})$$

is the subgroup of the Galois group of $\mathrm{Frac}(\overline{A})$ over $\mathrm{Frac}(A)$ that fixes $A_{K^{\mathrm{a}+}} := A \otimes_{K^+} K^{\mathrm{a}+}$. The (continuous) Galois cohomology

$$\mathscr{H}^\bullet(A) := H^\bullet_{\mathrm{cont}}(\Delta(A), \overline{A}^\wedge)$$

can be computed by the functorial simplicial complex $\mathscr{C}^\bullet(\Delta(A), \overline{A}^\wedge)$ such that

$$\mathscr{C}^n(\Delta(A), \overline{A}^\wedge) := \text{continuous maps } \Delta(A)^n \to \overline{A}^\wedge.$$

Let $\Delta(A)_\infty$ be the kernel of the natural surjective homomorphism:

$$\Delta(A) \to \mathrm{Gal}(A_\infty/A_{K^{\mathrm{a}+}}) \simeq \mathbb{Z}_p^{\oplus d}.$$

As a consequence of the foregoing almost vanishings, the Hochschild-Serre spectral sequence

$$E_2^{pq} : \quad H^p_{\mathrm{cont}}(\mathbb{Z}_p^{\oplus d}, H^q_{\mathrm{cont}}(\Delta(A)_\infty, \overline{A}^\wedge)) \Rightarrow \mathscr{H}^{p+q}(A)$$

degenerates up to m-torsion, and one obtains a natural $\mathrm{Gal}(K^{\mathrm{a}}/K)$-equivariant isomorphism

(1.2.3) $$\mathscr{H}^\bullet(A) \simeq A^\wedge_{K^{\mathrm{a}+}} \otimes_{\mathbb{Z}_p} \Lambda^\bullet_{\mathbb{Z}_p}((\mathbb{Z}_p(-1))^d) \oplus (\text{rest})$$

where $A^\wedge_{K^{\mathrm{a}+}}$ is the π-adic completion of $A_{K^{\mathrm{a}+}}$ and $\Lambda^\bullet_{\mathbb{Z}_p}$ denotes the exterior algebra of the free $\mathbb{Z}_p$-module that is the sum of d copies of $\mathbb{Z}_p(-1)$. The (rest) is a module annihilated by $p^{1/(p-1)}$.

Moreover, for any étale map $A \to B$ of small K^+-algebras, the induced map:

(1.2.4) $$\mathscr{H}^\bullet(A) \otimes_{A_{K^{\mathrm{a}+}}} B_{K^{\mathrm{a}+}} \to \mathscr{H}^\bullet(B)$$

---

[2]At the time of writing, there are still some obscure points in this proof

is an almost isomorphism.

Let X be a smooth projective K^+-scheme; we take an arbitrary (Zariski) hyper-covering $U_\bullet \to X$ consisting of *small* affine open subschemes, meaning that each U_i is the spectrum of a small K^+-algebra. By applying termwise the functor $\mathscr{C}^\bullet$ we deduce a bicosimplicial K^+-module $\mathscr{C}^\bullet(\Delta(U_\bullet), U_\bullet)$, whose diagonal is a cosimplicial complex that we denote by $\mathscr{D}^\bullet(X)$. The intermediate cohomology $\mathscr{H}^\bullet(X)$ is defined as the homology of $\mathscr{D}^\bullet(X) \otimes_{K^+} K$.

Using the fact that (1.2.4) is an almost isomorphism, a standard argument shows that $\mathscr{D}^\bullet(X)$ is independent – up to natural almost isomorphisms – on the choice of hypercovering. Functoriality on X is also clear, and one can even define cup products, Kunneth isomorphisms for products of varieties, as well as versions with torsion coefficients.

Finally, since étale cohomology is a globalization of Galois cohomology, it is not difficult to construct a natural transformation

$$(1.2.5) \qquad R\Gamma_{\text{ét}}(X_{K^a}, \mathbb{Z}_p) \otimes_{\mathbb{Z}_p} C^+ \to \mathscr{D}^\bullet(X)$$

(for this, one needs to know that X admits a basis of open subsets consisting of étale $K(\pi, 1)$-spaces, and this is also shown in [33]). The proof that (1.2.5) induces an almost isomorphism on cohomology is laborious, but not exceedingly difficult.

The relationship with Hodge-Tate cohomology is very direct, and can already be scented from (1.2.3); Faltings also spends a little extra effort to wring out some integral refinements (*i.e.*, to control the powers of p appearing in the denominators of the isomorphism map).

This is the basic outline of Faltings' proof; the method can even be extended to treat cohomology with not necessarily constant coefficients (see [34]), thereby providing the most comprehensive approach to p-adic Hodge theory found so far.

1.3. Contents of this book. Since each chapter is preceded by its own detailed introductory remarks, we will bound ourselves to a general overview of the organization of the monograph. The purpose of chapters 2 through 5 is to fully work out the foundations of "almost commutative algebra" outlined by Faltings; in the process we generalize and simplify considerably the theory, and also extend it in directions that were not explored in [33], [34].

It turns out that most of almost ring theory can be built up satisfactorily from a very slim and general set of assumptions: our basic setup, introduced in section 2.1, consists of a ring V and an ideal $\mathfrak{m} \subset V$ such that $\mathfrak{m} = \mathfrak{m}^2$; starting from (2.5.14) we also assume that $\mathfrak{m} \otimes_V \mathfrak{m}$ is a flat V-module : simple considerations show this to be a natural hypothesis, often verified in practice.

The V-modules killed by $\mathfrak{m}$ are the objects of a (full) Serre subcategory Σ of the category V-**Mod** of all V-modules, and the quotient V^a-**Mod** $:= V$-**Mod**$/\Sigma$ is an abelian category which we call the category of *almost V-modules*. It is easy to check that the usual tensor product of V-modules descends to a bifunctor $\otimes$ on almost V-modules, so that V^a-**Mod** is a monoidal abelian category in a natural way. Then an *almost ring* is just an almost V-module A endowed with a "multiplication" morphism $A \otimes A \to A$ satisfying certain natural axioms. Together with the obvious morphisms, these gadgets form a category V^a-**Alg**. Given any almost V-algebra A, one can then define the notion of A-module and A-algebra, just like for usual rings. The purpose of the game is to reconstruct in this new framework as much as possible (and useful) of classical linear and commutative algebra.

Essentially, this is the same as the ideology informing Deligne's paper [23], which sets out to develop algebraic geometry in the context of abstract tannakian categories. We could also claim an even earlier ancestry, in that some of the leading motifs resonating throughout our text, can be traced as far back as Gabriel's memoir [40] "Des catégories abéliennes". Furthermore, it has been recently pointed out to us that Roos, in a series of works dating from the mid-sixties, had already discovered much of the homological algebra that forms the backbone for our more systematic study of almost modules (see *e.g.* [68], [69]).

A pervasive theme – recurring throughout the text – is the study of deformations of various interesting objects, may they be almost algebras, almost modules, or almost group schemes. Especially, the analysis of *nilpotent* deformations of étale almost algebras and of almost projective modules is important for the proof of the almost purity theorem; whereas Faltings used Hochschild cohomology to this aim, we employ the cotangent complex; this gives us shorter proofs of essentially stronger results (and answers some questions of Faltings). Nilpotent deformations usher the way to *formal* deformations, hence to the definition of adic topologies on almost rings and modules; this in turns leads straight to the study of *henselian* deformations and to the notion of henselian pair in almost ring theory.

Another important thread is *descent theory* for almost algebras; faithfully flat descent is easy, but in [34] one needs also some cases of non-flat descent, so we give a comprehensive treatment of the latter.

Closely related to descent is the problem of constructing quotients under flat equivalence relations, and we dedicate section 4.5 to this question.

The third main ingredient in the newer proof of the almost purity theorem is the Frobenius endomorphism of almost algebras of positive characteristic: this is investigated in section 3.5.

In many instances, our results go well beyond what is strictly necessary in order to justify Faltings' proof of almost purity; this is mainly because our emphasis is on supplying natural frameworks and – as much as possible – pure thought arguments, rather than choosing the most economical presentation. Another reason is that we plan to discuss and extend the almost purity theorem in a future work: some of the extra generality gained here will pay off then.

Chapters 6 and 7 are dedicated to applications, respectively to valuation theory and to p-adic analytic geometry: especially the reader will find there our own contributions to almost purity. Much in these two chapters pertains to subjects that border on – but have, strictly speaking, no intersection with – almost ring theory; however, at some crucial junctures, the methods developed in the previous chapters intervene in an essential way and link up the discussion with almost mathematics. Two notable examples are:

(a) theorem 6.3.23, that generalizes the relationship between the module of differentials and the different ideal; this is classical in the case of a finite separable extension of discrete valuation rings, but new for general valuations of rank one, where the map is not any longer finite, but only *almost finite*;

(b) proposition 7.5.15 on the semicontinuity of the discriminant function, for a finite étale map of rings; again, this is a result of (usual) commutative algebra, whose statement and proof would both be very ackward without the machinery of almost rings.

We hope to demonstrate with these samples that almost ring theory has something to offer even to mathematicians that are not directly involved with p-adic Hodge theory.

We close with an appendix collecting some miscellanea: a sketch of a theory of the fundamental group for almost algebras, and the construction of the derived functors of some standard non-additive functors defined on almost modules.

1.4. The view from above. In evoking Deligne's and Gabriel's works, we have unveiled another source of motivation whose influence has steadily grown throughout the long gestation of our paper. Namely, we have come to view almost ring theory as a contribution to that expanding body of research of still uncertain range and shifting boundaries, that we could call "abstract algebraic geometry". We would like to encompass under this label several heterogeneous developments: notably, it should include various versions of non-commutative geometry that have been proposed in the last twenty years (*e.g.* [70]), but also the relative schemes of [47], as well as Deligne's ideas for algebraic geometry over symmetric monoidal categories.

The running thread loosely unifying these works is the realization that "geometric spaces" do not necessarily consist of set-theoretical points, and – perhaps more importantly – functions on such "spaces" do not necessarily form (sheaves of) commutative rings. Much effort has been devoted to extending the reach of geometric intuition to non-commutative algebras; alternatively, one can retain commutativity, but allow "structure sheaves" which take values in tensor categories other than the category of rings.

As a case in point, to any given almost ring A one can attach its *spectrum* Spec A, which is just A viewed as an object of the opposite of the category V^a-**Alg**. Spec A has even a natural *flat topology*, which allows to define more general *almost schemes* by gluing (*i.e.* taking colimits of) diagrams of affine spectra; all this is explained in section 5.7, where we also introduce *quasi-projective* almost schemes and investigate some basic properties of the *smooth locus* of a quasi-projective almost scheme.

By way of illustration, these generalities are applied in section 5.8 in order to solve a deformation problem for torsors over affine almost group schemes; let us stress that the problem in question is stated purely in terms of affine objects (*i.e.* almost rings and "almost Hopf algebras"), but the solution requires the introduction of certain auxiliary almost schemes that are not affine.

1.5. **Acknowledgements.** The second author is very much indebted to Gerd Faltings for many patient explanations on the method of almost étale extensions.

Next he would like to acknowledge several interesting discussions with Ioannis Emmanouil. He is also much obliged to Pierre Deligne, for a useful list of critical remarks.

Finally, he owes a special thank to Roberto Ferretti, who has read the first tentative versions of this work, has corrected many slips and has made several valuable suggestions.

The authors are pleased to thank Michel Mendès-France, who has lent his artistic talents to design the "prescano" that graces the cover of this volume.

This project began in 1997 while the second author was supported by the IHES, and has been completed in 2002, while the second author was a guest of the Laboratoire de Théorie des Nombres of the University of Paris VI.

2. HOMOLOGICAL THEORY

As explained in the introduction, in order to define a category of almost modules one requires a pair $(V, \mathfrak{m})$ consisting of a ring V and an ideal $\mathfrak{m} \subset V$ such that $\mathfrak{m} = \mathfrak{m}^2$. In section 2.1 we collect a few useful ring-theoretic preliminaries concerning such pairs. In section 2.2 we introduce the category V^a-**Mod** of *almost modules* : it is a quotient V-**Mod**$/\Sigma$ of the category of V-modules, where Σ is the thick subcategory of the V-modules killed by $\mathfrak{m}$. V^a-**Mod** is an abelian tensor category and its commutative unitary monoids, called *almost algebras*, are the chief objects of study in this work. The first useful observation is that the localization functor V-**Mod** $\to$ V^a-**Mod** admits both left and right adjoints. Taken together, these functors exhibit the kind of exactness properties that one associates to open embeddings of topoi, perhaps a hint of some deeper geometrical structure, still to be unearthed.

After these generalities, we treat in section 2.3 the question of finiteness conditions for almost modules. Let A denote an almost algebra, fixed for the rest of this introduction. It is certainly possible to define as usual a notion of finitely generated A-module, however this turns out to be too restrictive a class for applications. The main idea here is to define a uniform structure on the set of isomorphism classes of A-modules; then we will say that an A-module is *almost finitely generated* if its isomorphism class lies in the topological closure of the subspace of finitely generated A-modules. Similarly we define *almost finitely presented* A-modules. The uniform structure also comes handy when we want to construct operators on almost modules: if one can show that the operator in question is uniformly continuous on a class $\mathscr{C}$ of almost modules, then its definition extends right away by continuity to the topological closure $\overline{\mathscr{C}}$ of $\mathscr{C}$. This is exemplified by the construction of the (almost) Fitting ideals for A-modules, at the end of section 2.3.

In section 2.4 we introduce the basic toolkit of homological algebra, beginning with the notion of flat almost module, which poses no problem, since we do have a tensor product in our category. The notion of projectivity is more subtle : it turns out that the category of A-modules usually does *not* have enough projectives. The useful notion is *almost projectivity*: simply one uses the standard definition, except that the role of the Hom functor is played by the internal alHom functor. The scarcity of projectives should not be regarded as surprising or pathological: it is quite analogous to the lack of enough projective objects in the category of quasi-coherent $\mathscr{O}_X$-modules on a non-affine scheme X.

Section 2.5 introduces the cotangent complex of a morphism of almost rings, and establishes its usual properties, such as transitivity and Tor-independent base change theorems. These foundations will be put to use in chapter 3, to study infinitesimal deformations of almost algebras.

2.1. Some ring-theoretic preliminaries.
Unless otherwise stated, every ring is commutative with unit. This section collects some results of general nature that will be used throughout this work.

2.1.1.   Our *basic setup* consists of a fixed base ring V containing an ideal $\mathfrak{m}$ such that $\mathfrak{m}^2 = \mathfrak{m}$. Starting from (2.5.14), we will also assume that $\widetilde{\mathfrak{m}} := \mathfrak{m} \otimes_V \mathfrak{m}$ is a flat V-module.

Example 2.1.2.   (i) The main example is given by a non-discrete valuation ring $(V, |\cdot|)$ of rank one; in this case $\mathfrak{m}$ will be the maximal ideal.

(ii) Take $\mathfrak{m} := V$. This is the "classical limit". In this case almost ring theory reduces to usual ring theory. Thus, all the discussion that follows specializes to, and sometimes gives alternative proofs for, statements about rings and their modules.

2.1.3.   Let M be a given V-module. We say that M is *almost zero* if $\mathfrak{m}M = 0$. A map ϕ of V-modules is an *almost isomorphism* if both $\mathrm{Ker}\,\phi$ and $\mathrm{Coker}\,\phi$ are almost zero V-modules.

Remark 2.1.4.   (i) It is easy to check that a V-module M is almost zero if and only if $\mathfrak{m} \otimes_V M = 0$. Similarly, a map $M \to N$ of V-modules is an almost isomorphism if and only if the induced map $\widetilde{\mathfrak{m}} \otimes_V M \to \widetilde{\mathfrak{m}} \otimes_V N$ is an isomorphism. Notice also that, if $\mathfrak{m}$ is flat, then $\mathfrak{m} \simeq \widetilde{\mathfrak{m}}$.

(ii) Let $V \to W$ be a ring homomorphism. For a V-module M set $M_W := W \otimes_V M$. We have an exact sequence

$$(2.1.5) \qquad\qquad 0 \to K \to \mathfrak{m}_W \to \mathfrak{m}W \to 0$$

where $K := \mathrm{Tor}_1^V(V/\mathfrak{m}, W)$ is an almost zero W-module. By (i) it follows that $\mathfrak{m} \otimes_V K \simeq (\mathfrak{m}W) \otimes_W K \simeq 0$. Then, applying $\mathfrak{m}_W \otimes_W -$ and $- \otimes_W (\mathfrak{m}W)$ to (2.1.5) we derive

$$\mathfrak{m}_W \otimes_W \mathfrak{m}_W \simeq \mathfrak{m}_W \otimes_W (\mathfrak{m}W) \simeq (\mathfrak{m}W) \otimes_W (\mathfrak{m}W)$$

i.e. $\widetilde{\mathfrak{m}}_W \simeq (\mathfrak{m}W)^\sim$. In particular, if $\widetilde{\mathfrak{m}}$ is a flat V-module, then $(\mathfrak{m}W)^\sim$ is a flat W-module. This means that our basic assumptions on the pair $(V, \mathfrak{m})$ are stable under arbitrary base extension. Notice that the flatness of $\mathfrak{m}$ does not imply the flatness of $\mathfrak{m}W$. This partly explains why we insist that $\widetilde{\mathfrak{m}}$, rather than $\mathfrak{m}$, be flat.

2.1.6.   Before moving on, we want to analyze in some detail how our basic assumptions relate to certain other natural conditions that can be postulated on the pair $(V, \mathfrak{m})$. Indeed, let us consider the following two hypotheses:

(A) $\mathfrak{m} = \mathfrak{m}^2$ and $\mathfrak{m}$ is a filtered union of principal ideals.

(B) $\mathfrak{m} = \mathfrak{m}^2$ and, for all $k > 1$, the k-th powers of elements of $\mathfrak{m}$ generate $\mathfrak{m}$.

Clearly **(A)** implies **(B)**. Less obvious is the following result.

Proposition 2.1.7.   (i) **(A)** *implies that* $\widetilde{\mathfrak{m}}$ *is flat.*

(ii) *If* $\widetilde{\mathfrak{m}}$ *is flat then* **(B)** *holds.*

Proof. Suppose that **(A)** holds, so that $\mathfrak{m} = \mathrm{colim}_{\alpha \in I} Vx_\alpha$, where I is a directed set parametrizing elements $x_\alpha \in \mathfrak{m}$ (and $\alpha \leq \beta \Leftrightarrow Vx_\alpha \subset Vx_\beta$). For any $\alpha \in I$ we have natural isomorphisms

$$(2.1.8) \qquad\qquad Vx_\alpha \simeq V/\mathrm{Ann}_V(x_\alpha) \simeq (Vx_\alpha) \otimes_V (Vx_\alpha).$$

For $\alpha \leq \beta$, let $j_{\alpha\beta} : Vx_\alpha \hookrightarrow Vx_\beta$ be the imbedding; we have a commutative diagram

$$
\begin{CD}
V @>{\mu_{z^2}}>> V \\
@V{\pi_\alpha}VV @VV{\pi_\beta}V \\
(Vx_\alpha) \otimes_V (Vx_\alpha) @>{j_{\alpha\beta} \otimes j_{\alpha\beta}}>> (Vx_\beta) \otimes_V (Vx_\beta)
\end{CD}
$$

where $z \in V$ is such that $x_\alpha = z \cdot x_\beta$, μ_{z^2} is multiplication by z^2 and π_α is the projection induced by (2.1.8) (and similarly for π_β). Since $\mathfrak{m} = \mathfrak{m}^2$, for all $\alpha \in I$ we can find β such that x_α is a multiple of x_β^2. Say $x_\alpha = y \cdot x_\beta^2$; then we can take $z := y \cdot x_\beta$, so z^2 is a multiple of x_α and in the above diagram $\operatorname{Ker} \pi_\alpha \subset \operatorname{Ker} \mu_{z^2}$. Hence one can define a map $\lambda_{\alpha\beta} : (Vx_\alpha) \otimes_V (Vx_\alpha) \to V$ such that $\pi_\beta \circ \lambda_{\alpha\beta} = j_{\alpha\beta} \otimes j_{\alpha\beta}$ and $\lambda_{\alpha\beta} \circ \pi_\alpha = \mu_{z^2}$. It now follows that for every V-module N, the induced morphism $\operatorname{Tor}_1^V (N, (Vx_\alpha) \otimes_V (Vx_\alpha)) \to \operatorname{Tor}_1^V (N, (Vx_\beta) \otimes_V (Vx_\beta))$ is the zero map. Taking the colimit we derive that $\widetilde{\mathfrak{m}}$ is flat. This shows (i). In order to show (ii) we consider, for any prime number p, the following condition

$(*_p)$ $\mathfrak{m}/p \cdot \mathfrak{m}$ is generated (as a V-module) by the p-th powers of its elements.

Clearly (**B**) implies $(*_p)$ for all p. In fact we have :

Claim 2.1.9. (**B**) holds if and only if $(*_p)$ holds for every prime p.

Proof of the claim. Suppose that $(*_p)$ holds for every prime p. The polarization identity

$$
k! \cdot x_1 \cdot x_2 \cdot \ldots \cdot x_k = \sum_{I \subset \{1,2,\ldots,k\}} (-1)^{k-|I|} \cdot \left(\sum_{i \in I} x_i \right)^k
$$

shows that if $N := \sum_{x \in \mathfrak{m}} Vx^k$ then $k! \cdot \mathfrak{m} \subset N$. To prove that $N = \mathfrak{m}$ it then suffices to show that for every prime p dividing $k!$ we have $\mathfrak{m} = p \cdot \mathfrak{m} + N$. Let $\phi : V/pV \to V/pV$ be the Frobenius ($x \mapsto x^p$); we can denote by $(V/pV)^\phi$ the ring V/pV seen as a V/pV-algebra via the homomorphism ϕ. Also set $\phi^*M := M \otimes_{V/pV} (V/pV)^\phi$ for a V/pV-module M. Then the map $\phi^*(\mathfrak{m}/p \cdot \mathfrak{m}) \to (\mathfrak{m}/p \cdot \mathfrak{m})$ (defined by raising to p-th power) is surjective by $(*_p)$. Hence so is $(\phi^r)^*(\mathfrak{m}/p \cdot \mathfrak{m}) \to (\mathfrak{m}/p \cdot \mathfrak{m})$ for every $r > 0$, which says that $\mathfrak{m} = p \cdot \mathfrak{m} + N$ when $k = p^r$, hence for every k. $\Diamond$

Next recall (see [6, Exp. XVII 5.5.2]) that, if M is a V-module, the module of symmetric tensors $\operatorname{TS}^k(M)$ is defined as $(\otimes_V^k M)^{S_k}$, the invariants under the natural action of the symmetric group S_k on $\otimes_V^k M$. We have a natural map $\Gamma^k(M) \to \operatorname{TS}^k(M)$ that is an isomorphism when M is flat (see *loc. cit.* 5.5.2.5; here Γ^k denotes the k-th graded piece of the divided power algebra).

Claim 2.1.10. The group S_k acts trivially on $\otimes_V^k \mathfrak{m}$ and the map $\widetilde{\mathfrak{m}} \otimes_V \mathfrak{m} \to \widetilde{\mathfrak{m}}$ ($x \otimes y \otimes z \mapsto x \otimes yz$) is an isomorphism.

Proof of the claim. The first statement is reduced to the case of transpositions and to $k = 2$. There we can compute : $x \otimes yz = xy \otimes z = y \otimes xz = yz \otimes x$. For the second statement note that the imbedding $\mathfrak{m} \hookrightarrow V$ is an almost isomorphism, and apply remark 2.1.4(i).   ◊

Suppose now that $\tilde{\mathfrak{m}}$ is flat and pick a prime p. Then S_p acts trivially on $\otimes_V^p \tilde{\mathfrak{m}}$. Hence

$$(2.1.11) \qquad \Gamma^p(\tilde{\mathfrak{m}}) \simeq \otimes_V^p \tilde{\mathfrak{m}} \simeq \tilde{\mathfrak{m}}.$$

But $\Gamma^p(\tilde{\mathfrak{m}})$ is spanned as a V-module by the products $\gamma_{i_1}(x_1) \cdot \ldots \cdot \gamma_{i_k}(x_k)$ (where $x_i \in \tilde{\mathfrak{m}}$ and $\sum_j i_j = p$). Under the isomorphism (2.1.11) these elements map to $\binom{p}{i_1,\ldots,i_k} \cdot x_1^{i_1} \cdot \ldots \cdot x_k^{i_k}$; but such an element vanishes in $\tilde{\mathfrak{m}}/p \cdot \tilde{\mathfrak{m}}$ unless $i_l = p$ for some l. Therefore $\tilde{\mathfrak{m}}/p \cdot \tilde{\mathfrak{m}}$ is generated by p-th powers, so the same is true for $\mathfrak{m}/p \cdot \mathfrak{m}$, and by the above, **(B)** holds, which shows (ii).   □

Theorem 2.1.12. *Let $(\varepsilon_i \mid i \in I)$ be a set of generators of $\mathfrak{m}$ and, for every subset $S \subset I$, denote by $\mathfrak{m}_S \subset \mathfrak{m}$ the subideal generated by $(\varepsilon_i \mid i \in S)$. Then we have:*

(i) *Every countable subset $S \subset I$ is contained in an $S' \subset I$ such that:*
   (a) *S' is still countable.*
   (b) *$\mathfrak{m}_{S'}^2 = \mathfrak{m}_{S'}$.*
   (c) *If $\tilde{\mathfrak{m}}$ is a flat V-module, the same holds for $\mathfrak{m}_{S'} \otimes_V \mathfrak{m}_{S'}$.*
(ii) *Suppose that $\mathfrak{m}$ is countably generated as a V-module. Then :*
   (a) *$\tilde{\mathfrak{m}}$ is countably presented as a V-module.*
   (b) *If $\tilde{\mathfrak{m}}$ is a flat V-module, then it is of homological dimension ≤ 1.*

Proof. For every $i \in I$, we can write $\varepsilon_i = \sum_j x_{ij} \varepsilon_j$, for certain $x_{ij} \in \mathfrak{m}$. For any $i, j \in I$ such that $x_{ij} \neq 0$, let us write $x_{ij} = \sum_k x_{ijk} \varepsilon_k$ for some $x_{ijk} \in V$. We say that a subset $S \subset I$ is *saturated* if the following holds: whenever $i \in S$ and $x_{ijk} \neq 0$, we have $j, k \in S$.

Claim 2.1.13. Every countable subset $S \subset I$ is contained in a countable saturated subset $S_\infty \subset I$.

Proof of the claim. We let $S_0 := S$ and define recursively S_n for every $n > 0$ as follows. Suppose S_n has already been given; then we set

- $T_n := \{i \in I \mid$ there exists $a \in S_n, b \in I$ such that either $x_{aib} \neq 0$ or $x_{abi} \neq 0\}$
- $S_{n+1} := S_n \cup T_n$.

Notice that, for every $i \in I$ we have $x_{ij} = 0$ for all but finitely many $j \in I$, hence $x_{ijk} = 0$ for all but finitely many $j, k \in I$. It follows easily that $S_\infty := \bigcup_{n \in \mathbb{N}} S_n$ will do.   ◊

(i.a) and (i.b) are now straightforward, when one remarks that $\mathfrak{m}_S = \mathfrak{m}_S^2$ for every saturated subset S. Next, let $S \subset I$ be any saturated subset. Clearly the family $(\varepsilon_i \otimes \varepsilon_j \mid i, j \in S)$ generates $\tilde{\mathfrak{m}}_S := \mathfrak{m}_S \otimes_V \mathfrak{m}_S$ and $\varepsilon_i \cdot \varepsilon_j \cdot (\varepsilon_k \otimes \varepsilon_l) = \varepsilon_k \cdot \varepsilon_l \cdot (\varepsilon_i \otimes \varepsilon_j)$ for all $i, j, k, l \in S$. Let $F(S)$ be the V-module defined by generators $(e_{ij})_{i,j \in S}$, subject to the relations:

$$\varepsilon_i \cdot \varepsilon_j \cdot e_{kl} = \varepsilon_k \cdot \varepsilon_l \cdot e_{ij} \qquad e_{ik} = \sum_j x_{ij} e_{jk} \qquad \text{for all } i, j, k, l \in S.$$

We get an epimorphism $\pi_S : F(S) \to \widetilde{\mathfrak{m}}_S$ by $e_{ij} \mapsto \varepsilon_i \otimes \varepsilon_j$. The relations imply that, if $x := \sum_{k,l} y_{kl} e_{kl} \in \mathrm{Ker}\, \pi_S$, then $\varepsilon_i \cdot \varepsilon_j \cdot x = 0$, so $\mathfrak{m}_S \cdot \mathrm{Ker}\, \pi_S = 0$. Whence $\mathfrak{m}_S \otimes_V \mathrm{Ker}\, \pi_S = 0$ and $\mathbf{1}_{\mathfrak{m}_S} \otimes_V \pi_S$ is an isomorphism. We consider the diagram

$$
\begin{array}{ccc}
\mathfrak{m}_S \otimes_V F(S) & \xrightarrow{\;\sim\;} & \mathfrak{m}_S \otimes_V \widetilde{\mathfrak{m}}_S \\
\phi \downarrow & & \downarrow \psi \\
F(S) & \xrightarrow{\;\pi_S\;} & \widetilde{\mathfrak{m}}_S
\end{array}
$$

where ϕ and ψ are induced by scalar multiplication. We already know that ψ is an isomorphism, and since $F(S) = \mathfrak{m}_S \cdot F(S)$, we see that ϕ is an epimorphism, so π_S is an isomorphism. If now I is countable, this shows that (ii.a) holds. Now (ii.b) follows from (ii.a) and the following lemma 2.1.16. In order to show (i.c) we will use the following well known criterion.

Claim 2.1.14. ([55, Ch.I, Th.1.2]). Let R be a ring, M an R-module. Then M is R-flat if and only if, for every finitely presented R-module N, every morphism $N \to M$ factors as a composition $N \to L \to M$ where L is a free R-module of finite rank.

In order to apply claim 2.1.14 we show:

Claim 2.1.15. Let $S \subset I$ be a countable saturated subset. Then there exists a countable saturated subset $\sigma(S) \subset I$ containing S, with the following property. For every finitely presented V-module N and every morphism $f : N \to \widetilde{\mathfrak{m}}_S$, we can find a commutative diagram:

$$
\begin{array}{ccc}
N & \xrightarrow{\;f\;} & \widetilde{\mathfrak{m}}_S \\
\downarrow & & \downarrow j \\
L & \longrightarrow & \widetilde{\mathfrak{m}}_{\sigma(S)}
\end{array}
$$

where L is free V-module of finite rank and j is the natural map.

Proof of the claim. In view of (ii.a), we can write $\widetilde{\mathfrak{m}}_S \simeq \underset{\alpha \in A}{\mathrm{colim}}\, M_\alpha$, where A is some filtered countable set and every M_α is a finitely presented V-module. Given f as in the claim, we can find $\alpha \in A$ such that f factors through the natural map $\iota_\alpha : M_\alpha \to \widetilde{\mathfrak{m}}_S$, so we are reduced to prove the claim for $N = M_\alpha$ and $f = \iota_\alpha$. However, by assumption $\widetilde{\mathfrak{m}}$ is flat, so by claim 2.1.14 the composition $M_\alpha \overset{\iota_\alpha}{\to} \widetilde{\mathfrak{m}}_S \to \widetilde{\mathfrak{m}}$ factors through a map $L_\alpha \to \widetilde{\mathfrak{m}}$, with L_α free of finite rank over V. Furthermore, thanks to claim 2.1.13 we have $\widetilde{\mathfrak{m}} = \underset{J}{\mathrm{colim}}\, \widetilde{\mathfrak{m}}_J$, where J runs over the family of all countable saturated subsets of I. It follows that, for some countable saturated $S_\alpha \supset S$, the map $L_\alpha \to \widetilde{\mathfrak{m}}$ factors through $\widetilde{\mathfrak{m}}_{S_\alpha}$, and we can increase S_α so that the required diagram commutes. Clearly $\sigma(S) := \bigcup_{\alpha \in A} S_\alpha$ will do. $\Diamond$

Finally, for any countable subset $S \subset I$, let us set $S' := \bigcup_{n \in \mathbb{N}} \sigma^n(S_\infty)$, where S_∞ is the saturation of S as in claim 2.1.13. One verifies easily that (i.c) holds for this choice of S'. $\square$

The proof of following well known lemma is due to D.Lazard ([55, Ch.I, Th.3.2]), up to some slight imprecisions which were corrected in [59, pp.49-50]. For the convenience of the reader we reproduce the argument.

Lemma 2.1.16. *Let R be a (not necessarily commutative) ring. Any flat countably presented left R-module has homological dimension ≤ 1.*

Proof. Let M be flat and countable presented over R, and choose a presentation $F_1 \xrightarrow{\phi} F_0 \to M \to 0$ with F_0 and F_1 free R-modules of (infinite) countable rank. Let $(e_j \mid j \in \mathbb{N})$ be a basis of F_0 and $(f_i \mid i \in \mathbb{N})$ a basis of F_1. Say that $\phi(f_i) = \sum_i x_{ij} e_j$ with $x_{ij} \in R$, for every $i \in \mathbb{N}$. For every $i \in \mathbb{N}$ we define the finite set $S_i := \{j \in \mathbb{N} \mid x_{ij} \neq 0\}$; also, let $S'_n := \bigcup_{i \leq n}(\{i\} \cup S_i)$. Define R-modules $G_n := \oplus_{i \leq n} R f_i$, $H_n := \oplus_{j \in S'_n} R e_j$; the restriction of ϕ induces maps $\phi_n : G_n \to H_n$ and we have $M \simeq \operatorname*{colim}_{n \in \mathbb{N}} \operatorname{Coker} \phi_n$. By claim 2.1.14, the natural map $\operatorname{Coker} \phi_n \to M$ factors as a composition $\operatorname{Coker} \phi_n \xrightarrow{\alpha_n} L_n \xrightarrow{\psi_n} M$, where L_n is a free R-module of finite rank. Then ψ_n factors through a map $\psi'_n : L_n \to \operatorname{Coker} \phi_{t(n)}$ for some $t(n) \in \mathbb{N}$. We further define $\psi''_n : L_n \to \operatorname{Coker} \phi_{k(n)}$ as the composition of ψ'_n with the natural transition map $\operatorname{Coker} \phi_{t(n)} \to \operatorname{Coker} \phi_{k(n)}$, where $k(n)$ is suitably chosen, so that $k(n) > \max(n, t(n))$ and the composition $\psi''_n \circ \alpha_n : \operatorname{Coker} \phi_n \to \operatorname{Coker} \phi_{k(n)}$ is the natural transition map. We define by induction on $n \in \mathbb{N}$ a direct system of maps $\tau_n : L_{h_n} \to L_{h_{n+1}}$, as follows. Set $h_0 := 0$; next, suppose that $h_n \in \mathbb{N}$ has already been given up to some $n \in \mathbb{N}$; we let $h_{n+1} := k(h_n)$ and $\tau_n := \alpha_{h_{n+1}} \circ \psi''_{h_n}$. Clearly $M \simeq \operatorname*{colim}_{n \in \mathbb{N}} L_{h_n}$; set $L := \oplus_{n \in \mathbb{N}} L_{h_n}$ and let $\tau : L \to L$ be the map given by the rule $\tau(x_n \mid n \in \mathbb{N}) := (x_n - \tau_n(x_{n-1}) \mid n \in \mathbb{N})$. We derive a short exact sequence: $0 \to L \xrightarrow{\tau} L \to M \to 0$, whence the claim. $\square$

2.2. Categories of almost modules and algebras. If $\mathscr{C}$ is a category, and X, Y two objects of $\mathscr{C}$, we will usually denote by $\operatorname{Hom}_{\mathscr{C}}(X, Y)$ the set of morphisms in $\mathscr{C}$ from X to Y and by 1_X the identity morphism of X. Moreover we denote by $\mathscr{C}^o$ the opposite category of $\mathscr{C}$ and by $s.\mathscr{C}$ the category of simplicial objects in $\mathscr{C}$, that is, functors $\Delta^o \to \mathscr{C}$, where Δ is the category whose objects are the ordered sets $[n] := \{0, ..., n\}$ for each integer $n \geq 0$ and where a morphism $\phi : [p] \to [q]$ is a non-decreasing map. A morphism $f : X \to Y$ in $s.\mathscr{C}$ is a sequence of morphisms $f_{[n]} : X[n] \to Y[n]$, $n \geq 0$ such that the obvious diagrams commute. We can imbed $\mathscr{C}$ in $s.\mathscr{C}$ by sending each object X to the "constant" object $s.X$ such that $s.X[n] = X$ for all $n \geq 0$ and $s.X[\phi] = 1_X$ for all morphisms ϕ in Δ.

2.2.1. If $\mathscr{C}$ is an abelian category, $\mathsf{D}(\mathscr{C})$ will denote the derived category of $\mathscr{C}$. As usual we have also the full subcategories $\mathsf{D}^+(\mathscr{C}), \mathsf{D}^-(\mathscr{C})$ of complexes of objects of $\mathscr{C}$ that are exact for sufficiently large negative (resp. positive) degree. If R is a ring, the category of R-modules (resp. R-algebras) will be denoted by R-**Mod** (resp. R-**Alg**). Most of the times we will write $\operatorname{Hom}_R(M, N)$ instead of $\operatorname{Hom}_{R\text{-}\mathbf{Mod}}(M, N)$.

We denote by **Set** the category of sets. The symbol $\mathbb{N}$ denotes the set of non-negative integers; in particular $0 \in \mathbb{N}$.

2.2.2.   The full subcategory Σ of V-**Mod** consisting of all V-modules that are almost isomorphic to 0 is clearly a Serre subcategory and hence we can form the quotient category V-**Mod**$/\Sigma$. There is a localization functor

$$V\text{-}\mathbf{Mod} \to V\text{-}\mathbf{Mod}/\Sigma \qquad M \mapsto M^a$$

that takes a V-module M to the same module, seen as an object of V-**Mod**$/\Sigma$. In particular, we have the object V^a associated to V; it seems therefore natural to use the notation V^a-**Mod** for the category V-**Mod**$/\Sigma$, and an object of V^a-**Mod** will be indifferently referred to as "a V^a-module" or "an almost V-module". In case we need to stress the dependance on the ideal $\mathfrak{m}$, we can write $(V, \mathfrak{m})^a$-**Mod**.

Since the almost isomorphisms form a multiplicative system (see *e.g.* [77, Exerc.10.3.2]), it is possible to describe the morphisms in V^a-**Mod** via a calculus of fractions, as follows. Let V-**al.Iso** be the category that has the same objects as V-**Mod**, but such that $\mathrm{Hom}_{V\text{-}\mathbf{al.Iso}}(M,N)$ consists of all almost isomorphisms $M \to N$. If M is any object of V-**al.Iso** we write $(V\text{-}\mathbf{al.Iso}/M)$ for the category of objects of V-**al.Iso** over M (*i.e.* morphisms $\phi : X \to M$). If $\phi_i : X_i \to M$ $(i = 1, 2)$ are two objects of $(V\text{-}\mathbf{al.Iso}/M)$ then $\mathrm{Hom}_{(V\text{-}\mathbf{al.Iso}/M)}(\phi_1, \phi_2)$ consists of all morphisms $\psi : X_1 \to X_2$ in V-**al.Iso** such that $\phi_1 = \phi_2 \circ \psi$. For any two V-modules M, N we define a functor $\mathscr{F}_N : (V\text{-}\mathbf{al.Iso}/M)^o \to V\text{-}\mathbf{Mod}$ by associating to an object $\phi : P \to M$ the V-module $\mathrm{Hom}_V(P, N)$ and to a morphism $\alpha : P \to Q$ the map $\mathrm{Hom}_V(Q, N) \to \mathrm{Hom}_V(P, N) : \beta \mapsto \beta \circ \alpha$. Then we have

$$(2.2.3) \qquad \mathrm{Hom}_{V^a\text{-}\mathbf{Mod}}(M^a, N^a) = \operatorname*{colim}_{(V\text{-}\mathbf{al.Iso}/M)^o} \mathscr{F}_N.$$

However, formula (2.2.3) can be simplified considerably by remarking that for any V-module M, the natural morphism $\widetilde{\mathfrak{m}} \otimes_V M \to M$ is an initial object of $(V\text{-}\mathbf{al.Iso}/M)$. Indeed, let $\phi : N \to M$ be an almost isomorphism; the diagram

$$
\begin{array}{ccc}
\widetilde{\mathfrak{m}} \otimes_V N & \xrightarrow{\ \sim\ } & \widetilde{\mathfrak{m}} \otimes_V M \\
\downarrow & & \downarrow \\
N & \xrightarrow{\ \ \phi\ \ } & M
\end{array}
$$

(cp. remark 2.1.4(i)) allows one to define a morphism $\psi : \widetilde{\mathfrak{m}} \otimes_V M \to N$ over M. We need to show that ψ is unique. But if $\psi_1, \psi_2 : \widetilde{\mathfrak{m}} \otimes_V M \to N$ are two maps over M, then $\mathrm{Im}(\psi_1 - \psi_2) \subset \mathrm{Ker}(\phi)$ is almost zero, hence $\mathrm{Im}(\psi_1 - \psi_2) = 0$, since $\widetilde{\mathfrak{m}} \otimes_V M = \mathfrak{m} \cdot (\widetilde{\mathfrak{m}} \otimes_V M)$. Consequently, (2.2.3) boils down to

$$(2.2.4) \qquad \mathrm{Hom}_{V^a\text{-}\mathbf{Mod}}(M^a, N^a) = \mathrm{Hom}_V(\widetilde{\mathfrak{m}} \otimes_V M, N).$$

In particular $\mathrm{Hom}_{V^a\text{-}\mathbf{Mod}}(M, N)$ has a natural structure of V-module for any two V^a-modules M, N, *i.e.* $\mathrm{Hom}_{V^a\text{-}\mathbf{Mod}}(-, -)$ is a bifunctor that takes values in the category V-**Mod**.

Remark 2.2.5. More generally, the definition of the category V^a-**Mod** makes sense whenever V is a unitary ring and $\mathfrak{m}$ is a two-sided ideal with $\mathfrak{m} = \mathfrak{m}^2$. The abelian categories arising in this way have been characterized by J.-E.Roos as the Grothendieck categories in which axioms (AB4*) and (AB6) hold (see [68] : more precisely, the bilocalizing subcategories of V-**Mod** – *i.e.* the Serre subcategories for which the localization functor admits left and right adjoints – are the subcategories $(V/\mathfrak{m})$-**Mod** for $\mathfrak{m}$ idempotent two-sided ideal).

2.2.6.   One checks easily (for instance using (2.2.4)) that the usual tensor product induces a bifunctor $-\otimes_V-$ on almost V-modules, which, in the jargon of [24] makes of V^a-**Mod** an *abelian tensor category*. Then an *almost V-algebra* is just a commutative unitary monoid in the tensor category V^a-**Mod**. Let us recall what this means. Quite generally, let $(\mathscr{C}, \otimes, U, u)$ be any abelian tensor category, so that $\otimes$: $\mathscr{C} \times \mathscr{C} \to \mathscr{C}$ is a biadditive functor, U is the identity object of $\mathscr{C}$ (see [24, p.105]) and for any two objects M and N in $\mathscr{C}$ we have a "commutativity constraint" (*i.e.* a functorial isomorphism $\theta_{M|N} : M \otimes N \to N \otimes M$ that "switches the two factors") and a functorial isomorphism $\nu_M : U \otimes M \to M$. Then a $\mathscr{C}$-monoid A is an object of $\mathscr{C}$ endowed with a morphism $\mu_A : A \otimes A \to A$ (the "multiplication" of A) satisfying the associativity condition

$$\mu_A \circ (1_A \otimes \mu_A) = \mu_A \circ (\mu_A \otimes 1_A).$$

We say that A is *unitary* if additionally A is endowed with a "unit morphism" $\underline{1}_A$: $U \to A$ satisfying the (left and right) unit property :

$$\mu_A \circ (\underline{1}_A \otimes 1_A) = \nu_A \qquad \mu_A \circ (\underline{1}_A \otimes 1_A) \circ \theta_{A|U} = \mu_A \circ (1_A \otimes \underline{1}_A).$$

Finally A is *commutative* if $\mu_A = \mu_A \circ \theta_{A|A}$ (to be rigorous, in all of the above one should indicate the associativity constraints, which we have omitted : see [24]). A commutative unitary monoid will also be simply called an *algebra*. With the morphisms defined in the obvious way, the $\mathscr{C}$-monoids form a category; furthermore, given a $\mathscr{C}$-monoid A, a *left A-module* is an object M of $\mathscr{C}$ endowed with a morphism $\sigma_{M/A} : A \otimes M \to M$ such that $\sigma_{M/A} \circ (1_A \otimes \sigma_{M/A}) = \sigma_{M/A} \circ (\mu_A \otimes 1_M)$. Similarly one defines right A-modules and A-bimodules. In the case of bimodules we have left and right morphisms $\sigma_{M,l} : A \otimes M \to M$, $\sigma_{M,r} : M \otimes A \to M$ and one imposes that they "commute", *i.e.* that

$$\sigma_{M,r} \circ (\sigma_{M,l} \otimes 1_A) = \sigma_{M,l} \circ (1_A \otimes \sigma_{M,r}).$$

Clearly the (left resp. right) A-modules (and the A-bimodules) form an additive category with *A-linear morphisms* defined as one expects. One defines the notion of a submodule as an equivalence class of monomorphisms $N \to M$ such that the composition $A \otimes N \to A \otimes M \to M$ factors through N. Especially, a *two-sided ideal* of A is an A-sub-bimodule $I \to A$. For given submodules I, J of A one denotes $IJ := \mathrm{Im}(I \otimes J \to A \otimes A \xrightarrow{\mu_A} A)$. For an A-module M, the *annihilator* $\mathrm{Ann}_A(M)$ of M is the largest ideal $j : I \to A$ of A such that the composition $I \otimes M \xrightarrow{j \otimes 1_M} A \otimes M \xrightarrow{\sigma_{M/A}} M$ is the zero morphism.

2.2.7.   If $f : M \to N$ is a morphism of left A-modules, then $\mathrm{Ker}(f)$ exists in the underlying abelian category $\mathscr{C}$ and one checks easily that it has a unique structure of left A-module which makes it a submodule of M. *If moreover $\otimes$ is right exact* when either argument is fixed, then also $\mathrm{Coker}\, f$ has a unique A-module structure for which $N \to \mathrm{Coker}\, f$ is A-linear. In this case the category of left A-modules is abelian. Similarly, if A is a unitary $\mathscr{C}$-monoid, then one defines the notion of *unitary* left A-module by requiring that $\sigma_{M/A} \circ (\underline{1}_A \otimes 1_M) = \nu_M$ and these form an abelian category when $\otimes$ is right exact.

2.2.8. Specializing to our case we obtain the category V^a-**Alg** of almost V-algebras and, for every almost V-algebra A, the category A-**Mod** of unitary left A-modules. Clearly the localization functor restricts to a functor V-**Alg** $\to V^a$-**Alg** and for any V-algebra R we have a localization functor R-**Mod** $\to R^a$-**Mod**.

Next, if A is an almost V-algebra, we can define the category A-**Alg** of A-algebras. It consists of all the morphisms $A \to B$ of almost V-algebras.

2.2.9. Let again $(\mathscr{C}, \otimes, U, u)$ be any abelian tensor category. By [24, p.119], the endomorphism ring $\mathrm{End}_{\mathscr{C}}(U)$ of U is commutative. For any object M of $\mathscr{C}$, denote $M_* = \mathrm{Hom}_{\mathscr{C}}(U, M)$; then $M \mapsto M_*$ defines a functor $\mathscr{C} \to \mathrm{End}_{\mathscr{C}}(U)$-**Mod**. Moreover, if A is a $\mathscr{C}$-monoid, A_* is an associative $\mathrm{End}_{\mathscr{C}}(U)$-algebra, with multiplication given as follows. For $a, b \in A_*$ let $a \cdot b := \mu_A \circ (a \otimes b) \circ \nu_U^{-1}$. Similarly, if M is an A-module, M_* is an A_*-module in a natural way, and in this way we obtain a functor from A-modules and A-linear morphisms to A_*-modules and A_*-linear maps. Using [24, Prop.1.3], one can also check that $\mathrm{End}_{\mathscr{C}}(U) = U_*$ as $\mathrm{End}_{\mathscr{C}}(U)$-algebras, where U is viewed as a $\mathscr{C}$-monoid using ν_U.

2.2.10. All this applies especially to our categories of almost modules and almost algebras : in this case we call $M \mapsto M_*$ the *functor of almost elements*. So, if M is an almost module, an almost element of M is just an honest element of M_*. Using (2.2.4) one can show easily that for every V-module M the natural map $M \to (M^a)_*$ is an almost isomorphism.

2.2.11. Let A be a V^a-algebra, M, N two A-modules; the set $\mathrm{Hom}_{A\text{-}\mathbf{Mod}}(M, N)$ has a natural structure of A_*-module and we obtain an internal Hom functor by letting

$$\mathrm{alHom}_A(M, N) := \mathrm{Hom}_{A\text{-}\mathbf{Mod}}(M, N)^a.$$

This is the functor of *almost homomorphisms* from M to N.

2.2.12. For any A-module M we have also a functor of tensor product $M \otimes_A -$ on A-modules which, in view of the following proposition 2.2.14 can be shown to be a left adjoint to the functor $\mathrm{alHom}_A(M, -)$. It can be defined as $M \otimes_A N := (M_* \otimes_{A_*} N_*)^a$. A special case of this adjunction yields a natural identification:

$$\mathrm{alHom}_A(M, N)_* = \mathrm{Hom}_{A\text{-}\mathbf{Mod}}(M, N).$$

With this tensor product, A-**Mod** is an abelian tensor category as well, and A-**Alg** could also be described as the category of $(A$-**Mod**)-algebras. Under this equivalence, a morphism $\phi : A \to B$ of almost V-algebras becomes the unit morphism $\underline{1}_B : A \to B$ of the corresponding monoid. We will sometimes drop the subscript and write simply $\underline{1}$.

Remark 2.2.13. Let $V \to W$ be a map of base rings, W taken with the extended ideal $\mathfrak{m}W$. Then W^a is an almost V-algebra so we have defined the category W^a-**Mod** using base ring V and the category $(W, \mathfrak{m}W)^a$-**Mod** using base W. One shows easily that they are equivalent: we have an obvious functor $(W, \mathfrak{m}W)^a$-**Mod** $\to W^a$-**Mod** and a quasi-inverse is provided by $M \mapsto M_*^a$. Similar base comparison statements hold for the categories of almost algebras.

Proposition 2.2.14. *Let A be a V^a-algebra, R a V-algebra.*

(i) *There is a natural isomorphism $A \simeq A_*^a$ of almost V-algebras.*

(ii) *The functor $M \mapsto M_*$ from R^a-**Mod** to R-**Mod** (resp. from R^a-**Alg** to R-**Alg**) is right adjoint to the localization functor R-**Mod** $\to R^a$-**Mod** (resp. R-**Alg** $\to R^a$-**Alg**).*

(iii) *The counit of the adjunction $M_*^a \to M$ is a natural isomorphism from the composition of the two functors to the identity functor $1_{R^a\text{-}\mathbf{Mod}}$ (resp. $1_{R^a\text{-}\mathbf{Alg}}$).*

Proof. (i) has already been remarked. We show (ii). In light of remark 2.2.13 (applied with $W = R$) we can assume that $V = R$. Let M be a V-module and $N = N_0^a$ an almost V-module; we have natural bijections

$$\mathrm{Hom}_{V^a\text{-}\mathbf{Mod}}(M^a, N) \simeq \mathrm{Hom}_{V^a\text{-}\mathbf{Mod}}(M^a, N_0^a) \simeq \mathrm{Hom}_V(\tilde{\mathfrak{m}} \otimes_V M, N_0)$$
$$\simeq \mathrm{Hom}_V(M, \mathrm{Hom}_V(\tilde{\mathfrak{m}}, N_0))$$
$$\simeq \mathrm{Hom}_V(M, \mathrm{Hom}_{V^a\text{-}\mathbf{Mod}}(V, N_0^a))$$
$$\simeq \mathrm{Hom}_V(M, N_*)$$

which proves (ii). Now (iii) follows from (2.2.10), or by [40, Ch.III Prop.3]. □

Remark 2.2.15. (i) Let M_1, M_2 be two A-modules. By proposition 2.2.14(iii) it is clear that a morphism $\phi : M_1 \to M_2$ of A-modules is uniquely determined by the induced morphism $M_{1*} \to M_{2*}$. On this basis, we will very often define morphisms of A-modules (or A-algebras) by saying how they act on almost elements.

(ii) It is a bit tricky to deal with preimages of almost elements under morphisms: for instance, if $\phi : M_1 \to M_2$ is an epimorphism (by which we mean that Coker $\phi \simeq 0$) and $m_2 \in M_{2*}$, then it is not true in general that we can find an almost element $m_1 \in M_{1*}$ such that $\phi_*(m_1) = m_2$. What remains true is that for arbitrary $\varepsilon \in \mathfrak{m}$ we can find m_1 such that $\phi_*(m_1) = \varepsilon \cdot m_2$.

(iii) The existence of the right adjoint $M \mapsto M_*$ follows also directly from [40, Chap.III §3 Cor.1 or Chap.V §2].

Corollary 2.2.16. *The categories A-**Mod** and A-**Alg** are both complete and cocomplete.*

Proof. We recall that the categories A_*-**Mod** and A_*-**Alg** are both complete and cocomplete. Now let I be any small indexing category and $M : I \to A$-**Mod** be any functor. Denote by $M_* : I \to A_*$-**Mod** the composed functor $i \mapsto M(i)_*$. We claim that $\underset{I}{\mathrm{colim}}\, M = (\underset{I}{\mathrm{colim}}\, M_*)^a$. The proof is an easy application of proposition 2.2.14(iii). A similar argument also works for limits and for the category A-**Alg**. □

Example 2.2.17. Let I be any set; as a special case of corollary 2.2.16 we deduce the existence of the module $A^{(I)}$, defined as a colimit over the discrete category associated to I in the usual way. The A_*-module $(A^{(I)})_*$ is the submodule of $(A_*)^I$ consisting of all the sequences $a := (a_i \mid i \in I)$ such that $\varepsilon a \in (A_*)^{(I)}$ for every $\varepsilon \in \mathfrak{m}$. We leave the details to the reader.

2.2.18. For any V^a-algebra A, The abelian category A-**Mod** satisfies axiom (AB5) (see *e.g.* [77, §A.4]) and it has a generator, namely the object A itself. It then follows by a general result that A-**Mod** has enough injectives.

Corollary 2.2.19. *The functor* $M \mapsto M_*$ *from* R^a-**Mod** *to* R-**Mod** *sends injectives to injectives and injective envelopes to injective envelopes.*

Proof. The functor $M \mapsto M_*$ is right adjoint to an exact functor, hence it preserves injectives. Now, let J be an injective envelope of M; to show that J_* is an injective envelope of M_*, it suffices to show that J_* is an essential extension of M_*. However, if $N \subset J_*$ and $N \cap M_* = 0$, then $N^a \cap M = 0$, hence $\mathfrak{m}N = 0$, but J_* does not contain $\mathfrak{m}$-torsion, thus $N = 0$. □

2.2.20. Note that the essential image of $M \mapsto M_*$ is closed under limits. Next recall that the forgetful functor A_*-**Alg** → **Set** (resp. A_*-**Mod** → **Set**) has a left adjoint $A_*[-] :$ **Set** → A_*-**Alg** (resp. $A^{(-)} :$ **Set** → A_*-**Mod**) that assigns to a set S the free A_*-algebra $A_*[S]$ (resp. the free A_*-module $A_*^{(S)}$) generated by S. If S is any set, it is natural to write $A[S]$ (resp. $A^{(S)}$) for the A-algebra $(A_*[S])^a$ (resp. for the A-module $(A_*^{(S)})^a$. This yields a left adjoint, called the *free A-algebra* functor **Set** → A-**Alg** (resp. the *free A-module* functor **Set** → A-**Mod**) to the "forgetful" functor A-**Alg** → **Set** (resp. A-**Mod** → **Set**) $B \mapsto B_*$.

2.2.21. Now let R be any V-algebra; we want to construct a left adjoint to the localization functor R-**Mod** → R^a-**Mod**. For a given R^a-module M, let

(2.2.22) $$M_! := \widetilde{\mathfrak{m}} \otimes_V (M_*).$$

We have the natural map (unit of adjunction) $R \to R_*^a$, so that we can view $M_!$ as an R-module.

Proposition 2.2.23. *Let R be a V-algebra.*
  (i) *The functor* R^a-**Mod** → R-**Mod** *defined by (2.2.22) is left adjoint to localization.*
  (ii) *The unit of the adjunction $M \to M_!^a$ is a natural isomorphism from the identity functor $1_{R^a\text{-}\mathbf{Mod}}$ to the composition of the two functors.*

Proof. (i) follows easily from (2.2.4) and (ii) follows easily from (i). □

Corollary 2.2.24. *Suppose that $\widetilde{\mathfrak{m}}$ is a flat V-module. Then we have :*
  (i) *the functor $M \mapsto M_!$ is exact;*
  (ii) *the localization functor R-**Mod** → R^a-**Mod** sends injectives to injectives.*

Proof. By proposition 2.2.23 it follows that $M \mapsto M_!$ is right exact. To show that it is also left exact when $\widetilde{\mathfrak{m}}$ is a flat V-module, it suffices to remark that $M \mapsto M_*$ is left exact. Now, by (i), the functor $M \mapsto M^a$ is right adjoint to an exact functor, so (ii) is clear. □

2.2.25.   Let B be any A-algebra. The multiplication on B_* is inherited by $B_!$, which is therefore a non-unital ring in a natural way. We endow the V-module $V \oplus B_!$ with the ring structure determined by the rule:

$$(v, b) \cdot (v', b') := (v \cdot v', v \cdot b' + v' \cdot b + b \cdot b') \quad \text{for all } v, v' \in V \text{ and } b, b' \in B_!.$$

Then $V \oplus B_!$ is a (unital) ring. We notice that the V-submodule generated by all the elements of the form $(x \cdot y, -x \otimes y \otimes \underline{1})$ (for arbitrary $x, y \in \mathfrak{m}$) forms an ideal I of $V \oplus B_!$. Set $B_{!!} := (V \oplus B_!)/I$. Thus we have a sequence of V-modules

(2.2.26) $$0 \to \widetilde{\mathfrak{m}} \to V \oplus B_! \to B_{!!} \to 0$$

which in general is only right exact.

Definition 2.2.27. We say that B is an *exact V^a-algebra* if the sequence (2.2.26) is exact.

Remark 2.2.28. Notice that if $\widetilde{\mathfrak{m}} \xrightarrow{\sim} \mathfrak{m}$ (*e.g.* when $\mathfrak{m}$ is flat), then all V^a-algebras are exact. In the general case, if B is any A-algebra, then $V^a \times B$ is always exact. Indeed, we have $(V^a \times B)_* \simeq V_*^a \times B_*$ and, by remark 2.1.4(i), $\widetilde{\mathfrak{m}} \otimes_V V_*^a \simeq \widetilde{\mathfrak{m}}$.

Clearly we have a natural isomorphism $B \simeq B_{!!}^a$.

Proposition 2.2.29. *The functor $B \mapsto B_{!!}$ is left adjoint to the localization functor $A_{!!}$-**Alg** $\to A$-**Alg**.*

Proof. Let B be an A-algebra, C an $A_{!!}$-algebra and $\phi : B \to C^a$ a morphism of A-algebras. By proposition 2.2.23 we obtain a natural A_*-linear morphism $B_! \to C$. Together with the structure morphism $V \to C$ this yields a map $\widetilde{\phi} : V \oplus B_! \to C$ which is easily seen to be a ring homomorphism. It is equally clear that the ideal I defined above is mapped to zero by $\widetilde{\phi}$, hence the latter factors through a map of $A_{!!}$-algebras $B_{!!} \to C$. Conversely, such a map induces a morphism of A-algebras $B \to C^a$ just by taking localization. It is easy to check that the two procedures are inverse to each other, which shows the assertion. □

Remark 2.2.30. (i) The functor of almost elements commutes with arbitrary limits, because all right adjoints do. It does not in general commute with colimits, not even with arbitrary infinite direct sums. Dually, the functors $M \mapsto M_!$ and $B \mapsto B_{!!}$ commute with all colimits. In particular, the latter commutes with tensor products.

(ii) Resume the notation of remark 2.2.13. The change of setup functor

$$(W, \mathfrak{m}W)^a\text{-}\mathbf{Mod} \to W^a\text{-}\mathbf{Mod}$$

commutes with the operations $M \mapsto M_*$ and $M \mapsto M_!$. The corresponding functor $F : (W, \mathfrak{m}W)^a\text{-}\mathbf{Alg} \to W^a\text{-}\mathbf{Alg}$ satisfies the identity: $F(B)_{!!} \otimes_{W_{!!}^a} W = B_{!!}$ for every $(W, \mathfrak{m}W)^a$-algebra B.

2.3. Uniform spaces of almost modules. Let A be a V^a-algebra. For any cardinal number c, we let $\mathcal{M}_c(A)$ be the set of isomorphism classes of A-modules which admit a set of generators of cardinality $\leq c$. In the following we fix some (very) large infinite cardinality ω such that $\mathfrak{m}$ is generated by at most ω elements, and sometimes we write $\mathcal{M}(A)$ instead of $\mathcal{M}_\omega(A)$.

Definition 2.3.1. Let A be a V^a-algebra and M an A-module.

(i) We define a uniform structure on the set $\mathcal{I}_A(M)$ of A-submodules of M, as follows. For every finitely generated ideal $\mathfrak{m}_0 \subset \mathfrak{m}$, the subset of $\mathcal{I}_A(M) \times \mathcal{I}_A(M)$ given by $E_M(\mathfrak{m}_0) := \{(M_0, M_1) \mid \mathfrak{m}_0 M_0 \subset M_1 \text{ and } \mathfrak{m}_0 M_1 \subset M_0\}$ is an entourage for the uniform structure, and the subsets of this kind form a fundamental system of entourages.

(ii) We define a uniform structure on $\mathcal{M}(A)$ as follows. For every finitely generated ideal $\mathfrak{m}_0 \subset \mathfrak{m}$ and every integer $n \geq 0$ we define the entourage $E_{\mathcal{M}}(\mathfrak{m}_0) \subset \mathcal{M}(A) \times \mathcal{M}(A)$, which consists of all pairs of A-modules (M, M') such that there exist a third module N and morphisms $\phi : N \to M, \psi : N \to M'$, such that $\mathfrak{m}_0$ annihilates the kernel and cokernel of ϕ and ψ. We declare that the $E_{\mathcal{M}}(\mathfrak{m}_0)$ form a fundamental system of entourages for the uniform structure of $\mathcal{M}(A)$. This defines a uniform structure because $E_{\mathcal{M}}(\mathfrak{m}_0) \circ E_{\mathcal{M}}(\mathfrak{m}_0) \subset E_{\mathcal{M}}(\mathfrak{m}_0^2)$, and similarly in (i).

Remark 2.3.2. Notice that the entourage $E_{\mathcal{M}}(\mathfrak{m}_0)$ can be defined equivalently by all the pairs of A-modules (M, M') such that there exists a third module L and morphisms $\phi' : M \to L, \psi' : M' \to L$ such that $\mathfrak{m}_0$ annihilates the kernel and cokernel of ϕ' and ψ'. Indeed, given a pair $(M, M') \in E_{\mathcal{M}}(\mathfrak{m}_0)$, and a datum (N, ϕ, ψ) as in definition 2.3.1(ii), a datum (L, ϕ', ψ') satisfying the above condition is obtained from the push out diagram

(2.3.3)
$$
\begin{array}{ccc}
N & \xrightarrow{\phi} & M \\
\psi \downarrow & & \downarrow \phi' \\
M' & \xrightarrow{\psi'} & L.
\end{array}
$$

Conversely, given a datum (L, ϕ', ψ'), one obtains another diagram as (2.3.3), by letting N be the fibred product of M and M' over L.

2.3.4. We will also need occasionally a notion of "Cauchy product" : let $\prod_{n=0}^{\infty} I_n$ be a formal infinite product of ideals $I_n \subset A$. We say that the formal product *satisfies the Cauchy condition* (or briefly : *is a Cauchy product*) if, for every neighborhood $\mathcal{U}$ of A in $\mathcal{I}_A(A)$ there exists $n_0 \geq 0$ such that $\prod_{m=n}^{n+p} I_m \in \mathcal{U}$ for all $n \geq n_0$ and all $p \geq 0$.

Lemma 2.3.5. *Let M be an A-module.*

(i) *$\mathcal{I}_A(M)$ is complete and separated for the uniform structure of definition 2.3.1.*

(ii) *The following maps are uniformly continuous :*

(a) *$\mathcal{I}_A(M) \times \mathcal{I}_A(M) \to \mathcal{I}_A(M) : (M', M'') \mapsto M' \cap M''.$*

(b) $\mathscr{I}_A(M) \times \mathscr{I}_A(M) \to \mathscr{I}_A(M)$: $(M', M'') \mapsto M' + M''$.

(c) $\mathscr{I}_A(A) \times \mathscr{I}_A(A) \to \mathscr{I}_A(A)$: $(I, J) \mapsto IJ$.

(iii) *For any A-linear morphism $\phi : M \to N$, the following maps are uniformly continuous:*

(a) $\mathscr{I}_A(M) \to \mathscr{I}_A(N)$: $M' \mapsto \phi(M')$.

(b) $\mathscr{I}_A(N) \to \mathscr{I}_A(M)$: $N' \mapsto \phi^{-1}(N')$.

Proof. (i) : The separation property is easily verified. We show that $\mathscr{I}_A(M)$ is complete. Therefore, suppose that $\mathscr{F}$ is some Cauchy filter of $\mathscr{I}_A(M)$. Concretely, this means that for every finitely generated $\mathfrak{m}_0 \subset \mathfrak{m}$, there exists $F(\mathfrak{m}_0) \in \mathscr{F}$ such that $\mathfrak{m}_0 I \subset J$ for every $I, J \in F(\mathfrak{m}_0)$. Let $L := \bigcup_{F \in \mathscr{F}} (\bigcap_{I \in F} I)$. We claim that L is the limit of our filter. Indeed, for a given finitely generated $\mathfrak{m}_0 \subset \mathfrak{m}$, we have $\mathfrak{m}_0 I \subset \bigcap_{J \in F(\mathfrak{m}_0)} J$, for every $I \in F(\mathfrak{m}_0)$, whence $\mathfrak{m}_0 I \subset L$. On the other hand, if $I \in F \subset F(\mathfrak{m}_0)$, we can write: $\mathfrak{m}_0 L = \bigcup_{F' \subset F} \mathfrak{m}_0 (\bigcap_{J \in F'} J) \subset \bigcup_{F' \subset F} (\bigcap_{J \in F'} \mathfrak{m}_0 J) \subset \bigcup_{F' \subset F} I = I$ (where F' runs over all the subsets $F' \in \mathscr{F}$ such that $F' \subset F$). This shows that $(L, I) \in E_M(\mathfrak{m}_0)$ whenever $I \in F(\mathfrak{m}_0)$, which implies the claim. (ii) and (iii) are easy and will be left to the reader. $\square$

Remark 2.3.6. In general, the uniform space $\mathscr{M}(A)$ is not separated. For instance, consider the case where V is a valuation ring whose value group Γ is a dense proper subgroup of $\mathbb{R}_{>0}$. The invertible V^a-modules (see definition 2.4.23(vi)) are pairwise arbitrarily close with respect to the uniform structure of $\mathscr{M}(V^a)$; however, the set of isomorphism classes of such V^a-modules is in bijection with $\mathbb{R}_{>0}/\Gamma$ (cp. remark 6.1.17).

Lemma 2.3.7. *Let $\phi : M \to N$ be an A-linear morphism, B an A-algebra. The following maps are uniformly continuous :*

(i) $\mathscr{M}(A) \to \mathscr{I}_A(A)$: $M \mapsto \mathrm{Ann}_A(M)$.

(ii) $\mathscr{I}_A(M) \times \mathscr{I}_A(N) \to \mathscr{M}_\eta(A)$: $(M', N') \mapsto (\phi(M') + N')/\phi(M')$ *(for η large enough).*

(iii) $\mathscr{M}_\omega(A) \times \mathscr{M}_\omega(A) \to \mathscr{M}_\eta(A)$: $(M', M'') \mapsto \mathrm{alHom}_A(M', M'')$ *(for η large enough so that the map is defined).*

(iv) $\mathscr{M}(A) \times \mathscr{M}(A) \to \mathscr{M}(A)$: $(M', M'') \mapsto M' \otimes_A M''$.

(v) $\mathscr{M}(A) \to \mathscr{M}(B)$: $M \mapsto B \otimes_A M$.

(vi) $\mathscr{M}(A) \to \mathscr{M}(A)$: $M \mapsto \Lambda_A^r M$ *for any $r \geq 0$, provided* **(B)** *holds.*

Proof. We show (iv) and leave the others to the reader. By symmetry, we reduce to verifying that, if $(M', M'') \in E_{\mathscr{M}}(\mathfrak{m}_0)$ and N is an arbitrary A-module, then $(N \otimes_A M', N \otimes_A M'') \in E_{\mathscr{M}}(\mathfrak{m}_0^2)$. Then we can further assume that there is a morphism $\phi : M' \to M''$ with $\mathfrak{m}_0 \cdot \mathrm{Ker}\,\phi = \mathfrak{m}_0 \cdot \mathrm{Coker}\,\phi = 0$. We factor $\phi_!$ as an epimorphism followed by a monomorphism $M'_! \xrightarrow{\phi_1} \mathrm{Im}\,\phi_! \xrightarrow{\phi_2} M''_!$, and then we reduce to checking that the kernels and cokernels of both $1_N \otimes_A \phi_1$ and $1_N \otimes_A \phi_2$ are killed by $\mathfrak{m}_0$. This is clear for $1_N \otimes_A \phi_1$, and it follows easily for $1_N \otimes_A \phi_2$ as well, by using the Tor sequences. $\square$

Definition 2.3.8. For a subset S of a topological space T, let $\overline{S}$ denote the closure of S in T. Let M be an A-module.

(i) M is *finitely generated* if its isomorphism class lies in $\bigcup_{n \in \mathbb{N}} \mathscr{M}_n(A)$.

(ii) M is said to be *almost finitely generated* if its isomorphism class lies in $\overline{\bigcup_{n \in \mathbb{N}} \mathscr{M}_n(A)}$.

(iii) M is said to be *uniformly almost finitely generated* if there exists an integer $n \geq 0$ such that the isomorphism class of M lies in $\mathscr{U}_n(A) := \overline{\mathscr{M}_n(A)}$. Then we will say that n is a *uniform bound* for M.

(iv) M is said to be *finitely presented* if it is isomorphic to the cokernel of a morphism of free finitely generated A-modules. We denote by $\mathscr{F}\mathscr{P}(A) \subset \mathscr{M}(A)$ the subset of the isomorphism classes of finitely presented A-modules.

(v) M is said to be *almost finitely presented* if its isomorphism class lies in $\overline{\mathscr{F}\mathscr{P}(A)}$.

Remark 2.3.9. (i) Under condition (**A**), an A-module M lies in $\mathscr{U}_n(A)$ if and only if, for every $\varepsilon \in \mathfrak{m}$ there exists an A-linear morphism $A^n \to M$ whose cokernel is killed by ε.

(ii) One can check that if $\omega < \eta$, then $\mathscr{M}_\omega(A)$ is a closed subspace of $\mathscr{M}_\eta(A)$, so the notion of an almost finitely generated module is independent of the choice of the cardinal number ω.

Proposition 2.3.10. *Let M be an A-module.*

(i) *M is almost finitely generated if and only if for every finitely generated ideal $\mathfrak{m}_0 \subset \mathfrak{m}$ there exists a finitely generated submodule $M_0 \subset M$ such that $\mathfrak{m}_0 M \subset M_0$.*

(ii) *The following conditions are equivalent:*

(a) *M is almost finitely presented.*

(b) *for arbitrary $\varepsilon, \delta \in \mathfrak{m}$ there exist positive integers $n = n(\varepsilon)$, $m = m(\varepsilon)$ and a three term complex $A^m \xrightarrow{\psi_\varepsilon} A^n \xrightarrow{\phi_\varepsilon} M$ with $\varepsilon \cdot \operatorname{Coker}(\phi_\varepsilon) = 0$ and $\delta \cdot \operatorname{Ker}\phi_\varepsilon \subset \operatorname{Im}\psi_\varepsilon$.*

(c) *For every finitely generated ideal $\mathfrak{m}_0 \subset \mathfrak{m}$ there is a complex $A^m \xrightarrow{\psi} A^n \xrightarrow{\phi} M$ with $\mathfrak{m}_0 \cdot \operatorname{Coker}\phi = 0$ and $\mathfrak{m}_0 \cdot \operatorname{Ker}\phi \subset \operatorname{Im}\psi$.*

Proof. (i): Let M be an almost finitely generated A-module, and $\mathfrak{m}_0 \subset \mathfrak{m}$ a finitely generated subideal. Choose a finitely generated subideal $\mathfrak{m}_1 \subset \mathfrak{m}$ such that $\mathfrak{m}_0 \subset \mathfrak{m}_1^3$; by hypothesis, there exist A-modules M' and M'', where M'' is finitely generated, and morphisms $f : M' \to M$, $g : M' \to M''$ whose kernels and cokernels are annihilated by $\mathfrak{m}_1$. We get morphisms

$$\mathfrak{m}_1 \otimes_V M'' \to \operatorname{Im}(g) \qquad \text{and} \qquad \mathfrak{m}_1 \otimes_V \operatorname{Im}(g) \to M'$$

hence a composed morphism $\phi : \mathfrak{m}_1 \otimes_V \mathfrak{m}_1 \otimes_V M'' \to M'$; it is easy to check that $\operatorname{Coker}(f \circ \phi)$ is annihilated by $\mathfrak{m}_1^3$, hence $M_0 := \operatorname{Im}(f \circ \phi)$ will do.

To show (ii) we will need the following :

Claim 2.3.11. Let F_1 be a finitely generated A-module and suppose that we are given $a, b \in V$ and a (not necessarily commutative) diagram such that $q \circ \phi = a \cdot p$, $p \circ \psi = b \cdot q$. Let $I \subset V$ be an ideal such that $\operatorname{Ker} q$ has a finitely generated submodule containing $I \cdot \operatorname{Ker} q$. Then $\operatorname{Ker} p$ has a finitely generated submodule containing $ab \cdot I \cdot \operatorname{Ker} p$.

Proof of the claim. Let R be the submodule of $\operatorname{Ker} q$ given by the assumption. We have $\operatorname{Im}(\psi \circ \phi - ab \cdot 1_{F_1}) \subset \operatorname{Ker} p$ and $\psi(R) \subset \operatorname{Ker} p$. We take $R_1 := \operatorname{Im}(\psi \circ \phi - ab \cdot 1_{F_1}) + \psi(R)$. Clearly $\phi(\operatorname{Ker} p) \subset \operatorname{Ker} q$, so $I \cdot \phi(\operatorname{Ker} p) \subset R$, hence $I \cdot \psi \circ \phi(\operatorname{Ker} p) \subset \psi(R)$ and finally $ab \cdot I \cdot \operatorname{Ker} p \subset R_1$. $\Diamond$

Claim 2.3.12. If M satisfies condition (ii.b), and $\phi : F \to M$ is a morphism with $F \simeq A^n$, then for every finitely generated ideal $\mathfrak{m}_1 \subset \mathfrak{m} \cdot \operatorname{Ann}_V(\operatorname{Coker} \phi)$ there is a finitely generated submodule of $\operatorname{Ker} \phi$ containing $\mathfrak{m}_1 \cdot \operatorname{Ker} \phi$.

Proof of the claim. Now, let $\delta \in \operatorname{Ann}_V(\operatorname{Coker} \phi)$ and $\varepsilon_1, \varepsilon_2, \varepsilon_3, \varepsilon_4 \in \mathfrak{m}$. By assumption there is a complex $A^r \xrightarrow{t} A^s \xrightarrow{q} M$ with $\varepsilon_1 \cdot \operatorname{Coker} q = 0$, $\varepsilon_2 \cdot \operatorname{Ker} q \subset \operatorname{Im} t$. Letting $F_1 := F$, $F_2 := A^s$, $a := \varepsilon_1 \cdot \varepsilon_3$, $b := \varepsilon_4 \cdot \delta$, one checks easily that ψ and ϕ can be given such that all the assumptions of claim 2.3.11 are fulfilled. So, with $I := \varepsilon_2 \cdot V$ we see that $\varepsilon_1 \cdot \varepsilon_2 \cdot \varepsilon_3 \cdot \varepsilon_4 \cdot \delta \cdot \operatorname{Ker} \phi$ lies in a finitely generated submodule of $\operatorname{Ker} \phi$. But $\mathfrak{m}_1$ is contained in an ideal generated by finitely many such products $\varepsilon_1 \cdot \varepsilon_2 \cdot \varepsilon_3 \cdot \varepsilon_4 \cdot \delta$. $\Diamond$

Now, it is clear that (c) implies (a) and (b). To show that (b) implies (c), take a finitely generated ideal $\mathfrak{m}_1 \subset \mathfrak{m}$ such that $\mathfrak{m}_0 \subset \mathfrak{m} \cdot \mathfrak{m}_1$, pick a morphism $\phi : A^n \to M$ whose cokernel is annihilated by $\mathfrak{m}_1$, and apply claim 2.3.12. We show that (a) implies (c). For a given finitely generated subideal $\mathfrak{m}_0 \subset \mathfrak{m}$, pick another finitely generated $\mathfrak{m}_1 \subset \mathfrak{m}$ such that $\mathfrak{m}_0 \subset \mathfrak{m}_1^3$; find morphisms $f : M' \to M$ and $g : M' \to M''$ whose kernels and cokernels are annihilated by $\mathfrak{m}_1$, and such that M'' is finitely presented. Let $\underline{\varepsilon} := (\varepsilon_1, ..., \varepsilon_r)$ be a finite sequence of generators of $\mathfrak{m}_1$, and denote by $K_\bullet := K_\bullet(\underline{\varepsilon})$ the Koszul complex of V-modules associated to the sequence $\underline{\varepsilon}$. Set $\mathfrak{m}_1' := \operatorname{Coker}(K_2 \to K_1)$; we derive a natural surjection $\partial : \mathfrak{m}_1' \to \mathfrak{m}_1$ and, for every $i = 1, ..., r$, maps $e_i : V \to \mathfrak{m}_1'$ such that the compositions

$$\mathfrak{m}_1' \xrightarrow{\partial} \mathfrak{m}_1 \hookrightarrow V \xrightarrow{e_i} \mathfrak{m}_1' \qquad V \xrightarrow{e_i} \mathfrak{m}_1' \xrightarrow{\partial} \mathfrak{m}_1 \hookrightarrow V$$

are both scalar multiplication by ε_i. Hence, for every V^a-module N, the kernel and cokernel of the natural morphism $\mathfrak{m}_1' \otimes_V N \to N$ are annihilated by $\mathfrak{m}_1$. Let now ϕ be as in the proof of (i); notice that the diagram:

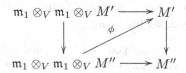

commutes. It follows that the composed morphism

$$\mathfrak{m}_1' \otimes_V \mathfrak{m}_1' \otimes_V M'' \to \mathfrak{m}_1 \otimes_V \mathfrak{m}_1 \otimes_V M'' \xrightarrow{\phi} M' \to M$$

has kernel and cokernel annihilated by $\mathfrak{m}_1^3$, so the claim follows.   □

Corollary 2.3.13. *Let R be a V-algebra and M an R-module. Then M^a is an almost finitely generated (resp. almost finitely presented) R^a-module if and only if for every finitely generated subideal $\mathfrak{m}_0 \subset \mathfrak{m}$ there is a finitely generated (resp. finitely presented) R-module N and a homomorphism $N \to M$ whose kernel and cokernel are annihilated by $\mathfrak{m}_0$.*

Proof. The direction $\Leftarrow$ is clear, hence suppose that M^a is almost finitely generated. By proposition 2.3.10(i) there is a finitely generated submodule L of M^a containing $\mathfrak{m}_0 M^a$. One can lift $\mathfrak{m}_0 L$ to a finitely generated submodule of M containing $\mathfrak{m}_0^3 M$.

Finally, suppose that M^a is almost finitely presented. We already know that there is $n \in \mathbb{N}$ and $\phi : R^n \to M$ with $\mathfrak{m}_0 \cdot \text{Coker } \phi = 0$. By claim 2.3.12, Ker ϕ^a has a finitely generated submodule L containing $\mathfrak{m}_0^2 \cdot$ Ker ϕ^a. One can lift $\mathfrak{m}_0 L$ to a finitely generated submodule of Ker ϕ containing $\mathfrak{m}_0^4 \cdot$ Ker ϕ.   □

2.3.14. Suppose that $\mathfrak{m} = \bigcup_{\lambda \in \Lambda} \mathfrak{m}_\lambda$, where $(\mathfrak{m}_\lambda \mid \lambda \in \Lambda)$ is a filtered family of subideals such that $\mathfrak{m}_\lambda^2 = \mathfrak{m}_\lambda$ for every $\lambda \in \Lambda$. Let R be a V-algebra, M an R-module, and denote by R_λ^a (resp. R^a) the $(V, \mathfrak{m}_\lambda)^a$-algebra (resp. $(V, \mathfrak{m})^a$-algebra) associated to R; define similarly M^a and M_λ^a, for all $\lambda \in \Lambda$.

Lemma 2.3.15. *With the notation of (2.3.14), the following are equivalent:*

(i) *M^a is an almost finitely generated (resp. almost finitely presented) R^a-module.*

(ii) *M_λ^a is an almost finitely generated (resp. almost finitely presented) R_λ^a-module for all $\lambda \in \Lambda$.*

Proof. It is an easy consequence of corollary 2.3.13.   □

The following proposition generalizes a well-known characterization of finitely presented modules over usual rings.

Proposition 2.3.16. *Let M be an A-module.*

(i) *M is almost finitely generated if and only if, for every filtered system $(N_\lambda, \phi_{\lambda\mu})$ (indexed by a directed set Λ) the natural morphism*

(2.3.17) $$\operatorname*{colim}_\Lambda \text{alHom}_A(M, N_\lambda) \to \text{alHom}_A(M, \operatorname*{colim}_\Lambda N_\lambda)$$

is a monomorphism.

(ii) *M is almost finitely presented if and only if for every filtered inductive sytem as above, (2.3.17) is an isomorphism.*

Proof. The "only if" part in (i) (resp. (ii)) is first checked when M is finitely generated (resp. finitely presented) and then extended to the general case. We leave the details to the reader and we proceed to verify the "if" part. For (i), choose a set I and an epimorphism $p : A^{(I)} \to M$. Let Λ be the directed set of finite subsets of I, ordered by inclusion. For $S \in \Lambda$, let $M_S := p(A^S)$. Then $\operatorname*{colim}_{\Lambda}(M/M_S) = 0$, so the assumption gives $\operatorname*{colim}_{\Lambda} \text{alHom}_A(M, M/M_S) = 0$, i.e. $\operatorname*{colim}_{\Lambda} \text{Hom}_A(M, M/M_S) = 0$ is almost zero, so, for every $\varepsilon \in \mathfrak{m}$, the image of $\varepsilon \cdot 1_M$ in the above colimit is 0, i.e. there exists $S \in \Lambda$ such that $\varepsilon M \subset M_S$, which proves the contention. For (ii), we present M as a filtered colimit $\operatorname*{colim}_{\Lambda} M_\lambda$, where each M_λ is finitely presented (this can be done e.g. by taking such a presentation of the A_*-module M_* and applying $N \mapsto N^a$). The assumption of (ii) gives that $\operatorname*{colim}_{\Lambda} \text{Hom}_A(M, M_\lambda) \to \text{Hom}_A(M, M)$ is an almost isomorphism, hence, for every $\varepsilon \in \mathfrak{m}$ there is $\lambda \in \Lambda$ and $\phi_\varepsilon : M \to M_\lambda$ such that $p_\lambda \circ \phi_\varepsilon = \varepsilon \cdot 1_M$, where $p_\lambda : M_\lambda \to M$ is the natural morphism to the colimit. If such a ϕ_ε exists for λ, then it exists for every $\mu \geq \lambda$. Hence, if $\mathfrak{m}_0 \subset \mathfrak{m}$ is a finitely generated subideal, say $\mathfrak{m}_0 = \sum_j^k V\varepsilon_j$, then there exist $\lambda \in \Lambda$ and $\phi_i : M \to M_\lambda$ such that $p_\lambda \circ \phi_i = \varepsilon_i \cdot 1_M$ for $i = 1, ..., k$. Hence $\text{Im}(\phi_i \circ p_\lambda - \varepsilon_i \cdot 1_{M_\lambda})$ is contained in $\text{Ker}\, p_\lambda$ and contains $\varepsilon_i \cdot \text{Ker}\, p_\lambda$. Hence $\text{Ker}\, p_\lambda$ has a finitely generated submodule L containing $\mathfrak{m}_0 \cdot \text{Ker}\, p_\lambda$. Choose a presentation $A^m \to A^n \xrightarrow{\pi} M_\lambda$. Then one can lift $\mathfrak{m}_0 L$ to a finitely generated submodule L' of A^n. Then $\text{Ker}(\pi) + L'$ is a finitely generated submodule of $\text{Ker}(p_\lambda \circ \pi)$ containing $\mathfrak{m}_0^2 \cdot \text{Ker}(p_\lambda \circ \pi)$. Since we also have $\mathfrak{m}_0 \cdot \text{Coker}(p_\lambda \circ \pi) = 0$ and $\mathfrak{m}_0$ is arbitrary, the conclusion follows from proposition 2.3.10. □

Lemma 2.3.18. *Let $0 \to M' \to M \to M'' \to 0$ be an exact sequence of A-modules. Then:*

(i) *If M', M'' are almost finitely generated (resp. presented) then so is M.*

(ii) *If M is almost finitely presented, then M'' is almost finitely presented if and only if M' is almost finitely generated.*

(iii) *If M is almost finitely generated and M'' is almost finitely presented, then M' is almost finitely generated.*

Proof. These facts can be deduced from proposition 2.3.16 and remark 2.4.12(iii), or proved directly. □

Lemma 2.3.19. *Let $(M_n ; \phi_n : M_n \to M_{n+1} \mid n \in \mathbb{N})$ be a direct system of A-modules and suppose there exist sequences $(\varepsilon_n \mid n \in \mathbb{N})$ and $(\delta_n \mid n \in \mathbb{N})$ of ideals of V such that*

(i) $\lim_{n \to \infty} \varepsilon_n^a = V^a$ *(for the uniform structure of definition 2.3.1) and $\prod_{j=0}^{\infty} \delta_j^a$ is a Cauchy product (see (2.3.4));*

(ii) *for all $n \in \mathbb{N}$ there exist integers $N(n)$ and morphisms of A-modules $\psi_n : A^{N(n)} \to M_n$ such that $\varepsilon_n \cdot \text{Coker}\, \psi_n = 0$;*

(iii) $\delta_n \cdot \text{Coker}\, \phi_n = 0$ *for all $n \in \mathbb{N}$.*

Then $\operatorname*{colim}_{n \in \mathbb{N}} M_n$ is an almost finitely generated A-module.

Proof. Let $M := \operatorname*{colim}_{n \in \mathbb{N}} M_n$. For any $n \in \mathbb{N}$ let $a_n = \bigcap_{m \geq 0}(\prod_{j=n}^{n+m} \delta_j)$. Then $\lim_{n \to \infty} a_n = V$. For $m > n$ set $\phi_{n,m} = \phi_m \circ ... \circ \phi_{n+1} \circ \phi_n : M_n \to M_{m+1}$ and let

$\phi_{n,\infty} : M_n \to M$ be the natural morphism. An easy induction shows that $\prod_{j=n}^{m} \delta_j \cdot$ Coker $\phi_{n,m} = 0$ for all $m > n \in \mathbb{N}$. Since Coker $\phi_{n,\infty} = \operatorname*{colim}_{m \in \mathbb{N}} \operatorname{Coker} \phi_{n,m}$ we obtain $a_n \cdot \operatorname{Coker} \phi_{n,\infty} = 0$ for all $n \in \mathbb{N}$. Therefore $\varepsilon_n a_n \cdot \operatorname{Coker}(\phi_{n,\infty} \circ \psi_n) = 0$ for all $n \in \mathbb{N}$. Since $\lim_{n \to \infty} \varepsilon_n a_n = V$, the claim follows. $\qquad\square$

In the remaining of this section *we assume that condition* (**B**) *of* (2.1.6) *holds.* We wish to define the Fitting ideals of an arbitrary uniformly almost finitely generated A-module M. This will be achieved in two steps: first we will see how to define the Fitting ideals of a finitely generated module, then we will deal with the general case. We refer to [54, Ch.XIX] for the definition of the Fitting ideals $F_i(M)$ of a finitely generated module over an arbitrary commutative ring R.

Lemma 2.3.20. *Let R be a V-algebra and M, N two finitely generated R-modules with an R^a-linear isomorphism $M^a \simeq N^a$. Then $F_i(M)^a = F_i(N)^a$ for all $i \geq 0$.*

Proof. By the usual arguments, for every $\varepsilon \in \mathfrak{m}$ we have morphisms $\alpha : M \to N$, $\beta : N \to M$ with kernels and cokernels killed by ε^2. Then we have: $F_i(N) \supset F_i(\operatorname{Im} \alpha) \cdot F_0(\operatorname{Coker} \alpha)$. If N is generated by k elements, then the same holds for Coker α, whence $\operatorname{Ann}_R(\operatorname{Coker} \alpha)^k \subset F_0(\operatorname{Coker} \alpha)$, therefore $\varepsilon^{2k} R \subset F_0(\operatorname{Coker} \alpha)$, and consequently $F_i(N) \supset \varepsilon^{2k} F_i(\operatorname{Im} \alpha)$. Since $\operatorname{Im}(\alpha)$ is a quotient of M, it is clear that $F_i(M) \subset F_i(\operatorname{Im} \alpha)$, so finally $\varepsilon^{2k} F_i(M) \subset F_i(N)$. Arguing symmetrically with β one has $\varepsilon^{2k} F_i(N) \subset F_i(M)$. Since we assume (**B**), the claim follows. $\qquad\square$

2.3.21.   Let M be a finitely generated A-module. In light of lemma 2.3.20, the Fitting ideals $F_i(M)$ are well defined as ideals in A.

Lemma 2.3.22. *Let $\mathfrak{m}_0 \subset \mathfrak{m}$ be a finitely generated subideal and $n \in \mathbb{N}$. Pick $\varepsilon_1, ..., \varepsilon_k \in \mathfrak{m}$ such that $\mathfrak{m}_0 \subset (\varepsilon_1^{3n}, ..., \varepsilon_k^{3n})$ and set $\mathfrak{m}_1 := (\varepsilon_1, ..., \varepsilon_k)$. Then $(F_i(M), F_i(M')) \in E_A(\mathfrak{m}_0)$ for every $(M, M') \in E_{\mathscr{M}}(\mathfrak{m}_1)$ such that M and M' are generated by at most n of their almost elements.*

Proof. Let M, M' be as in the lemma. By hypothesis, there exist an A-module N and morphisms $\phi : N \to M$ and $\psi : N \to M'$ such that $\mathfrak{m}_1$ annihilates the kernel and cokernel of ϕ and ψ. By symmetry, it suffices to show that $\varepsilon_i^{3n} F_i(M) \subset F_i(M')$ for every $i = 1, ..., k$. Now, for every $i \leq k$, the morphism $M \to M : x \mapsto \varepsilon_i \cdot x$ factors through a morphism $\alpha : M \to \phi(N)$, and similarly, scalar multiplication by ε_i on N factors through a morphism $\beta : \phi(N) \to N$. Then $\eta := \psi \circ \beta \circ \alpha : M \to M'$ has kernel and cokernel annihilated by ε_i^3. Pick A_*-modules $L \subset M_*$, $L' \subset M'_*$ generated by n elements, such that $L^a = M$ and $L'^a = M'$, and set $L'' := L' + \eta_*(L)$. Then $F_i(M) = F_i(L)^a$, $F_i(L'')^a = F_i(M')$ and $F_i(L'') \supset F_i(L/(L \cap \operatorname{Ker} \eta_*)) \cdot F_0(L''/\eta_* L)$. Since $L''/\eta_*(L)$ is generated by at most n elements and is annihilated by $\varepsilon_i^3 \cdot \mathfrak{m}$, we have $\varepsilon_i^{3n} \cdot \mathfrak{m} \subset F_0(L''/(\eta_* L))$. Furthermore $F_i(L/(L \cap \operatorname{Ker} \eta_*)) \supset F_i(L)$, so the claim follows. $\qquad\square$

Proposition 2.3.23. *For every $i, n \geq 0$, the map $F_i : \mathscr{M}_n(A) \to \mathscr{I}_A(A)$ is uniformly continuous and therefore it extends uniquely to a uniformly continuous map $F_i : \mathscr{U}_n(A) \to \mathscr{I}_A(A)$.*

Proof. The uniform continuity follows readily from lemma 2.3.22. Since $\mathscr{I}_A(A)$ is complete, it follows that F_i extends to the whole of $\mathscr{U}_n(A)$. Finally, the extension is unique because $\mathscr{I}_A(A)$ is separated. □

Definition 2.3.24. Let M be a uniformly almost finitely generated A-module. We call $F_i(M)$ the *i-th Fitting ideal* of M.

Proposition 2.3.25. *Let $0 \to M' \xrightarrow{\phi} M \xrightarrow{\psi} M'' \to 0$ be a short exact sequence of A-modules, such that M is uniformly almost finitely generated.*

  (i) *M'' is uniformly almost finitely generated and $F_i(M) \subset F_i(M'')$ for every $i \in \mathbb{N}$.*

  (ii) *If M' is also uniformly almost finitely generated, then*

$$\sum_{j+k=i} F_j(M') \cdot F_k(M'') \subset F_i(M) \qquad \textit{for every } i \geq 0.$$

  (iii) *For any A-algebra B and any $i \geq 0$ we have $F_i(B \otimes_A M) = F_i(M) \cdot B$.*

Proof. (i) is easy and shall be left to the reader.

(ii): Let n be a uniform bound for M and M'; by remark 2.3.2 we can find, for every subideal $\mathfrak{m}_0 \subset \mathfrak{m}$, A-modules M_0, M_0', L, L' and morphisms

$$M \xrightarrow{\alpha} L \xleftarrow{\beta} M_0 \qquad M' \xrightarrow{\alpha'} L' \xleftarrow{\beta'} M_0'$$

whose kernels and cokernels are annihilated by $\mathfrak{m}_0$, and such that M_0 and M_0' are generated by n almost elements. Let N be defined by the push-out diagram

$$
\begin{array}{ccc}
M' & \xrightarrow{\phi} M \xrightarrow{\alpha} & L \\
{\scriptstyle \alpha'}\downarrow & & \downarrow{\scriptstyle \gamma} \\
L' & \xrightarrow{\gamma'} & N.
\end{array}
$$

Furthermore set

$$M_1' := \mathrm{Im}(\gamma' \circ \beta' : M_0' \to N)$$
$$M_1 := \mathrm{Im}((\gamma \circ \beta) \oplus (\gamma' \circ \beta') : M_0 \oplus M_0' \to N)$$

and let M_1'' be the cokernel of the induced monomorphism $M_1' \to M_1$. We deduce a commutative diagram with short exact rows:

(2.3.26)
$$
\begin{array}{ccccccccc}
0 & \longrightarrow & M' & \xrightarrow{\phi} & M & \xrightarrow{\psi} & M'' & \longrightarrow & 0 \\
 & & \downarrow & & \downarrow & & \downarrow & & \\
0 & \longrightarrow & \mathrm{Im}\,\gamma' & \longrightarrow & N & \longrightarrow & \mathrm{Coker}\,\gamma' & \longrightarrow & 0 \\
 & & \uparrow & & \uparrow & & \uparrow & & \\
0 & \longrightarrow & M_1' & \longrightarrow & M_1 & \longrightarrow & M_1'' & \longrightarrow & 0.
\end{array}
$$

One checks easily that the kernels and cokernels of all the vertical arrows in (2.3.26) are annihilated by $\mathfrak{m}_0^2$, *i.e.* $(M, M_1), (M', M_1'), (M'', M_1'') \in E_{\mathscr{M}}(\mathfrak{m}_0^2)$.

Let $x_1, ..., x_n \in M_{0*}$ (resp. $x'_1, ..., x'_n \in M'_{0*}$) be a set of generators for M_0 (resp. for M'_0). For every $i = 1, ..., n$, let $z_i := \gamma \circ \beta(x_i)$ and $z'_i := \gamma' \circ \beta'(x'_i)$. Let $Q' \subset N_*$ (resp $Q \subset N_*$) be the A_*-module generated by the z'_i (resp. and by the z_i). It is clear that the bottom row of (2.3.26) is naturally isomorphic to the short exact sequence $(0 \rightarrow Q' \rightarrow Q \rightarrow Q/Q' \rightarrow 0)^a$. It is well-known that $F_i(Q') \cdot F_j(Q/Q') \subset F_{i+j}(Q)$ for every $i, j \in \mathbb{N}$; by lemma 2.3.5(ii.c) and proposition 2.3.23 all the operations under considerations are uniformly continuous, so we deduce $F_i(M') \cdot F_j(M/M') \subset F_{i+j}(M)$, which is (i).

(iii): Since the identity is known for usual finitely generated modules over rings, the claim follows easily from proposition 2.3.23 and lemma 2.3.7(v).   □

2.4. Almost homological algebra. In this section we fix an almost V-algebra A and we consider various constructions in the category of A-modules.

2.4.1. By corollary 2.2.16 any inverse system $(M_n \mid n \in \mathbb{N})$ of A-modules has an (inverse) limit $\lim_{n \in \mathbb{N}} M$. As usual, we denote by $\lim^1$ the right derived functor of the inverse limit functor. Notice that [77, Cor. 3.5.4] holds in the almost case since axiom (AB4*) holds in A-**Mod**; on the other hand, [77, Lemma 3.5.3] does not hold under (AB4*), (the proof given there uses elements : for a counterexample in an exotic abelian category, see [62]).

Lemma 2.4.2. *Let* $(M_n ; \phi_n : M_n \rightarrow M_{n+1} \mid n \in \mathbb{N})$ *(resp.* $(N_n ; \psi_n : N_{n+1} \rightarrow N_n \mid n \in \mathbb{N}))$ *be a direct (resp. inverse) system of A-modules and morphisms and* $(\varepsilon_n \mid n \in \mathbb{N})$ *a sequence of ideals of V^a converging to V^a (for the uniform structure of definition* 2.3.1).

(i) *If* $\varepsilon_n \cdot M_n = 0$ *for all* $n \in \mathbb{N}$ *then* $\operatorname*{colim}_{n \in \mathbb{N}} M_n \simeq 0$.

(ii) *If* $\varepsilon_n \cdot N_n = 0$ *for all* $n \in \mathbb{N}$ *then* $\lim_{n \in \mathbb{N}} N_n \simeq 0 \simeq \lim^1_{n \in \mathbb{N}} N_n$.

(iii) *If* $\varepsilon_n \cdot \operatorname{Coker} \psi_n = 0$ *for all* $n \in \mathbb{N}$ *and* $\prod_{j=0}^{\infty} \varepsilon_j$ *is a Cauchy product, then* $\lim^1_{n \in \mathbb{N}} N_n \simeq 0$.

Proof. (i) and (ii): We remark only that $\lim^1_{n \in \mathbb{N}} N_n \simeq \lim^1_{n \in \mathbb{N}} N_{n+p}$ for all $p \in \mathbb{N}$ and leave the details to the reader. We prove (iii). From [77, Cor. 3.5.4] it follows easily that $(\lim^1_{n \in \mathbb{N}} N_{n*})^a \simeq \lim^1_{n \in \mathbb{N}} N_n$. It then suffices to show that $\lim^1_{n \in \mathbb{N}} N_{n*}$ is almost zero. We have $\varepsilon_n^2 \cdot \operatorname{Coker} \psi_{n*} = 0$ and the product $\prod_{j=0}^{\infty} \varepsilon_j^2$ is again a Cauchy product. Next let $N'_n := \bigcap_{p \geq 0} \operatorname{Im}(N_{n+p*} \rightarrow N_{n*})$. If $J_n := \bigcap_{p \geq 0} (\varepsilon_n \cdot \varepsilon_{n+1} \cdot ... \cdot \varepsilon_{n+p})^2$ then $J_n N_{n*} \subset N'_n$ and $\lim_{n \rightarrow \infty} J_n^a = V^a$. In view of (ii), $\lim^1_{n \in \mathbb{N}} N_{n*}/N'_n$ is almost zero, hence we reduce to showing that $\lim^1_{n \in \mathbb{N}} N'_n$ is almost zero. But

$$J_{n+p+q} \cdot N'_n \subset \operatorname{Im}(N'_{n+p+q} \rightarrow N'_n) \subset \operatorname{Im}(N'_{n+p} \rightarrow N'_n)$$

for all $n, p, q \in \mathbb{N}$. On the other hand, since the ideals J_n^a converge to V^a, we get $\bigcup_{q=0}^{\infty} \mathfrak{m} \cdot J_{n+p+q} = \mathfrak{m}$, hence $\mathfrak{m} N'_n \subset \operatorname{Im}(N'_{n+p} \rightarrow N'_n)$ and finally $\mathfrak{m} N'_n =$

$\mathfrak{m}^2 N_n' \subset \mathrm{Im}(\mathfrak{m} N_{n+p}' \to \mathfrak{m} N_n')$ which means that $\{\mathfrak{m} N_n'\}$ is a surjective inverse system, so its $\lim^1$ vanishes and the result follows. □

Example 2.4.3. Let $(V, \mathfrak{m})$ be as in example 2.1.2. Then every finitely generated ideal in V is principal, so in the situation of lemma 2.4.2 we can write $\varepsilon_j = (x_j)$ for some $x_j \in V$. Then the hypothesis in (iii) means that there exists $c \in \mathbb{N}$ such that $x_j \neq 0$ for all $j \geq c$ and the sequence $n \mapsto \prod_{j=c}^n |x_j|$ is Cauchy in Γ.

Definition 2.4.4. Let M be an A-module.

(i) We say that M is *flat* (resp. *faithfully flat*) if the functor $N \mapsto M \otimes_A N$, from the category of A-modules to itself is exact (resp. exact and faithful).

(ii) We say that M is *almost projective* if the functor $N \mapsto \mathrm{alHom}_A(M, N)$ is exact.

For euphonic reasons, we will use the expression "almost finitely generated projective" to denote an A-module which is almost projective and almost finitely generated. This convention does not give rise to ambiguities, since we will never consider projective almost modules : indeed, the following example 2.4.5 explains why the categorical notion of projectivity is useless in the setting of almost ring theory.

Example 2.4.5. First of all we remark that the functor $M \mapsto M_!$ preserves (categorical) projectivity, since it is left adjoint to an exact functor. Moreover, if $P_!$ is a projective V-module P is a projective V^a-module, as one checks easily using the fact the functor $M \mapsto M_!$ is right exact. Hence one has an equivalence from the full subcategory of projective V^a-modules, to the full subcategory of projective V-modules P such that the counit of the adjunction $\tilde{\mathfrak{m}} \otimes_V P \to P$ is an isomorphism. The latter condition is equivalent to $P = \mathfrak{m} P$; indeed, as P is flat, we have $\mathfrak{m} P = \mathfrak{m} \otimes_V P$.

As an example, suppose that V is local; then every projective V-module is free, so if $\mathfrak{m} \neq V$ the condition $P = \mathfrak{m} P$ implies $P = 0$, *ergo*, there are no non-trivial projective V^a-modules. For more results along these lines (including a non-commutative generalization), see [69], especially Théorème 2 and Exemple 4 of *loc.cit.*

Lemma 2.4.6. *Let M be an almost finitely generated A-module and B a flat A-algebra. Then $\mathrm{Ann}_B(B \otimes_A M) = B \otimes_A \mathrm{Ann}_A(M)$.*

Proof. Using lemma 2.3.7(i),(v) we reduce to the case of a finitely generated A-module M. Then, let $x_1, ..., x_k \in M_*$ be a finite set of generators for M; we have:

$$\mathrm{Ann}_A(M) = \mathrm{Ker}(\phi : A \to M^k)$$

where ϕ is defined by the rule: $a \mapsto (ax_1, ..., ax_k)$ for every $a \in A_*$. Since B is flat, we have $\mathrm{Ker}(1_B \otimes_A \phi) \simeq B \otimes_A \mathrm{Ker} \phi$, whence the claim. □

Lemma 2.4.7. *Let $\mathbf{P}$ be one of the properties : "flat", "almost projective", "almost finitely generated", "almost finitely presented". If B is a $\mathbf{P}$ A-algebra, and M is a $\mathbf{P}$ B-module, then M is $\mathbf{P}$ as an A-module.*

Proof. Left to the reader. □

2.4.8.   Let R be a V-algebra and M a flat (resp. faithfully flat) R-module (in the usual sense, see [58, p.45]). Then M^a is a flat (resp. faithfully flat) R^a-module. Indeed, the functor $M \otimes_R -$ preserves the Serre subcategory of almost zero modules, so by general facts it induces an exact functor on the localized categories (cp. [40, p.369]). For the faithfulness we have to show that an R-module N is almost zero whenever $M \otimes_R N$ is almost zero. However, $M \otimes_R N$ is almost zero $\Leftrightarrow M \otimes_R (\mathfrak{m} \otimes_V N) = 0 \Leftrightarrow \mathfrak{m} \otimes_V N = 0 \Leftrightarrow N$ is almost zero. It is clear that A-**Mod** has enough almost projective (resp. flat) objects.

2.4.9.   Suppose that $\tilde{\mathfrak{m}}$ is a flat V-module and let R be a V-algebra. The localization functor induces a functor

$$G : \mathrm{D}(R\text{-}\mathbf{Mod}) \to \mathrm{D}(R^a\text{-}\mathbf{Mod})$$

and, in view of corollary 2.2.24, $M \mapsto M_!$ induces a functor $F : \mathrm{D}(R^a\text{-}\mathbf{Mod}) \to \mathrm{D}(R\text{-}\mathbf{Mod})$. We have a natural isomorphism $G \circ F \simeq 1_{\mathrm{D}(R^a\text{-}\mathbf{Mod})}$ and a natural transformation $F \circ G \to 1_{\mathrm{D}(R\text{-}\mathbf{Mod})}$. These satisfy the triangular identities of [57, p.83], so F is a left adjoint to G. If Σ denotes the multiplicative set of morphisms in $\mathrm{D}(R\text{-}\mathbf{Mod})$ which induce almost isomorphisms on the cohomology modules, then Σ is locally small and the localized category $\Sigma^{-1}\mathrm{D}(R\text{-}\mathbf{Mod})$ exists (see *e.g.* [77, Th.10.3.7]); then by the same argument we get an equivalence of categories:

$$\Sigma^{-1}\mathrm{D}(R\text{-}\mathbf{Mod}) \simeq \mathrm{D}(R^a\text{-}\mathbf{Mod}).$$

2.4.10.   Given an A-module M, we can derive the functors $M \otimes_A -$ (resp. $\mathrm{alHom}_A(M, -)$, resp. $\mathrm{alHom}_A(-, M)$) by taking flat (resp. injective, resp. almost projective) resolutions : one remarks that bounded above exact complexes of flat (resp. almost projective) A-modules are acyclic for the functor $M \otimes_A -$ (resp. $\mathrm{alHom}_A(-, M)$) (recall the standard argument: if $F_\bullet$ is a bounded above exact complex of flat A-modules, let $\Phi_\bullet$ be a flat resolution of M; then $\mathrm{Tot}(\Phi_\bullet \otimes_A F_\bullet) \to M \otimes_A F_\bullet$ is a quasi-isomorphism since it is so on rows, and $\mathrm{Tot}(\Phi_\bullet \otimes_A F_\bullet)$ is acyclic since its columns are; similarly, if $P_\bullet$ is a complex of almost projective objects, one considers the double complex $\mathrm{alHom}_A(P_\bullet, J^\bullet)$ where $J^\bullet$ is an injective resolution of M; cp. [77, §2.7]); then one uses the construction detailed in [77, Th.10.5.9]. We denote by $\mathrm{Tor}_i^A(M, -)$ (resp. $\mathrm{alExt}_A^i(M, -)$, resp. $\mathrm{alExt}_A^i(-, M)$) the corresponding derived functors. If $A := R^a$ for some V-algebra R, we obtain easily natural isomorphisms

(2.4.11) $\mathrm{Tor}_i^R(M, N)^a \simeq \mathrm{Tor}_i^A(M^a, N^a)$

for all R-modules M, N. Similarly:

$$\mathrm{Ext}_R^i(M, N)^a \simeq \mathrm{alExt}_A^i(M^a, N^a) \simeq \mathrm{Ext}_A^i(M^a, N^a)^a.$$

Remark 2.4.12.   (i) Clearly, an A-module M is flat (resp. almost projective) if and only if $\mathrm{Tor}_i^A(M, N) = 0$ (resp. $\mathrm{alExt}_A^i(M, N) = 0$) for all A-modules N and all $i > 0$. In particular, an almost projective A-module is flat, because for every $\varepsilon \in \mathfrak{m}$ the scalar multiplication by $\varepsilon : M \to M$ factors through a free module.

(ii) Let M, N be two flat (resp. almost projective) A-modules. Then $M \otimes_A N$ is a flat (resp. almost projective) A-module and for any A-algebra B, the B-module $B \otimes_A M$ is flat (resp. almost projective).

(iii) Resume the notation of proposition 2.3.16. If M is almost finitely presented, then one has also that the natural morphism

$$\operatorname*{colim}_\Lambda \operatorname{alExt}^1_A(M, N_\lambda) \to \operatorname{alExt}^1_A(M, \operatorname*{colim}_\Lambda N_\lambda)$$

is a monomorphism. This is deduced from proposition 2.3.16(ii), using the fact that (N_λ) can be injected into an inductive system (J_λ) of injective A-modules (e.g. $J_\lambda = E^{\operatorname{Hom}_A(N_\lambda, E)}$, where E is an injective cogenerator for A-**Mod**), and by applying alExt sequences.

Lemma 2.4.13. *Resume the notation of* (2.3.14). *The following conditions are equivalent:*

(i) M^a *is a flat (resp. almost projective) R^a-module.*

(ii) M^a_λ *is a flat (resp. almost projective) R^a_λ-module for every $\lambda \in \Lambda$.*

Proof. It is a direct consequence of (2.4.11) and its analogue for $\operatorname{Ext}^i_R(M, N)$. □

Lemma 2.4.14. *Suppose that $\widetilde{\mathfrak{m}}$ is a flat V-module and let R be a V-algebra.*

(i) *There is a natural isomorphism:*

$$\operatorname{Ext}^i_R(M_!, N) \simeq \operatorname{Ext}^i_{R^a}(M, N^a)$$

for every R^a-module M, every R-module N and every $i \in \mathbb{N}$.

(ii) *If P is an almost projective R^a-module, then:*

$$\operatorname{hom.dim}_R P_! \leq \operatorname{hom.dim}_V \widetilde{\mathfrak{m}}.$$

Proof. (Here hom.dim denotes homological dimension.) (i) is a straightforward consequence of the existence of the adjunction (F, G) of (2.4.9). Next we consider, for arbitrary R-modules M and N, the spectral sequence:

$$E^{p,q}_2 := \operatorname{Ext}^p_V(\widetilde{\mathfrak{m}}, \operatorname{Ext}^q_R(M, N)) \Rightarrow \operatorname{Ext}^{p+q}_R(\widetilde{\mathfrak{m}} \otimes_V M, N).$$

(This is constructed *e.g.* from the double complex $\operatorname{Hom}_V(F_p, \operatorname{Hom}_R(F'_q, N))$ where $F_\bullet$ (resp. $F'_\bullet$) is a projective resolution of $\widetilde{\mathfrak{m}}$ (resp. of M).) If now we let $M := P_!$, we deduce from (i) that $\operatorname{Ext}^p_V(\widetilde{\mathfrak{m}}, \operatorname{Ext}^q_R(P_!, N)) \simeq \operatorname{Ext}^p_{V^a}(V^a, \operatorname{Ext}^q_R(P_!, N)^a) \simeq 0$ for every $p \in \mathbb{N}$ and every $q > 0$. Since $\widetilde{\mathfrak{m}} \otimes_V P_! \simeq P_!$, assertion (ii) follows easily. □

Lemma 2.4.15. *Let M be an almost finitely generated A-module. Then M is almost projective if and only if, for arbitrary $\varepsilon \in \mathfrak{m}$, there exist $n(\varepsilon) \in \mathbb{N}$ and A-linear morphisms*

(2.4.16)
$$M \xrightarrow{u_\varepsilon} A^{n(\varepsilon)} \xrightarrow{v_\varepsilon} M$$

such that $v_\varepsilon \circ u_\varepsilon = \varepsilon \cdot 1_M$.

Proof. Let morphisms as in (2.4.16) be given. Pick any A-module N and apply the functor $\mathrm{alExt}_A^i(-, N)$ to (2.4.16) to get morphisms

$$\mathrm{alExt}_A^i(M, N) \to \mathrm{alExt}_A^i(A^{n(\varepsilon)}, N) \to \mathrm{alExt}_A^i(M, N)$$

whose composition is again the scalar multiplication by ε; hence $\varepsilon \cdot \mathrm{alExt}_A^i(M, N) = 0$ for all $i > 0$. Since ε is arbitrary, it follows from remark 2.4.12(i) that M is almost projective. Conversely, suppose that M is almost projective; by hypothesis, for arbitrary $\varepsilon \in \mathfrak{m}$ we can find $n := n(\varepsilon)$ and a morphism $\phi_\varepsilon : A^n \to M$ such that $\varepsilon \cdot \mathrm{Coker}\, \phi_\varepsilon = 0$. Let M_ε be the image of ϕ_ε, so that ϕ_ε factors as $A^{n(\varepsilon)} \xrightarrow{\psi_\varepsilon} M_\varepsilon \xrightarrow{j_\varepsilon} M$. Also $\varepsilon \cdot 1_M : M \to M$ factors as $M \xrightarrow{\gamma_\varepsilon} M_\varepsilon \xrightarrow{j_\varepsilon} M$. Since by hypothesis M is almost projective, the natural morphism induced by ψ_ε :

$$\mathrm{alHom}_A(M, A^n) \xrightarrow{\psi_\varepsilon^*} \mathrm{alHom}_A(M, M_\varepsilon)$$

is an epimorphism. Then for arbitrary $\delta \in \mathfrak{m}$ the morphism $\delta \cdot \gamma_\varepsilon$ is in the image of ψ_ε^*, in other words, there exists an A-linear morphism $u_{\varepsilon\delta} : M \to A^n$ such that $\psi_\varepsilon \circ u_{\varepsilon\delta} = \delta \cdot \gamma_\varepsilon$. If now we take $v_{\varepsilon\delta} := \phi_\varepsilon$, it is clear that $v_{\varepsilon\delta} \circ u_{\varepsilon\delta} = \varepsilon \cdot \delta \cdot 1_M$. This proves the claim. $\square$

Lemma 2.4.17. *Let R be any ring, M any R-module and $C := \mathrm{Coker}(\phi : R^n \to R^m)$ any finitely presented (left) R-module. Let $C' := \mathrm{Coker}(\phi^* : R^m \to R^n)$ be the cokernel of the transpose of the map ϕ. Then there is a natural isomorphism*

$$\mathrm{Tor}_1^R(C', M) \simeq \mathrm{Hom}_R(C, M)/\mathrm{Im}(\mathrm{Hom}_R(C, R) \otimes_R M).$$

Proof. We have a spectral sequence :

$$E_{ij}^2 := \mathrm{Tor}_i^R(H_j(\mathrm{Cone}\,\phi^*), M) \Rightarrow H_{i+j}(\mathrm{Cone}(\phi^*) \otimes_R M).$$

On the other hand we have also natural isomorphisms

$$\mathrm{Cone}(\phi^*) \otimes_R M \simeq \mathrm{Hom}_R(\mathrm{Cone}\,\phi, R)[1] \otimes_R M \simeq \mathrm{Hom}_R(\mathrm{Cone}\,\phi, M)[1].$$

Hence :

$$E_{10}^2 \simeq E_{10}^\infty \simeq H_1(\mathrm{Cone}(\phi^*) \otimes_R M)/E_{01}^\infty \simeq H^0(\mathrm{Hom}_R(\mathrm{Cone}\,\phi, M))/\mathrm{Im}(E_{01}^2)$$
$$\simeq \mathrm{Hom}_R(C, M)/\mathrm{Im}(\mathrm{Hom}_R(C, R) \otimes_R M)$$

which is the claim. $\square$

Proposition 2.4.18. *Let A be a V^a-algebra.*

   (i) *Every almost finitely generated projective A-module is almost finitely presented.*

   (ii) *Every almost finitely presented flat A-module is almost projective.*

Proof. (ii): Let M be such an A-module. Let $\varepsilon, \delta \in \mathfrak{m}$ and pick a three term complex

$$A^m \xrightarrow{\psi} A^n \xrightarrow{\phi} M$$

such that $\varepsilon \cdot \mathrm{Coker}\, \phi = \delta \cdot \mathrm{Ker}(\phi)/\mathrm{Im}(\psi) = 0$. Set $P := \mathrm{Coker}\, \psi_*$; this is a finitely presented A_*-module and ϕ_* factors through a morphism $\overline{\phi}_* : P \to M_*$. Let $\gamma \in \mathfrak{m}$; from lemma 2.4.17 we see that $\gamma \cdot \overline{\phi}$ is the image of some element $\sum_{j=1}^n \phi_j \otimes m_j \in \mathrm{Hom}_{A_*}(P, A_*) \otimes_{A_*} M_*$. If we define $L := A_*^n$ and $v : P \to L$,

$w : L \to M_*$ by $v(x) := (\phi_1(x), ..., \phi_n(x))$ and $w(y_1, ..., y_n) := \sum_{j=1}^{n} y_j \cdot m_j$, then clearly $\gamma \cdot \overline{\phi} = w \circ v$. Let $K := \mathrm{Ker}\, \overline{\phi}_*$. Then $\delta \cdot K^a = 0$ and the map $\delta \cdot 1_{P^a}$ factors through a morphism $\sigma : (P/K)^a \to P^a$. Similarly the map $\varepsilon \cdot 1_M$ factors through a morphism $\lambda : M \to (P/K)^a$. Let $\alpha := v^a \circ \sigma \circ \lambda : M \to L^a$ and $\beta := w^a : L^a \to M$. The reader can check that $\beta \circ \alpha = \varepsilon \cdot \delta \cdot \gamma \cdot 1_M$. By lemma 2.4.15 the claim follows.

(i): Let P be such an almost finitely generated projective A-module. For any finitely generated ideal $\mathfrak{m}_0 \subset \mathfrak{m}$ pick a morphism $\phi : A^r \to P$ such that $\mathfrak{m}_0 \cdot \mathrm{Coker}\, \phi = 0$. If $\varepsilon_1, ..., \varepsilon_k$ is a set of generators for $\mathfrak{m}_0$, a standard argument shows that, for any $i \leq k$, $\varepsilon_i \cdot 1_P$ lifts to a morphism $\psi_i : P \to A^r/\mathrm{Ker}\, \phi$; then, since P is almost projective, $\varepsilon_j \psi_i$ lifts to a morphism $\psi_{ij} : P \to A^r$. Now claim 2.3.11 applies with $F_1 := A^r$, $F_2 := M = P$, $p := \phi$, $q := 1_P$ and $\psi := \psi_{ij}$ and shows that $\mathrm{Ker}\, \phi$ has a finitely generated submodule M_{ij} containing $\varepsilon_i \varepsilon_j \cdot \mathrm{Ker}\, \phi$. Then the span of all such M_{ij} is a finitely generated submodule of $\mathrm{Ker}\, \phi$ containing $\mathfrak{m}_0^2 \cdot \mathrm{Ker}\, \phi$. By proposition 2.3.10(ii), the claim follows. □

In general a flat almost finitely generated A-module is not necessarily almost finitely presented, but one can give the following criterion, which extends [17, §1, Exerc. 13].

Proposition 2.4.19. *If $A \to B$ is a monomorphism of V^a-algebras and M is an almost finitely generated flat A-module such that $B \otimes_A M$ is almost finitely presented over B, then M is almost finitely presented over A.*

Proof. Let $\mathfrak{m}_0$ be a finitely generated subideal of $\mathfrak{m}$ and an A-linear morphism $\phi : A^n \to M$ such that $\mathfrak{m}_0 \cdot \mathrm{Coker}\, \phi = 0$. By assumption and claim 2.3.12, we can find a finitely generated B-submodule R of $\mathrm{Ker}(1_B \otimes_A \phi)$ such that

$$(2.4.20) \qquad \mathfrak{m}_0^2 \cdot \mathrm{Ker}(1_B \otimes_A \phi) \subset R.$$

By a Tor sequence we have $\mathfrak{m}_0 \cdot \mathrm{Coker}(1_B \otimes_A \mathrm{Ker}(\phi) \to \mathrm{Ker}(1_B \otimes_A \phi)) = 0$, hence $\mathfrak{m} \cdot \mathfrak{m}_0 \cdot \mathrm{Coker}(1_{B_*} \otimes_{A_*} \mathrm{Ker}(\phi)_* \to \mathrm{Ker}(1_B \otimes_A \phi)_*) = 0$, therefore there exists a finitely generated submodule R_0 of $\mathrm{Ker}\, \phi$ such that

$$(2.4.21) \qquad \mathfrak{m}_0^2 R \subset B \cdot \mathrm{Im}(R_0 \to \mathrm{Ker}(1_B \otimes_A \phi)).$$

By lemma 2.4.17, for every $\varepsilon \in \mathfrak{m}$ the morphism $\varepsilon \cdot \overline{\phi} : A^n/R_0 \to M$ factors through a morphism $\psi : A^n/R_0 \to F$, where F is a finitely generated free A-module. Since $F \subset B \otimes_A F$, we deduce easily $\mathrm{Ker}(A^n/R_0 \to B^n/B \cdot R_0) \subset \mathrm{Ker}\, \psi$; on the other hand, by (2.4.20) and (2.4.21) we derive $\mathfrak{m}_0^4 \cdot \mathrm{Ker}\, \phi \subset \mathrm{Ker}(A^n/R_0 \to B^n/B \cdot R_0)$. Thus ψ factors through $M' := A^n/(R_0 + \mathfrak{m}_0^4 \cdot \mathrm{Ker}\, \phi)$. Clearly $\mathfrak{m}_0^4 \cdot \mathrm{Ker}(M' \to M) = \mathfrak{m}_0 \cdot \mathrm{Coker}(M' \to M) = 0$; hence, for every A-module N, the kernel of the induced morphism

$$(2.4.22) \qquad \mathrm{alExt}_A^1(M, N) \to \mathrm{alExt}_A^1(M', N)$$

is annihilated by $\mathfrak{m}_0^5$; however $\varepsilon \cdot (2.4.22)$ factors through $\mathrm{alExt}_A^1(F, N) = 0$, therefore $\mathfrak{m}_0^6 \cdot \mathrm{alExt}_A^1(M, N) = 0$ for every A-module N. This shows that M is almost projective, which is equivalent to the conclusion, in view of proposition 2.4.18(i). □

Definition 2.4.23. Let M be an A-module, $f : A \to M$ an almost element of M.

(i) The *dual A-module* of M is the A-module $M^* := \mathrm{alHom}_A(M, A)$.

(ii) The *evaluation morphism* is the morphism

$$\mathrm{ev}_{M/A} : M \otimes_A M^* \to A \; : \; m \otimes \phi \mapsto \phi(m).$$

(iii) The *evaluation ideal* of M is the ideal $\mathscr{E}_{M/A} := \mathrm{Im}\, \mathrm{ev}_{M/A}$.

(iv) The *evaluation ideal* of f is the ideal

$$\mathscr{E}_{M/A}(f) := \mathrm{Im}(\mathrm{ev}_{M/A} \circ (f \otimes_A 1_{M^*})).$$

(v) We say that M is *reflexive* if the natural morphism

$$(2.4.24) \qquad M \to (M^*)^* \qquad m \mapsto (f \mapsto f(m))$$

is an isomorphism of A-modules.

(vi) We say that M is *invertible* if $M \otimes_A M^* \simeq A$.

Remark 2.4.25. Let B be an A-algebra, N any A-module and M any B-module.

(i) By "restriction of scalars" M is also an A-module and the A-module $\mathrm{alHom}_A(M, N)$ has a natural structure of B-module. This is defined by the rule $(b \cdot f)(m) := f(b \cdot m)$ (for all $b \in B_*$, $m \in M_*$ and $f \in \mathrm{Hom}_A(M, N)$). Especially, the dual A-module M^* of M is a B-module. With respect to this structure (2.4.24) becomes a B-linear morphism. Incidentally, notice that the two meanings of "M_*^*" coincide, *i.e.* $(M_*)^* \simeq (M^*)_*$.

(ii) Furthermore, the functor

$$A\text{-}\mathbf{Mod} \to B\text{-}\mathbf{Mod} \qquad : \qquad N \mapsto \mathrm{alHom}_A(B, N)$$

is right adjoint to the forgetful functor $B\text{-}\mathbf{Mod} \to A\text{-}\mathbf{Mod}$. More precisely, one has a natural B_*-linear isomorphism

$$\mathrm{Hom}_A(M, N) \xrightarrow{\sim} \mathrm{Hom}_B(M, \mathrm{alHom}_A(B, N))$$

defined by the rule: $f \mapsto (x \mapsto (b \mapsto f(b \cdot x)))$ for every $f : M \to N$, $x \in M_*$, $b \in B_*$. Its inverse is the morphism given by the rule: $g \mapsto (x \mapsto g(x)(1))$ for every $g : M \to \mathrm{alHom}_A(B, M)$ and $x \in M_*$. We leave to the reader the easy verification.

2.4.26. If E, F and N are A-modules, there is a natural morphism :

$$(2.4.27) \qquad E \otimes_A \mathrm{alHom}_A(F, N) \to \mathrm{alHom}_A(F, E \otimes_A N).$$

Let P be an A-module. As a special case of (2.4.27) we have the morphism:

$$\omega_{P/A} : P \otimes_A P^* \to \mathrm{End}_A(P)^a := \mathrm{alHom}_A(P, P)$$

such that $\omega_{P/A}(p \otimes \phi)(q) := p \cdot \phi(q)$ for every $p, q \in P_*$ and $\phi : P \to A$.

Proposition 2.4.28. *Let P be an almost projective A-module and $f \in P_*$.*

(i) *For every morphism of algebras $A \to B$ we have $\mathscr{E}_{B \otimes_A P/B} = \mathscr{E}_{P/A} \cdot B$.*

(ii) *$\mathscr{E}_{P/A} = \mathscr{E}_{P/A}^2$.*

(iii) *$P = 0$ if and only if $\mathscr{E}_{P/A} = 0$.*

(iv) *P is faithfully flat if and only if $\mathscr{E}_{P/A} = A$.*

(v) *$\mathscr{E}_{P/A}(f)$ is the smallest of the ideals $J \subset A$ such that $f \in (JP)_*$.*

Proof. Pick an indexing set I large enough, and an epimorphism $\phi : F := A^{(I)} \to P$. For every $i \in I$ we have the standard morphisms $A \xrightarrow{e_i} F \xrightarrow{\pi_i} A$ such that $\pi_i \circ e_j = \delta_{ij} \cdot 1_A$ and $\sum_{i \in I} e_i \circ \pi_i = 1_F$. For every $x \in \mathfrak{m}$ choose $\psi_x \in \mathrm{Hom}_A(P, F)$ such that $\phi \circ \psi_x = x \cdot 1_P$. It is easy to check that $\mathscr{E}_{P/A}$ is generated by the almost elements $\pi_i \circ \psi_x \circ \phi \circ e_j$ $(i, j \in I, x \in \mathfrak{m})$. (i) follows already. For (iii), the "only if" is clear; if $\mathscr{E}_{P/A} = 0$, then $\psi_x \circ \phi = 0$ for all $x \in \mathfrak{m}$, hence $\psi_x = 0$ and therefore $x \cdot 1_P = 0$, *i.e.* $P = 0$. Next, notice that (i) and (iii) imply $P/\mathscr{E}_{P/A}P = 0$, *i.e.* $P = \mathscr{E}_{P/A}P$, so (ii) follows directly from the definition of $\mathscr{E}_{P/A}$. Since P is flat, to show the "if" claim of (iv) we have only to verify that the functor $M \mapsto P \otimes_A M$ is faithful. To this purpose, it suffices to check that $P \otimes_A A/J \neq 0$ for every proper ideal J of A. This follows easily from (i) and (iii). Conversely, if P is faithfully flat, (i) and (iii) imply easily that $\mathscr{E}_{P/A} = A$. Finally, it clear that $\mathscr{E}_{P/A}(f) \subset J$ for every ideal $J \subset A$ with $f \in (JP)_*$. Conversely, suppose first that $P = A^{(I)}$; by example 2.2.17, for every $x \in \mathfrak{m}$ we can represent xf as an element of $(A_*)^{(I)}$, so $xf = \sum_{i \in I} \pi_i(xf)e_i \in \mathscr{E}_{P/A}(xf)_*P_* \subset (\mathscr{E}_{P/A}(f)P)_*$, whence the claim in this case. For general P, choose an epimorphism $\phi : F \to P$ and morphisms ψ_x $(x \in \mathfrak{m})$ as in the foregoing; we deduce that $\psi_x(f) \in (\mathscr{E}_{F/A}(\psi_x(f))F)_* \subset (\mathscr{E}_{P/A}(f)F)_*$, so $xf = \phi \circ \psi_x(f) \in (\mathscr{E}_{P/A}(f)P)_*$ which implies the claim. $\square$

Lemma 2.4.29. *Let E, F, N be three A-modules.*

(i) *The morphism (2.4.27) is an isomorphism in the following cases :*
   (a) *when E is flat and F is almost finitely presented;*
   (b) *when either E or F is almost finitely generated projective;*
   (c) *when F is almost projective and E is almost finitely presented;*
   (d) *when E is almost projective and F is almost finitely generated.*

(ii) *The morphism (2.4.27) is a monomorphism in the following cases :*
   (a) *when E is flat and F is almost finitely generated;*
   (b) *when E is almost projective.*

(iii) *The morphism (2.4.27) is an epimorphism when F is almost projective and E is almost finitely generated.*

Proof. If $F \simeq A^{(I)}$ for some finite set I, then $\mathrm{alHom}_A(F, N) \simeq N^{(I)}$ and the claims are obvious. More generally, if F is almost finitely generated projective, for any $\varepsilon \in \mathfrak{m}$ there exists a finite set $I := I(\varepsilon)$ and morphisms

(2.4.30) $$F \xrightarrow{u_\varepsilon} A^{(I)} \xrightarrow{v_\varepsilon} F$$

such that $v_\varepsilon \circ u_\varepsilon = \varepsilon \cdot 1_F$. We apply the natural transformation

$$E \otimes_A \mathrm{alHom}_A(-, N) \to \mathrm{alHom}_A(-, E \otimes_A N)$$

to (2.4.30) : an easy diagram chase allows then to conclude that the kernel and cokernel of (2.4.27) are killed by ε. As ε is arbitrary, it follows that (2.4.27) is an isomorphism in this case. An analogous argument works when E is almost finitely generated projective, so we get (i.b). If F is only almost projective, then we still have morphisms of the type (2.4.30), but now $I(\varepsilon)$ is no longer necessarily finite. However, the cokernels of the induced morphisms $1_E \otimes u_\varepsilon$ and $\mathrm{alHom}_A(v_\varepsilon, E \otimes_A N)$ are still annihilated by ε. Hence, to show (iii) (resp. (i.c)) it suffices to consider the case when F is free and E is almost finitely generated (resp. presented). By passing to almost elements, we can further reduce to the analogous question for usual rings and modules, and by the usual juggling we can even replace E by a finitely generated (resp. presented) A_*-module and F by a free A_*-module. This case is easily dealt with, and (iii) and (i.c) follow. Case (i.d) (resp. (ii.b)) is similar : one considers almost elements and replaces E_* by a free A_*-module (resp. and F_* by a finitely generated A_*-module). In case (ii.a) (resp. (i.a)), for every finitely generated submodule $\mathfrak{m}_0$ of $\mathfrak{m}$ we can find, by proposition 2.3.10, a finitely generated (resp. presented) A-module F_0 and a morphism $F_0 \to F$ whose kernel and cokernel are annihilated by $\mathfrak{m}_0$. It follows easily that we can replace F by F_0 and suppose that F is finitely generated (resp. presented). Then the argument in [16, Ch.I §2 Prop.10] can be taken over *verbatim* to show (ii.a) (resp. (i.a)). □

Lemma 2.4.31. *Let B be an A-algebra.*

  (i) *Let P be an A-module. If either P or B is almost finitely generated projective as an A-module, the natural morphism*

(2.4.32) $$B \otimes_A \mathrm{alHom}_A(P, N) \to \mathrm{alHom}_B(B \otimes_A P, B \otimes_A N)$$

  is an isomorphism for all A-modules N.

  (ii) *Every almost finitely generated projective A-module is reflexive.*

  (iii) *If P is an almost finitely generated projective B-module, the morphism*

(2.4.33) $$\mathrm{alHom}_B(P, B) \otimes_B B^* \to \mathrm{alHom}_A(P, A) \; : \; \phi \otimes \psi \mapsto \psi \circ \phi$$

  is an isomorphism of B-modules.

Proof. (i) is an easy consequence of lemma 2.4.29(i.b). To prove (ii), we apply the natural transformation (2.4.24) to (2.4.30) : by diagram chase one sees that the kernel and cokernel of the morphism $F \to (F^*)^*$ are killed by ε. (iii) is analogous: one applies the natural transformation (2.4.33) to (2.4.30). □

Lemma 2.4.34. *Let $(M_n \, ; \, \phi_n : M_n \to M_{n+1} \mid n \in \mathbb{N})$ be a direct system of A-modules and suppose there exist sequences $(\varepsilon_n \mid n \in \mathbb{N})$ and $(\delta_n \mid n \in \mathbb{N})$ of ideals of V such that*

  (i) $\lim_{n \to \infty} \varepsilon_n^a = V^a$ *and* $\prod_{j=0}^{\infty} \delta_j^a$ *is a Cauchy product (see (2.3.4));*

  (ii) $\varepsilon_n \cdot \mathrm{alExt}_A^i(M_n, N) = \delta_n \cdot \mathrm{alExt}_A^i(\mathrm{Coker}\, \phi_n, N) = 0$ *for all A-modules N, all $i > 0$ and all $n \in \mathbb{N}$;*

  (iii) $\delta_n \cdot \mathrm{Ker}\, \phi_n = 0$ *for all $n \in \mathbb{N}$.*

Then $\mathrm{colim}_{n \in \mathbb{N}} M_n$ *is an almost projective A-module.*

Proof. Let $M := \mathrm{colim}_{n \in \mathbb{N}} M_n$. In view of remark 2.4.12(i), it suffices to show that $\mathrm{alExt}_A^i(M, N)$ vanishes for all $i > 0$ and all A-modules N. The morphisms ϕ_n assemble to $\phi : \oplus_n M_n \to \oplus_n M_n$ and we have a short exact sequence $0 \to \oplus_n M_n \xrightarrow{1-\phi} \oplus_n M_n \longrightarrow M \to 0$. Applying the long exact alExt sequence one obtains a short exact sequence (cp. [77, 3.5.10])

$$0 \to \varprojlim_{n\in\mathbb{N}}{}^1 \text{alExt}_A^{i-1}(M_n, N) \to \text{alExt}_A^i(M, N) \to \varprojlim_{n\in\mathbb{N}} \text{alExt}_A^i(M_n, N) \to 0.$$

Then lemma 2.4.2(ii) implies that $\text{alExt}_A^i(M, N) \simeq 0$ for all $i > 1$ and moreover $\text{alExt}_A^1(M, N)$ is isomorphic to $\varprojlim_{n\in\mathbb{N}}{}^1 \text{alHom}_A(M_n, N)$. Let

$$\phi_n^* : \text{alHom}_A(M_{n+1}, N) \to \text{alHom}_A(M_n, N) \quad : \quad f \mapsto f \circ \phi_n$$

be the transpose of ϕ_n and write ϕ_n as a composition $M_n \xrightarrow{p_n} \text{Im}(\phi_n) \xrightarrow{q_n} M_{n+1}$, so that $\phi_n^* = q_n^* \circ p_n^*$, the composition of the respective transposed morphisms. We have monomorphisms

$$\text{Coker}\, p_n^* \hookrightarrow \text{alHom}_A(\text{Ker}\, \phi_n, N)$$
$$\text{Coker}\, q_n^* \hookrightarrow \text{alExt}_A^1(\text{Coker}\, \phi_n, N)$$

for all $n \in \mathbb{N}$. Hence $\delta_n^2 \cdot \text{Coker}\, \phi_n^* = 0$ for all $n \in \mathbb{N}$. Since $\prod_{n=0}^{\infty} \delta_n^2$ is a Cauchy product, lemma 2.4.2(iii) shows that $\varprojlim_{n\in\mathbb{N}}{}^1 \text{alHom}_A(M_n, N) \simeq 0$ and the assertion follows.   □

Proposition 2.4.35. *For any V-algebra R the functor $R^a\text{-}\mathbf{Mod} \to R\text{-}\mathbf{Mod}$: $M \mapsto M_!$ commutes with tensor products. If $\tilde{\mathfrak{m}}$ is a flat V-module, then this functor takes flat R^a-modules to flat R-modules.*

Proof. Let M be a flat R^a-module and $N \hookrightarrow N'$ an injective map of R-modules. Denote by K the kernel of the induced map $M_! \otimes_R N \to M_! \otimes_R N'$; we have $K^a \simeq 0$. We obtain an exact sequence $0 \to \tilde{\mathfrak{m}} \otimes_V K \to \tilde{\mathfrak{m}} \otimes_V M_! \otimes_R N \to \tilde{\mathfrak{m}} \otimes_V M_! \otimes_R N'$. But one sees easily that $\tilde{\mathfrak{m}} \otimes_V K = 0$ and $\tilde{\mathfrak{m}} \otimes_V M_! \simeq M_!$, which shows that $M_!$ is a flat R-module. Similarly, let M, N be two R^a-modules. Then the natural map $M_* \otimes_R N_* \to (M \otimes_{R^a} N)_*$ is an almost isomorphism and the assertion follows from remark 2.1.4(i).   □

2.5. Almost homotopical algebra.

The formalism of abelian tensor categories provides a minimal framework wherein the rudiments of deformation theory can be developed.

2.5.1. Let $(\mathscr{C}, \otimes, U, u)$ be an abelian tensor category; we assume henceforth that $\otimes$ is a right exact functor. Let A be a given $\mathscr{C}$-monoid. For any two-sided ideal I of A, the quotient A/I in the underlying abelian category $\mathscr{C}$ has a unique $\mathscr{C}$-monoid structure such that $A \to A/I$ is a morphism of monoids. A/I is unitary if A is. If I is a two-sided ideal of A such that $I^2 = 0$, then, using the right exactness of $\otimes$ one checks that I has a natural structure of an A/I-bimodule, unitary when A is.

Definition 2.5.2. A $\mathscr{C}$-*extension* of a $\mathscr{C}$-monoid B by a B-bimodule I is a short exact sequence of objects of $\mathscr{C}$

(2.5.3) $$X : \qquad 0 \to I \to C \xrightarrow{p} B \to 0$$

such that C is a $\mathscr{C}$-monoid, p is a morphism of $\mathscr{C}$-monoids, I is a square zero two-sided ideal in C and the E/I-bimodule structure on I coincides with the given bimodule structure on I. The $\mathscr{C}$-extensions form a category $\mathbf{Exmon}_{\mathscr{C}}$. The morphisms are commutative diagrams with exact rows

$$
\begin{array}{ccccccccc}
X: & 0 & \longrightarrow & I & \longrightarrow & E & \xrightarrow{\ p\ } & B & \longrightarrow & 0 \\
& \downarrow & & \downarrow{\scriptstyle f} & & \downarrow{\scriptstyle g} & & \downarrow{\scriptstyle h} & & \\
X': & 0 & \longrightarrow & I' & \longrightarrow & E' & \xrightarrow{\ p'\ } & B' & \longrightarrow & 0
\end{array}
$$

such that g and h are morphisms of $\mathscr{C}$-monoids. We let $\mathbf{Exmon}_{\mathscr{C}}(B, I)$ be the subcategory of $\mathbf{Exmon}_{\mathscr{C}}$ consisting of all $\mathscr{C}$-extensions of B by I, where the morphisms are all short exact sequences as above such that $f := 1_I$ and $h := 1_B$.

2.5.4. We have also the variant in which all the $\mathscr{C}$-monoids in (2.5.3) are required to be unitary (resp. to be algebras) and I is a unitary B-bimodule (resp. a bimodule whose left and right B-module actions coincide, *i.e.* are switched by composition with the "commutativity constraints" $\theta_{B|I}$ and $\theta_{I|B}$, see (2.2.6)); we will call $\mathbf{Exun}_{\mathscr{C}}$ (resp. $\mathbf{Exal}_{\mathscr{C}}$) the corresponding category.

2.5.5. For a morphism $\phi : C \to B$ of $\mathscr{C}$-monoids, and a $\mathscr{C}$-extension X in $\mathbf{Exmon}_{\mathscr{C}}(B, I)$, we can pullback X via ϕ to obtain an exact sequence $X * \phi$ with a morphism $\phi^* : X * \phi \to X$; one checks easily that there exists a unique structure of $\mathscr{C}$-extension on $X * \phi$ such that ϕ^* is a morphism of $\mathscr{C}$-extension; then $X * \phi$ is an object in $\mathbf{Exmon}_{\mathscr{C}}(C, I)$. Similarly, given a B-linear morphism $\psi : I \to J$, we can push out X and obtain a well defined object $\psi * X$ in $\mathbf{Exmon}_{\mathscr{C}}(B, J)$ with a morphism $X \to \psi * X$ of $\mathbf{Exmon}_{\mathscr{C}}$. In particular, if I_1 and I_2 are two B-bimodules, the functors $p_i *$ ($i = 1, 2$) associated to the natural projections $p_i : I_1 \oplus I_2 \to I_i$ establish an equivalence of categories

(2.5.6) $\mathbf{Exmon}_{\mathscr{C}}(B, I_1 \oplus I_2) \xrightarrow{\sim} \mathbf{Exmon}_{\mathscr{C}}(B, I_1) \times \mathbf{Exmon}_{\mathscr{C}}(B, I_2)$

whose quasi-inverse is given by $(E_1, E_2) \mapsto (E_1 \oplus E_2) * \delta$, where $\delta : B \to B \oplus B$ is the diagonal morphism. A similar statement holds for $\mathbf{Exal}$ and $\mathbf{Exun}$. These operations can be used to induce an abelian group structure on the set $\mathrm{Exmon}_{\mathscr{C}}(B, I)$ of isomorphism classes of objects of $\mathbf{Exmon}_{\mathscr{C}}(B, I)$ as follows. For any two objects X, Y of $\mathbf{Exmon}_{\mathscr{C}}(B, I)$ we can form $X \oplus Y$ which is an object of $\mathbf{Exmon}_{\mathscr{C}}(B \oplus B, I \oplus I)$. Let $\alpha : I \oplus I \to I$ be the addition morphism of I. Then we set

$$X + Y := \alpha * (X \oplus Y) * \delta.$$

One can check that $X + Y \simeq Y + X$ for any X, Y and that the trivial split $\mathscr{C}$-extension $B \oplus I$ is a neutral element for $+$. Moreover every isomorphism class has an inverse $-X$. The functors $X \mapsto X * \phi$ and $X \mapsto \psi * X$ commute with the operation thus defined, and induce group homomorphisms

$$*\phi : \mathrm{Exmon}_{\mathscr{C}}(B, I) \to \mathrm{Exmon}_{\mathscr{C}}(C, I)$$
$$\psi* : \mathrm{Exmon}_{\mathscr{C}}(B, I) \to \mathrm{Exmon}_{\mathscr{C}}(B, J).$$

2.5.7. We will need the variant $\mathrm{Exal}_{\mathscr{C}}(B, I)$ defined in the same way, starting from the category $\mathbf{Exal}_{\mathscr{C}}(B, I)$. For instance, if A is an almost algebra (resp. a commutative ring), we can consider the abelian tensor category $\mathscr{C} = A\text{-}\mathbf{Mod}$. In this case the $\mathscr{C}$-extensions will be called simply A-extensions, and we will write $\mathbf{Exal}_A$ rather than $\mathbf{Exal}_{\mathscr{C}}$. In fact the commutative unitary case will soon become prominent in our work, and the more general setup is only required for technical reasons, in the proof of proposition 2.5.13 below, which is the abstract version of a well-known result on the lifting of idempotents over nilpotent ring extensions.

2.5.8. Let A be a $\mathscr{C}$-monoid. We form the biproduct $A^\dagger := U \oplus A$ in $\mathscr{C}$. We denote by p_1, p_2 the associated projections from $A^\dagger$ onto U and respectively A. Also, let i_1, i_2 be the natural monomorphisms from U, resp. A to $A^\dagger$. $A^\dagger$ is equipped with a unitary monoid structure

$$\mu^\dagger := i_2 \circ \mu \circ (p_2 \otimes p_2) + i_2 \circ \ell_A^{-1} \circ (p_1 \otimes p_2) + i_2 \circ r_A^{-1} \circ (p_2 \otimes p_1) + i_1 \circ u^{-1} \circ (p_1 \otimes p_1)$$

where ℓ_A, r_A are the natural isomorphisms provided by [24, Prop. 1.3] and $u : U \rightarrow U \otimes U$ is as in [24, §1]. In terms of the ring $A_*^\dagger \simeq U_* \oplus A_*$ this is the multiplication

$$(u_1, b_1) \cdot (u_2, b_2) := (u_1 \cdot u_2, b_1 \cdot b_2 + b_1 \cdot u_2 + u_1 \cdot b_2).$$

Then i_2 is a morphism of monoids and one verifies that the "restriction of scalars" functor i_2^* defines an equivalence from the category $A^\dagger$-**Uni.Mod** of unitary $A^\dagger$-modules to the category A-**Mod** of all A-modules; let j denote the inverse functor. A similar discussion applies to bimodules.

2.5.9. Similarly, we derive equivalences of categories

$$\mathbf{Exun}_\mathscr{C}(A^\dagger, j(M)) \underset{(-)^\dagger}{\overset{*i_2}{\rightleftarrows}} \mathbf{Exmon}_\mathscr{C}(A, M)$$

for all A-bimodules M.

2.5.10. Next we specialize to $A := U$: for a given U-module M let $e_M := \sigma_{M/U} \circ \ell_M : M \rightarrow M$; working out the definitions one finds that the condition that $(M, \sigma_{M/U})$ is a module structure is equivalent to $e_M^2 = e_M$. Let $U \times U$ be the product of U by itself in the category of $\mathscr{C}$-monoids. There is an isomorphism of unitary $\mathscr{C}$-monoids $\zeta : U^\dagger \rightarrow U \times U$ given by $\zeta := i_1 \circ p_1 + i_2 \circ p_1 + i_2 \circ p_2$. Another isomorphism is $\tau \circ \zeta$, where τ is the flip $i_1 \circ p_2 + i_2 \circ p_1$. Hence we get equivalences of categories

$$U\text{-}\mathbf{Mod} \underset{i_2^*}{\overset{j}{\rightleftarrows}} U^\dagger\text{-}\mathbf{Uni.Mod} \underset{(\tau \circ \zeta)^*}{\overset{(\zeta^{-1})^*}{\rightleftarrows}} (U \times U)\text{-}\mathbf{Uni.Mod}.$$

The composition $i_2^* \circ (\zeta^{-1} \circ \tau \circ \zeta)^* \circ j$ defines a self-equivalence of U-**Mod** which associates to a given U-module M the new U-module M^{flip} whose underlying object in $\mathscr{C}$ is M and such that $e_{M^{\text{flip}}} = 1_M - e_M$. The same construction applies to U-bimodules and finally we get equivalences

(2.5.11) $\mathbf{Exmon}_\mathscr{C}(U, M) \overset{\sim}{\rightarrow} \mathbf{Exmon}_\mathscr{C}(U, M^{\text{flip}})$ $X \mapsto X^{\text{flip}}$

for all U-bimodules M. If $X := (0 \rightarrow M \rightarrow E \overset{\pi}{\rightarrow} U \rightarrow 0)$ is an extension and $X^{\text{flip}} := (0 \rightarrow M^{\text{flip}} \rightarrow E^{\text{flip}} \rightarrow U \rightarrow 0)$, then one verifies that there is a natural isomorphism $X^{\text{flip}} \rightarrow X$ of complexes in $\mathscr{C}$ inducing -1_M on M, the identity on U and carrying the multiplication morphism on E^{flip} to

$$-\mu_E + \ell_E^{-1} \circ (\pi \otimes 1_E) + r_E^{-1} \circ (1_E \otimes \pi) : E \otimes E \rightarrow E.$$

In terms of the associated rings, this corresponds to replacing the given multiplication $(x, y) \mapsto x \cdot y$ of E_* by the new operation $(x, y) \mapsto \pi_*(x) \cdot y + \pi_*(y) \cdot x - x \cdot y$.

Lemma 2.5.12. *If M is a U-bimodule whose left and right actions coincide, then every extension of U by M splits uniquely.*

Proof. Using the idempotent e_M we get a U-linear decomposition $M \simeq M_1 \oplus M_2$ where the bimodule structure on M_1 is given by the zero morphisms and the bimodule structure on M_2 is given by ℓ_M^{-1} and r_M^{-1}. We have to prove that $\mathbf{Exmon}_{\mathscr{C}}(U, M)$ is equivalent to a one-point category. By (2.5.6) we can assume that $M = M_1$ or $M = M_2$. By (2.5.11) we have $\mathbf{Exmon}_{\mathscr{C}}(U, M_2) \simeq \mathbf{Exmon}_{\mathscr{C}}(U, M_2^{\mathrm{flip}})$ and on M_2^{flip} the bimodule actions are the zero morphisms. So it is enough to consider $M = M_1$. In this case, if $X := (0 \to M \to E \to U \to 0)$ is any extension, $\mu_E : E \otimes E \to E$ factors through a morphism $U \otimes U \to E$ and composing with $u : U \to U \otimes U$ we get a right inverse of $E \to U$, which shows that X is the split extension. Then it is easy to see that X does not have any non-trivial automorphisms, which proves the assertion. $\square$

Proposition 2.5.13. *Let $X := (0 \to I \to A \xrightarrow{p} A' \to 0)$ be a $\mathscr{C}$-extension.*

(i) *Let $e' \in A'_*$ be an idempotent element whose left action on the A'-bimodule I coincides with its right action. Then there exists a unique idempotent $e \in A_*$ such that $p_*(e) = e'$.*

(ii) *Especially, if A' is unitary and I is a unitary A'-bimodule, then every extension of A' by I is unitary.*

Proof. (i): The hypothesis $e'^2 = e'$ implies that $e' : U \to A'$ is a morphism of (non-unitary) $\mathscr{C}$-monoids. We can then replace X by $X * e'$ and thereby assume that $A' = U$, $p : A \to U$ and I is a (non-unitary) U-bimodule and the right and left actions on I coincide. The assertion to prove is that $\underline{1}_U$ lifts to a unique idempotent $e \in A_*$. However, this follows easily from lemma 2.5.12. To show (ii), we observe that, by (i), the unit $\underline{1}_{A'}$ of A'_* lifts uniquely to an idempotent $e \in A_*$. We have to show that e is a unit for A. Let us show the left unit property. Via $e : U \to A$ we can view the extension X as an exact sequence of left U-modules. We can then split X as the direct sum $X_1 \oplus X_2$ where X_1 is a sequence of unitary U-modules and X_2 is a sequence of U-modules with trivial actions. But by hypothesis, on I and on A' the U-module structure is unitary, so $X = X_1$ and this is the left unit property. $\square$

2.5.14.   So much for the general nonsense; we now return to almost algebras. As already announced, *from here on, we assume throughout that $\widetilde{\mathfrak{m}}$ is a flat V-module.* As an immediate consequence of proposition 2.5.13 we get natural equivalences of categories

(2.5.15) $\mathbf{Exal}_{A_1}(B_1, M_1) \times \mathbf{Exal}_{A_2}(B_2, M_2) \xrightarrow{\sim} \mathbf{Exal}_{A_1 \times A_2}(B_1 \times B_2, M)$

whenever A_1, A_2 are V^a-algebras, B_i is a A_i-algebra, M_i is a (unitary) B_i-module, $i = 1, 2$ and $M := M_1 \oplus M_2$, viewed as a $B_1 \times B_2$-module.

2.5.16. If R is a V-algebra, S (resp. J) is a R-algebra (resp. an S-module) and X is any object of $\mathbf{Exal}_R(S, J)$, then by applying termwise the localization functor we get an object X^a of $\mathbf{Exal}_{R^a}(S^a, J^a)$. With this notation we have:

Lemma 2.5.17. *Let B be any A-algebra and I a B-module.*

(i) *The natural functor*

$$(2.5.18) \qquad \mathbf{Exal}_{A_{!!}}(B_{!!}, I_*) \to \mathbf{Exal}_A(B, I) \quad : \quad X \mapsto X^a$$

is an equivalence of categories.

(ii) *The equivalence (2.5.18) induces a group isomorphism*

$$\mathrm{Exal}_{A_{!!}}(B_{!!}, I_*) \xrightarrow{\sim} \mathrm{Exal}_A(B, I)$$

functorial in all arguments.

Proof. Of course (ii) is an immediate consequence of (i). To show (i), let $X := (0 \to I \to E \to B \to 0)$ be any object of $\mathbf{Exal}_A(B, I)$. Using corollary 2.2.24 one sees easily that the sequence $X_! := (0 \to I_! \to E_{!!} \to B_{!!} \to 0)$ is right exact; $X_!$ won't be exact in general, unless B (and therefore E) is an exact algebra. In any case, the kernel of $I_! \to E_{!!}$ is almost zero, so we get an extension of $B_{!!}$ by a quotient of $I_!$ which maps to I_*. In particular we get by pushout an extension $X_{!*}$ by I_*, *i.e.* an object of $\mathbf{Exal}_{A_{!!}}(B_{!!}, I_*)$ and in fact the assignment $X \mapsto X_{!*}$ is a quasi-inverse for the functor (2.5.18). $\qquad \square$

Remark 2.5.19. By inspecting the proof, we see that one can replace I_* by $I_{!*} := \mathrm{Im}(I_! \to I_*)$ in (i) and (ii) above. When B is exact, also $I_!$ (or any $A_{!!}$-module representing I) will do.

In [50, II.1.2] it is shown how to associate to any ring homomorphism $R \to S$ a natural simplicial complex of S-modules denoted $\mathbb{L}_{S/R}$ and called the cotangent complex of S over R.

Definition 2.5.20. Let $A \to B$ be a morphism of almost V-algebras. The *almost cotangent complex* of B over A is the simplicial $B_{!!}$-module

$$\mathbb{L}_{B/A} := B_{!!} \otimes_{(V^a \times B)_{!!}} \mathbb{L}_{(V^a \times B)_{!!}/(V^a \times A)_{!!}}.$$

2.5.21. Usually we will want to view $\mathbb{L}_{B/A}$ as an object of the derived category $\mathrm{D}(s.B_{!!}\text{-}\mathbf{Mod})$ of simplicial $B_{!!}$-modules. Indeed, the hyperext functors computed in this category relate the cotangent complex to a number of important invariants. Recall that, for any simplicial ring R and any two R-modules E, F the hyperext of E and F is the abelian group defined as

$$\mathbb{Ext}_R^p(E, F) := \underset{n \geq -p}{\mathrm{colim}} \, \mathrm{Hom}_{\mathrm{D}(R\text{-}\mathbf{Mod})}(\sigma^n E, \sigma^{n+p} F)$$

(where σ is the suspension functor of [50, I.3.2.1.4]).

Let us fix an almost algebra A. First we want to establish the relationship with differentials.

Definition 2.5.22. Let B be any A-algebra, M any B-module.

(i) An *A-derivation* of B with values in M is an A-linear morphism $\partial : B \to M$ such that $\partial(b_1 \cdot b_2) = b_1 \cdot \partial(b_2) + b_2 \cdot \partial(b_1)$ for $b_1, b_2 \in B_*$. The set of all M-valued A-derivations of B forms a V-module $\mathrm{Der}_A(B, M)$ and the almost V-module $\mathrm{Der}_A(B, M)^a$ has a natural structure of B-module.

(ii) We reserve the notation $I_{B/A}$ for the ideal $\mathrm{Ker}(\mu_{B/A} : B \otimes_A B \to B)$. The *module of relative differentials* of ϕ is defined as the (left) B-module $\Omega_{B/A} := I_{B/A}/I_{B/A}^2$. It is endowed with a natural A-derivation $\delta : B \to \Omega_{B/A}$ defined by $b \mapsto \underline{1} \otimes b - b \otimes \underline{1}$ for all $b \in B_*$. The assignment $(A \to B) \mapsto \Omega_{B/A}$ defines a functor

(2.5.23) $$\Omega : V^a\text{-}\mathbf{Alg.Morph} \to V^a\text{-}\mathbf{Alg.Mod}$$

from the category of morphisms $A \to B$ of almost V-algebras to the category denoted V^a-**Alg.Mod**, consisting of all pairs (B, M) where B is an almost V-algebra and M is a B-module. The morphisms in V^a-**Alg.Morph** are the commutative squares; the morphisms $(B, M) \to (B', M')$ in V^a-**Alg.Mod** are all pairs (ϕ, f) where $\phi : B \to B'$ is a morphism of almost V-algebras and $f : B' \otimes_B M \to M'$ is a morphism of B'-modules.

2.5.24. The module of relative differentials enjoys the familiar universal properties that one expects. In particular $\Omega_{B/A}$ represents the functor $\mathrm{Der}_A(B, -)$, *i.e.* for any (left) B-module M the morphism

(2.5.25) $$\mathrm{Hom}_B(\Omega_{B/A}, M) \to \mathrm{Der}_A(B, M) \qquad f \mapsto f \circ \delta$$

is an isomorphism. As an exercise, the reader can supply the proof for this claim and for the following standard proposition.

Proposition 2.5.26. *Let B and C be two A-algebras.*

(i) *There is a natural isomorphism:*
$$\Omega_{C \otimes_A B/C} \simeq C \otimes_A \Omega_{B/A}.$$

(ii) *Suppose that C is a B-algebra. Then there is a natural exact sequence of C-modules:*
$$C \otimes_B \Omega_{B/A} \to \Omega_{C/A} \to \Omega_{C/B} \to 0.$$

(iii) *Let I be an ideal of B and let $C := B/I$ be the quotient A-algebra. Then there is a natural exact sequence:* $I/I^2 \to C \otimes_B \Omega_{B/A} \to \Omega_{C/A} \to 0$.

(iv) *The functor (2.5.23) commutes with all colimits.* $\qquad\qquad\square$

We supplement these generalities with one more statement which is in the same vein as lemma 2.3.22 and which will be useful in section 6.3 to calculate the Fitting ideals of modules of differentials.

Lemma 2.5.27. *Let $\phi : B \to B'$ be a morphism of A-algebras such that $I \cdot \mathrm{Ker}(\phi) = I \cdot \mathrm{Coker}(\phi) = 0$ for an ideal $I \subset A$. Let $d\phi : \Omega_{B/A} \otimes_B B' \to \Omega_{B'/A}$ be the natural morphism. Then $I \cdot \mathrm{Coker}\, d\phi = 0$ and $I^4 \cdot \mathrm{Ker}\, d\phi = 0$.*

Proof. We will use the standard presentation

(2.5.28) $$H(B/A) \; : \; B \otimes_A B \otimes_A B \xrightarrow{\partial} B \otimes_A B \xrightarrow{d} \Omega_{B/A} \to 0$$

where d is defined by : $b_1 \otimes b_2 \mapsto b_1 \cdot db_2$ and ∂ is the differential of the Hochschild complex :

$$b_1 \otimes b_2 \otimes b_3 \mapsto b_1 b_2 \otimes b_3 - b_1 \otimes b_2 b_3 + b_1 b_3 \otimes b_2.$$

By naturality of $H(B)$, we deduce a morphism of complexes : $B' \otimes_B H(B/A) \to H(B'/A)$. Then, by snake lemma, we derive an exact sequence : $\mathrm{Ker}(1_{B'} \otimes_A \phi) \to \mathrm{Ker}\, d\phi \to X$, where X is a quotient of $B' \otimes_A \mathrm{Coker}(\phi \otimes_A \phi)$. Using the Tor exact sequences we see that $\mathrm{Ker}(1_{B'} \otimes_A \phi)$ is annihilated by I^2. It follows easily that I^4 annihilates $\mathrm{Ker}\, d\phi$. Similarly, $\mathrm{Coker}\, d\phi$ is a quotient of $\mathrm{Coker}(1_{B'} \otimes_A \phi)$, so $I \cdot \mathrm{Coker}\, d\phi = 0$. $\qquad\square$

Lemma 2.5.29. *For any A-algebra B there is a natural $B_{!!}$-linear isomorphism:*

$$(\Omega_{B/A})_! \simeq \Omega_{B_{!!}/A_{!!}}.$$

Proof. Using the adjunction (2.5.25) we are reduced to showing that the natural map

$$\phi_M : \mathrm{Der}_{A_{!!}}(B_{!!}, M) \to \mathrm{Der}_A(B, M^a)$$

is a bijection for all $B_{!!}$-modules M. Given $\partial : B \to M^a$ we construct $\partial_! : B_! \to M_!^a \to M$. We extend $\partial_!$ to $V \oplus B_!$ by setting it equal to zero on V. Then it is easy to check that the resulting map descends to $B_{!!}$, hence giving an A-derivation $B_{!!} \to M$. This procedure yields a right inverse ψ_M to ϕ_M. To show that ϕ_M is injective, suppose that $\partial : B_{!!} \to M$ is an almost zero A-derivation. Composing with the natural A-linear map $B_! \to B_{!!}$ we obtain an almost zero map $\partial' : B_! \to M$. But $\mathfrak{m} \cdot B_! = B_!$, hence $\partial' = 0$. This implies that in fact $\partial = 0$, and the assertion follows. $\qquad\square$

Proposition 2.5.30. *For any A-algebra B and any $B_{!!}$-module M we have a natural $B_{!!}$-linear isomorphism:*

$$\mathbb{Ext}^0_{B_{!!}}(\mathbb{L}_{B/A}, M) \simeq \mathrm{Der}_A(B, M^a).$$

Equivalently, $H_0(\mathbb{L}_{B/A})$ is naturally isomorphic to $(\Omega_{B/A})_!$.

Proof. To ease notation, set $\widetilde{A} := V^a \times A$ and $\widetilde{B} := V^a \times B$. We have:

$$\begin{aligned}
\mathbb{Ext}^0_{B_{!!}}(\mathbb{L}_{B/A}, M) &\simeq \mathbb{Ext}^0_{\widetilde{B}_{!!}}(\mathbb{L}_{\widetilde{B}_{!!}/\widetilde{A}_{!!}}, M) && \text{by [50, I.3.3.4.4]} \\
&\simeq \mathrm{Der}_{\widetilde{A}_{!!}}(\widetilde{B}_{!!}, M) && \text{by [50, II.1.2.4.2]} \\
&\simeq \mathrm{Der}_{\widetilde{A}}(\widetilde{B}, M^a) && \text{by lemma 2.5.29.}
\end{aligned}$$

But it is easy to see that the natural map $\mathrm{Der}_A(B, M^a) \to \mathrm{Der}_{\widetilde{A}}(\widetilde{B}, M^a)$ is an isomorphism. $\qquad\square$

Theorem 2.5.31. *For B, M as in proposition 2.5.30, there is a natural isomorphism:*

(2.5.32) $$\mathrm{Exal}_A(B, M^a) \xrightarrow{\sim} \mathbb{Ext}^1_{B_{!!}}(\mathbb{L}_{B/A}, M).$$

Proof. With the notation of the proof of proposition 2.5.30 we have:

$$\mathrm{Ext}^1_{B_{!!}}(\mathbb{L}_{B/A}, M) \simeq \mathrm{Ext}^1_{\widetilde{B}_{!!}}(\mathbb{L}_{\widetilde{B}_{!!}/\widetilde{A}_{!!}}, M) \quad \text{by [50, I.3.3.4.4]}$$
$$\simeq \mathrm{Exal}_{\widetilde{A}_{!!}}(\widetilde{B}_{!!}, M) \quad \text{by [50, III.1.2.3]}$$
$$\simeq \mathrm{Exal}_{\widetilde{A}}(\widetilde{B}, M^a)$$

where the last isomorphism follows directly from lemma 2.5.17(ii) and the subsequent remark 2.5.19. Finally, (2.5.15) shows that $\mathrm{Exal}_{\widetilde{A}}(\widetilde{B}, M^a) \simeq \mathrm{Exal}_A(B, M^a)$, as required. □

Moreover we have the following transitivity theorem as in [50, II.2.1.2].

Theorem 2.5.33. *Let $A \to B \to C$ be a sequence of morphisms of almost V-algebras. There exists a natural distinguished triangle of $\mathsf{D}(s.C_{!!}\text{-}\mathbf{Mod})$*

$$C_{!!} \otimes_{B_{!!}} \mathbb{L}_{B/A} \xrightarrow{u} \mathbb{L}_{C/A} \xrightarrow{v} \mathbb{L}_{C/B} \to C_{!!} \otimes_{B_{!!}} \sigma\mathbb{L}_{B/A}$$

where the morphisms u and v are obtained by functoriality of $\mathbb{L}$.

Proof. It follows directly from *loc. cit.* □

Proposition 2.5.34. *Let $(A_\lambda \to B_\lambda)_{\lambda \in I}$ be a system of morphisms of V^a-algebra indexed by a small filtered category I. Then there is a natural isomorphism in $\mathsf{D}(s.\mathrm{colim}_{\lambda \in I} B_{\lambda !!}\text{-}\mathbf{Mod})$*

$$\mathop{\mathrm{colim}}_{\lambda \in I} \mathbb{L}_{B_\lambda/A_\lambda} \simeq \mathbb{L}_{\mathop{\mathrm{colim}}_{\lambda \in I} B_\lambda/\mathop{\mathrm{colim}}_{\lambda \in I} A_\lambda}.$$

Proof. Remark 2.2.30(i) gives an isomorphism : $\mathop{\mathrm{colim}}_{\lambda \in I} A_{\lambda !!} \xrightarrow{\sim} (\mathop{\mathrm{colim}}_{\lambda \in I} A_\lambda)_{!!}$ (and likewise for $\mathop{\mathrm{colim}}_{\lambda \in I} B_\lambda$). Then the claim follows from [50, II.1.2.3.4]. □

Next we want to prove the almost version of the flat base change theorem [50, II.2.2.1]. To this purpose we need some preparation.

Proposition 2.5.35. *Let B and C be two A-algebras and set $T_i := \mathrm{Tor}_i^{A_{!!}}(B_{!!}, C_{!!})$.*

(i) *If A, B, C and $B \otimes_A C$ are all exact, then for every $i > 0$ the natural morphism $\widetilde{\mathfrak{m}} \otimes_V T_i \to T_i$ is an isomorphism.*

(ii) *If furthermore, $\mathrm{Tor}_i^A(B, C) \simeq 0$ for some $i > 0$, then T_i vanishes.*

Proof. (i): For any almost V-algebra D we let k_D denote the complex of $D_{!!}$-modules $[\widetilde{\mathfrak{m}} \otimes_V D_{!!} \to D_{!!}]$ placed in degrees $-1, 0$; we have a distiguished triangle

$$\mathscr{T}(D) : \widetilde{\mathfrak{m}} \otimes_V D_{!!} \to D_{!!} \to k_D \to \widetilde{\mathfrak{m}} \otimes_V D_{!!}[1].$$

By assumption, the natural map $k_A \to k_B$ is a quasi-isomorphism and $\widetilde{\mathfrak{m}} \otimes_V B_{!!} \simeq B_!$. On the other hand, for all $i \in \mathbb{N}$ we have

$$\mathrm{Tor}_i^{A_{!!}}(k_B, C_{!!}) \simeq \mathrm{Tor}_i^{A_{!!}}(k_A, C_{!!}) \simeq H^{-i}(k_A \otimes_{A_{!!}} C_{!!}) = H^{-i}(k_C).$$

In particular $\mathrm{Tor}_i^{A_{!!}}(k_B, C_{!!}) = 0$ for all $i > 1$. As $\widetilde{\mathfrak{m}}$ is flat over V, we have $\widetilde{\mathfrak{m}} \otimes_V T_i \simeq \mathrm{Tor}_i^{A_{!!}}(\widetilde{\mathfrak{m}} \otimes_V B_{!!}, C_{!!})$. Then by the long exact Tor sequence associated to $\mathscr{T}(B) \overset{\mathrm{L}}{\otimes}_{A_{!!}} C_{!!}$ we get the assertion for all $i > 1$. Next we consider the natural map of distinguished triangles $\mathscr{T}(A) \overset{L}{\otimes}_{A_{!!}} A_{!!} \to \mathscr{T}(B) \overset{L}{\otimes}_{A_{!!}} C_{!!}$; writing down the associated morphism of long exact Tor sequences, we obtain a diagram with exact rows :

$$
\begin{array}{ccccccc}
0 & \longrightarrow & \mathrm{Tor}_1^{A_{!!}}(k_A, A_{!!}) & \overset{\partial}{\longrightarrow} & (\widetilde{\mathfrak{m}} \otimes_V A_{!!}) \otimes_{A_{!!}} A_{!!} & \overset{i}{\longrightarrow} & A_{!!} \otimes_{A_{!!}} A_{!!} \\
& & \downarrow & & \downarrow & & \downarrow \\
& & \mathrm{Tor}_1^{A_{!!}}(k_B, C_{!!}) & \overset{\partial'}{\longrightarrow} & (\widetilde{\mathfrak{m}} \otimes_V B_{!!}) \otimes_{A_{!!}} C_{!!} & \overset{i'}{\longrightarrow} & B_{!!} \otimes_{A_{!!}} C_{!!}.
\end{array}
$$

By the above, the leftmost vertical map is an isomorphism; moreover, the assumption gives $\mathrm{Ker}\, i \simeq \mathrm{Ker}(\widetilde{\mathfrak{m}} \to V) \simeq \mathrm{Ker}\, i'$. Then, since ∂ is injective, also ∂' must be injective, which implies our assertion for the remaining case $i = 1$. (ii) follows directly from (i). □

Theorem 2.5.36. *Let B, A' be two A-algebras. Suppose that the natural morphism $B \overset{L}{\otimes}_A A' \to B' := B \otimes_A A'$ is an isomorphism in $\mathsf{D}(s.A\text{-}\mathbf{Mod})$. Then the natural morphisms*

$$
B'_{!!} \otimes_{B_{!!}} \mathbb{L}_{B/A} \to \mathbb{L}_{B'/A'}
$$
$$
(B'_{!!} \otimes_{B_{!!}} \mathbb{L}_{B/A}) \oplus (B'_{!!} \otimes_{A'_{!!}} \mathbb{L}_{A'/A}) \to \mathbb{L}_{B'/A}
$$

are quasi-isomorphisms.

Proof. Let us remark that the functor $D \mapsto V^a \times D : A\text{-}\mathbf{Alg} \to (V^a \times A)\text{-}\mathbf{Alg}$ commutes with tensor products; hence the same holds for the functor $D \mapsto (V^a \times D)_{!!}$ (see remark 2.2.30(i)). Then, in view of proposition 2.5.35(ii), the theorem is reduced immediately to [50, II.2.2.1]. □

As an application we obtain the vanishing of the almost cotangent complex for a certain class of morphisms.

Theorem 2.5.37. *Let $R \to S$ be a morphism of almost algebras such that*

$$
\mathrm{Tor}_i^R(S, S) \simeq 0 \simeq \mathrm{Tor}_i^{S \otimes_R S}(S, S) \qquad \text{for all } i > 0
$$

(for the $S \otimes_R S$-module structure induced by $\mu_{S/R}$). Then $\mathbb{L}_{S/R} \simeq 0$ in $\mathsf{D}(S_{!!}\text{-}\mathbf{Mod})$.

Proof. Since $\mathrm{Tor}_i^R(S, S) = 0$ for all $i > 0$, theorem 2.5.36 applies (with $A := R$ and $B := A' := S$), giving the natural isomorphisms

$$
(2.5.38) \qquad \begin{aligned} (S \otimes_R S)_{!!} \otimes_{S_{!!}} \mathbb{L}_{S/R} &\simeq \mathbb{L}_{S \otimes_R S/S} \\ ((S \otimes_R S)_{!!} \otimes_{S_{!!}} \mathbb{L}_{S/R}) \oplus ((S \otimes_R S)_{!!} \otimes_{S_{!!}} \mathbb{L}_{S/R}) &\simeq \mathbb{L}_{S \otimes_R S/R}. \end{aligned}
$$

Since $\mathrm{Tor}_i^{S \otimes_R S}(S, S) = 0$, the same theorem also applies with $A := S \otimes_R S$, $B := S$, $A' := S$, and we notice that in this case $B' \simeq S$; hence we have

$$
(2.5.39) \qquad \mathbb{L}_{S/S \otimes_R S} \simeq S_{!!} \otimes_{S_{!!}} \mathbb{L}_{S/S \otimes_R S} \simeq \mathbb{L}_{S/S} \simeq 0.
$$

Next we apply transitivity to the sequence $R \to S \otimes_R S \to S$, to obtain (thanks to (2.5.39))

$$
(2.5.40) \qquad S_{!!} \otimes_{S \otimes_R S_{!!}} \mathbb{L}_{S \otimes_R S/R} \simeq \mathbb{L}_{S/R}.
$$

Applying $S_{!!} \otimes_{S \otimes_R S_{!!}} -$ to the second isomorphism (2.5.38) we obtain

(2.5.41) $$\mathbb{L}_{S/R} \oplus \mathbb{L}_{S/R} \simeq S_{!!} \otimes_{S \otimes_R S_{!!}} \mathbb{L}_{S \otimes_R S/R}.$$

Finally, composing (2.5.40) and (2.5.41) we derive

(2.5.42) $$\mathbb{L}_{S/R} \oplus \mathbb{L}_{S/R} \xrightarrow{\sim} \mathbb{L}_{S/R}.$$

However, by inspection, the isomorphism (2.5.42) is the sum map. Consequently $\mathbb{L}_{S/R} \simeq 0$, as claimed. □

The following proposition shows that $\mathbb{L}_{B/A}$ is already determined by $\mathbb{L}_{B/A}^a$.

Proposition 2.5.43. *Let $A \to B$ be a morphism of exact almost V-algebras. Then the natural map $\widetilde{\mathfrak{m}} \otimes_V \mathbb{L}_{B_{!!}/A_{!!}} \to \mathbb{L}_{B_{!!}/A_{!!}}$ is a quasi-isomorphism.*

Proof. By transitivity we may assume $A = V^a$. Let $P_\bullet := P_V(B_{!!})$ be the standard resolution of $B_{!!}$ (see [50, II.1.2.1]). Each $P[n]^a$ contains V^a as a direct summand, hence it is exact, so that we have an exact sequence of simplicial V-modules $0 \to s.\widetilde{\mathfrak{m}} \to s.V \oplus (P_\bullet^a)_! \to (P_\bullet^a)_{!!} \to 0$. The augmentation $(P_\bullet^a)_! \to (B_{!!}^a)_{!!} \simeq B_!$ is a quasi-isomorphism and we deduce that $(P_\bullet^a)_{!!} \to B_{!!}$ is a quasi-isomorphism; hence $(P_\bullet^a)_{!!} \to P_\bullet$ is a quasi-isomorphism as well. We have $P[n] \simeq \mathrm{Sym}(F_n)$ for a free V-module F_n and the map $(P[n]^a)_{!!} \to P[n]$ is identified with $\mathrm{Sym}(\widetilde{\mathfrak{m}} \otimes_V F_n) \to \mathrm{Sym}(F_n)$, whence $\Omega_{P[n]_{!!}^a/V} \otimes_{P[n]_{!!}^a} P[n] \to \Omega_{P[n]/V}$ is identified with $\widetilde{\mathfrak{m}} \otimes_V \Omega_{P[n]/V} \to \Omega_{P[n]/V}$. By [50, II.1.2.6.2] the map $\mathbb{L}_{(P_\bullet^a)_{!!}/V}^\Delta \to \mathbb{L}_{P_\bullet/V}^\Delta$ is a quasi-isomorphism. In view of [50, II.1.2.4.4] we derive that $\Omega_{(P_\bullet^a)_{!!}/V} \to \Omega_{P_\bullet/V}$ is a quasi-isomorphism, *i.e.* $\widetilde{\mathfrak{m}} \otimes_V \Omega_{P_\bullet/V} \to \Omega_{P_\bullet/V}$ is a quasi-isomorphism. Since $\widetilde{\mathfrak{m}}$ is flat and $\Omega_{P_\bullet/V} \to \Omega_{P_\bullet/V} \otimes_{P_\bullet} B_{!!} = \mathbb{L}_{B_{!!}/V}$ is a quasi-isomorphism, we get the desired conclusion. □

Finally we have a fundamental spectral sequence as in [50, III.3.3.2].

Theorem 2.5.44. *Let $\phi : A \to B$ be a morphism of almost algebras such that $B \otimes_A B \simeq B$ (e.g. such that B is a quotient of A). Then there is a first quadrant homology spectral sequence of bigraded almost algebras*

$$E_{pq}^2 := H_{p+q}(\mathrm{Sym}_B^q(\mathbb{L}_{B/A}^a)) \Rightarrow \mathrm{Tor}_{p+q}^A(B, B).$$

Proof. We replace ϕ by $1_{V^a} \times \phi$ and apply the functor $B \mapsto B_{!!}$ (which commutes with tensor products by remark 2.2.30(i)) thereby reducing the assertion to [50, III.3.3.2]. □

3. Almost ring theory

With this chapter we begin in earnest the study of almost commutative algebra: in section 3.1 the classes of flat, unramified and étale morphisms are defined, together with some variants. In section 3.2 we derive the infinitesimal lifting theorems for étale algebras (theorem 3.2.18) and for almost projective modules (theorem 3.2.28).

In section 3.4 we turn to study some cases of non-flat descent; when we specialize to usual rings, we recover known theorems (of course, standard commutative algebra is a particular case of almost ring theory). But if the result is not new, the argument is : indeed, we believe that our treatment, even when specialized to usual rings, improves upon the method found in the literature.

The last section of chapter 3 calls on stage the Frobenius endomorphism of an almost algebra of positive characteristic. The main results are invariance of étale morphisms under pull-back by Frobenius maps (theorem 3.5.13) and theorem 3.5.28, which can be interpreted as a purity theorem. Perhaps the most remarkable aspect of the latter result is how cheap it is : in positive characteristic, the availability of the Frobenius map allows for a quick and easy proof. Philosophically, this proof is not too far removed from the method devised by Faltings for his more recent proof of purity in mixed characteristic.

Taken together, the above-mentioned four sections leave us with a decent understanding of the morphisms "of relative dimension zero" (this expression should be taken with a grain of salt, since we do not try to define the dimension of an almost algebra). On one hand, a good hold on the case of relative dimension zero is all that is required for the applications currently in sight (especially for the proof of the almost purity theorem, but also for the needs of our chapters 6 and 7); on the other hand, having reached this stage, one cannot help wondering what lies ahead, for instance whether there is a good notion of smooth morphism of almost algebras. The full answer to this question shall be delayed until chapter 5 : there we will introduce a class of morphisms that generalize "in higher dimension" the class of weakly étale morphisms, and that specialize to formally smooth morphisms in the "classical limit" $V = \mathfrak{m}$. We will present evidence that our notion of smoothness is well behaved and worthwhile; however we shall also see that smoothness "in higher dimension" exhibits some extra twists that have no analogue in standard commutative algebra, and cannot be easily guessed just by extrapolating from the case of étale morphisms of V^a-algebras (which, after all, reproduce very faithfully the behaviour of the classical notion defined in EGA).

Such extra twists are already foreshadowed by our results on the nilpotent deformation of almost projective modules : we show that such deformations exist, but are not unique; however, any two such deformations are "very close" in a precise sense (proposition 3.2.30).

In (usual) algebraic geometry one can also regard projective modules of finite rank n as GL_n-torsors (say for the Zariski topology); in almost ring theory this description carries through at least for the class of almost projective modules *of finite rank* (to be defined in section 4.3). From this vantage, one is naturally led to ask how much of the standard deformation theory for torsors over arbitrary group schemes generalizes to the almost world. We take up this question in section 3.3, focusing especially on the case of *smooth affine almost group schemes*, as a warm up to the later study of general smooth morphisms. Having committed seriously to almost group schemes and almost torsors, it is only a short while before one grows impatient at the limited expressive range afforded by the purely algebraic terminology introduced thus far, which by this point starts feeling a little like a linguistic straightjacket. That is why we find ourselves compelled to introduce a more geometric language : for our purpose an *affine almost scheme* X is just an object of the opposite of the category of V^a-algebras; we define also the category of $\mathcal{O}_X$-modules on X, as well as some suggestive notation to go with them, mimicking the standard usage in algebraic geometry.

Once these preliminaries are in place, the theory proceeds as in [51] : the techniques are rather sophisticated, but all the hard work has already been done in the previous sections, and we can just adapt Illusie's treatise without much difficulty.

3.1. **Flat, unramified and étale morphisms.** Let $A \to B$ be a morphism of almost V-algebras. Using the natural "multiplication" morphism of A-algebras $\mu_{B/A} : B \otimes_A B \to B$ we can view B as a $B \otimes_A B$-algebra.

Definition 3.1.1. Let $\phi : A \to B$ be a morphism of almost V-algebras.

(i) We say that ϕ is a *flat* (resp. *faithfully flat*, resp. *almost projective*) *morphism* if B is a flat (resp. faithfully flat, resp. almost projective) A-module.

(ii) We say that ϕ is *(uniformly) almost finite* (resp. *finite*) if B is a (uniformly) almost finitely generated (resp. finitely generated) A-module.

(iii) We say that ϕ is *weakly unramified* (resp. *unramified*) if B is a flat (resp. almost projective) $B \otimes_A B$-module (via the morphism $\mu_{B/A}$ defined above).

(iv) ϕ is *weakly étale* (resp. *étale*) if it is flat and weakly unramified (resp. unramified).

Furthermore, in analogy with definition 2.4.4, we shall write "(uniformly) almost finite projective" to denote a morphism ϕ which is both (uniformly) almost finite and almost projective.

Lemma 3.1.2. *Let $\phi : A \to B$ and $\psi : B \to C$ be morphisms of almost V-algebras.*

(i) *Let $A \to A'$ be any morphism of V^a-algebras; if ϕ is flat (resp. almost projective, resp. faithfully flat, resp. almost finite, resp. weakly unramified, resp. unramified, resp. weakly étale, resp. étale) then the same holds for $\phi \otimes_A 1_{A'}$.*

(ii) *If both ϕ and ψ are flat (resp. almost projective, resp. faithfully flat, resp. almost finite, resp. weakly unramified, resp. unramified, resp. weakly étale, resp. étale), then so is $\psi \circ \phi$.*

(iii) *If ϕ is flat and $\psi \circ \phi$ is faithfully flat, then ϕ is faithfully flat.*

(iv) *If ϕ is weakly unramified and $\psi \circ \phi$ is flat (resp. weakly étale), then ψ is flat (resp. weakly étale).*

(v) *If ϕ is unramified and $\psi \circ \phi$ is étale, then ψ is étale.*

   (vi) ϕ *is faithfully flat if and only if it is a monomorphism and* B/A *is a flat A-module.*

   (vii) *If* ϕ *is almost finite and weakly unramified, then* ϕ *is unramified.*

   (viii) *If* ψ *is faithfully flat and* $\psi \circ \phi$ *is flat (resp. weakly unramified), then* ϕ *is flat (resp. weakly unramified).*

Proof. For (vi) use the Tor sequences. In view of proposition 2.4.18(ii), to show (vii) it suffices to know that B is an almost finitely presented $B \otimes_A B$-module; but this follows from the existence of an epimorphism of $B \otimes_A B$-modules $(B \otimes_A B) \otimes_A B \rightarrow \text{Ker } \mu_{B/A}$ defined by $x \otimes b \mapsto x \cdot (\underline{1} \otimes b - b \otimes \underline{1})$. Of the remaining assertions, only (iv) and (v) are not obvious, but the proof is just the "almost version" of a well-known argument. Let us show (v); the same argument applies to (iv). We remark that $\mu_{B/A}$ is an étale morphism, since ϕ is unramified. Define $\Gamma_\psi := 1_C \otimes_B \mu_{B/A}$. By (i), Γ_ψ is étale. Define also $p := (\psi \circ \phi) \otimes_A 1_B$. By (i), p is flat (resp. étale). The claim follows by remarking that $\psi = \Gamma_\psi \circ p$ and applying (ii). $\qquad\square$

Remark 3.1.3. (i) Suppose we work in the classical limit case, that is, $\mathfrak{m} := V$ (cp. example 2.1.2(ii)). Then we caution the reader that our notion of "étale morphism" is more general than the usual one, as defined in [44]. The relationship between the usual notion and ours is discussed in the digression (3.4.44).

   (ii) The naive hope that the functor $A \mapsto A_{!!}$ might preserve flatness is crushed by the following counterexample. Let $(V, \mathfrak{m})$ be as in example 2.1.2(i) and let k be the residue field of V. Consider the flat map $V \times V \rightarrow V$ defined as $(x, y) \mapsto x$. We get a flat morphism $V^a \times V^a \rightarrow V^a$ in V^a-**Alg**; applying the left adjoint to localization yields a map $V \times_k V \rightarrow V$ that is not flat. On the other hand, faithful flatness *is* preserved. Indeed, let $\phi : A \rightarrow B$ be a morphism of almost algebras. Then ϕ is a monomorphism if and only if $\phi_{!!}$ is injective; moreover, $B_{!!}/\text{Im}(A_{!!}) \simeq B_!/A_!$, which is flat over $A_{!!}$ if and only if B/A is flat over A, by proposition 2.4.35.

   We will find useful to study certain "almost idempotents", as in the following proposition.

Proposition 3.1.4. *A morphism* $\phi : A \rightarrow B$ *is unramified if and only if there exists an almost element* $e_{B/A} \in B \otimes_A B_*$ *such that*

   (i) $e_{B/A}^2 = e_{B/A}$;

   (ii) $\mu_{B/A}(e_{B/A}) = \underline{1}$;

   (iii) $x \cdot e_{B/A} = 0$ *for all* $x \in I_{B/A*}$.

Proof. Suppose that ϕ is unramified. We start by showing that for every $\varepsilon \in \mathfrak{m}$ there exist almost elements e_ε of $B \otimes_A B$ such that

$$(3.1.5) \qquad e_\varepsilon^2 = \varepsilon \cdot e_\varepsilon \qquad \mu_{B/A}(e_\varepsilon) = \varepsilon \cdot \underline{1} \qquad I_{B/A*} \cdot e_\varepsilon = 0.$$

Since B is an almost projective $B \otimes_A B$-module, for every $\varepsilon \in \mathfrak{m}$ there exists an "approximate splitting" for the epimorphism $\mu_{B/A} : B \otimes_A B \rightarrow B$, i.e. a $B \otimes_A B$-linear morphism $u_\varepsilon : B \rightarrow B \otimes_A B$ such that $\mu_{B/A} \circ u_\varepsilon = \varepsilon \cdot 1_B$. Set $e_\varepsilon := u_\varepsilon \circ \underline{1} : A \rightarrow B \otimes_A B$. We see that $\mu_{B/A}(e_\varepsilon) = \varepsilon \cdot \underline{1}$. To show that $e_\varepsilon^2 = \varepsilon \cdot e_\varepsilon$ we use the $B \otimes_A B$-linearity of u_ε to compute

$$e_\varepsilon^2 = e_\varepsilon \cdot u_\varepsilon(\underline{1}) = u_\varepsilon(\mu_{B/A}(e_\varepsilon) \cdot \underline{1}) = u_\varepsilon(\mu_{B/A}(e_\varepsilon)) = \varepsilon \cdot e_\varepsilon.$$

Next take any almost element x of $I_{B/A}$ and compute

$$x \cdot e_\varepsilon = x \cdot u_\varepsilon(\underline{1}) = u_\varepsilon(\mu_{B/A}(x) \cdot \underline{1}) = 0.$$

This establishes (3.1.5). Next let us take any other $\delta \in \mathfrak{m}$ and a corresponding almost element e_δ. Both $\varepsilon \cdot \underline{1} - e_\varepsilon$ and $\delta \cdot \underline{1} - e_\delta$ are elements of $I_{B/A*}$, hence we have $(\delta \cdot \underline{1} - e_\delta) \cdot e_\varepsilon = 0 = (\varepsilon \cdot \underline{1} - e_\varepsilon) \cdot e_\delta$ which implies

(3.1.6) $$\delta \cdot e_\varepsilon = \varepsilon \cdot e_\delta \qquad \text{for all } \varepsilon, \delta \in \mathfrak{m}.$$

Let us define a map $e_{B/A} : \mathfrak{m} \otimes_V \mathfrak{m} \to B \otimes_A B_*$ by the rule

(3.1.7) $$\varepsilon \otimes \delta \mapsto \delta \cdot e_\varepsilon \qquad \text{for all } \varepsilon, \delta \in \mathfrak{m}.$$

To show that (3.1.7) does indeed determine a well defined morphism, we need to check that $\delta \cdot v \cdot e_\varepsilon = \delta \cdot e_{v \cdot \varepsilon}$ and $\delta \cdot e_{\varepsilon + \varepsilon'} = \delta \cdot (e_\varepsilon + e_{\varepsilon'})$ for all $\varepsilon, \varepsilon', \delta \in \mathfrak{m}$ and all $v \in V$. However, both identities follow easily by a repeated application of (3.1.6). It is easy to see that $e_{B/A}$ defines an almost element with the required properties.

Conversely, suppose an almost element $e_{B/A}$ of $B \otimes_A B$ is given with the stated properties. We define $u : B \to B \otimes_A B$ by $b \mapsto e_{B/A} \cdot (1 \otimes b)$ $(b \in B_*)$ and $v := \mu_{B/A}$. Then (iii) says that u is a $B \otimes_A B$-linear morphism and (ii) shows that $v \circ u = 1_B$. Hence, by lemma 2.4.15, ϕ is unramified. $\qquad\square$

Remark 3.1.8. The proof of proposition 3.1.4 shows that if I is an ideal in an almost V-algebra A, then A/I is almost projective over A if and only if I is generated by an idempotent of A_*. This idempotent is uniquely determined.

Corollary 3.1.9. *Under the hypotheses and notation of the proposition, the ideal $I_{B/A}$ has a natural structure of $B \otimes_A B$-algebra, with unit morphism given by $\underline{1} := 1_{B \otimes_A B/A} - e_{B/A}$ and whose multiplication is the restriction of $\mu_{B \otimes_A B/A}$ to $I_{B/A}$. Moreover the natural morphism*

$$B \otimes_A B \to I_{B/A} \oplus B \qquad x \mapsto (x \cdot \underline{1} \oplus \mu_{B/A}(x))$$

is an isomorphism of $B \otimes_A B$-algebras.

Proof. Left to the reader as an exercise. $\qquad\square$

3.2. Nilpotent deformations of almost algebras and modules.

Throughout the following, the terminology "epimorphism of V^a-algebras" will refer to a morphism of V^a-algebras that induces an epimorphism on the underlying V^a-modules.

Lemma 3.2.1. *Let $A \to B$ be an epimorphism of almost V-algebras with kernel I. Let U be the A-extension $0 \to I/I^2 \to A/I^2 \to B \to 0$. Then the assignment $f \mapsto f * U$ defines a natural isomorphism*

(3.2.2) $$\operatorname{Hom}_B(I/I^2, M) \xrightarrow{\sim} \operatorname{Exal}_A(B, M).$$

Proof. Let $X := (0 \to M \to E \xrightarrow{p} B \to 0)$ be any A-extension of B by M. The composition $g : A \to E \xrightarrow{p} B$ of the structural morphism for E followed by p coincides with the projection $A \to B$. Therefore $g(I) \subset M$ and $g(I^2) = 0$. Hence g factors through A/I^2; the restriction of g to I/I^2 defines a morphism

$f \in \mathrm{Hom}_B(I/I^2, M)$ and a morphism of A-extensions $f * U \to X$. In this way we obtain an inverse for (3.2.2). $\qquad\square$

3.2.3.   Now consider any morphism of A-extensions

$$(3.2.4) \qquad
\begin{array}{ccccccccc}
\widetilde{B}: & & 0 & \longrightarrow & I & \longrightarrow & B & \longrightarrow & B_0 & \longrightarrow & 0 \\
\downarrow\widetilde{f} & & & & \downarrow{u} & & \downarrow{f} & & \downarrow{f_0} & & \\
\widetilde{C}: & & 0 & \longrightarrow & J & \longrightarrow & C & \longrightarrow & C_0 & \longrightarrow & 0.
\end{array}$$

The morphism u induces by adjunction a morphism of C_0-modules

$$(3.2.5) \qquad\qquad C_0 \otimes_{B_0} I \to J$$

whose image is the ideal $I \cdot C$, so that the square diagram of almost algebras defined by $\widetilde{f}$ is cofibred (*i.e.* $C_0 \simeq C \otimes_B B_0$) if and only if (3.2.5) is an epimorphism.

Lemma 3.2.6. *Let* $\widetilde{f} : \widetilde{B} \to \widetilde{C}$ *be a morphism of A-extensions as above, such that the corresponding square diagram of almost algebras is cofibred. Then the morphism $f : B \to C$ is flat if and only if $f_0 : B_0 \to C_0$ is flat and (3.2.5) is an isomorphism.*

Proof. It follows directly from the (almost version of the) local flatness criterion (see [58, Th. 22.3]). $\qquad\square$

We are now ready to put together all the work done so far and begin the study of deformations of almost algebras.

3.2.7.   The morphism $u : I \to J$ is an element in $\mathrm{Hom}_{B_0}(I, J)$; by lemma 3.2.1 the latter group is naturally isomorphic to $\mathrm{Exal}_B(B_0, J)$. On the other hand, in view of proposition 2.5.43 and lemma 2.4.14(i) we have natural isomorphisms:

$$(3.2.8) \qquad \mathbb{Ext}^i_{C_{0!!}}(\mathbb{L}_{C_0/B_0}, M_!) \simeq \mathbb{Ext}^i_{C_0}(\mathbb{L}^a_{C_0/B_0}, M)$$

for every $i \in \mathbb{N}$ and every C_0-module M. By applying transitivity (theorem 2.5.33) to the sequence of morphisms $B \to B_0 \xrightarrow{f_0} C_0$ we deduce an exact sequence of abelian groups

$$\mathrm{Exal}_{B_0}(C_0, J) \to \mathrm{Exal}_B(C_0, J) \to \mathrm{Hom}_{B_0}(I, J) \xrightarrow{\partial} \mathbb{Ext}^2_{C_0}(\mathbb{L}^a_{C_0/B_0}, J).$$

Hence we can form the element $\omega(\widetilde{B}, f_0, u) := \partial(u) \in \mathbb{Ext}^2_{C_0}(\mathbb{L}^a_{C_0/B_0}, J)$. The proof of the next result goes exactly as in [50, Ch.III, Prop.2.1.2.3].

Proposition 3.2.9. *Let the A-extension $\widetilde{B}$, the B_0-linear morphism $u : I \to J$ and the morphism of A-algebras $f_0 : B_0 \to C_0$ be given as above.*

(i) *There exists an A-extension $\widetilde{C}$ and a morphism $\widetilde{f} : \widetilde{B} \to \widetilde{C}$ completing diagram (3.2.4) if and only if $\omega(\widetilde{B}, f_0, u) = 0$. (I.e. $\omega(\widetilde{B}, f_0, u)$ is the obstruction to the lifting of $\widetilde{B}$ over f_0.)*

(ii) *Assume that the obstruction $\omega(\widetilde{B}, f_0, u)$ vanishes. Then the set of isomorphism classes of A-extensions $\widetilde{C}$ as in* (i) *forms a torsor under the group:*

$$\mathrm{Exal}_{B_0}(C_0, J) \simeq \mathbf{Ext}^1_{C_0}(\mathbb{L}^a_{C_0/B_0}, J).$$

(iii) *The group of automorphisms of an A-extension $\widetilde{C}$ as in* (i) *is naturally isomorphic to* $\mathrm{Der}_{B_0}(C_0, J)$ $(\simeq \mathbf{Ext}^0_{C_0}(\mathbb{L}^a_{C_0/B_0}, J))$. $\square$

3.2.10.   The obstruction $\omega(\widetilde{B}, f_0, u)$ depends functorially on u. More exactly, if we denote by

$$\omega(\widetilde{B}, f_0) \in \mathbf{Ext}^2_{C_0}(\mathbb{L}^a_{C_0/B_0}, C_0 \otimes_{B_0} I)$$

the obstruction corresponding to the natural morphism $I \to C_0 \otimes_{B_0} I$, then for any other morphism $u : I \to J$ we have

$$\omega(\widetilde{B}, f_0, u) = v_! \circ \omega(\widetilde{B}, f_0)$$

where v is the morphism (3.2.5). Taking lemma 3.2.6 into account we deduce

Corollary 3.2.11. *Suppose that $B_0 \to C_0$ is flat. Then*

(i) *The class $\omega(\widetilde{B}, f_0)$ is the obstruction to the existence of a flat deformation of C_0 over B, i.e. of a B-extension $\widetilde{C}$ as in* (3.2.4) *such that C is flat over B and $C \otimes_B B_0 \to C_0$ is an isomorphism.*

(ii) *If $\omega(\widetilde{B}, f_0) = 0$, then the set of isomorphism classes of flat deformations of C_0 over B forms a torsor under the group $\mathrm{Exal}_{B_0}(C_0, C_0 \otimes_{B_0} I)$.*

(iii) *The group of automorphisms of a given flat deformation of C_0 over B is naturally isomorphic to $\mathrm{Der}_{B_0}(C_0, C_0 \otimes_{B_0} I)$.* $\square$

3.2.12.   Now, suppose we are given two A-extensions $\widetilde{C}^1, \widetilde{C}^2$ with morphisms of A-extensions

$$
\begin{array}{ccccccccc}
\widetilde{B}: & & 0 & \longrightarrow & I & \longrightarrow & B & \longrightarrow & B_0 & \longrightarrow 0 \\
& \downarrow{\scriptstyle \widetilde{f}^i} & & & \downarrow{\scriptstyle u^i} & & \downarrow{\scriptstyle f^i} & & \downarrow{\scriptstyle f_0^i} & \\
\widetilde{C}^i: & & 0 & \longrightarrow & J^i & \longrightarrow & C^i & \longrightarrow & C_0^i & \longrightarrow 0
\end{array}
$$

and morphisms $v : J^1 \to J^2$, $g_0 : C_0^1 \to C_0^2$ such that

(3.2.13) $u^2 = v \circ u^1$ and $f_0^2 = g_0 \circ f_0^1$.

We consider the problem of finding a morphism of A-extensions

(3.2.14)
$$
\begin{array}{ccccccccc}
\widetilde{C}^1: & & 0 & \longrightarrow & J^1 & \longrightarrow & C^1 & \longrightarrow & C_0^1 & \longrightarrow 0 \\
& \downarrow{\scriptstyle \widetilde{g}} & & & \downarrow{\scriptstyle v} & & \downarrow{\scriptstyle g} & & \downarrow{\scriptstyle g_0} & \\
\widetilde{C}^2: & & 0 & \longrightarrow & J^2 & \longrightarrow & C^2 & \longrightarrow & C_0^2 & \longrightarrow 0
\end{array}
$$

such that $\widetilde{f}^2 = \widetilde{g} \circ \widetilde{f}^1$. Let us denote by $e(\widetilde{C}^i) \in \mathbf{Ext}^1_{C_0^i}(\mathbb{L}^a_{C_0^i/B}, J^i)$ the classes defined by the B-extensions $\widetilde{C}^1, \widetilde{C}^2$ via the isomorphisms (2.5.32), (3.2.8), and by

$$v* : \operatorname{Ext}^1_{C^1_0}(\mathbb{L}^a_{C^1_0/B}, J^1) \to \operatorname{Ext}^1_{C^1_0}(\mathbb{L}^a_{C^1_0/B}, J^2)$$
$$*g_0 : \operatorname{Ext}^1_{C^2_0}(\mathbb{L}^a_{C^2_0/B}, J^2) \to \operatorname{Ext}^1_{C^2_0}(C^2_0 \otimes_{C^1_0} \mathbb{L}^a_{C^1_0/B}, J^2)$$

the canonical morphisms defined by v and g_0. Using the natural isomorphism

$$\operatorname{Ext}^1_{C^1_0}(\mathbb{L}^a_{C^1_0/B}, J^2) \simeq \operatorname{Ext}^1_{C^2_0}(C^2_0 \otimes_{C^1_0} \mathbb{L}_{C^1_0/B}, J^2)$$

we can identify the target of both $v*$ and $*g$ with $\operatorname{Ext}^1_{C^1_0}(\mathbb{L}^a_{C^1_0/B}, J^2)$. It is clear that the problem admits a solution if and only if the A-extensions $v * \widetilde{C}^1$ and $\widetilde{C}^2 * g_0$ coincide, *i.e.* if and only if $v * e(\widetilde{C}^1) - e(\widetilde{C}^2) * g_0 = 0$. By applying transitivity to the sequence of morphisms $B \to B_0 \to C^1_0$ we obtain an exact sequence

$$\operatorname{Ext}^1_{C^1_0}(\mathbb{L}^a_{C^1_0/B_0}, J^2) \hookrightarrow \operatorname{Ext}^1_{C^1_0}(\mathbb{L}^a_{C^1_0/B}, J^2) \to \operatorname{Hom}_{C^1_0}(C^1_0 \otimes_{B_0} I, J^2).$$

It follows from (3.2.13) that the image of $v * e(\widetilde{C}^1) - e(\widetilde{C}^2) * g_0$ in the group $\operatorname{Hom}_{C^1_0}(C^1_0 \otimes_{B_0} I, J^2)$ vanishes, therefore

(3.2.15) $$v * e(\widetilde{C}^1) - e(\widetilde{C}^2) * g_0 \in \operatorname{Ext}^1_{C^1_0}(\mathbb{L}^a_{C^1_0/B_0}, J^2).$$

In conclusion, we derive the following result as in [50, Ch.III, Prop.2.2.2].

Proposition 3.2.16. *With the above notations:*

(i) *The class (3.2.15) is the obstruction to the existence of a morphism of A-extensions $\widetilde{g} : \widetilde{C}^1 \to \widetilde{C}^2$ as in (3.2.14) such that $\widetilde{f}^2 = \widetilde{g} \circ \widetilde{f}^1$.*

(ii) *If the obstruction vanishes, the set of such morphisms forms a torsor under the group $\operatorname{Der}_{B_0}(C^1_0, J^2)$ (which is identified with $\operatorname{Ext}^0_{C^2}(C^2_0 \otimes_{C^1_0} \mathbb{L}^a_{C^1_0/B_0}, J^2))$.* □

3.2.17.   For a given almost V-algebra A, we define the category A-**w.Ét** (resp. A-**Ét**) as the full subcategory of A-**Alg** consisting of all weakly étale (resp. étale) A-algebras. Notice that, by lemma 3.1.2(iv) all morphisms in A-**w.Ét** are weakly étale.

Theorem 3.2.18. *Let A be a V^a-algebra.*

(i) *Let B be a weakly étale A-algebra, C any A-algebra and $I \subset C$ a nilpotent ideal. Then the natural morphism*

$$\operatorname{Hom}_{A\text{-}\mathbf{Alg}}(B, C) \to \operatorname{Hom}_{A\text{-}\mathbf{Alg}}(B, C/I)$$

is bijective.

(ii) *Let $I \subset A$ a nilpotent ideal and $A' := A/I$. Then the natural functor*

$$A\text{-}\mathbf{w.Ét} \to A'\text{-}\mathbf{w.Ét} \quad (\phi : A \to B) \mapsto (1_{A'} \otimes_A \phi : A' \to A' \otimes_A B)$$

is an equivalence of categories.

(iii) *The equivalence of (ii) restricts to an equivalence A-Ét $\to A'$-Ét.*

Proof. By induction we can assume $I^2 = 0$. Then (i) follows directly from proposition 3.2.16 and theorem 2.5.37. We show (ii): by corollary 3.2.11 (and again theorem 2.5.37) a given weakly étale morphism $\phi' : A' \to B'$ can be lifted to a *unique* flat morphism $\phi : A \to B$. We need to prove that ϕ is weakly étale, *i.e.*

that B is $B \otimes_A B$-flat. However, it is clear that $\mu_{B'/A'} : B' \otimes_{A'} B' \to B'$ is weakly étale, hence it has a flat lifting $\tilde{\mu} : B \otimes_A B \to C$. Then the composition $A \to B \otimes_A B \to C$ is flat and it is a lifting of ϕ'. We deduce that there is an isomorphism of A-algebras $\alpha : B \to C$ lifting $1_{B'}$ and moreover the morphisms $b \mapsto \tilde{\mu}(b \otimes \underline{1})$ and $b \mapsto \tilde{\mu}(\underline{1} \otimes b)$ coincide with α. Claim (ii) follows. To show (iii), suppose that $A' \to B'$ is étale and let $I_{B'/A'}$ denote as usual the kernel of $\mu_{B'/A'}$. By corollary 3.1.9 there is a natural morphism of almost algebras $B' \otimes_{A'} B' \to I_{B'/A'}$ which is clearly étale. Hence $I_{B'/A'}$ lifts to a weakly étale $B \otimes_A B$-algebra C, and the isomorphism $B' \otimes_{A'} B' \simeq I_{B'/A'} \oplus B'$ lifts to an isomorphism $B \otimes_A B \simeq C \oplus B$ of $B \otimes_A B$-algebras. It follows that B is an almost projective $B \otimes_A B$-module, i.e. $A \to B$ is étale, as claimed. □

We conclude with some results on deformations of almost modules. These can be established independently of the theory of the cotangent complex, along the lines of [50, Ch.IV, §3.1.12].

3.2.19. We begin by recalling some notation from *loc. cit.* Let R be a ring and $J \subset R$ an ideal with $J^2 = 0$. Set $R' := R/J$; an extension of R-modules $\underline{M} := (0 \to K \to M \xrightarrow{p} M' \to 0)$ where K and M' are killed by J, defines a natural morphism of R'-modules $u(\underline{M}) : J \otimes_{R'} M' \to K$ such that $u(\underline{M})(x \otimes m') = xm$ for $x \in J$, $m \in M$ and $p(m) = m'$. By the local flatness criterion ([58, Th. 22.3]) M is flat over R if and only if M' is flat over R' and $u(\underline{M})$ is an isomorphism. One can then show the following.

Proposition 3.2.20. *(cp. [50, Ch.IV, Prop.3.1.5]) With the notation of (3.2.19) :*

(i) *Given R'-modules M' and K and a morphism $u' : J \otimes_{R'} M' \to K$ there exists an obstruction $\omega(R, u') \in \mathrm{Ext}^2_{R'}(M', K)$ whose vanishing is necessary and sufficient for the existence of an extension of R-modules $\underline{M}$ of M' by K such that $u(\underline{M}) = u'$.*

(ii) *When $\omega(R, u') = 0$, the set of isomorphism classes of such extensions $\underline{M}$ forms a torsor under $\mathrm{Ext}^1_{R'}(M', K)$; the group of automorphisms of such an extension is isomorphic to $\mathrm{Hom}_{R'}(M', K)$.* □

Lemma 3.2.21. *Let R be a ring, M a finitely generated R-module such that $\mathrm{Ann}_R M$ is a nilpotent ideal. Then R admits a filtration $0 = J_m \subset ... \subset J_1 \subset J_0 = R$ such that each J_i/J_{i+1} is a quotient of a direct sum of copies of M.*

Proof. This is [45, 1.1.5]; for the convenience of the reader we reproduce the proof. Let $I := F_0(M) \subset R$; if M is generated by k elements, we have $(\mathrm{Ann}_R M)^k \subset I \subset \mathrm{Ann}_R M$, especially I is nilpotent.

Claim 3.2.22. It suffices to show that $\overline{R} := R/I$ admits a filtration as above.

Proof of the claim. Indeed, if $0 = J'_0 \subset J'_1 \subset ... \subset J'_{n-1} \subset J'_n = R/I$ is such a filtration, we deduce filtrations $0 \subset J'_1(I^t/I^{t+1}) \subset ... \subset J'_{n-1}(I^t/I^{t+1}) \subset (I^t/I^{t+1})$ for every $t \in \mathbb{N}$; the graded module associated to this filtration is a direct sum of quotients of modules of the form $(J'_k/J'_{k+1}) \otimes_R (I^t/I^{t+1})$, so the claim follows easily. ◇

Let $F_0(M/IM)$ be the 0-th Fitting ideal of the $\overline{R}$-module M/IM; we have $F_0(M/IM) = F_0(M) \cdot \overline{R} = 0$, and $(\text{Ann}_{\overline{R}}M/IM)^k \subset F_0(M/IM)$, i.e. $\text{Ann}_{\overline{R}}M/IM$ is a nilpotent ideal. Thus, thanks to claim 3.2.22 we can replace R and M by $\overline{R}$ and M/IM, and thereby reduce to the case where $F_0(M) = 0$. We claim that the filtration $0 = F_0(M) \subset F_1(M) \subset ... \subset F_k(M) = R$ will do in this case. Indeed, let $L_1 \overset{\phi}{\to} L_0 \to M \to 0$ be a presentation of M, where L_0 and L_1 are free R-modules and the rank of L_0 equals k; by definition $F_j(M)$ is the image of the map $\Lambda_R^{k-j}L_1 \otimes_R \Lambda_R^j L_0 \to \Lambda_R^k L_0 \simeq R$ defined by the rule $x \otimes y \mapsto \Lambda_R^{k-j}\phi(x) \wedge y$. We deduce easily that for $1 \leq j \leq k$ the induced surjection $\Lambda_R^{k-j}L_1 \otimes_R \Lambda_R^j L_0 \to F_j(M)/F_{j-1}(M)$ factors through the module $\Lambda_R^{k-j}L_1 \otimes_R \Lambda_R^j M$; however the latter is a quotient of sums of M, so the claim follows. □

Lemma 3.2.23. *Let $A \to B$ be a finite morphism of almost algebras with nilpotent kernel. There exists $m \geq 0$ such that the following holds. For every A-linear morphism $\phi : M \to N$, set $\phi_B := \phi \otimes_A 1_B : M \otimes_A B \to N \otimes_A B$; then :*

    (i) $\text{Ann}_A(\text{Coker}\,\phi_B)^m \subset \text{Ann}_A(\text{Coker}\,\phi)$.

    (ii) $(\text{Ann}_A(\text{Ker}\phi_B) \cdot \text{Ann}_A(\text{Tor}_1^A(B,N))) \cdot \text{Ann}_A(\text{Coker}\phi))^m \subset \text{Ann}_A\text{Ker}\phi$.

If $B = A/I$ for an ideal I such that $I^n = 0$, then we can take $m = n$ in (i) and (ii).

Proof. Under the assumptions, we can find a finitely generated A_*-module Q such that $\mathfrak{m} \cdot B_* \subset Q \subset B_*$. By lemma 3.2.21 there exists a finite filtration $0 = J_m \subset ... \subset J_1 \subset J_0 = A_*$ such that each J_i/J_{i+1} is a quotient of a direct sum of copies of Q. This implies that

$$(3.2.24) \qquad \text{Ann}_A(M \otimes_A B)^m \subset \text{Ann}_A(M)$$

for every A-module M; (i) follows easily. Notice that if $B = A/I$ and $I^n = 0$, then we can take $m = n$ in (3.2.24). For (ii) let $C^\bullet := \text{Cone}\,\phi$. We estimate $H := H^{-1}(C^\bullet \overset{\mathbf{L}}{\otimes}_A B)$ in two ways. By the first spectral sequence of hyperhomology we have an exact sequence $\text{Tor}_1^A(N,B) \to H \to \text{Ker}\,\phi_B$. By the second spectral sequence for hyperhomology we have an exact sequence $\text{Tor}_2^A(\text{Coker}\,\phi,B) \to \text{Ker}(\phi) \otimes_A B \to H$. Hence $\text{Ker}(\phi) \otimes_A B$ is annihilated by the product of the three annihilators in (ii) and the result follows by applying (3.2.24) with $M := \text{Ker}\,\phi$. □

Lemma 3.2.25. *Keep the assumptions of lemma 3.2.23 and let M be an A-module.*

    (i) *If $A \to B$ is an epimorphism, M is flat and $M_B := B \otimes_A M$ is almost projective over B, then M is almost projective over A.*

    (ii) *If M_B is an almost finitely generated B-module then M is an almost finitely generated A-module.*

    (iii) *If $\text{Tor}_1^A(B,M) = 0$ and M_B is almost finitely presented over B, then M is almost finitely presented over A.*

Proof. (i): We have to show that $\text{Ext}_A^1(M,N)$ is almost zero for every A-module N. Let $I := \text{Ker}(A \to B)$; by assumption I is nilpotent, so by the usual devissage we may assume that $IN = 0$. If $\chi \in \text{Ext}_A^1(M,N)$ is represented by an extension

$0 \to N \to Q \to M \to 0$ then after tensoring by B and using the flatness of M we get an exact sequence of B-modules $0 \to N \to B \otimes_A Q \to M_B \to 0$. Thus χ comes from an element of $\operatorname{Ext}_B^1(M_B, N)$ which is almost zero by assumption.

(ii): For a given finitely generated subideal $\mathfrak{m}_0 \subset \mathfrak{m}$, let $N \subset M_B$ be a finitely generated B-submodule such that $\mathfrak{m}_0 M_B \subset N$. Since the induced map $M_* \otimes_{A_*} B_* \to (M_B)_*$ is almost surjective, we can find a finitely generated A-submodule $N_0 \subset M$ such that $\mathfrak{m}_0 N \subset \operatorname{Im}((N_0)_B \to M_B)$; by lemma 3.2.23(i) it follows that $\mathfrak{m}_0^{2n}(M/N_0) = 0$ for some $n \geq 0$ depending only on B, whence the claim.

(iii): Let $\mathfrak{m}_0$ be as above. By (ii), M is almost finitely generated over A, so we can choose a morphism $\phi : A^r \to M$ such that $\mathfrak{m}_0 \cdot \operatorname{Coker} \phi = 0$. Consider $\phi_B := \phi \otimes_A 1_B : B^r \to M_B$. By claim 2.3.12, there is a finitely generated submodule N of $\operatorname{Ker} \phi_B$ containing $\mathfrak{m}_0^2 \cdot \operatorname{Ker} \phi_B$. Notice that $\operatorname{Ker}(\phi) \otimes_A B$ maps onto $\operatorname{Ker}(B^r \to \operatorname{Im}(\phi) \otimes_A B)$ and $\operatorname{Ker}(\operatorname{Im}(\phi) \otimes_A B \to M_B) \simeq \operatorname{Tor}_1^A(B, \operatorname{Coker} \phi)$ is annihilated by $\mathfrak{m}_0$. Hence $\mathfrak{m}_0 \cdot \operatorname{Ker} \phi_B$ is contained in the submodule generated by the image of $\operatorname{Ker} \phi$ and therefore a finite generating set $\{x_1', ..., x_n'\}$ for $\mathfrak{m}_0^2 N$ is contained in the B_*-module generated by the images of almost elements $\{x_1, ..., x_m\}$ of $\operatorname{Ker} \phi$. If we quotient A^r by the span of these x_i, we get a finitely presented A-module F with a morphism $\overline{\phi} : F \to M$ such that $\operatorname{Ker}(\overline{\phi} \otimes_A B)$ is annihilated by $\mathfrak{m}_0^4$ and $\operatorname{Coker} \overline{\phi}$ is annihilated by $\mathfrak{m}_0$. By lemma 3.2.23(ii) we derive $\mathfrak{m}_0^{5m} \cdot \operatorname{Ker} \overline{\phi} = 0$ for some $m \geq 0$. Since $\mathfrak{m}_0$ is arbitrary, this proves the result. $\qquad\square$

Remark 3.2.26. (i) Inspecting the proof, one sees that parts (ii) and (iii) of lemma 3.2.25 hold whenever (3.2.24) holds. For instance, if $A \to B$ is any faithfully flat morphism, then (3.2.24) holds with $m := 1$.

(ii) Hence, if $A \to B$ is faithfully flat and M is an A-module such that M_B is flat (resp. almost finitely generated, resp. almost finitely presented) over B, then M is flat (resp. almost finitely generated, resp. almost finitely presented) over A.

(iii) On the other hand, we do not know whether a general faithfully flat morphism $A \to B$ descends almost projectivity. However, using (ii) and proposition 2.4.18 we see that if the B-module M_B is almost finitely generated projective, then M has the same property.

(iv) Furthermore, if B is faithfully flat and almost finitely presented as an A-module, then $A \to B$ does descend almost projectivity, as can be easily deduced from lemma 2.4.31(i) and proposition 2.4.18(ii).

3.2.27.   We denote by $A\text{-}\acute{\mathbf{Et}}_{\mathrm{afp}}$ the full subcategory of $A\text{-}\acute{\mathbf{Et}}$ consisting of all étale A-algebras B such that B is almost finitely presented as an A-module.

Theorem 3.2.28. *Let $I \subset A$ be a nilpotent ideal, and set $A' := A/I$.*

  (i) *Suppose that $\widetilde{\mathfrak{m}}$ is a (flat) V-module of homological dimension ≤ 1. Let P' be an almost projective A'-module.*
    (a) *There is an almost projective A-module P with $A' \otimes_A P \simeq P'$.*
    (b) *If P' is almost finitely presented, then P is almost finitely presented.*
  (ii) *The equivalence of theorem 3.2.18(ii) restricts to an equivalence*

$$A\text{-}\acute{\mathbf{Et}}_{\mathrm{afp}} \to A'\text{-}\acute{\mathbf{Et}}_{\mathrm{afp}}.$$

Proof. (i.a): As usual we reduce to $I^2 = 0$. Then proposition 3.2.20(i) applies with $R := A_*$, $J := I_*$, $R' := A_*/I_*$, $M' := P'_!$, $K := I_* \otimes_{R'} P'_!$ and $u' := 1_K$. We obtain a class $\omega(A_*, u') \in \mathrm{Ext}^2_{R'}(P'_!, I_* \otimes_{R'} P'_!)$ which gives the obstruction to the existence of a flat A_*-module F lifting $P'_!$. Since P' is almost projective, lemma 2.4.14(ii) says that $\mathrm{Ext}^2_{R'}(P'_!, I_* \otimes_{R'} P'_!) = 0$, so such F can always be found, and then the A-module $P = F^a$ is a flat lifting of P'; by lemma 3.2.25(i) we see that P is almost projective. Now (i.b) follows from (i.a), lemma 3.2.25(ii) and proposition 2.4.18(i).

(ii): In view of theorem 3.2.18(iii), we only have to show that an étale A-algebra B is almost finitely presented as an A-module whenever $B \otimes_A A'$ is almost finitely presented as an A'-module. However, the assertion is a direct consequence of lemma 3.2.25(iii). $\qquad\square$

Remark 3.2.29. (i) According to proposition 2.1.12(ii.b), theorem 3.2.28(i) applies especially when $\mathfrak{m}$ is countably generated as a V-module.

(ii) For P and P' as in theorem 3.2.28(i.b) let $\sigma_P : P \to P'$ be the projection. It is natural to ask whether the pair (P, σ_P) is uniquely determined up to isomorphism, *i.e.* whether, for any other pair $(Q, \sigma_Q) : Q \to P')$ for which theorem 3.2.28(i.b) holds, there exists an A-linear isomorphism $\phi : P \to Q$ such that $\sigma_Q \circ \phi = \sigma_P$. The answer is negative in general. Consider the case $P' := A'$. Take $P := Q := A$ and let σ_P be the natural projection, while $\sigma_Q := (u' \cdot 1_{A'}) \circ \sigma_P$, where u' is a unit in A'_*. Then the uniqueness question amounts to whether every unit in A'_* lifts to a unit of A_*. The following counterexample is related to the fact that the completion of the algebraic closure $\overline{\mathbb{Q}}_p$ of $\mathbb{Q}_p$ is not maximally complete. Let $V := \overline{\mathbb{Z}}_p$, the integral closure of $\mathbb{Z}_p$ in $\overline{\mathbb{Q}}_p$. Then V is a non-discrete valuation ring of rank one, and we take for $\mathfrak{m}$ the maximal ideal of V, $A := (V/p^2V)^a$ and $A' := A/pA$. Choose a compatible system of roots of p. An almost element of A' is just a V-linear morphism $\phi : \mathrm{colim}_{n>0} p^{1/n!}V \to V/pV$. Such a ϕ can be represented (in a non-unique way) by an infinite series of the form $\sum_{n=1}^{\infty} a_n p^{1-1/n!}$ ($a_n \in V$). The meaning of this expression is as follows. For every $m > 0$, scalar multiplication by the element $\sum_{n=1}^{m} a_n p^{1-1/n!} \in V$ defines a morphism $\phi_m : p^{1/m!}V \to V/pV$. For $m' > m$, let $j_{m,m'} : p^{1/m!}V \to p^{1/m'!}V$ be the imbedding. Then we have $\phi_{m'} \circ j_{m,m'} = \phi_m$, so that we can define $\phi := \mathrm{colim}_{m>0} \phi_m$. Similarly, every almost element of A can be represented by an expression of the form $a_0 + \sum_{n=1}^{\infty} a_n p^{2-1/n!}$. Now, if $\sigma : A \to A'$ is the natural projection, the induced map $\sigma_* : A_* \to A'_*$ is given by: $a_0 + \sum_{n=1}^{\infty} a_n p^{2-1/n!} \mapsto a_0$. In particular, its image is the subring $V/p \subset (V/p)_* = A'_*$. (This is a proper subring since A'_* is uncountable.) For instance, the unit $\sum_{n=1}^{\infty} p^{1-1/n!}$ of A'_* does not lie in the image of this map.

In the light of remark 3.2.29, the best one can achieve in general is the following.

Proposition 3.2.30. *Assume* **(A)** *(see (2.1.6)) and keep the notation of theorem 3.2.28(i). Suppose that $(Q, \sigma_Q : Q \to P')$ and $(P, \sigma_P : P \to P')$ are two pairs as in remark 3.2.29(ii). Then for every $\varepsilon \in \mathfrak{m}$ there exist A-linear morphisms $t_\varepsilon : P \to Q$ and $s_\varepsilon : Q \to P$ such that*

$$\mathbf{PQ}(\varepsilon) \qquad \begin{matrix} \sigma_Q \circ t_\varepsilon = \varepsilon \cdot \sigma_P & \sigma_P \circ s_\varepsilon = \varepsilon \cdot \sigma_Q \\ s_\varepsilon \circ t_\varepsilon = \varepsilon^2 \cdot \mathbf{1}_P & t_\varepsilon \circ s_\varepsilon = \varepsilon^2 \cdot \mathbf{1}_Q. \end{matrix}$$

Proof. Since both Q and P are almost projective and σ_P, σ_Q are epimorphisms, there exist morphisms $\bar{t}_\varepsilon : P \to Q$ and $\bar{s}_\varepsilon : Q \to P$ such that $\sigma_Q \circ \bar{t}_\varepsilon = \varepsilon \cdot \sigma_P$ and $\sigma_P \circ \bar{s}_\varepsilon = \varepsilon \cdot \sigma_Q$. Then we have $\sigma_P \circ (\bar{s}_\varepsilon \circ \bar{t}_\varepsilon - \varepsilon^2 \cdot \mathbf{1}_P) = 0$ and $\sigma_Q \circ (\bar{t}_\varepsilon \circ \bar{s}_\varepsilon - \varepsilon^2 \circ \mathbf{1}_Q) = 0$, *i.e.* the morphism $u_\varepsilon := \varepsilon^2 \cdot \mathbf{1}_P - \bar{s}_\varepsilon \circ \bar{t}_\varepsilon$ (resp. $v_\varepsilon := \varepsilon^2 \cdot \mathbf{1}_Q - \bar{t}_\varepsilon \circ \bar{s}_\varepsilon$) has image contained in the almost submodule IP (resp. IQ). Since $I^m = 0$ this implies $u_\varepsilon^m = 0$ and $v_\varepsilon^m = 0$. Hence

$$\varepsilon^{2m} \cdot \mathbf{1}_P = (\varepsilon^2 \mathbf{1}_P)^m - u_\varepsilon^m = \Big(\sum_{a=0}^{m-1} \varepsilon^{2a} u_\varepsilon^{m-1-a} \Big) \circ \bar{s}_\varepsilon \circ \bar{t}_\varepsilon.$$

Define $\bar{s}_{(2m-1)\varepsilon} := \big(\sum_{a=0}^{m-1} \varepsilon^{2a} u_\varepsilon^{m-1-a} \big) \circ \bar{s}_\varepsilon$. Notice that $\bar{s}_{(2m-1)\varepsilon} = \bar{s}_\varepsilon \circ \big(\sum_{a=0}^{m-1} \varepsilon^{2a} v_\varepsilon^{m-1-a} \big)$. This implies the equalities $\bar{s}_{(2m-1)\varepsilon} \circ \bar{t}_\varepsilon = \varepsilon^{2m} \cdot \mathbf{1}_P$ and $\bar{t}_\varepsilon \circ \bar{s}_{(2m-1)\varepsilon} = \varepsilon^{2m} \cdot \mathbf{1}_Q$. Then the pair $(\bar{s}_{(2m-1)\varepsilon}, \varepsilon^{2(m-1)} \cdot \bar{t}_\varepsilon)$ satisfies $\mathbf{PQ}(\varepsilon^{2m-1})$. Under **(A)**, every element of $\mathfrak{m}$ is a multiple of an element of the form ε^{2m-1}, therefore the claim follows for arbitrary $\varepsilon \in \mathfrak{m}$. $\qquad\square$

3.3. Nilpotent deformations of torsors. For the considerations that follow it will be convenient to extend yet further our basic setup. Namely, suppose that T is any topos; we can define a *basic setup relative to* T as a pair $(V, \mathfrak{m})$ consisting of a T-ring V and an ideal $\mathfrak{m} \subset V$ satisfying the usual assumptions (2.1.1). Then most of the discussion of chapter 2 extends *verbatim* to this relative setting. Accordingly, we generally continue to use the same notation as in *loc.cit.*; however, if it is desirable for clarity's sake, we may sometimes stress the dependance on T by mentioning it explicitly. For instance, an almost T-ring is an object of the category V^a-**Alg** of associative, commutative and unitary monoids of the abelian tensor category of V^a-modules, and sometimes the same category is denoted $(T, V, \mathfrak{m})^a$-**Alg**.

3.3.1. Let T be a topos and $(V, \mathfrak{m})$ a basic setup relative to T. For every object U of T, we can consider the restriction $(V_{/U}, \mathfrak{m}_{/U})$ of $(V, \mathfrak{m})$ to $T_{/U}$, which is a basic setup relative to the latter topos. As usual, for every object $X \to U$ of $T_{/U}$, one defines $V_{/U}(X) := \mathrm{Hom}_{T_{/U}}(X, V)$ and similarly for $\mathfrak{m}_{/U}(X)$. The restriction functor $V\text{-}\mathbf{Mod} \to V_{/U}\text{-}\mathbf{Mod}$ clearly preserves almost isomorphisms, whence a restriction functor

$$(T, V, \mathfrak{m})^a\text{-}\mathbf{Mod} \to (T_{/U}, V_{/U}, \mathfrak{m}_{/U})^a\text{-}\mathbf{Mod} \qquad M \mapsto M_{/U}.$$

Similar functors exist for the categories of V^a-algebras, and more generally, for A-algebras, where A is any $(T, V, \mathfrak{m})^a$-algebra.

3.3.2. For every almost T-module M, we define a functor $M_* : T^o \to \mathbb{Z}\text{-}\mathbf{Mod}$ by the rule: $U \mapsto \mathrm{Hom}_{V_{/U}^a}(V_{/U}^a, M_{/U})$. Let N be a V-module; using the natural isomorphism

$$\mathrm{Hom}_{V_{/U}}(\widetilde{\mathfrak{m}}_{/U}, N_{/U}) \xrightarrow{\sim} N_*^a(U)$$

(cp. (2.2.4)) one sees that M_* is a sheaf for the canonical topology of T (with small values), hence it is representable by an abelian group object of T, which we denote by the same name. It is then easy to check that M_* is a V-module, and that the functor $M \mapsto M_*$ is right adjoint to the localization functor $V\text{-}\mathbf{Mod} \to V^a\text{-}\mathbf{Mod}$. We can then generalize to the case of the localization functors $A\text{-}\mathbf{Mod} \to A^a\text{-}\mathbf{Mod}$ and $A\text{-}\mathbf{Alg} \to A^a\text{-}\mathbf{Alg}$, for an arbitrary V-algebra A. Likewise, the left adjoint functors to localization $M \mapsto M_!$ and $B \mapsto B_{!!}$ are obtained in the same way as in the earlier treatment of the one-point topos.

3.3.3. Let T and $(V, \mathfrak{m})$ be as in (3.3.1) and let R be any V-algebra. An *affine R-scheme* is an object of the category $R\text{-}\mathbf{Alg}^o$. An *affine almost R-scheme* (or an *affine R^a-scheme*) is an object of the category $R^a\text{-}\mathbf{Alg}^o$. If X is an affine R^a-scheme, then we may write $\mathscr{O}_X$ in place of the R^a-algebra X^o, and an X^o-module will also be called a *$\mathscr{O}_X$-module*. A morphism $\phi : X \to Y$ of affine almost schemes is the same as the morphism of almost algebras $\phi^\sharp := \phi^o : \mathscr{O}_Y \to \mathscr{O}_X$; moreover, ϕ induces pullback (*i.e.* tensor product $- \otimes_{\mathscr{O}_Y} \mathscr{O}_X$) and direct image functors (*i.e.* restriction of scalars), which we will denote

(3.3.4) $\phi^* : \mathscr{O}_Y\text{-}\mathbf{Mod} \to \mathscr{O}_X\text{-}\mathbf{Mod}$ and $\phi_* : \mathscr{O}_X\text{-}\mathbf{Mod} \to \mathscr{O}_Y\text{-}\mathbf{Mod}$

respectively. In the same vein, if A is any R^a-algebra, we may write $\operatorname{Spec} A$ to denote the object A^o represented by A in the opposite category $R^a\text{-}\mathbf{Alg}^o$. We say that X is *flat* over R^a if $\mathscr{O}_X$ is a flat R^a-algebra. Clearly the category of affine R^a-schemes admits arbitrary products. Hence, we can define an *affine R^a-group scheme* as a group object in the category of affine R^a-schemes (and likewise for the notion of affine R-group scheme).

3.3.5. The functors $B \mapsto B_*$ and $B \mapsto B_{!!}$ induce functors on almost schemes, which we denote in the same way. Notice that the functor $X \mapsto X_*$ (resp. $X \mapsto X_{!!}$) from affine R^a-schemes to R-schemes (resp. to $R^a_{!!}$-schemes) commutes with all colimits (resp. with all limits). Especially, if G is an affine R^a-group scheme, then $G_{!!}$ is an affine $R^a_{!!}$-group scheme.

3.3.6. Throughout the rest of this section we fix a V^a-algebra A and let $S :=\operatorname{Spec} A$. Let X an affine A-scheme, G an affine A-group scheme. A *right action* of G on X is a morphism of S-schemes $\rho : X \times_S G \to X$ fulfilling the usual conditions. To the datum (X, G, ρ) one assigns its *nerve* $G^\bullet_X$ which is a simplicial affine S-scheme whose component in degree n is $G^n_X := X \times_S G^n$, and whose face and degeneracy morphisms

$$\partial_i : G^{n+1}_X \to G^n_X \qquad \sigma_j : G^n_X \to G^{n+1}_X \qquad i = 0, ..., n+1; j = 0, ..., n$$

are defined for every $n \in \mathbb{N}$ as in [51, Ch.VI, §2.5].

3.3.7. Let (X, G, ρ) be as in (3.3.6), and let M be an $\mathscr{O}_X$-module. A *G-action* on M is a morphism of $\mathscr{O}_{G^1_X}$-modules

(3.3.8) $\beta : \partial^*_0 M \to \partial^*_1 M$

such that the following diagram commutes:

$$(3.3.9) \quad \begin{array}{ccc} \partial_0^* \partial_0^* M & \xrightarrow{\partial_0^* \beta} & \partial_0^* \partial_1^* M \Longrightarrow \partial_2^* \partial_0^* M \\ \| & & \downarrow{\partial_2^* \beta} \\ \partial_1^* \partial_0^* M & \xrightarrow{\partial_1^* \beta} & \partial_1^* \partial_1^* M \Longrightarrow \partial_2^* \partial_1^* M \end{array}$$

and such that

$$(3.3.10) \qquad\qquad \sigma_0^* \beta = 1_M.$$

One also says that (M, β) is a G-equivariant $\mathscr{O}_X$-module. One defines in the obvious way the morphisms of G-equivariant $\mathscr{O}_X$-modules, and we denote by $\mathscr{O}_X$-$\mathbf{Mod}_G$ the category of G-equivariant $\mathscr{O}_X$-modules.

Lemma 3.3.11. *For every G-equivariant $\mathscr{O}_X$-module M, the morphism (3.3.8) is an isomorphism.*

Proof. We let $\tau : G_X^1 \rightarrow G_X^2$ be the morphism given on T-points by the rule: $(g, x) \mapsto (g, g^{-1}, x)$. Working out the identifications, one checks easily that $\tau^*(3.3.9)$ boils down to the diagram

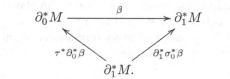

However, $\partial_1^* \sigma_0^* \beta = 1_{\partial_1^* M}$, in view of (3.3.10), hence β is an epimorphism and $\tau^* \partial_0^* \beta$ is an monomorphism. Since $\partial_0 \circ \tau$ is an isomorphism, we deduce that β is already a monomorphism, whence the claim. $\qquad\square$

Lemma 3.3.12. *If G is a flat affine S-group scheme acting on an affine S-scheme X, then the category of G-equivariant $\mathscr{O}_X$-modules is abelian.*

Proof. We need to verify that every (G-equivariant) morphism $\phi : M \rightarrow N$ of G-equivariant $\mathscr{O}_X$-modules admits a kernel and a cokernel (and then it will be clear that the kernel of $N \rightarrow \mathrm{Coker}\, \phi$ equals the cokernel of $\mathrm{Ker}\, \phi \rightarrow M$, since the same holds for $\mathscr{O}_X$-modules). The obvious candidates are the kernel K and cokernel C taken in the category of $\mathscr{O}_X$-modules, and one has only to show that the G-actions of M and N induce G-actions on K and C. This is always the case for C (even when G is not flat). To deal with K, one remarks that both morphisms $\partial_0, \partial_1 : G_X^1 \rightarrow X$ are flat; indeed, this is clear for ∂_1. Then the same holds for $\partial_0 := \partial_1 \circ \omega$, where $\omega := (\rho, i) : G_X^1 \rightarrow G_X^1$ is the isomorphism deduced from the action $\rho : G_X^1 \rightarrow X$ and the inverse map $i : G \rightarrow G$. It follows that $\mathrm{Ker}\,(\partial_j^* \phi) \simeq \partial_j^*(\mathrm{Ker}\, \phi)$ for $j = 0, 1$, whence the claim. $\qquad\square$

3.3.13. Let $\phi : X \rightarrow Y$ be a morphism of affine S-schemes. A *square zero deformation* of X over Y is a datum $(j : X \rightarrow X', \mathscr{I}, \beta)$, consisting of:

    (a) a morphism of Y-schemes $j : X \rightarrow X'$ such that the induced morphism of $\mathscr{O}_Y$-algebras $j^\sharp : \mathscr{O}_{X'} \rightarrow \mathscr{O}_X$ is an epimorphism and $\mathscr{I} := \mathrm{Ker}\, j^\sharp$ is a square zero ideal, and

(b) an $\mathscr{O}_X$-module $\mathscr{I}$ with an isomorphism $\beta : j^* \mathscr{J} \xrightarrow{\sim} \mathscr{I}$ of $\mathscr{O}_X$-modules.

The square zero deformations form a category $\mathbf{Exal}_Y(X, \mathscr{I})$, with morphisms defined in the obvious way. As in the case of the one-point topos, we can compute the isomorphism classes of square zero deformations of X in terms of an appropriate cotangent complex. And, just as in the earlier treatment, we have to make sure that we are dealing with exact algebras, hence the right definition of the cotangent complex of ϕ is:

$$\mathbb{L}_{X/Y} := \iota^* \mathbb{L}_{(X \amalg \operatorname{Spec}V^a)_{!!}/(Y \amalg \operatorname{Spec}V^a)_{!!}}$$

where $\iota : X_{!!} \to (X \amalg \operatorname{Spec}V^a)_{!!}$ is the natural morphism of schemes. This is an object of $\mathsf{D}(\mathscr{O}_{X_{!!}}\text{-}\mathbf{Mod})$.

3.3.14.   Next, let G be a flat affine S-group scheme acting on X and Y, in such a way that ϕ is G-equivariant; it follows from lemma 3.3.12 that $\mathscr{J}$ has a natural G-action. A *G-equivariant square zero deformations* of X over Y, is a datum $(j : X \to X', \mathscr{I}, \beta)$ as above, such that, additionally, X' and $\mathscr{I}$ are endowed with a G-action and both j and β are G-equivariant. Let us denote by $\mathbf{Exal}_Y(X/G, \mathscr{I})$ the category of such G-equivariant deformations. We aim to classify the isomorphism classes of objects of $\mathbf{Exal}_Y(X/G, \mathscr{I})$, and more generally, study the G-equivariant deformation theory of X by means of an appropriate cotangent complex cohomology. This is achieved by the following device.

3.3.15.   Let I be a (small) category; recall ([51, Ch.VI, §5.1]) that a *fibred topos over I* is a pseudo-functor X of I in the 2-category of topoi, *i.e.* the datum of:

(a) for every object i of I, a topos X_i

(b) for every arrow $f : i \to j$ in I, a morphism of topoi $X_f : X_i \to X_j$ (sometimes denoted f) with $X_{1_i} = 1_{X_i}$ for every object i of I

(c) for every composition $i \xrightarrow{f} j \xrightarrow{g} k$, a *transitivity isomorphism* $X_{f,g} : X_g \circ X_f \xrightarrow{\sim} X_{gf}$, submitted to certain compatibility conditions (in practice, we will omit from the notation the transitivity isomorphisms).

Given a fibred topos X over I, one denotes by $\operatorname{Top}(X)$ the following category, which is easily seen to be a topos ([51, Ch.VI, §5.2]). An object of E of $\operatorname{Top}(X)$ is the datum of

(a) for every $i \in I$, an object E_i of X_i

(b) for every arrow $f : i \to j$, a morphism $E_f : f^* E_j \to E_i$ such that $E_{1_i} = 1_{E_i}$ for every object i of I, and such that for every composition $i \xrightarrow{f} j \xrightarrow{g} k$, one has $E_{gf} = E_f \circ f^* E_g$, provided one identifies $(gf)^* E_k$ with $f^* g^* E_k$ via $X_{f,g}$.

As an example we have the topos $s.T$ whose objects are the cosimplicial objects of T; indeed this is the topos $\operatorname{Top}(F)$ associated to a fibred topos F, whose indexing category is the category Δ^o defined in (2.2) : to every object $[n]$ of Δ^o one assigns $F_{[n]} := T$ and for every morphism $f : [n] \to [m]$ of Δ^o, one takes $F_f := 1_T$, the identity morphism of the topos T.

There is an obvious functor $T \to s.T$ that assigns to any object Z of T the constant cosimplicial T-object $s.Z$ associated to Z. Especially, if $(T, V, \mathfrak{m})$ is a basic setup for T, then $(s.T, s.V, s.\mathfrak{m})$ is a basic setup relative to $s.T$.

3.3.16. Suppose now that $M_\bullet$ is a cosimplicial V^a-module. By applying termwise the functor $N \mapsto N_*$ we deduce a cosimplicial V-module $(M_\bullet)_*$, whence an object of $s.T$ which we denote by the same name. Clearly $(M_\bullet)_*$ is a $s.V$-module and we can therefore take its image in the localized category $(s.V)^a$-**Mod**. This defines a functor

(3.3.17) $$ s.(V^a\text{-}\mathbf{Mod}) \to (s.V)^a\text{-}\mathbf{Mod} $$

and it is not difficult to see that (3.3.17) is an equivalence of categories. Similar equivalences then follow for categories of cosimplicial V^a-algebras (a.k.a. simplicial V^a-schemes) and the like. For instance, for (X, G, ρ) as in (3.3.6) we can regard the nerve $G_X^\bullet$ as an affine $s.S$-scheme.

3.3.18. More generally, let $X_\bullet$ be any simplicial S-scheme. An $\mathcal{O}_{X_\bullet}$-module is the same as a cosimplicial $\mathcal{O}_S$-module $M_\bullet$, such that M_n is an $\mathcal{O}_{X_n}$-module for every $n \in \mathbb{N}$, and the coface (resp. codegeneracy) morphisms

$$ \partial^i : M_n \to M_{n+1} \qquad (\text{resp.} \quad \sigma^j : M_n \to M_{n-1}) $$

are ∂_i-linear (resp. σ_j-linear), i.e. they induce $\mathcal{O}_{X_n}$-linear morphisms $M_n \to \partial_{i*} M_{n+1}$ (resp. $M_n \to \sigma_{j*} M_{n-1}$) (notation of (3.3.4)). It is convenient to introduce the following notation: for every $n \in \mathbb{N}$ and $0 \le i \le n+1$ (resp. $0 \le j \le n-1$) we let

$$ \overline{\partial}^i : \partial_i^* M_n \to M_{n+1} \qquad (\text{resp.} \quad \overline{\sigma}^j : \sigma_j^* M_n \to M_{n-1}) $$

be the $\mathcal{O}_{X_{n+1}}$-linear (resp. $\mathcal{O}_{X_{n-1}}$-linear) morphism deduced from ∂^i (resp. σ^j) by extension of scalars.

3.3.19. For every S-scheme X on which G acts, and for every $n \in \mathbb{N}$, set $\pi_{X,n} := \partial_1 \circ \partial_2 \circ \ldots \circ \partial_n : G_X^n \to X$. For any G-equivariant $\mathcal{O}_X$-module M, we define a quasi-coherent $\mathcal{O}_{G_X^\bullet}$-module $\pi_X^* M$ as follows. According to (3.3.16), this is the same as defining a module $\pi_X^* M$ over the cosimplicial almost algebra $\mathcal{O}_{G_X^\bullet}$; then we set $\pi_X^* M_n := \pi_{X,n}^* M$ for every object $[n]$ of Δ^o. Next, we remark that $\pi_{X,n-1} \circ \partial_i = \pi_{X,n}$ for every $0 < i \le n$, hence we have natural isomorphisms $\pi_X^* M_n \xrightarrow{\sim} \partial_i^* \pi_X^* M_{n-1}$, from which we deduce the coface morphisms $\partial^i : \pi_X^* M_{n-1} \to \pi_X^* M_n$ for every $0 < i \le n$. Finally we use the morphism (3.3.8) and the cartesian diagram:

$$
\begin{array}{ccc}
G_X^n & \xrightarrow{\partial_0} & G_X^{n-1} \\
\scriptstyle \tau := \partial_2 \circ \ldots \circ \partial_n \downarrow & & \downarrow \scriptstyle \pi_{X,n-1} \\
G_X^1 & \xrightarrow{\partial_0} & X
\end{array}
$$

to define ∂^0 as the composition:

$$\pi_X^* M_{n-1} \longrightarrow \partial_0^* \pi_X^* M_{n-1} = \partial_0^* \pi_{X,n-1}^* M = \tau^* \partial_0^* M \xrightarrow{\tau^* \beta} \pi_{X,n}^* M.$$

We leave to the reader the task of defining the codegeneracy morphisms; using the cocycle relation encoded in (3.3.9) one can then verify the required cosimplicial identities. In this way we obtain a functor

$$(3.3.20) \qquad \mathscr{O}_X\text{-}\mathbf{Mod}_G \to \mathscr{O}_{G_X^\bullet}\text{-}\mathbf{Mod} \qquad M \mapsto \pi_X^* M.$$

Proposition 3.3.21. *The functor* (3.3.20) *is fully faithful, and its essential image is the full subcategory of all* $\mathscr{O}_{G_X^\bullet}$-modules $(M_n \mid n \in \mathbb{N})$ *such that* $\overline{\partial}^n : \partial_n^* M_{n-1} \to M_n$ *is an isomorphism for every* $n \in \mathbb{N}$ *(notation of* (3.3.18)*).*

Proof. Let $\phi_\bullet : \pi_X^* M \to \phi_X^* N$ be a morphism of $\mathscr{O}_{G_X^\bullet}$-modules. Since $M_n = \pi_{X,n}^* M$ for every $n \in \mathbb{N}$, we see that $\phi_\bullet$ is already determined by its component $\phi_0 : M \to N$, whence the full faithfulness of (3.3.20). On the other hand, let $M_\bullet$ be an $\mathscr{O}_{G_X^\bullet}$-module satisfying the condition of the lemma, and set $M := M_0$. We define a morphism $\beta : \partial_0^* M \to \partial_1^* M$ as the composition

$$\partial_0^* M \xrightarrow{\overline{\partial}^0} M_1 \xrightarrow{(\overline{\partial}^1)^{-1}} \partial_1^* M.$$

Claim 3.3.22. β defines an action of G on M.

Proof of the claim. The identity $\sigma_0^* \beta = \mathbf{1}_M$ is immediate, hence it suffices to show that (3.3.9) commutes. This is the same as showing the commutativity of the following diagram:

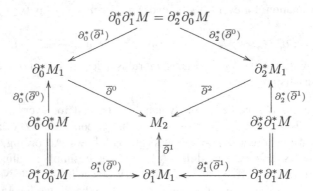

which can be checked separately on each of its three quadrangular subdiagrams. Let us verify for instance the commutativity of the bottom left diagram. By linearity, we can simplify down to the diagram:

$$\begin{array}{ccc} M_1 & \xrightarrow{\partial^0} & M_2 \\ {\scriptstyle \partial^0} \uparrow & & \uparrow {\scriptstyle \partial^1} \\ M & \xrightarrow{\partial_0} & M_1 \end{array}$$

whose commutativity expresses one of the identities defining the cosimplicial module $M_\bullet$. $\diamond$

To conclude the proof, it suffices to exhibit an isomorphism $\gamma_\bullet : \pi_X^* M \xrightarrow{\sim} M_\bullet$. For every $n \in \mathbb{N}$, let $\gamma_n : \pi_{X,n}^* M \to M_n$ be the morphism induced by the composition $\partial^n \circ \partial^{n-1} \circ \ldots \circ \partial^1$; under our assumption, γ_n is an isomorphism, and we leave to the reader the verification that $(\gamma_n \mid n \in \mathbb{N})$ defines a morphism of cosimplicial modules. $\qquad\qquad\square$

3.3.23.   In case G is flat over S, $\mathscr{I}$ is G-equivariant, and $\phi : X \to Y$ is an equivariant morphism of affine S-schemes on which G acts, we deduce a natural functor:

$$(3.3.24) \qquad \mathbf{Exal}_Y(X/G, \mathscr{I}) \to \mathbf{Exal}_{G_Y^\bullet}(G_X^\bullet, \pi_X^* \mathscr{I}).$$

Namely, to any G-equivariant square zero deformation $D := (j : X \to X', \mathscr{I}, \beta)$ one assigns the datum $G_D := (G_j : G_X^\bullet \to G_{X'}^\bullet, \pi_X^* \mathscr{I}, \pi_X^* \beta)$. The flatness of G ensures that $\operatorname{Ker}(\mathscr{O}_{G_{X'}^n} \to \mathscr{O}_{G_X^n}) \simeq \pi_{X,n}^* \mathscr{I}$ for every $n \in \mathbb{N}$, which means that G_D is indeed a deformation of $G_X^\bullet$ over $G_Y^\bullet$.

The basic observation is contained in the following:

Lemma 3.3.25. *Under the assumptions of (3.3.23), the functor (3.3.24) is an equivalence of categories.*

Proof. Let $(j : G_X^\bullet \to X'_\bullet, \pi_X^* \mathscr{I}, \beta)$ be a deformation of $G_X^\bullet$. For every $n \in \mathbb{N}$ we have a commutative diagram of affine S-schemes:

$$
\begin{array}{ccc}
X'_n & \xrightarrow{\ \partial\ } & X'_0 \\
\downarrow & & \downarrow \\
G_Y^n & \xrightarrow{\ \pi_Y\ } & Y
\end{array}
$$

where $\partial := \partial_1 \circ \partial_2 \circ \ldots \circ \partial_n$ and $\pi_{Y,n}$ is the natural projection as in (3.3.19). We deduce a unique morphism $\alpha_n : X'_n \to G_Y^n \times_Y X'_0 \simeq G_{X'_0}^n$. By construction, $\alpha_n^\sharp$ fits into a commutative diagram

$$
\begin{array}{ccccccccc}
0 & \longrightarrow & \mathscr{O}_{G^n} \otimes_{\mathscr{O}_S} \mathscr{I} & \longrightarrow & \mathscr{O}_{G^n} \otimes_{\mathscr{O}_S} \mathscr{O}_{X'_0} & \longrightarrow & \mathscr{O}_{G^n} \otimes_{\mathscr{O}_S} \mathscr{O}_X & \longrightarrow & 0 \\
& & \| & & \downarrow{\scriptstyle \alpha_n^\sharp} & & \| & & \\
0 & \longrightarrow & \mathscr{O}_{G^n} \otimes_{\mathscr{O}_S} \mathscr{I} & \longrightarrow & \mathscr{O}_{X'_n} & \longrightarrow & \mathscr{O}_{G^n} \otimes_{\mathscr{O}_S} \mathscr{O}_X & \longrightarrow & 0
\end{array}
$$

hence α_n is an isomorphism for every $n \in \mathbb{N}$. To conclude, it suffices to verify that the system of morphisms $(\alpha_n \mid n \in \mathbb{N})$ defines a morphism of simplicial S-schemes: $\alpha_\bullet : X'_\bullet \to G_{X'_0}^\bullet$. This amounts to showing that the α_n commute with the face and degeneracy morphisms, which however is easily checked from the definition. $\qquad\square$

3.3.26.   Combining lemmata 3.3.25 and 2.5.17 (which holds *verbatim* in the present context) we derive a natural equivalence of categories:

$$\mathbf{Exal}_Y(X/G, \mathscr{I}) \to \mathbf{Exal}_{G_{Y!!}^\bullet}(G_{X!!}^\bullet, (\pi_X^* \mathscr{I})_*).$$

This enables us to use the usual theory of the cotangent complex to classify the G-equivariant deformations of X.

3.3.27.   According to [51, Ch.VI, §5.3] we have a morphism of topoi

$$[n]_T : T \to s.T$$

for every $n \in \mathbb{N}$, called *restriction to the n-th level*. It is given by a pair of adjoint functors $([n]_T^*, [n]_{T*})$ such that $[n]_T^* : s.T \to T$ is the functor that assigns to any cosimplicial object $(X_{[k]} \mid k \in \mathbb{N})$ the object $X_{[n]}$ of T. For every $k \in \mathbb{N}$ set $N_{X,k} := G_X^k \amalg \operatorname{Spec} V^a$; the system $(N_{X,k} \mid k \in \mathbb{N})$ defines a simplicial $\operatorname{Spec} V^a$-scheme N_X (and likewise we define N_Y). In view of [50, Ch.II, (1.2.3.5)], we deduce natural isomorphisms of simplicial complexes of flat $\mathscr{O}_{N_{X,n!!}}$-modules:

$$[n]_T^* \mathbb{L}_{N_{X!!}/N_{Y!!}} \xrightarrow{\sim} \mathbb{L}_{N_{X,n!!}/N_{Y,n!!}} \qquad \text{for every } n \in \mathbb{N}$$

whence natural isomorphisms of simplicial complexes of flat $\mathscr{O}_{G_{X!!}^n}$-modules:

$$[n]_T^* \mathbb{L}_{G_X^\bullet/G_Y^\bullet} := [n]_T^* \iota_\bullet^* \mathbb{L}_{N_{X!!}/N_{Y!!}} \xrightarrow{\quad\sim\quad} \iota_{[n]}^* [n]_T^* \mathbb{L}_{N_{X!!}/N_{Y!!}}$$

$$\downarrow \sim$$

$$\iota_{[n]}^* \mathbb{L}_{N_{X,n!!}/N_{Y,n!!}} =: \mathbb{L}_{G_X^n/G_Y^n}$$

where $\iota_\bullet : G_{X!!}^\bullet \to N_{X!!}$ is the morphism of simplicial schemes defined as in (3.3.13). In other words, $\mathbb{L}_{G_X^\bullet/G_Y^\bullet}$ is a mixed simplicial-cosimplicial module $\mathbb{L}_{\bullet\bullet}$ whose rows $\mathbb{L}_{\bullet n}$ are the cotangent complexes of the morphisms $G_\phi : G_X^n \to G_Y^n$. Furthermore, for every $n \in \mathbb{N}$ and every $i \leq n+1$ we have a cartesian diagram

$$\begin{array}{ccc}
G_X^{n+1} & \xrightarrow{\ \partial_i\ } & G_X^n \\
{\scriptstyle G_\phi^{n+1}}\downarrow & & \downarrow{\scriptstyle G_\phi^n} \\
G_Y^{n+1} & \xrightarrow{\ \partial_i\ } & G_Y^n
\end{array}$$

whose horizontal arrows are flat morphisms (since G is S-flat by assumption). Whence, taking into account theorem 2.5.36, a natural isomorphism:

$$\overline{\partial}_{\mathbb{L}}^i : \partial_i^* \mathbb{L}_{G_X^n/G_Y^n} \xrightarrow{\sim} \mathbb{L}_{G_X^{n+1}/G_Y^{n+1}}$$

in $\mathsf{D}(\mathscr{O}_{N_{X,n+1!!}}\text{-}\mathbf{Mod})$ and, by unwinding the definitions, one sees that $\overline{\partial}_{\mathbb{L}}^i$ is induced by the i-th coface morphism $\mathbb{L}_{\bullet n} \to \mathbb{L}_{\bullet n+1}$ of the double complex $\mathbb{L}_{\bullet\bullet}$.

Definition 3.3.28. Let G be an affine S-group scheme, $\phi : X \to Y$ a G-equivariant morphism of affine S-schemes on which G acts (on the right). We say that ϕ is a *G-torsor over Y* if the action of G on Y is trivial (*i.e.* $\rho : G \times_S Y \to Y$ is the natural projection) and there exists a cartesian diagram of affine S-schemes

(3.3.29)
$$\begin{array}{ccc}
G \times_S Z & \xrightarrow{\ g\ } & X \\
{\scriptstyle p_Z}\downarrow & & \downarrow{\scriptstyle \phi} \\
Z & \xrightarrow{\ f\ } & Y
\end{array}$$

such that f is faithfully flat, g is G-equivariant for the action on $G \times_S Z$ deduced from G, and p_Z is the natural projection.

3.3.30.   Suppose that G is a flat affine S-group scheme and $X \to Y$ is a G-torsor over Y. Then the groupoid $G \times_S X \rightrightarrows X$ defined by the G-action on X is an effective equivalence relation (cp. (4.5.1)) and $Y \simeq X/G$. Furthermore, let $X^{\times_Y n} := X \times_Y X \times_Y \dots \times_Y X$ (the n-th fold cartesian power of X over Y); there are natural identifications $G_X^n \simeq X^{\times_Y n+1}$, amounting to an isomorphism of (semi-)simplicial S-schemes augmented over Y:

$$G_X^\bullet \xrightarrow{\sim} [Y|X].$$

(cp. the notation of [5, Exp.V^{bis}, (1.2.7)]). We can regard the cosimplicial T-ring $\mathscr{O}_{G_X^\bullet}$ as a ring of the topos $\Gamma(\Delta \times T)$ deduced from the simplicial topos $\Delta \times T$ (notation of [5, Exp.V^{bis}, (1.2.8)]), whence an augmentation of fibred topoi

$$(3.3.31) \qquad \theta : (\Delta \times T, \mathscr{O}_{G_X^\bullet}) \to (\Delta \times T, \mathscr{O}_Y)$$

and it follows from the foregoing and from remark 3.1.3 that (3.3.31) is an *augmentation of 2-cohomological descent* (defined as in [5, Exp.V^{bis}, Déf.2.2.6]). Denote by

$$\bar{\theta}^* : (T, \mathscr{O}_Y)\text{-}\mathbf{Mod} \to (\Gamma(\Delta \times T), \mathscr{O}_{G_X^\bullet})\text{-}\mathbf{Mod}$$

the morphism obtained as the composition of the constant functor

$$\varepsilon^* : (T, \mathscr{O}_Y)\text{-}\mathbf{Mod} \to (\Gamma(\Delta \times T), \mathscr{O}_Y)\text{-}\mathbf{Mod}$$

and the functor $\Gamma(\theta^*) : (\Gamma(\Delta \times T), \mathscr{O}_Y)\text{-}\mathbf{Mod} \to \mathscr{O}_{G_X^\bullet}\text{-}\mathbf{Mod}$. Since the morphisms $\partial_{\mathbb{L}}^i$ are isomorphisms, it then follows from general cohomological descent ([5, Exp.V^{bis}, Prop.2.2.7]) that, for every $k \in \mathbb{N}$, the truncated system $(\tau_{[-k}\mathbb{L}_{G_X^n/G_Y^n} \mid n \in \mathbb{N})$ is in the essential image of the functor $L^+\bar{\theta}^*$. In the following we will be only interested in the case where the cotangent complex is concentrated in degree zero, in which case one can avoid the recourse to cohomological descent, and rather appeal to more down-to-earth faithfully flat descent. In any case, the foregoing shows that there exists a uniquely determined pro-object $(\mathbb{L}_{X/Y,k}^G \mid k \in \mathbb{N})$ of $\mathsf{D}(\mathscr{O}_Y\text{-}\mathbf{Mod})$ such that

$$L^+\bar{\theta}^*(\mathbb{L}_{X/Y,k}^G) \simeq \tau_{[-k}\mathbb{L}_{G_X^\bullet/G_Y^\bullet}^a \qquad \text{for every } k \in \mathbb{N}.$$

By trivial duality there follow natural isomorphisms for every $k \in \mathbb{N}$ and every complex $K^\bullet \in \mathsf{D}^+(\mathscr{O}_{G_X^\bullet}\text{-}\mathbf{Mod})$ such that $K = \tau_{[-k}K$:

$$(3.3.32) \qquad \mathbf{Ext}_{\mathscr{O}_{G_X^\bullet}}^k (\mathbb{L}_{G_X^\bullet/G_Y^\bullet}^a, K^\bullet) \simeq \mathbf{Ext}_{\mathscr{O}_Y}^k (\mathbb{L}_{X/Y,k}^G, R^+\bar{\theta}_*K^\bullet).$$

Definition 3.3.33. Let G be an affine S-scheme and $e : S \to G$ its unit section. The *co-Lie complex* of G is the complex of $\mathscr{O}_S$-modules $\ell_{G/S} := e^*\mathbb{L}_{G/S}^a$.

Proposition 3.3.34. *Let G be a flat S-group scheme, $\phi : X \to Y$ be a G-torsor over Y and $\pi_Y : Y \to S$ the structure morphism. Then, for every $k \in \mathbb{N}$, the complex $\mathbb{L}_{X/Y,k}^G$ is locally isomorphic to $\pi_Y^*\tau_{[-k}\ell_{G/S}$ in the fpqc topology of Y.*

Proof. We will exhibit more precisely a faithfully flat morphism $f : Z \to Y$ such that, for every $k \in \mathbb{N}$ there exist isomorphisms $R^+f^*\mathbb{L}_{X/Y,k}^G \simeq R^+f^*\pi_Y^*\tau_{[-k}\ell_{G/S}$ in $\mathsf{D}(\mathscr{O}_Z\text{-}\mathbf{Mod})$. Let $\pi_G : G \to S$ be the structure morphism; first of all we remark:

Claim 3.3.35. $\mathbb{L}^a_{G/S} \simeq \pi^*_G \ell_{G/S}$.

Proof of the claim. Indeed, π_G is trivially a G-torsor over S, hence we have a compatible system of isomorphisms $\tau_{[-k}\mathbb{L}^a_{G/S} \simeq \pi^*_Y \mathbb{L}^G_{G/S,k}$. If $e : S \to G$ is the unit section, we deduce: $\tau_{[-k}\ell_{G/S} \simeq e^* \tau_{[-k}\mathbb{L}^a_{G/S} \simeq \mathbb{L}^G_{G/S,k}$ for every $k \in \mathbb{N}$. After taking π^*_G of the two sides, the claim follows. $\diamondsuit$

Let now $f : Z \to Y$ be a faithfully flat morphism that trivializes the given G-torsor $\phi : X \to Y$, so we have a cartesian diagram (3.3.29). Denoting by $p_G : G \times_S Z \to G$ the natural projection, we deduce (since G is S-flat) an isomorphism $g^* \mathbb{L}_{X/Y} \simeq p^*_G \mathbb{L}_{G/S}$. Whence, in view of claim 3.3.35 :

$$\begin{aligned}(3.3.36) \qquad p^*_Z f^* \mathbb{L}^G_{X/Y,k} &\simeq p^*_G \pi^*_G \tau_{[-k}\ell_{G/S} \\ &\simeq p^*_Z \pi^*_Z \tau_{[-k}\ell_{G/S} \simeq p^*_Z f^* \pi^*_Y \tau_{[-k}\ell_{G/S}.\end{aligned}$$

Let $e_Z := e \times_S 1_Z : Z \to G \times_S Z$; the claim follows after applying e^*_Z to (3.3.36). $\square$

3.3.37. Finally we can wrap up this section with a discussion of equivariant deformations of torsors. Hence, let $\phi : X \to Y$ be a G-torsor over Y, and $j_Y : Y \to Y'$ a morphism of affine S-schemes such that $\mathscr{I} := \text{Ker}(j^\sharp_Y : \mathscr{O}_{Y'} \to \mathscr{O}_Y)$ is a square zero ideal of $\mathscr{O}_{Y'}$. We wish to classify the *square zero deformations of the torsor* ϕ *over* Y', that is, the isomorphism classes of cartesian diagrams

$$\begin{array}{ccc} X & \xrightarrow{\;j_X\;} & X' \\ {\scriptstyle \phi}\downarrow & & \downarrow{\scriptstyle \phi'} \\ Y & \xrightarrow{\;j_Y\;} & Y' \end{array}$$

such that ϕ' is a G-torsor over Y' and j_X is G-equivariant.

Theorem 3.3.38. *Suppose that G is flat over S, that $H_i(\ell_{G/S}) = 0$ for $i > 0$ and that $H_0(\ell_{G/S})$ is an almost finitely generated projective $\mathscr{O}_S$-module. Furthermore, suppose that the homological dimension of $\widetilde{\mathfrak{m}}$ is ≤ 1. Then, in the situation of (3.3.37), we have:*

(i) *The pro-object $(\mathbb{L}^G_{X/Y,k} \mid k \in \mathbb{N})$ is constant, isomorphic to a complex of $\mathsf{D}(\mathscr{O}_Y\text{-}\mathbf{Mod})$ concentrated in degree zero, that we shall denote by $\mathbb{L}^G_{X/Y}$.*

(ii) *The set of isomorphism classes of square zero deformations of the torsor $\phi : X \to Y$ over Y' is a torsor under the group $\text{Ext}^1_{\mathscr{O}_Y}(\mathbb{L}^G_{X/Y}, \mathscr{I})$.*

(iii) *The group of automorphisms of a square zero deformation $\phi' : X' \to Y'$ as in (3.3.37) is naturally isomorphic to $\text{Ext}^0_{\mathscr{O}_Y}(\mathbb{L}^G_{X/Y}, \mathscr{I})$.*

Proof. (i) follows easily from proposition 3.3.34; moreover it follows from remark 3.2.26(iii) that $H_0(\mathbb{L}^G_{X/Y})^a$ is an almost finitely generated projective $\mathscr{O}_Y$-module. Let $\mathscr{J} := \text{Ker}(j^\sharp : \mathscr{O}_{X'} \to \mathscr{O}_X)$; by flatness, the natural morphism $\phi^* \mathscr{I} \to \mathscr{J}$ is a G-equivariant isomorphism. We notice that, for every $\mathscr{O}_Y$-module M, there is a natural isomorphism

$$\bar{\theta}^* M_! \xrightarrow{\sim} (\pi^*_X \phi^* M)_!.$$

By cohomological descent (or else, by plain old-fashioned faithfully flat descent) it follows that the counit of the adjunction:

$$\mathscr{I}_! \to R^+\bar{\theta}_* \pi_X^* \mathscr{I}_!$$

is an isomorphism, whence, in light of (3.3.32), natural isomorphisms:

(3.3.39) $\mathrm{Ext}^k_{\mathscr{O}_{G_X^\bullet}}(\mathbb{L}^a_{G_X^\bullet/G_Y^\bullet}, \pi_X^* \mathscr{I}) \simeq \mathrm{Ext}^k_{\mathscr{O}_Y}(H_0(\mathbb{L}^G_{X/Y}), \mathscr{I})$ for every $k \in \mathbb{N}$.

Combining (3.3.32) and lemma 2.4.14(i),(ii), we deduce that

$$\mathrm{Ext}^2_{\mathscr{O}_{G_X^\bullet}}(\mathbb{L}^a_{G_X^\bullet/G_Y^\bullet}, \pi_X^* \mathscr{I}) = 0.$$

However, by (3.3.26) and the usual arguments (cp. section 3.2) one knows that the obstruction to deforming the torsor ϕ is a class in the latter group; since the obstruction vanishes, one deduces (ii). (iii) follows in the same way. □

3.4. **Descent.** Faithfully flat descent in the almost setting presents no particular surprises: since the functor $A \mapsto A_{!!}$ preserves faithful flatness of morphisms (see remark 3.1.3) many well-known results for usual rings and modules extend *verbatim* to almost algebras.

3.4.1. So for instance, faithfully flat morphisms are of universal effective descent for the fibred categories

$$F : V^a\text{-}\mathbf{Alg.Mod}^o \to V^a\text{-}\mathbf{Alg}^o \quad \text{and} \quad G : V^a\text{-}\mathbf{Alg.Morph}^o \to V^a\text{-}\mathbf{Alg}^o$$

(see definition 2.5.22: for an almost V-algebra B, the fibre F_B (resp. G_B) is the opposite of the category of B-modules (resp. B-algebras)). Then, using remark 3.2.26, we deduce also universal effective descent for the fibred subcategories of flat (resp. almost finitely generated, resp. almost finitely presented, resp. almost finitely generated projective) modules. Likewise, a faithfully flat morphism is of universal effective descent for the fibred subcategories $\text{Ét}^o \to V^a\text{-}\mathbf{Alg}^o$ of étale (resp. **w.Ét**$^o \to V^a\text{-}\mathbf{Alg}^o$ of weakly étale) algebras.

3.4.2. More generally, since the functor $A \mapsto A_{!!}$ preserves pure morphisms in the sense of [64], and since, by a theorem of Olivier (*loc. cit.*), pure morphisms are of universal effective descent for modules, the same holds for pure morphisms of almost algebras.

3.4.3. Non-flat descent is more delicate. Our results are not as complete here as it could be wished, but nevertheless, they suffice for the further study of étale and unramified morphisms that shall be taken up in chapter 5 (and they also cover the cases needed in [34]). Our first statement is the almost version of a theorem of Gruson and Raynaud ([46, Part II, Th. 1.2.4]).

Proposition 3.4.4. *A finite monomorphism of almost algebras descends flatness.*

Proof. Let $\phi : A \to B$ be such a morphism. Under the assumption, we can find a finite A_*-module Q such that $\mathfrak{m}B_* \subset Q \subset B_*$.

Claim 3.4.5. If $(0 \to N \to L \to P \to 0)$ is an exact sequence of A_*-modules with L flat, such that $\mathrm{Im}(N \otimes_{A_*} Q)$ is a pure submodule of $L \otimes_{A_*} Q$, then P is flat.

Proof of the claim. One sees easily that Q is a faithful A_*-module, so the claim follows from [46, Part II, Th. 1.2.4 and lemma 1.2.2],   ◇

Now let M be an A-module such that $M \otimes_A B$ is flat. Pick an epimorphism $p : F \to M$ with F free over A. Then $\underline{Y} := (0 \to \mathrm{Ker}(p \otimes_A 1_B) \to F \otimes_A B \to M \otimes_A B \to 0)$ is universally exact over B, hence over A. Consider the sequence $\underline{X} := (0 \to \mathrm{Im}(\mathrm{Ker}(p)_! \otimes_{A_*} Q) \to F_! \otimes_{A_*} Q \to M_! \otimes_{A_*} Q \to 0)$. Clearly $\underline{X}^a \simeq \underline{Y}$. However, it is easy to check that a sequence $\underline{E}$ of A-modules is universally exact if and only if the sequence $\underline{E}_!$ is universally exact over A_*. We conclude that $\underline{X} = (\underline{X}^a)_!$ is a universally exact sequence of A_*-modules, hence, by claim (3.4.5), $M_!$ is flat over A_*, *i.e.* M is flat over A as required.   □

Corollary 3.4.6. *Let $A \to B$ be a finite morphism of almost algebras, with nilpotent kernel. If C is a flat A-algebras such that $C \otimes_A B$ is weakly étale (resp. étale) over B, then C is weakly étale (resp. étale) over A.*

Proof. In the weakly étale case, we have to show that the multiplication morphism $\mu : C \otimes_A C \to C$ is flat. As $N := \mathrm{Ker}(A \to B)$ is nilpotent, the local flatness criterion reduces the question to the situation over A/N. So we may assume that $A \to B$ is a monomorphism. Then $C \otimes_A C \to (C \otimes_A C) \otimes_A B$ is a monomorphism, but $\mu \otimes_{C \otimes_A C} 1_{(C \otimes_A C) \otimes_A B}$ is the multiplication morphism of $C \otimes_A B$, which is flat by assumption. Therefore, by proposition 3.4.4, μ is flat.

For the étale case, we have to show that C is almost finitely presented as a $C \otimes_A C$-module. By hypothesis $C \otimes_A B$ is almost finitely presented as a $C \otimes_A C \otimes_A B$-module and we know already that C is flat as a $C \otimes_A C$-module, so by lemma 3.2.25(iii) (applied to the finite morphism $C \otimes_A C \to C \otimes_A C \otimes_A B$) the claim follows.   □

3.4.7.   Suppose now that we are given a cartesian diagram $\mathscr{D}$ of almost algebras

(3.4.8)
$$
\begin{array}{ccc}
A_0 & \xrightarrow{f_2} & A_2 \\
f_1 \downarrow & & \downarrow g_2 \\
A_1 & \xrightarrow{g_1} & A_3
\end{array}
$$

such that one of the morphisms $A_i \to A_3$ ($i = 1, 2$) is an epimorphism. Diagram $\mathscr{D}$ induces an essentially commutative diagram for the corresponding categories A_i-**Mod**, where the arrows are given by the "extension of scalars" functors. We define the category of $\mathscr{D}$-*modules* as the 2-fibre product $\mathscr{D}$-**Mod** := A_1-**Mod** $\times_{A_3\text{-Mod}} A_2$-**Mod**. Recall (see [8, Ch.VII §3] or [47, Ch.I] for generalities on 2-categories and 2-fibre products) that $\mathscr{D}$-**Mod** is the category whose objects are the triples (M_1, M_2, ξ), where M_i is an A_i-module ($i = 1, 2$) and $\xi : A_3 \otimes_{A_1} M_1 \xrightarrow{\sim} A_3 \otimes_{A_2} M_2$ is an A_3-linear isomorphism. There follows a natural functor

$$\pi : A_0\text{-}\mathbf{Mod} \to \mathscr{D}\text{-}\mathbf{Mod}.$$

Given an object (M_1, M_2, ξ) of $\mathscr{D}$-**Mod**, let us denote $M_3 := A_3 \otimes_{A_2} M_2$; we have a natural morphism $M_2 \to M_3$, and ξ gives a morphism $M_1 \to M_3$, so we can form the fibre product $T(M_1, M_2, \xi) := M_1 \times_{M_3} M_2$. In this way we obtain a functor $T : \mathscr{D}\text{-}\mathbf{Mod} \to A_0\text{-}\mathbf{Mod}$, and we leave to the reader the verification that T is right adjoint to π. Let us denote by $\varepsilon : 1_{\mathscr{M}_0} \to T \circ \pi$ and $\eta : \pi \circ T \to 1_{\mathscr{D}\text{-}\mathbf{Mod}}$ the unit and counit of the adjunction.

Lemma 3.4.9. *The functor π restricts to an equivalence from the full subcategory of A_0-$\mathbf{Mod}$ consisting of all objects X such that ε_X is an isomorphism, to the full subcategory of $\mathscr{D}$-$\mathbf{Mod}$ consisting of all objects Y such that η_Y is an isomorphism. Furthermore, T restricts to a quasi-inverse for this equivalence.*

Proof. General nonsense. □

Lemma 3.4.10. *Let M be any A_0-module.*
  (i) *ε_M is an epimorphism.*
  (ii) *If $\mathrm{Tor}_1^{A_0}(M, A_3) = 0$, then ε_M is an isomorphism.*

Proof. Indeed, $\varepsilon_M : M \to (A_1 \otimes_{A_0} M) \times_{A_3 \otimes_{A_0} M} (A_2 \otimes_{A_0} M)$ is the natural morphism. So, the assertions follow by applying $- \otimes_{A_0} M$ to the short exact sequence of A_0-modules

$$(3.4.11) \qquad\qquad 0 \to A_0 \xrightarrow{f} A_1 \oplus A_2 \xrightarrow{g} A_3 \to 0$$

where $f(a) := (f_1(a), f_2(a))$ and $g(a, b) := g_1(a) - g_2(b)$. □

3.4.12.   There is another case of interest, in which ε_M is an isomorphism. Namely, suppose that one of the morphisms $A_i \to A_3$ ($i = 1, 2$), say $A_1 \to A_3$, has a section. Then also the morphism $A_0 \to A_2$ gains a section $s : A_2 \to A_0$ and we have the following :

Lemma 3.4.13. *In the situation of (3.4.12), suppose that the A_0-module M arises by extension of scalars from an A_2-module M', via the section $s : A_2 \to A_0$. Then ε_M is an isomorphism.*

Proof. Indeed, in this case (3.4.11) is split exact as a sequence of A_2-modules, and it remains such after tensoring by M'. □

Lemma 3.4.14. *$\eta_{(M_1, M_2, \xi)}$ is an isomorphism for all objects (M_1, M_2, ξ).*

Proof. To fix ideas, suppose that $A_1 \to A_3$ is an epimorphism. Consider any $\mathscr{D}$-module (M_1, M_2, ξ). Let $M := T(M_1, M_2, \xi)$; we deduce a natural morphism

$$\phi : (M \otimes_{A_0} A_1) \times_{M \otimes_{A_0} A_3} (M \otimes_{A_0} A_2) \to M_1 \times_{M_3} M_2$$

such that $\phi \circ \varepsilon_M = 1_M$. It follows that ε_M is injective, hence it is an isomorphism, by lemma 3.4.10. We derive a commutative diagram with exact rows :

$$
\begin{array}{ccccccccc}
0 & \longrightarrow & M & \longrightarrow & (M \otimes_{A_0} A_1) \oplus (M \otimes_{A_0} A_2) & \longrightarrow & M \otimes_{A_0} A_3 & \longrightarrow & 0 \\
& & \| & & \downarrow{\scriptstyle \phi_1 \oplus \phi_2} & & \downarrow{\scriptstyle \phi_3} & & \\
0 & \longrightarrow & M & \longrightarrow & M_1 \oplus M_2 & \longrightarrow & M_3 & \longrightarrow & 0.
\end{array}
$$

From the snake lemma we deduce

$$(*) \qquad \mathrm{Ker}(\phi_1) \oplus \mathrm{Ker}(\phi_2) \simeq \mathrm{Ker}(\phi_3)$$
$$(**) \qquad \mathrm{Coker}(\phi_1) \oplus \mathrm{Coker}(\phi_2) \simeq \mathrm{Coker}(\phi_3).$$

Since $M_3 \simeq M_1 \otimes_{A_1} A_3$ we have $A_3 \otimes_{A_1} \mathrm{Coker}\, \phi_1 \simeq \mathrm{Coker}\, \phi_3$. But by assumption $A_1 \to A_3$ is an epimorphism, so also $\mathrm{Coker}\, \phi_1 \to \mathrm{Coker}\, \phi_3$ is an epimorphism. Then $(**)$ implies that $\mathrm{Coker}\, \phi_2 = 0$. But $\phi_3 = 1_{A_3} \otimes_{A_2} \phi_2$, thus $\mathrm{Coker}\, \phi_3 = 0$ as well. We look at the exact sequence $0 \to \mathrm{Ker}\, \phi_1 \to M \otimes_{A_0} A_1 \xrightarrow{\phi_1} M_1 \to 0$: applying $A_3 \otimes_{A_1} -$ we obtain an epimorphism $A_3 \otimes_{A_1} \mathrm{Ker}\, \phi_1 \to \mathrm{Ker}\, \phi_3$. From $(*)$ it follows that $\mathrm{Ker}\, \phi_2 = 0$. In conclusion, ϕ_2 is an isomorphism. Hence the same is true for $\phi_3 = 1_{A_3} \otimes_{A_2} \phi_2$, and again $(*)$, $(**)$ show that ϕ_1 is an isomorphism as well, which implies the claim. $\square$

Lemma 3.4.15. *In the situation of* (3.4.7), *let M be any A_0-module and $n \geq 1$ an integer. The following conditions are equivalent:*

(a) $\mathrm{Tor}_j^{A_0}(M, A_i) = 0$ *for every $1 \leq j \leq n$ and $i = 1, 2, 3$.*

(b) $\mathrm{Tor}_j^{A_i}(A_i \otimes_{A_0} M, A_3) = 0$ *for every $1 \leq j \leq n$ and $i = 1, 2$.*

Proof. There is a base change spectral sequence

$$(3.4.16) \qquad E_{pq}^2 := \mathrm{Tor}_p^{A_i}(\mathrm{Tor}_q^{A_0}(M, A_i), A_3) \Rightarrow \mathrm{Tor}_{p+q}^{A_0}(M, A_3).$$

If now (a) holds, we deduce that $\mathrm{Tor}_p(A_i \otimes_{A_0} M, A_3) \simeq \mathrm{Tor}_p(M, A_3)$ whenever $p \leq n$, and then (b) follows. Conversely, suppose that (b) holds; we show (a) by induction on n. Say that the morphism $A_1 \to A_3$ is an epimorphism, so that the same holds for the morphism $A_0 \to A_2$, and denote by I the common kernel of these morphisms. For $n = 1$ and $i = 1$, the assumption is equivalent to saying that the natural morphism $I \otimes_{A_1} (A_1 \otimes_{A_0} M) \to (A_1 \otimes_{A_0} M)$ is injective. It follows that the same holds for the morphism $I \otimes_{A_0} M \to M$, which already shows that $\mathrm{Tor}_1^{A_0}(M, A_2) = 0$. Next, the assumption for $i = 2$ means that the term E_{10}^2 of the spectral sequence (3.4.16) vanishes, whence an isomorphism $\mathrm{Tor}_1^{A_0}(M, A_3) \simeq \mathrm{Tor}_1^{A_0}(M, A_2) \otimes_{A_2} A_3 \simeq 0$. Finally, we use the long exact Tor sequence arising from the short exact sequence (3.4.11) to deduce that also $\mathrm{Tor}_1^{A_0}(M, A_1)$ vanishes. Let now $n > 1$ and suppose that assertion (a) is already known for $1 \leq j < n$. We choose a presentation

$$(3.4.17) \qquad 0 \to R \to F \to M \to 0$$

with F flat over A_0; using the long exact Tor sequence we deduce that

$$\mathrm{Tor}_j^{A_0}(M, A_i) \simeq \mathrm{Tor}_{j-1}^{A_0}(R, A_i) \quad \text{for every } j > 1 \text{ and every } i \leq 3.$$

Moreover, assertion (a) taken with $j = 1$ shows that the sequence $A_i \otimes_{A_0}$ (3.4.17) is again exact for $i \leq 3$, which by the Tor sequences gives that the A_0-module R fulfills condition (b) for every $1 \leq j < n$, hence the inductive assumption shows that $\mathrm{Tor}_j^{A_0}(R, A_i) = 0$ for every $1 \leq j < n$ and $i = 1, 2, 3$; in turns this implies the sought vanishing for M. $\square$

The following lemma 3.4.18(iii) generalizes a result of D.Ferrand ([35, lemma]).

Lemma 3.4.18. *Let M be any A_0-module. We have:*

(i) $\mathrm{Ann}_{A_0}(M \otimes_{A_0} A_1) \cdot \mathrm{Ann}_{A_0}(M \otimes_{A_0} A_2) \subset \mathrm{Ann}_{A_0}(M)$.

(ii) M admits a three-step filtration $0 \subset \mathrm{Fil}_0 M \subset \mathrm{Fil}_1 M \subset \mathrm{Fil}_2 M = M$ such that $\mathrm{Fil}_0 M$ and $\mathrm{gr}_2 M$ are A_2-modules and $\mathrm{gr}_1 M$ is an A_1-module.

(iii) If $(A_1 \times A_2) \otimes_{A_0} M$ is flat over $A_1 \times A_2$, then M is flat over A_0.

Proof. To fix ideas, suppose that $A_1 \to A_3$ is an epimorphism, and let I be its kernel; let also $M_i := A_i \otimes_{A_0} M$ for $i = 1, 2, 3$.

(i): Clearly $I \simeq \mathrm{Ker}(A_0 \to A_2)$, hence $aM \subset IM$ for every $a \in \mathrm{Ann}_{A_0}(M_2)$. On the other hand, the natural morphism $I \otimes_{A_1} M_1 \xrightarrow{\sim} I \otimes_{A_0} M \to IM$ is obviously an epimorphism, whence the assertion.

(ii): We set $\mathrm{Fil}_0 M := \mathrm{Ker}(\varepsilon_M : M \to M_1 \times_{M_3} M_2)$; using the short exact sequence (3.4.16) we see that $\mathrm{Fil}_0 M \simeq \mathrm{Tor}_1^{A_0}(M, A_3)/\mathrm{Tor}_1^{A_0}(A_1 \oplus A_2, M)$. Obviously I annihilates $\mathrm{Fil}_0 M$ (since it annihilates already $\mathrm{Tor}_1^{A_0}(M, A_3)$), hence the latter is an A_2-module. By lemma 3.4.10, we have $M' := M/\mathrm{Fil}_0 M \simeq M_1 \times_{M_3} M_2$. We can then filter the latter module by defining $\mathrm{Fil}_0 M' := 0$, $\mathrm{Fil}_1 M' := \mathrm{Ker}(M' \to M_2) \simeq \mathrm{Ker}(M_1 \to M_3)$, which is a A_1-module, and $\mathrm{Fil}_2 M' := M'$. Since $\mathrm{gr}_2 M' \simeq M_2$, the assertion follows.

(iii): In view of lemma 3.4.15, it suffices to show the following:

Claim 3.4.19. M is flat over A_0 if and only if M_i is A_i-flat and $\mathrm{Tor}_1^{A_0}(M, A_i) = 0$ for $i \leq 2$.

Proof of the claim. It suffices to prove the non-obvious implication, and in view of (ii) we are reduced to showing that $\mathrm{Tor}_1^{A_0}(M, L) = 0$ whenever L is an A_i-module, for $i = 1, 2$. However, for any A_i-module L we have a base change spectral sequence $E_{pq}^2 := \mathrm{Tor}_p^{A_i}(\mathrm{Tor}_q^{A_0}(M, A_i), L) \Rightarrow \mathrm{Tor}_{p+q}^{A_0}(M, L)$. If $\mathrm{Tor}_1^{A_0}(M, A_i) = 0$, this yields $\mathrm{Tor}_1^{A_0}(M, L) \simeq \mathrm{Tor}_1^{A_i}(M_i, L)$, which vanishes when M_i is A_i-flat. $\square$

3.4.20.   For any V^a-algebra A, let $A\text{-}\mathbf{Mod}_{\mathrm{fl}}$ (resp. $A\text{-}\mathbf{Mod}_{\mathrm{proj}}$, resp. $A\text{-}\mathbf{Mod}_{\mathrm{afpr}}$) denote the full subcategory of $A\text{-}\mathbf{Mod}$ consisting of all flat (resp. almost projective, resp. almost finitely generated projective) A-modules. For any integer $n \geq 1$, let $A_0\text{-}\mathbf{Mod}_n$ be the full subcategory of all A_0-modules satisfying condition (a) of lemma 3.4.15; let also $A_i\text{-}\mathbf{Mod}_n$ (for $i = 1, 2$) be the full subcategory of $A_i\text{-}\mathbf{Mod}$ consisting of all A_i-modules M such that $\mathrm{Tor}_j^{A_i}(M, A_3) = 0$ for every $j \leq n$. Finally, Let $A\text{-}\mathbf{Alg}_{\mathrm{fl}}$ be the full subcategory of $A\text{-}\mathbf{Alg}$ consisting of all flat A-algebras.

Proposition 3.4.21. *In the situation of (3.4.7), the natural essentially commutative diagram:*

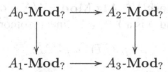

is 2-cartesian (i.e. cartesian in the 2-category of categories, cp. [47, Ch.I]) whenever ? = "fl" *or* ? = "proj" *or* ? = "afpr", *or* ? = n, *for any integer* $n \geq 1$.

Proof. The assertion for flat almost modules follows directly from lemmata 3.4.9, 3.4.10, 3.4.14 and 3.4.18(iii). Similarly the assertion for the categories A_i-$\mathbf{Mod}_n$ follows from the same lemmata and from lemma 3.4.15. Set $B := A_1 \times A_2$. To establish the assertion for projective modules, it suffices to show that, if P is an A_0-module such that $B \otimes_{A_0} P$ is almost projective over B, then P is almost projective over A_0, or which is the same, that $\mathrm{alExt}^i_{A_0}(P, N) \simeq 0$ for all $i > 0$ and any A_0-module N. We know already that P is flat. Let M be any A_0-module and N any B-module. The standard isomorphism $R\mathrm{Hom}_B(B \overset{L}{\otimes}_{A_0} M, N) \simeq R\mathrm{Hom}_{A_0}(M, N)$ yields a natural isomorphism $\mathrm{alExt}^i_B(B \otimes_{A_0} M, N) \simeq \mathrm{alExt}^i_{A_0}(M, N)$, whenever $\mathrm{Tor}^{A_0}_j(B, M) = 0$ for every $j > 0$. In particular, we have $\mathrm{alExt}^i_{A_0}(P, N) \simeq 0$ whenever N comes from either an A_1-module, or an A_2-module. In view of lemma 3.4.18(ii), we deduce that the sought vanishing holds in fact for every A_0-module N. Finally, suppose that $P \otimes_A B$ is almost finitely generated projective over B; we have to show that P is almost finitely generated over A. To this aim, notice that $\mathrm{Ann}_{A_0}(P \otimes_A B)^2 \subset \mathrm{Ann}_{A_0}(P)$, in view of lemma 3.4.18(i); then the claim follows from remark 3.2.26(i). $\qquad\square$

Corollary 3.4.22. *In the situation of* (3.4.7), *the natural essentially commutative diagram:*

$$
\begin{array}{ccc}
A_0\text{-}\mathscr{C} & \longrightarrow & A_2\text{-}\mathscr{C} \\
\downarrow & & \downarrow \\
A_1\text{-}\mathscr{C} & \longrightarrow & A_3\text{-}\mathscr{C}
\end{array}
$$

is 2-cartesian whenever $\mathscr{C}$ is one of the categories $\mathbf{Alg}_{\mathrm{fl}}$, $\mathbf{\acute{E}t}$, $\mathbf{w.\acute{E}t}$, $\mathbf{\acute{E}t}_{\mathrm{afp}}$ *(notation of* (3.2.27)). $\qquad\square$

3.4.23.   Next we want to reinterpret the equivalences of proposition 3.4.21 in terms of descent data. If $F : \mathscr{C} \to V^a\text{-}\mathbf{Alg}^o$ is a fibred category over the category of affine almost schemes, and if $\phi : X \to Y$ is a given morphism of almost algebras, we shall denote by either $\mathbf{Desc}(\mathscr{C}, Y/X)$ or $\mathbf{Desc}(\mathscr{C}, \mathrm{Spec}\,\phi)$ the category of objects of the fibre category F_Y, endowed with a descent datum relative to the morphism $\mathrm{Spec}\,\phi : \mathrm{Spec}\,Y \to \mathrm{Spec}\,X$ (cp. [41, Ch.II §1]). In the arguments hereafter, we consider morphisms of almost algebras and modules, and one has to reverse the direction of the arrows to pass to morphisms in the relevant fibred category. Denote by $p_i : Y \to Y \otimes_X Y$ ($i = 1, 2$), resp. $p_{ij} : Y \otimes_X Y \to Y \otimes_X Y \otimes_X Y$ ($1 \le i < j \le 3$) the usual morphisms.

3.4.24.   As an example, $\mathbf{Desc}(V^a\text{-}\mathbf{Alg.Mod}^o, Y/X)$ consists of the pairs (M, β) where M is a Y-module and β is a $Y \otimes_X Y$-linear isomorphism

$$\beta : p_2^*(M) \overset{\sim}{\to} p_1^*(M)$$

such that

(3.4.25) $$p_{12}^*(\beta) \circ p_{23}^*(\beta) = p_{13}^*(\beta).$$

3.4.26.   Let now $I \subset X$ be an ideal, and set $\overline{X} := X/I, \overline{Y} := Y/IY$. For any $F : \mathscr{C} \to V^a\text{-}\mathbf{Alg}^o$ as in (3.4.23), one has an essentially commutative diagram:

(3.4.27)
$$
\begin{array}{ccc}
\mathbf{Desc}(\mathscr{C}, Y/X) & \longrightarrow & \mathbf{Desc}(\mathscr{C}, \overline{Y}/\overline{X}) \\
\downarrow & & \downarrow \\
F_Y & \longrightarrow & F_{\overline{Y}}.
\end{array}
$$

This induces a functor :

(3.4.28)
$$\mathbf{Desc}(\mathscr{C}, Y/X) \to \mathbf{Desc}(\mathscr{C}, \overline{Y}/\overline{X}) \times_{F_{\overline{Y}}} F_Y.$$

Lemma 3.4.29. *In the situation of* (3.4.26), *suppose moreover that the natural morphism $I \to IY$ is an isomorphism. Then diagram* (3.4.27) *is 2-cartesian whenever $\mathscr{C}$ is one of the fibred categories $V^a\text{-}\mathbf{Alg.Mod}^o$, $V^a\text{-}\mathbf{Alg.Morph}^o$, $\acute{\mathrm{E}}\mathrm{t}^o$, $\mathrm{w}.\acute{\mathrm{E}}\mathrm{t}^o$.*

Proof. For any $n > 0$, denote by $Y^{\otimes n}$ (resp. $\overline{Y}^{\otimes n}$) the n-fold tensor product of Y (resp. $\overline{Y}$) with itself over X (resp. $\overline{X}$), and by $\rho_n : Y^{\otimes n} \to \overline{Y}^{\otimes n}$ the natural morphism. First of all we claim that, for every $n > 0$, the natural diagram of almost algebras

(3.4.30)
$$
\begin{array}{ccc}
Y^{\otimes n} & \xrightarrow{\rho_n} & \overline{Y}^{\otimes n} \\
\mu_n \downarrow & & \downarrow \overline{\mu}_n \\
Y & \xrightarrow{\rho_1} & \overline{Y}
\end{array}
$$

is cartesian (where μ_n and $\overline{\mu}_n$ are n-fold multiplication morphisms). For this, we need to verify that, for every $n > 0$, the induced morphism $\operatorname{Ker} \rho_n \to \operatorname{Ker} \rho_1$ is an isomorphism. It then suffices to check that the natural morphism $\operatorname{Ker} \rho_n \to \operatorname{Ker} \rho_{n-1}$ (defined by multiplication of the first two factors) is an isomorphism for all $n > 1$. Indeed, consider the commutative diagram

$$
\begin{array}{ccccc}
I \otimes_X Y^{\otimes n-1} & \xrightarrow{\;p\;} & IY^{\otimes n-1} & \xrightarrow{\;i\;} & Y^{\otimes n-1} \\
\| & & \psi \downarrow & \phi \otimes 1_{Y^{\otimes n-1}} \downarrow & \searrow{\scriptstyle 1_{Y^{\otimes n-1}}} \\
I \otimes_X Y^{\otimes n-1} & \xrightarrow{\;p'\;} & \operatorname{Ker} \rho_n & \xrightarrow[i']{} & Y^{\otimes n} \xrightarrow[\mu_{Y/X} \otimes 1_{Y^{\otimes n-2}}]{} Y^{\otimes n-1}.
\end{array}
$$

From $IY = \phi(Y)$, it follows that p' is an epimorphism. Hence also ψ is an epimorphism. Since i is a monomorphism, it follows that ψ is also a monomorphism, hence ψ is an isomorphism and the claim follows easily.

We consider first the case $\mathscr{C} := V^a\text{-}\mathbf{Alg.Mod}^o$; we see that (3.4.30) is a diagram of the kind considered in (3.4.8), hence, for every $n > 0$, we have the associated functor $\pi_n : Y^{\otimes n}\text{-}\mathbf{Mod} \to \overline{Y}^{\otimes n}\text{-}\mathbf{Mod} \times_{\overline{Y}\text{-}\mathbf{Mod}} Y\text{-}\mathbf{Mod}$ and also its right adjoint T_n. Denote by $\overline{p}_i : \overline{Y} \to \overline{Y}^{\otimes 2}$ ($i = 1, 2$) the usual morphisms, and similarly define $\overline{p}_{ij} : \overline{Y}^{\otimes 2} \to \overline{Y}^{\otimes 3}$. Suppose there is given a descent datum $(\overline{M}, \overline{\beta})$ for $\overline{M}$, relative to $\overline{X} \to \overline{Y}$. The cocycle condition (3.4.25) implies easily that $\overline{\mu}_2^*(\overline{\beta})$ is the identity on $\overline{\mu}_2^*(\overline{p}_i^*\overline{M}) = \overline{M}$. It follows that the pair $(\overline{\beta}, 1_M)$ defines an isomorphism $\pi_2(p_1^* M) \xrightarrow{\sim} \pi_2(p_2^* M)$ in the category $\overline{Y}^{\otimes 2}$-**Mod** $\times_{\overline{Y}\text{-Mod}} Y$-**Mod**. Hence $T_2(\overline{\beta}, 1_M) : T_2 \circ \pi_2(p_1^* M) \to T_2 \circ \pi_2(p_2^* M)$ is an isomorphism. However, we remark that either morphism $\overline{p}_i$ yields a section for μ_2, hence we are in the situation contemplated in lemma 3.4.13, and we derive an isomorphism $\beta : p_2^*(M) \xrightarrow{\sim} p_1^*(M)$. We claim that (M, β) is an object of $\mathbf{Desc}(\mathscr{C}, Y/X)$, *i.e.* that β verifies the cocycle condition (3.4.25). Indeed, we can compute: $\pi_3(p_{ij}^* \beta) = (\rho_3^*(p_{ij}^* \beta), \mu_3^*(p_{ij}^* \beta))$ and by construction we have $\rho_3^*(p_{ij}^* \beta) = \overline{p}_{ij}^*(\overline{\beta})$ and $\mu_3^*(p_{ij}^* \beta) = \mu_2^*(\beta) = 1_M$. Therefore, the cocycle identity for $\overline{\beta}$ implies the equality $\pi_3(p_{12}^* \beta) \circ \pi_3(p_{23}^* \beta) = \pi_3(p_{13}^* \beta)$. If we now apply the functor T_3 to this equality, and then invoke again lemma 3.4.13, the required cocycle identity for β will ensue. Clearly β is the only descent datum on M lifting $\overline{\beta}$. This proves that (3.4.28) is essentially surjective. The same sort of argument also shows that the functor (3.4.28) induces bijections on morphisms, so the lemma follows in this case. Next, the case $\mathscr{C} := V^a$-**Alg.Morph**^{o} can be deduced formally from the previous case, by applying repeatedly natural isomorphisms of the kind $p_i^*(M \otimes_Y N) \simeq p_i^*(M) \otimes_{Y \otimes_X Y} p_i^*(N)$ $(i = 1, 2)$. Finally, the "étaleness" of an object of $\mathbf{Desc}(V^a$-**Alg.Morph**$^o, Y/X)$ can be checked on its projection onto Y-**Alg**^{o}, hence also the cases $\mathscr{C} := \text{w.}\mathbf{\acute{E}t}^o$ and $\mathscr{C} := \mathbf{\acute{E}t}^o$ follow directly. $\square$

3.4.31.   Now, let

$$B := A_1 \times A_2.$$

To an object (M, β) in $\mathbf{Desc}(V^a$-**Alg.Mod**$^o, B/A)$ we assign as follows a $\mathscr{D}$-module (M_1, M_2, ξ) (notation of (3.4.7)). Set $M_i := A_i \otimes_B M$ $(i = 1, 2)$ and $A_{ij} := A_i \otimes_{A_0} A_j$. We can write $B \otimes_{A_0} B = \prod_{i,j=1}^{2} A_{ij}$ and β gives rise to the A_{ij}-linear isomorphisms $\beta_{ij} : A_{ij} \otimes_{B \otimes_{A_0} B} p_2^*(M) \xrightarrow{\sim} A_{ij} \otimes_{B \otimes_{A_0} B} p_1^*(M)$. In other words, we obtain isomorphisms $\beta_{ij} : A_i \otimes_{A_0} M_j \to M_i \otimes_{A_0} A_j$. However, we have a natural isomorphism $A_{12} \simeq A_3$ (indeed, suppose that $A_1 \to A_3$ is an epimorphism with kernel I; then I is also an ideal of A_0 and $A_0/I \simeq A_2$; now the claim follows by remarking that $IA_1 = I$). Hence we can choose $\xi = \beta_{12}$. In this way we obtain a functor :

(3.4.32) $\mathbf{Desc}(V^a$-**Alg.Mod**$^o, B/A_0) \to \mathscr{D}$-**Mod**^{o}.

Proposition 3.4.33. *The functor* (3.4.32) *is an equivalence of categories.*

Proof. Let us say that $A_1 \to A_3$ is an epimorphism with kernel I. Then I is also an ideal of B and we have $B/I \simeq A_3 \times A_2$ and $A_0/I \simeq A_2$. We intend to apply lemma 3.4.29 to the morphism $A_0 \to B$. However, the induced morphism $\overline{B} := B/I \to \overline{A}_0 := A_0/I$ in V^a-**Alg**^{o} has a section, and hence it is of universal effective descent for every fibred category. Thus, we can replace in (3.4.28) the category $\mathbf{Desc}(V^a$-**Alg.Mod**$^o, \overline{B}/\overline{A}_0)$ by $\overline{A}_0$-**Mod**^{o}, and thereby, identify (up to equivalence) the target of (3.4.28) with the 2-fibred product $(A_1$-**Mod** $\times A_2$-**Mod**$)^o \times_{(A_3\text{-Mod} \times A_2\text{-Mod})^o} A_2$-**Mod**^{o}. The latter is equivalent to the category $\mathscr{D}$-**Mod**^{o} and the resulting functor $\mathbf{Desc}(V^a$-**Alg.Mod**$^o, B/A_0) \to \mathscr{D}$-**Mod**^{o} is canonically isomorphic to (3.4.32), whence the claim. $\square$

Combining propositions 3.4.21 and 3.4.33 we obtain the following :

Corollary 3.4.34. *In the situation of* (3.4.8), *the morphism* $A_0 \to A_1 \times A_2$ *is of effective descent for the fibred categories of flat modules and of almost projective modules.* □

3.4.35. Next we would like to give sufficient conditions to ensure that a morphism of almost algebras is of effective descent for the fibred category $\mathbf{w.\acute{E}t}^o \to V^a\text{-}\mathbf{Alg}^o$ of weakly étale algebras (resp. for étale algebras). To this aim we are led to the following :

Definition 3.4.36. A morphism $\phi : A \to B$ of almost algebras is said to be *strictly finite* if $\mathrm{Ker}\,\phi$ is nilpotent and $B \simeq R^a$, where R is a finite A_*-algebra.

Theorem 3.4.37. *Let* $\phi : A \to B$ *be a strictly finite morphism. Then :*

(i) *For every A-algebra C, the induced morphism $C \to C \otimes_A B$ is strictly finite.*

(ii) *If M is a flat A-module and $B \otimes_A M$ is almost projective over B, then M is almost projective over A.*

(iii) $\mathrm{Spec}\,B \to \mathrm{Spec}\,A$ *is of universal effective descent for the fibred categories of weakly étale (resp. étale) affine almost schemes.*

Proof. (i): Suppose that $B = R^a$ for a finite A_*-algebra R; then $S := C_* \otimes_{A_*} R$ is a finite C_*-algebra and $S^a \simeq C$. It remains to show that $\mathrm{Ker}(C \to C \otimes_A B)$ is nilpotent. Suppose that R is generated by n elements as an A_*-module and let $F_{A_*}(R)$ (resp. $F_{C_*}(S)$) be the Fitting ideal of R (resp.of S); we have $\mathrm{Ann}_{C_*}(S)^n \subset F_{C_*}(S) \subset \mathrm{Ann}_{C_*}(S)$ (see [54, Ch.XIX Prop.2.5]); on the other hand $F_{C_*}(S) = F_{A_*}(R) \cdot C_*$, so the claim is clear.

(iii): We shall consider the fibred category $F : \mathbf{w.\acute{E}t}^o \to V^a\text{-}\mathbf{Alg}^o$; the same argument applies also to étale almost schemes. We begin by establishing a very special case :

Claim 3.4.38. Assertion (iii) holds when $B = (A/I_1) \times (A/I_2)$, where I_1 and I_2 are ideals in A such that $I_1 \cap I_2$ is nilpotent.

Proof of the claim. First of all we remark that the situation considered in the claim is stable under arbitrary base change, therefore it suffices to show that $\mathrm{Spec}\,\phi$ is of F-2-descent in this case. The morphism ϕ factors through $\psi : A/\mathrm{Ker}\,\phi \to B$ and clearly we have a natural identification

$$\mathbf{Desc}(F, \mathrm{Spec}\,\phi) = \mathbf{Desc}(F, \mathrm{Spec}\,\psi).$$

By theorem 3.2.18, the pull-back functor $F_A \to F_{A/\mathrm{Ker}\,\phi}$ is an equivalence; hence we are reduced to showing that $\mathrm{Spec}\,\psi$ is of F-2-descent, which follows from corollaries 3.4.6 and 3.4.34. ◊

Claim 3.4.39. More generally, assertion (iii) holds when $B = \prod_{i=1}^n A/I_i$, where $I_1, ..., I_n$ are ideals of A, such that $\bigcap_{i=1}^n I_i$ is nilpotent.

Proof of the claim. We prove this by induction on n, the case $n = 2$ being covered by claim 3.4.38. Therefore, suppose that $n > 2$, and set $B' := A/(\bigcap_{i=1}^{n-1} I_j)$. By induction, the morphism $\amalg_{i=1}^{n-1} \operatorname{Spec} A/I_i \to \operatorname{Spec} B'$ is of universal F-2-descent. However, according to [41, Ch.II Prop.1.1.3], the sieves of universal F-2-descent form a topology on $V^a\text{-}\mathbf{Alg}^o$, and given any two affine almost schemes X, Y, the family $\{X, Y\}$ is covering for $X \amalg Y$ in this topology. Hence, since $\operatorname{Spec} B' \times (A/I_n) \to \operatorname{Spec} A$ is a covering morphism, $\{\operatorname{Spec} B', \operatorname{Spec} A/I_n\}$ is a covering family of $\operatorname{Spec} A$, and then, by composition of covering families, $\{\amalg_{i=1}^{n-1} \operatorname{Spec} A/I_i, \operatorname{Spec} A/I_n\}$ is a covering family of $\operatorname{Spec} A$, which is equivalent to the claim. $\Diamond$

Now, let $A \to B$ be a general strictly finite morphism, so that $B = R^a$ for some finite A_*-algebra R. Pick generators $f_1, ..., f_m$ of the A_*-module R, and monic polynomials $p_1(X), ..., p_m(X)$ such that $p_i(f_i) = 0$ for $i = 1, ..., m$.

Claim 3.4.40. There exists a finite and faithfully flat extension C of A_* such that the images in $C[X]$ of $p_1(X), ..., p_m(X)$ split as products of monic linear factors.

Proof of the claim. This extension C can be obtained as follows. It suffices to find, for each $i = 1, ..., m$, an extension C_i that splits $p_i(X)$, because then $C := C_1 \otimes_{A_*} ... \otimes_{A_*} C_m$ will split them all, so we can assume that $m = 1$ and $p_1(X) = p(X)$; moreover, by induction on the degree of $p(X)$, it suffices to find an extension C' such that $p(X)$ factors in $C'[X]$ as a product of the form $p(X) = (X - \alpha) \cdot q(X)$, where $q(X)$ is a monic polynomial of degree $\deg(p) - 1$. Clearly we can take $C' := A_*[T]/(p(T))$. $\Diamond$

Given a C as in claim 3.4.40, we remark that the morphism $\operatorname{Spec} C^a \to \operatorname{Spec} A$ is of universal F-2-descent. Considering again the topology of universal F-2-descent, it follows that $\operatorname{Spec} B \to \operatorname{Spec} A$ is of universal F-2-descent if and only if the same holds for the induced morphism $\operatorname{Spec} C^a \otimes_A B \to \operatorname{Spec} C^a$. Therefore, in proving assertion (iii) we can replace ϕ by $1_C \otimes_A \phi$ and assume from start that the polynomials $p_i(X)$ factor in $A_*[X]$ as product of linear factors. Now, let $d_i := \deg(p_i)$ and $p_i(X) := \prod_j^{d_i}(X - \alpha_{ij})$ (for $i = 1, ..., m$). We get a surjective homomorphism of A_*-algebras $D := A_*[X_1, ..., X_m]/(p_1(X_1), ..., p_m(X_m)) \to R$ by the rule $X_i \mapsto f_i$ ($i = 1, ..., m$). Moreover, any sequence

$$\underline{\alpha} := (\alpha_{1,j_1}, \alpha_{2,j_2}, ..., \alpha_{m,j_m})$$

yields a homomorphism $\psi_{\underline{\alpha}} : D \to A_*$, determined by the assignment $X_i \mapsto \alpha_{i,j_i}$. A simple combinatorial argument shows that $\prod_{\underline{\alpha}} \operatorname{Ker} \psi_{\underline{\alpha}} = 0$, where $\underline{\alpha}$ runs over all the sequences as above. Hence the product map $\prod_{\underline{\alpha}} \psi_{\underline{\alpha}} : D \to \prod_{\underline{\alpha}} A_*$ has nilpotent kernel. We notice that the A_*-algebra $(\prod_{\underline{\alpha}} A_*) \otimes_D R$ is a quotient of $\prod_{\underline{\alpha}} A_*$, hence it can be written as a product of rings of the form $A_*/I_{\underline{\alpha}}$, for various ideals $I_{\underline{\alpha}}$. By (i), the kernel of the induced homomorphism $R \to \prod_{\underline{\alpha}} A_*/I_{\underline{\alpha}}$ is nilpotent, hence the same holds for the kernel of the composition $A \to \prod_{\underline{\alpha}} A/I_{\underline{\alpha}}^a$, which is therefore of the kind considered in claim 3.4.39. Hence $\amalg_{\underline{\alpha}} \operatorname{Spec} A/I_{\underline{\alpha}}^a \to \operatorname{Spec} A$ is of universal F-2-descent. Since such morphisms form a topology, it follows that also $\operatorname{Spec} B \to \operatorname{Spec} A$ is of universal F-2-descent, so (iii) holds.

Finally, let M be as in (ii) and pick again C as in the proof of claim 3.4.40. By remark 3.2.26(iv), M is almost projective over A if and only if $C^a \otimes_A M$ is almost projective over C^a; hence we can replace ϕ by $1_{C^a} \otimes_A \phi$, and by arguing as in the proof of (iii), we can assume from start that $B = \prod_{j=1}^{n} A/I_j$ for ideals $I_j \subset A, j = 1, ..., n$ such that $I := \bigcap_{j=1}^{n} I_j$ is nilpotent. By an easy induction, we can furthermore reduce to the case $n = 2$. We factor ϕ as $A \to A/I \to B$; by proposition 3.4.21 it follows that $(A/I) \otimes_A M$ is almost projective over A/I, and then lemma 3.2.25(i) says that M itself is almost projective.   □

Remark 3.4.41. It is natural to ask whether theorem 3.4.37 holds if we replace everywhere "strictly finite" by "finite with nilpotent kernel" (or even by "almost finite with nilpotent kernel"). We do not know the answer to this question.

3.4.42.   On the category V^a-**Alg** (taken in some universe) consider the topologies τ_e, resp. τ_w of universal effective descent for the fibred category $\acute{\mathrm{E}}\mathrm{t}^o$, resp. w.$\acute{\mathrm{E}}\mathrm{t}^o$ (limited to the same universe). For a ring R denote by $\mathrm{Idemp}(R)$ the set of idempotents of R.

Proposition 3.4.43. *With the notation of (3.4.42) we have:*

(i) *The presheaf $A \mapsto \mathrm{Idemp}(A_*)$ is a sheaf for both τ_e and τ_w.*
(ii) *If $f : A \to B$ is an étale (resp. weakly étale) morphism of almost V-algebras and there is a covering family $\{\mathrm{Spec}\, A_\alpha \to \mathrm{Spec}\, A \mid \alpha \in I\}$ for τ_e (resp. τ_w) such that $A_\alpha \to A_\alpha \otimes_A B$ is an almost projective epimorphism for all $\alpha \in I$, then f is an almost projective epimorphism.*
(iii) *τ_e is finer than τ_w.*

Proof. (i): Use descent of morphisms and the bijection

$$\mathrm{Hom}_{A\text{-}\mathbf{Alg}}(A \times A, A) \xrightarrow{\sim} \mathrm{Idemp}(A_*) \qquad \phi \mapsto \phi_*(1, 0).$$

(ii): By remark 3.1.8, $\mathrm{Ker}(A_\alpha \to A_\alpha \otimes_A B)$ is generated by $e_\alpha \in \mathrm{Idemp}(A_{\alpha *})$. e_α and e_β agree in $(A_\alpha \otimes_A A_\beta)_*$, so by (i) there is an idempotent $e \in A_*$ that restricts to e_α in $\mathrm{Idemp}(A_{\alpha *})$, for each α. The A-algebras B and A/eA become isomorphic after applying $- \otimes_A A_\alpha$; these isomorphisms are unique and are compatible on $A_\alpha \otimes_A A_\beta$, hence they patch to an isomorphism $B \simeq A/eA$.

(iii): We have to show that if A is a V^a-algebra, R a sieve of $(V^a\text{-}\mathbf{Alg})^o/A$ and R is of universal w.$\acute{\mathrm{E}}\mathrm{t}^o$-2-descent, then R is of universal $\acute{\mathrm{E}}\mathrm{t}^o$-2-descent. Since the assumption is stable under base change, it suffices to show that R is of $\acute{\mathrm{E}}\mathrm{t}^o$-2-descent. Descent of morphisms is clear. Let R be the sieve generated by a family of morphisms $\{\mathrm{Spec}\, A_\alpha \to \mathrm{Spec}\, A \mid \alpha \in I\}$. Any descent datum consisting of étale A_α-algebras B_α and isomorphisms $A_\alpha \otimes_A B_\beta \simeq B_\alpha \otimes_A A_\beta$ satisfying the cocycle condition, becomes effective when we pass to w.$\acute{\mathrm{E}}\mathrm{t}^o$. So one has to verify that if B is a weakly étale A-algebra such that $B \otimes_A A_\alpha$ is étale over A_α for all α, then B is étale over A. Indeed, an application of (ii) gives that $B \otimes_A B \to B$ is almost projective.   □

3.4.44.   We conclude with a digression to explain the relationship between our results and known facts that can be extracted from the literature. So, we now place ourselves in the "classical limit" $\mathfrak{m} := V$ (cp. example 2.1.2(ii)). In this case, weakly étale morphisms had already been considered in some earlier work, and they were called "absolutely flat" morphisms. A ring homomorphism $A \to B$ is étale in the usual sense of [44] if and only if it is absolutely flat and of finite presentation. Let us denote by $\mathbf{u.\acute{E}t}^o$ the fibred category over $V\text{-}\mathbf{Alg}^o$, whose fibre over a V-algebra A is the opposite of the category of étale A-algebras in the usual sense. We claim that, if a morphism $A \to B$ of V-algebras is of universal effective descent for the fibred category $\mathbf{w.\acute{E}t}^o$ (resp. $\mathbf{\acute{E}t}^o$), then it is a morphism of universal effective descent for $\mathbf{u.\acute{E}t}^o$. Indeed, let C be an étale A-algebra (in the sense of definition 3.1.1) and such that $C \otimes_A B$ is étale over B in the usual sense. We have to show that C is étale in the usual sense, *i.e.* that it is of finite presentation over A. This amounts to showing that, for every filtered inductive system $(A_\lambda)_{\lambda \in \Lambda}$ of A-algebras, we have

$$\operatorname*{colim}_{\lambda \in \Lambda} \operatorname{Hom}_{A\text{-}\mathbf{Alg}}(C, A_\lambda) \simeq \operatorname{Hom}_{A\text{-}\mathbf{Alg}}(C, \operatorname*{colim}_{\lambda \in \Lambda} A_\lambda).$$

Since, by assumption, this is known after extending scalars to B and to $B \otimes_A B$, it suffices to show that, for any A-algebra D, the set $\operatorname{Hom}_{A\text{-}\mathbf{Alg}}(C, D)$ is the equalizer of the natural maps:

$$\operatorname{Hom}_{B\text{-}\mathbf{Alg}}(C_B, D_B) \rightrightarrows \operatorname{Hom}_{B \otimes_A B\text{-}\mathbf{Alg}}(C_{B \otimes_A B}, D_{B \otimes_A B}).$$

For this, note that $\operatorname{Hom}_{A\text{-}\mathbf{Alg}}(C, D) = \operatorname{Hom}_{D\text{-}\mathbf{Alg}}(C_D, D)$ (and similarly for the other terms) and by hypothesis $(D \to D \otimes_A B)^o$ is a morphism of 1-descent for the fibred category $\mathbf{w.\acute{E}t}^o$ (resp. $\mathbf{\acute{E}t}^o$).

As a consequence of these observations and of theorem 3.4.37, we see that any finite ring homomorphism $\phi : A \to B$ with nilpotent kernel is of universal effective descent for the fibred category of étale algebras. This fact was known as follows. A standard reduction allows to suppose that A and B are noetherian. Then, by [44, Exp.IX, 4.7], $\operatorname{Spec} \phi$ is of universal effective descent for the fibred category of separated étale morphisms of finite type. By [27, Ch.II, 6.7.1], if X is such a scheme over A, such that $X \otimes_A B$ is affine, then X is affine.

3.5. Behaviour of étale morphisms under Frobenius.

We consider the following category $\mathscr{B}$ of basic setups. The objects of $\mathscr{B}$ are the pairs $(V, \mathfrak{m})$, where V is a ring and $\mathfrak{m}$ is an ideal of V with $\mathfrak{m} = \mathfrak{m}^2$ and $\tilde{\mathfrak{m}}$ is flat. The morphisms $(V, \mathfrak{m}_V) \to (W, \mathfrak{m}_W)$ between two objects of $\mathscr{B}$ are the ring homomorphisms $f : V \to W$ such that $\mathfrak{m}_W = f(\mathfrak{m}_V) \cdot W$.

3.5.1.   We have a fibred and cofibred category $\mathscr{B}\text{-}\mathbf{Mod} \to \mathscr{B}$ (see [44, Exp.VI §5,6,10] for generalities on fibred categories). An object of $\mathscr{B}\text{-}\mathbf{Mod}$ (which we may call a "$\mathscr{B}$-module") consists of a pair $((V, \mathfrak{m}), M)$, where $(V, \mathfrak{m})$ is an object of $\mathscr{B}$ and M is a V-module. Given two objects $X := ((V, \mathfrak{m}_V), M)$ and $Y := ((W, \mathfrak{m}_W), N)$, the morphisms $X \to Y$ are the pairs (f, g), where $f : (V, \mathfrak{m}_V) \to (W, \mathfrak{m}_W)$ is a morphism in $\mathscr{B}$ and $g : M \to N$ is an f-linear map.

3.5.2. Similarly one has a fibred and cofibred category

$$\mathscr{B}\text{-}\mathbf{Alg} \to \mathscr{B}$$

of $\mathscr{B}$-algebras. We will also need to consider the fibred and cofibred category $\mathscr{B}\text{-}\mathbf{Mon} \to \mathscr{B}$ of non-unitary commutative $\mathscr{B}$-monoids: an object of $\mathscr{B}\text{-}\mathbf{Mon}$ is a pair $((V, \mathfrak{m}), A)$ where A is a V-module endowed with a morphism $A \otimes_V A \to A$ subject to associativity and commutativity conditions, as discussed in section 2.2. The fibre over an object $(V, \mathfrak{m})$ of $\mathscr{B}$, is the category of V-monoids denoted $(V, \mathfrak{m})\text{-}\mathbf{Mon}$ or simply $V\text{-}\mathbf{Mon}$.

3.5.3. The almost isomorphisms in the fibres of $\mathscr{B}\text{-}\mathbf{Mod} \to \mathscr{B}$ give a multiplicative system Σ in $\mathscr{B}\text{-}\mathbf{Mod}$, admitting a calculus of both left and right fractions. The "locally small" conditions are satisfied (see [77, p.381]), so that one can form the localized category

$$\mathscr{B}^a\text{-}\mathbf{Mod} := \Sigma^{-1}(\mathscr{B}\text{-}\mathbf{Mod}).$$

The fibres of the localized category over the objects of $\mathscr{B}$ are the previously considered categories of almost modules. Similar considerations hold for $\mathscr{B}\text{-}\mathbf{Alg}$ and $\mathscr{B}\text{-}\mathbf{Mon}$, and we get the fibred and cofibred categories

$$\mathscr{B}^a\text{-}\mathbf{Mod} \to \mathscr{B} \qquad \mathscr{B}^a\text{-}\mathbf{Alg} \to \mathscr{B} \qquad \mathscr{B}^a\text{-}\mathbf{Mon} \to \mathscr{B}.$$

In particular, for every object $(V, \mathfrak{m})$ of $\mathscr{B}$, we have an obvious notion of almost V-monoid and the category consisting of these is denoted $V^a\text{-}\mathbf{Mon}$.

3.5.4. The localization functors

$$\mathscr{B}\text{-}\mathbf{Mod} \to \mathscr{B}^a\text{-}\mathbf{Mod} : M \mapsto M^a \qquad \mathscr{B}\text{-}\mathbf{Alg} \to \mathscr{B}^a\text{-}\mathbf{Alg} : B \mapsto B^a$$

have left and right adjoints. These adjoints can be chosen as functors of categories over $\mathscr{B}$ such that the adjunction units and counits are morphisms over identity arrows in $\mathscr{B}$. On the fibres these induce the previously considered left and right adjoints $M \mapsto M_!$, $M \mapsto M_*$, $B \mapsto B_{!!}$, $B \mapsto B_*$. We will use the same notation for the corresponding functors on the larger categories. Then it is easy to check that the functor $M \mapsto M_!$ is cartesian and cocartesian (*i.e.* it sends cartesian arrows to cartesian arrows and cocartesian arrows to cocartesian arrows), $M \mapsto M_*$ and $B \mapsto B_*$ are cartesian, and $B \mapsto B_{!!}$ is cocartesian.

3.5.5. Let $\mathscr{B}/\mathbb{F}_p$ be the full subcategory of $\mathscr{B}$ consisting of all objects $(V, \mathfrak{m})$ where V is an $\mathbb{F}_p$-algebra. Define similarly $\mathscr{B}\text{-}\mathbf{Alg}/\mathbb{F}_p$, $\mathscr{B}\text{-}\mathbf{Mon}/\mathbb{F}_p$ as well as $\mathscr{B}^a\text{-}\mathbf{Alg}/\mathbb{F}_p$, $\mathscr{B}^a\text{-}\mathbf{Mon}/\mathbb{F}_p$, so that we have again fibred and cofibred categories

$$\mathscr{B}\text{-}\mathbf{Alg}/\mathbb{F}_p \to \mathscr{B}/\mathbb{F}_p \qquad \text{and} \qquad \mathscr{B}^a\text{-}\mathbf{Alg}/\mathbb{F}_p \to \mathscr{B}/\mathbb{F}_p$$

(and likewise for non-unitary monoids). We remark that the categories $\mathscr{B}^a\text{-}\mathbf{Alg}/\mathbb{F}_p$ and $\mathscr{B}^a\text{-}\mathbf{Mon}/\mathbb{F}_p$ have small limits and colimits, and these are preserved by the projection to $\mathscr{B}/\mathbb{F}_p$. Especially, if $A \to B$ and $A \to C$ are two morphisms in $\mathscr{B}^a\text{-}\mathbf{Alg}/\mathbb{F}_p$ or $\mathscr{B}^a\text{-}\mathbf{Mon}/\mathbb{F}_p$, we can define $B \otimes_A C$ as such a colimit.

3.5.6. If A is a (unitary or non-unitary) $\mathscr{B}$-monoid over $\mathbb{F}_p$, we denote by

$$\Phi_A : A \to A \quad : \quad x \mapsto x^p$$

the *Frobenius endomorphism*. If $(V, \mathfrak{m})$ is an object of $\mathscr{B}/\mathbb{F}_p$, it follows from proposition 2.1.7(ii) that $\Phi_V : (V, \mathfrak{m}) \to (V, \mathfrak{m})$ is a morphism in $\mathscr{B}$. If B is an object of $\mathscr{B}\text{-}\mathbf{Alg}/\mathbb{F}_p$ (resp. $\mathscr{B}\text{-}\mathbf{Mon}/\mathbb{F}_p$) over V, then the Frobenius map induces a morphism $\Phi_B : B \to B$ in $\mathscr{B}\text{-}\mathbf{Alg}/\mathbb{F}_p$ (resp. $\mathscr{B}\text{-}\mathbf{Mon}/\mathbb{F}_p$) over Φ_V. In this way we get a natural transformation from the identity functor of $\mathscr{B}\text{-}\mathbf{Alg}/\mathbb{F}_p$ (resp. $\mathscr{B}\text{-}\mathbf{Mon}/\mathbb{F}_p$) to itself that induces a natural transformation on the identity functor of $\mathscr{B}^a\text{-}\mathbf{Alg}/\mathbb{F}_p$ (resp. $\mathscr{B}^a\text{-}\mathbf{Mon}/F_p$).

3.5.7. Using the pull-back functors, any object B of $\mathscr{B}\text{-}\mathbf{Alg}$ over V defines new objects $B_{(m)}$ of $\mathscr{B}\text{-}\mathbf{Alg}$ ($m \in \mathbb{N}$) over V, where $B_{(m)} := (\Phi_V^m)^*(B)$, which is just B considered as a V-algebra via the homomorphism $V \xrightarrow{\Phi^m} V \to B$. These operations also induce functors $B \mapsto B_{(m)}$ on almost $\mathscr{B}$-algebras.

Definition 3.5.8. Let $(V, \mathfrak{m})$ be an object of $\mathscr{B}/\mathbb{F}_p$.

(i) We say that a morphism $f : A \to B$ of almost V-algebras (resp. almost V-monoids) is *invertible up to* Φ^m if there exists a morphism $f' : B \to A$ in $\mathscr{B}^a\text{-}\mathbf{Alg}$ (resp. $\mathscr{B}^a\text{-}\mathbf{Mon}$) over Φ_V^m, such that $f' \circ f = \Phi_A^m$ and $f \circ f' = \Phi_B^m$.

(ii) We say that an almost V-monoid I (*e.g.* an ideal in a V^a-algebra) is *Frobenius nilpotent* if Φ_I is nilpotent.

3.5.9. Notice that a morphism f of $V^a\text{-}\mathbf{Alg}$ (or $V^a\text{-}\mathbf{Mon}$) is invertible up to Φ^m if and only if $f_* : A_* \to B_*$ is so as a morphism of $\mathbb{F}_p$-algebras.

Lemma 3.5.10. *Let $(V, \mathfrak{m})$ be an object of $\mathscr{B}/\mathbb{F}_p$ and let $f : A \to B$, $g : B \to C$ be morphisms of almost V-algebras or almost V-monoids.*

(i) *If f (resp. g) is invertible up to Φ^n (resp. Φ^m), then $g \circ f$ is invertible up to Φ^{m+n}.*

(ii) *If f (resp. $g \circ f$) is invertible up to Φ^n (resp. Φ^m), then g is invertible up to Φ^{m+n}.*

(iii) *If g (resp. $g \circ f$) is invertible up to Φ^n (resp. Φ^m), then f is invertible up to Φ^{m+n}.*

(iv) *The Frobenius morphisms induce Φ_V-linear morphisms (i.e. morphisms in $\mathscr{B}^a\text{-}\mathbf{Mod}$ over Φ_V) $\Phi' : \mathrm{Ker}\, f \to \mathrm{Ker}\, f$ and $\Phi'' : \mathrm{Coker}\, f \to \mathrm{Coker}\, f$, and f is invertible up to some power of Φ if and only if both Φ' and Φ'' are nilpotent.*

(v) *Consider a map of short exact sequences of almost V-monoids :*

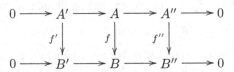

and suppose that two of the morphisms f', f, f'' are invertible up to a power of Φ. Then also the third morphism has this property.

Proof. (i): If f' is an inverse of f up to Φ^n and g' is an inverse of g up to Φ^m, then $f' \circ g'$ is an inverse of $g \circ f$ up to Φ^{m+n}.

(ii): Given an inverse f' of f up to Φ^n and an inverse h' of $h := g \circ f$ up to Φ^m, let $g' := \Phi_B^n \circ f \circ h'$. We compute :

$$g \circ g' = g \circ \Phi_B^n \circ f \circ h' = \Phi_C^n \circ g \circ f \circ h = \Phi_C^n \circ \Phi_C^m$$
$$g' \circ g = \Phi_B^n \circ f \circ h' \circ g = f \circ h' \circ g \circ \Phi_B^n = f \circ h' \circ g \circ f \circ f'$$
$$= f \circ \Phi_A^m \circ f' = \Phi_B^m \circ f \circ f' = \Phi_B^m \circ \Phi_B^n.$$

(iii) is similar and (iv) is an easy diagram chasing left to the reader. (v) follows from (iv) and the snake lemma. □

Lemma 3.5.11. *Let $(V, \mathfrak{m})$ be an object of $\mathscr{B}/\mathbb{F}_p$.*

(i) *If $f : A \to B$ is a morphism of almost V-algebras, invertible up to Φ^n, then so is $A' \to A' \otimes_A B$ for every morphism $A \to A'$ of almost algebras.*

(ii) *If $f : (V, \mathfrak{m}_V) \to (W, \mathfrak{m}_W)$ is a morphism in $\mathscr{B}/\mathbb{F}_p$, both functors*

$$f_* : (V, \mathfrak{m}_V)^a\text{-}\mathbf{Alg} \to (W, \mathfrak{m}_W)^a\text{-}\mathbf{Alg} \qquad f^* : (W, \mathfrak{m}_W)^a\text{-}\mathbf{Alg} \to (V, \mathfrak{m}_V)^a\text{-}\mathbf{Alg}$$

preserve the class of morphisms invertible up to Φ^n.

Proof. (i): Given $f' : B \to A_{(m)}$, construct a morphism $A' \otimes_A B \to A'_{(m)}$ using the morphism $A' \to A'_{(m)}$ coming from $\Phi_{A'}^m$ and f'.

(ii): The assertion for f^* is clear, and the assertion for f_* follows from (i). □

Remark 3.5.12. Statements like those of lemma 3.5.11 hold for the classes of flat, (weakly) unramified, (weakly) étale morphisms.

Theorem 3.5.13. *Let $(V, \mathfrak{m})$ be an object of $\mathscr{B}/\mathbb{F}_p$ and $f : A \to B$ a weakly étale morphism of almost V-algebras.*

(i) *If f is invertible up to Φ^n ($n \geq 0$), then it is an isomorphism.*

(ii) *For every integer $m \geq 0$ the natural square diagram*

(3.5.14)
$$\begin{array}{ccc} A & \xrightarrow{\ f\ } & B \\ {\scriptstyle \Phi_A^m} \downarrow & & \downarrow {\scriptstyle \Phi_B^m} \\ A_{(m)} & \xrightarrow{\ f_{(m)}\ } & B_{(m)} \end{array}$$

is cocartesian.

Proof. (i): We first show that f is faithfully flat. Since f is flat, it remains to show that if M is an A-module such that $M \otimes_A B = 0$, then $M = 0$. It suffice to do this for $M := A/I$, for an arbitrary ideal I of A. After base change by $A \to A/I$, we reduce to show that $B = 0$ implies $A = 0$. However, $A_* \to B_*$ is invertible up to Φ^n, so $\Phi_{A_*}^n = 0$ which means $A_* = 0$. In particular, f is a monomorphism, hence the proof is complete in case that f is an epimorphism. In general, consider the composition $B \xrightarrow{1_B \otimes f} B \otimes_A B \xrightarrow{\mu_{B/A}} B$. From lemma 3.5.11(i) it follows that $1_B \otimes f$ is invertible up to Φ^n; then lemma 3.5.10(ii) says that $\mu_{B/A}$ is invertible up to Φ^n. The latter is also weakly étale; by the foregoing we derive that it is an isomorphism.

Consequently $1_B \otimes f$ is an isomorphism, and finally, by faithful flatness, f itself is an isomorphism.

(ii): The morphisms Φ_A^m and Φ_B^m are invertible up to Φ^m. By lemma 3.5.11(i) it follows that $1_B \otimes \Phi_A^m : B \to B \otimes_A A_{(m)}$ is invertible up to Φ^m; hence, by lemma 3.5.10(ii), the morphism $h : B \otimes_A A_{(m)} \to B_{(m)}$ induced by Φ_B^m and $f_{(m)}$ is invertible up to Φ^{2m} (in fact one verifies that it is invertible up to Φ^m). But h is a morphism of weakly étale $A_{(m)}$-algebras, so it is weakly étale, so it is an isomorphism by (i). $\qquad\square$

Remark 3.5.15. Theorem 3.5.13(ii) extends a statement of Faltings ([34, p.195]) for his notion of almost étale extensions.

3.5.16.   We recall (cp. [41, Ch.0, 3.5]) that a morphism $f : X \to Y$ of objects in a site is called *bicovering* if the induced map of associated sheaves of sets is an isomorphism; if f is squarable ("quarrable" in French), this is equivalent to the condition that both f and the diagonal morphism $\Delta_f : X \to X \times_Y X$ are covering morphisms.

3.5.17.   Let $F \to E$ be a fibered category and $f : P \to Q$ a squarable morphism of E. Consider the following condition:

(3.5.18)   for every base change $P \times_Q Q' \to Q'$ of f, the inverse image functor $F_{Q'} \to F_{P \times_Q Q'}$ is an equivalence of categories.

Inspecting the arguments in [41, Ch.II, §1.1] one can show:

Lemma 3.5.19. *With the above notation, let τ be the topology of universal effective descent relative to $F \to E$. Then we have :*

(i) *if (3.5.18) holds, then f is a covering morphism for the topology τ.*

(ii) *Suppose that $\Delta_f : P \to P \times_Q P$ is squarable; then f is bicovering for τ if and only if (3.5.18) holds both for f and for Δ_f.* $\qquad\square$

Remark 3.5.20. In [41, Ch.II, 1.1.3(iv)] it is stated that "la réciproque est vraie si $i = 2$", meaning that (3.5.18) is equivalent to the condition that f is bicovering for τ. (Actually the cited statement is given in terms of presheaves, but one can show that (3.5.18) is equivalent to the corresponding condition for the fibered category $F^+ \to \widehat{E}_U$ considered in *loc.cit.*) However, this fails in general : as a counterexample we can give the following. Let E be the category of schemes of finite type over a field k; set $P = \mathbb{A}_k^1$, $Q = \operatorname{Spec} k$. Finally let $F \to E$ be the discretely fibered category defined by the presheaf $X \mapsto H^0(X, \mathbb{Z})$. Then it is easy to show that f satisfies (3.5.18) but the diagonal map does not, so f is not bicovering. The mistake in the proof is in [41, Ch.II, 1.1.3.5], where one knows that $F^+(d)$ is an equivalence of categories (notation of *loc.cit.*) but one needs it also after base changes of d.

Lemma 3.5.21. *Let $f : A \to B$ be a morphism of V^a-algebras.*

(i) *If f is invertible up to Φ^m, then the induced functors A-Ét $\to B$-Ét and A-w.Ét $\to B$-w.Ét are equivalences of categories.*

(ii) *If f is weakly étale and $C \to D$ is a morphism of A-algebras invertible up to Φ^m, then the induced map:* $\mathrm{Hom}_{A\text{-}\mathbf{Alg}}(B, C) \to \mathrm{Hom}_{A\text{-}\mathbf{Alg}}(B, D)$ *is bijective.*

Proof. We first consider (i) for the special case where $f := \Phi_A^m : A \to A_{(m)}$. The functor $(\Phi_V^m)^* : V^a\text{-}\mathbf{Alg} \to V^a\text{-}\mathbf{Alg}$ induces a functor $(-)_{(m)} : A\text{-}\mathbf{Alg} \to A_{(m)}\text{-}\mathbf{Alg}$, and by restriction (see remark 3.5.12) we obtain a functor $(-)_{(m)} : A\text{-}\acute{\mathrm{Et}} \to A_{(m)}\text{-}\acute{\mathrm{Et}}$; by theorem 3.5.13(ii), the latter is isomorphic to the functor $(\Phi^m)_* : A\text{-}\acute{\mathrm{Et}} \to A_{(m)}\text{-}\acute{\mathrm{Et}}$ of the lemma. Furthermore, from remark 2.1.4(ii) and (2.2.4) we derive a natural ring isomorphism $\omega : A_{(m)*} \xrightarrow{\sim} A_*$, hence an essentially commutative diagram

$$
\begin{array}{ccccc}
A\text{-}\acute{\mathrm{Et}} & \longrightarrow & A\text{-}\mathbf{Alg} & \xrightarrow{\ \alpha\ } & (A_*, \mathfrak{m} \cdot A_*)^a\text{-}\mathbf{Alg} \\
{\scriptstyle (\Phi^m)_*}\downarrow & & {\scriptstyle (-)_{(m)}}\downarrow & & {\scriptstyle \omega^*}\downarrow \\
A_{(m)}\text{-}\acute{\mathrm{Et}} & \longrightarrow & A_{(m)}\text{-}\mathbf{Alg} & \xrightarrow{\ \beta\ } & (A_{(m)*}, \mathfrak{m} \cdot A_{(m)*})^a\text{-}\mathbf{Alg}
\end{array}
$$

where α and β are the equivalences of remark 2.2.13. Clearly α and β restrict to equivalences on the corresponding categories of étale algebras, hence the lemma follows in this case.

For the general case of (i), let $f' : B \to A_{(m)}$ be a morphism as in definition 3.5.8. Diagram (3.5.14) – completed with f' – induces an essentially commutative diagram of the corresponding categories of algebras, so by the previous case, the functor $(f')_* : B\text{-}\acute{\mathrm{Et}} \to A_{(m)}\text{-}\acute{\mathrm{Et}}$ has both a left quasi-inverse and a right quasi-inverse; these quasi-inverses must be isomorphic, so f_* has a quasi-inverse as desired. Finally, we remark that the map in (ii) is the same as the map $\mathrm{Hom}_{C\text{-}\mathbf{Alg}}(B \otimes_A C, C) \to \mathrm{Hom}_{D\text{-}\mathbf{Alg}}(B \otimes_A D, D)$, and the latter is a bijection in view of (i). $\qquad\square$

Remark 3.5.22. Notice that lemma 3.5.21(ii) generalizes theorem 3.2.18(i) (in case V is an $\mathbb{F}_p$-algebra). Similarly, it follows from lemmata 3.5.21(i) and 3.5.10(iv) that, in case V is an $\mathbb{F}_p$-algebra, one can replace "nilpotent" in theorem 3.2.18(ii),(iii) by "Frobenius nilpotent".

3.5.23. In the following, τ will denote indifferently the topology of universal effective descent defined by either of the fibered categories $\mathrm{w.}\acute{\mathrm{Et}}^o \to V^a\text{-}\mathbf{Alg}^o$ or $\acute{\mathrm{Et}}^o \to V^a\text{-}\mathbf{Alg}^o$.

Proposition 3.5.24. *If $f : A \to B$ is a morphism of almost V-algebras which is invertible up to Φ^m, then f^o is bicovering for the topology τ.*

Proof. In light of lemmata 3.5.19(ii) and 3.5.21(i), it suffices to show that $\mu_{B/A}$ is invertible up to a power of Φ. For this, factor the identity morphism of B as $B \xrightarrow{1_B \otimes f} B \otimes_A B \xrightarrow{\mu_{B/A}} B$ and argue as in the proof of theorem 3.5.13. $\qquad\square$

Proposition 3.5.25. *Let $A \to B$ be a morphism of almost V-algebras, $I \subset A$ and $J \subset B$ ideals such that $IB \subset J$. Set $\overline{A} := A/I$ and $\overline{B} := B/J$. Suppose that either*

(a) $I \to J$ *is an epimorphism with nilpotent kernel, or*

(b) V *is an* $\mathbb{F}_p$-*algebra and* $I \to J$ *is invertible up to a power of* Φ.

Then we have :

(i) *conditions* (a) *and* (b) *are stable under any base change* $A \to C$.

(ii) $(A \to B)^o$ *is covering (resp. bicovering) for* τ *if and only if* $(\overline{A} \to \overline{B})^o$ *is.*

Proof. Suppose first that both morphisms $I \to IB \to J$ are isomorphism; in this case we claim that $IC \to I(C \otimes_A B)$ is an epimorphism and $\mathrm{Ker}(IC \to I(C \otimes_A B))^2 = 0$ for any A-algebra C. Indeed, since by assumption $I \simeq IB$, $C \otimes_A B$ acts on $C \otimes_A I$, hence $\mathrm{Ker}(C \to C \otimes_A B)$ annihilates $C \otimes_A I$, hence annihilates its image IC, whence the claim. If, moreover, V is an $\mathbb{F}_p$-algebra, lemma 3.5.10(iv) implies that $IC \to I(C \otimes_A B)$ is invertible up to a power of Φ.

In the general case, consider the intermediate almost V-algebra $A_1 := \overline{A} \times_{\overline{B}} B$ equipped with the ideal $I_1 := 0 \times_{\overline{B}} J$. In case (a), $I_1 = IA_1$ and $A \to A_1$ is an epimorphism with nilpotent kernel, hence it remains such after any base change $A \to C$. To prove (i) in case (a), it suffices then to consider the morphism $A_1 \to B$, hence we can assume from start that $I \to J$ is an isomorphism, which is the case already dealt with. To prove (i) in case (b), it suffices to consider the cases of $(A, I) \to (A_1, I_1)$ and $(A_1, I_1) \to (B, J)$. The second case is treated above. In the first case, we do not necessarily have $I_1 = IA_1$ and the assertion to be checked is that, for every A-algebra C, the morphism $IC \to I_1(A_1 \otimes_A C)$ is invertible up to a power of Φ. We apply lemma 3.5.10(v) to the commutative diagram with exact rows:

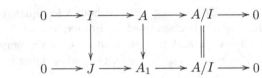

to deduce that $A \to A_1$ is invertible up to some power of Φ, hence so is $C \to A_1 \otimes_A C$, which implies the assertion.

As for (ii), we remark that the "only if" part is trivial in case (a), and follows in case (b) because $(B/IB \to B/J)^o$ is bicovering by proposition 3.5.24; we assume therefore that $(\overline{A} \to \overline{B})^o$ is τ-covering (resp. τ-bicovering). Consider first the assertion for "covering". We need to show that $(A \to B)^o$ is of universal effective descent for F, where F is either one of our two fibered categories. In light of (i), this is reduced to the assertion that $(A \to B)^o$ is of effective descent for F. We notice that $(A \to A_1)^o$ is bicovering for τ (in case (a) by theorem 3.2.18 and lemma 3.5.19(ii), in case (b) by proposition 3.5.24). As $(\overline{A} \to A_1/I_1)^o$ is an isomorphism, the assertion is reduced to the case where $I \to J$ is an isomorphism. In this case, by lemma 3.4.29, there is a natural equivalence: $\mathbf{Desc}(F, B/A) \xrightarrow{\sim} \mathbf{Desc}(F, \overline{B}/\overline{A}) \times_{F_{\overline{B}}} F_B$. Then the assertion follows easily from corollary 3.4.22. Finally suppose that $(\overline{A} \to \overline{B})^o$ is bicovering. The foregoing already says that $(A \to B)^o$ is covering, so it remains to show that $(B \otimes_A B \to B)^o$ is also covering. The above argument again reduces to the case where $I \to J$ is an isomorphism. Then, as in the proof of lemma 3.4.29, the induced morphism

$I(B \otimes_A B) \to IB$ is an isomorphism as well. Thus the assertion for "bicovering" is reduced to the assertion for "covering". $\qquad\square$

We conclude this section with a result of a more special nature, which can be interpreted as an easy case of almost purity in positive characteristic.

3.5.26. We suppose now that the basic setup $(V, \mathfrak{m})$ consists of a *perfect* $\mathbb{F}_p$-algebra V, *i.e.* such that the Frobenius endomorphism $\Phi_V : V \to V$ is bijective; moreover we assume that there exists a non-zero-divisor $\varepsilon \in \mathfrak{m}$ such that $\mathfrak{m} = \bigcup_{n>0} \varepsilon^{1/p^n} V$. Let us denote by $V^a\text{-}\acute{\text{E}}\text{t}_{\text{uafp}}$ (resp. $V[\varepsilon^{-1}]\text{-}\text{u}.\acute{\text{E}}\text{t}_{\text{fp}}$) the category of uniformly almost finite projective étale V^a-algebras (resp. of finite étale $V[\varepsilon^{-1}]$-algebra in the usual sense of [44]). We will be concerned with the functor:

$$(3.5.27) \qquad V^a\text{-}\acute{\text{E}}\text{t}_{\text{uafp}} \to V[\varepsilon^{-1}]\text{-}\text{u}.\acute{\text{E}}\text{t}_{\text{fp}} \quad : \quad A \mapsto A_*[\varepsilon^{-1}].$$

Theorem 3.5.28. *Under the assumptions of (3.5.26), the functor (3.5.27) is an equivalence of categories.*

Proof. Let R be a finite étale $V[\varepsilon^{-1}]$-algebra. Since $V[\varepsilon^{-1}]$ is perfect, the same holds for R, in view of theorem 3.5.13(ii) (applied in the classical limit case of example 2.1.2(ii)). Let us choose a finite V-algebra $R_0 \subset R$ such that $R_0[\varepsilon^{-1}] = R$ and define $R_1 := \bigcup_{n \in \mathbb{N}} \Phi_R^{-n}(R_0)$. (This is an increasing union.)

Claim 3.5.29. The V^a-algebra R_1^a does not depend on the choice of R_0.

Proof of the claim. Let $R_0' \subset R$ be another finite V-algebra such that $R_0'[\varepsilon^{-1}] = R$; clearly we have $\varepsilon^m R_0 \subset R_0' \subset \varepsilon^{-m} R_0$ for $m \in \mathbb{N}$ sufficiently large. It follows that $\varepsilon^{m/p^n} \Phi_R^{-n}(R_0) \subset \Phi_R^{-n}(R_0') \subset \varepsilon^{-m/p^n} \Phi_R^{-n}(R_0)$ for every $n \in \mathbb{N}$. The claim readily follows. $\qquad\diamond$

We set $J := \bigcup_{n \in \mathbb{N}} \text{Ann}_{R_1 \otimes_V R_1}(\varepsilon^n)$ and let $e_R \in R \otimes_{V[\varepsilon^{-1}]} R$ be the idempotent provided by proposition 3.1.4.

Claim 3.5.30. $\text{Ker}((R_1 \otimes_V R_1/J)^a \to R_1^a)$ is generated by an idempotent almost element whose image in $R \otimes_V R$ equals $1 - e_R$.

Proof of the claim. For $m \in \mathbb{N}$ large enough, $\varepsilon^m \cdot e_R$ is contained in the image of $R_0 \otimes_V R_0$. Hence, for every $n \in \mathbb{N}$, $\varepsilon^{m/p^n} \cdot e_R$ is in the image of $\Phi_R^{-n}(R_0) \otimes_V \Phi_R^{-n}(R_0)$, so, arguing as in the proof of proposition 3.1.4, e_R defines an idempotent almost element in $(R_1 \otimes_V R_1/J)_*^a$ and the claim follows. $\qquad\diamond$

Claim 3.5.31. R_1^a is a uniformly almost finite V^a-algebra.

Proof of the claim. For large enough $m \in \mathbb{N}$ we have: $R_0 \subset \Phi_R^{-1}(R_0) \subset \varepsilon^{-m} R_0$, therefore $\Phi_R^{-n}(R_0) \subset \Phi_R^{-(n+1)}(R_0) \subset \varepsilon^{-m/p^n} \Phi_R^{-n}(R_0)$ for every $n \in \mathbb{N}$. By an easy induction we deduce:

$$\Phi_R^{-(n+k)}(R_0) \subset \prod_{j=0}^{k} \varepsilon^{-m/p^{n+j}} \cdot \Phi_R^{-n}(R_0) \subset \varepsilon^{-m/p^{n-1}} \Phi_R^{-n}(R_0)$$

for every $n, k \in \mathbb{N}$. Finally, this implies that $R_1 \subset \varepsilon^{-m/p^{n-1}} \Phi_R^{-n}(R_0)$ for every $n \in \mathbb{N}$ and the claim follows. $\qquad\diamond$

Claim 3.5.32. Let S be the integral closure of V in R; then $R_1^a = S^a$.

Proof of the claim. Let us endow R with the unique ring topology τ such that the induced subspace topology on R_0 is the ε-adic topology and R_0 is open in R. It is easy to check that S consists of power-bounded elements of R relative to the topology τ. Since clearly $R_1 \subset S$, it suffices therefore to show that $(R_1^a)_* \subset R$ is the subring of all power-bounded elements of R. However, $(R_1^a)_*$ can be characterized as the subring of all $x \in R$ such that $\mathfrak{m} \cdot x \subset R_1$; this already implies that $(R_1^a)_*$ consists of power-bounded elements. On the other hand, if $x \in R$ is power-bounded, it follows that $\delta \cdot x$ is topologically nilpotent for every $\delta \in \mathfrak{m}$; since R_0 is open in R, it follows that, for every $\delta \in \mathfrak{m}$ there exists $n_0 \in \mathbb{N}$ such that $(\delta \cdot x)^n \in R_0$ for every $n > n_0$. By taking $n := p^k$ for sufficiently large $k \in \mathbb{N}$, we deduce that $\Phi_R^k(\delta \cdot x) \in R_0$, that is $\delta \cdot x \in R_1$, and the claim follows. $\Diamond$

Claim 3.5.33. R_1^a is an almost projective V^a-algebra.

Proof of the claim. As a special case of claim 3.5.32, let W be the integral closure of V in $V[\varepsilon^{-1}]$; then:

(3.5.34) $$W^a = V^a.$$

Next, let $\mathrm{Tr}_{R/V[\varepsilon^{-1}]} : R \to V[\varepsilon^{-1}]$ be the trace map of the finite étale extension $V[\varepsilon^{-1}] \to R$; recall that $\mathrm{Tr}_{R/V[\varepsilon^{-1}]}$ sends elements integral over V to elements integral over V (to see this, we can assume that R has constant rank n over $V[\varepsilon^{-1}]$; then the assertion can be checked after a faithfully flat base change $V[\varepsilon^{-1}] \to S$, so we can further suppose that $R \simeq V[\varepsilon^{-1}]^n$, in which case everything is clear); it then follows from claim 3.5.32 and (3.5.34) that $\mathrm{Tr}_{R/V[\varepsilon^{-1}]}^a$ restricts to a morphism $T : R_1^a \to V^a$. Furthermore, by claim 3.5.30, for every $\delta \in \mathfrak{m}$ we can write $\delta \cdot e_R = \sum_i^n x_i \otimes y_i$ for certain $x_i, y_i \in R_1$. By remark 4.1.17 (whose proof does not use theorem 3.5.28) we deduce the identity: $\delta \cdot b = \sum_i^n x_i \cdot T(b \cdot y_i)$ for every $b \in (R_1^a)_*$. This allows us to define morphisms

$$\alpha : R_1^a \to (V^a)^n \qquad \beta : (V^a)^n \to R_1^a$$

with $\beta \circ \alpha = \delta \cdot 1_{R_1^a}$, namely

$$\alpha(b) = (T(b \cdot y_1), ..., T(b \cdot y_n)) \quad \text{and} \quad \beta(v_1, ..., v_n) = \sum_i^n x_i \cdot v_i$$

for every $b \in R_1$ and $v_1, ..., v_n \in V$. By lemma 2.4.15, the claim follows. $\Diamond$

Claim 3.5.35. R_1^a is an unramified V^a-algebra.

Proof of the claim. In light of claim 3.5.33 we deduce that actually $J^a = 0$; the assertion then follows from claim 3.5.30 and proposition 3.1.4. $\Diamond$

Claim 3.5.36. The functor (3.5.27) is fully faithful.

Proof of the claim. First of all, it is clear that, for every flat V^a-algebras A, B, the natural map

(3.5.37) $$\mathrm{Hom}_{V^a\text{-}\mathbf{Alg}}(A, B) \to \mathrm{Hom}_{V[\varepsilon^{-1}]\text{-}\mathbf{Alg}}(A_*[\varepsilon^{-1}], B_*[\varepsilon^{-1}])$$

is injective, since $A_* \subset A_*[\varepsilon^{-1}]$ and similarly for B. Suppose now that A and B are étale and almost finite over V^a; then Φ_A and Φ_B are automorphisms, due to theorem 3.5.13(ii) and the assumption that V is perfect. Let $\psi : A_*[\varepsilon^{-1}] \to B_*[\varepsilon^{-1}]$ be any map of $V[\varepsilon^{-1}]$-algebras; since A is almost finite, we have $\psi(A_*) \subset \varepsilon^{-m}B_*$ for $m \in \mathbb{N}$ large enough. Since Frobenius commutes with every ring homomorphism, we deduce $\psi(A_*) = \psi(\Phi_{A_*}^{-n}(A_*)) \subset \varepsilon^{-m/p^n}\Phi_{B_*}^{-n}(B_*) = \varepsilon^{-m/p^n}B_*$ for every $n \in \mathbb{N}$, so ψ induces a morphism $\psi^a : A \to B$, which shows that (3.5.37) is surjective. $\diamond$

It now follows from claims 3.5.29, 3.5.31, 3.5.33 3.5.35 that the assignment $R \to R_1^a$ defines a quasi-inverse to (3.5.27); together with claim 3.5.36, this concludes the proof of the theorem. $\square$

Remark 3.5.38. (i) In the terminology of definition 8.2.27, claim 3.5.32 says that R_1^a is the integral closure of V^a in R^a. So a quasi-inverse to (3.5.27) is given by the functor

$$R \mapsto \mathrm{i.c.}(V^a, R^a).$$

(ii) The proof of theorem 3.5.28 shows as well that the natural base change functor

$$V^a\text{-}\mathbf{\acute{E}t}_{\mathrm{afp}} \to V[\varepsilon^{-1}]\text{-}\mathbf{u.\acute{E}t}_{\mathrm{fp}}$$

is an equivalence.

4. FINE STUDY OF ALMOST PROJECTIVE MODULES

An alternative title for this chapter could have been "Everything you can do with traces". Right at the outset we find the definition of the trace map of an almost projective almost finitely generated A-module. The whole purpose of the chapter is to showcase the versatility of this construction, a real swiss-knife of almost linear algebra. For instance, we apply it to characterize étale morphisms (theorem 4.1.14); more generally, it is used to define the *different ideal* of an almost finite A-algebra. In section 4.3 it is employed in an essential way to study the important class of A-modules of *finite rank*, *i.e.* those almost projective A-modules P such that $\Lambda_A^i P = 0$ for sufficiently large $i \in \mathbb{N}$. A rather complete and satisfactory description is achieved for such A-modules (proposition 4.3.27). This is further generalized in theorem 4.3.28, to arbitrary A-modules so called of *almost finite rank* (see definition 4.3.9(ii)). The interest of the latter class is that it contains basically all the almost projective modules found in nature; indeed, we cannot produce a single example of an almost projective module that is almost finitely generated but has not almost finite rank (but we suspect that they do exist). In any case, almost finitely generated modules whose rank is not almost finite are certainly rather weird beasts : some clue about their looks can be gained by analyzing the structure of *invertible* modules : we do this at the end of section 4.4.

The other main construction of chapter 4 is the *splitting algebra* of an almost projective module, introduced in section 4.4 : with its aid we show that A-modules of finite rank are locally free in the flat topology of A. It should be clear that this is a very pleasant and useful culmination for our study of almost projective modules; we put it to use right away in the following section 4.5, where we show that an étale groupoid of finite rank over the category of affine almost schemes (more prosaically, the opposite of the category of almost algebras) is universally effective, that is, it admits a good quotient, as in the classical algebro-geometric setting.

4.1. Almost traces. Let A be a V^a-algebra.

Definition 4.1.1. Let P be an almost finitely generated projective A-module. Then $\omega_{P/A}$ is an isomorphism by lemma 2.4.29(i.b). The *trace morphism* of P is the A-linear morphism

$$\mathrm{tr}_{P/A} := \mathrm{ev}_{P/A} \circ \omega_{P/A}^{-1} : \mathrm{End}_A(P)^a \to A.$$

We let ζ_P be the unique almost element of $P \otimes_A P^*$ such that $\omega_{P/A}(\zeta_P) = 1_P$.

Lemma 4.1.2. *Let M, N be almost finitely generated projective A-modules, and $\phi : M \to N$, $\psi : N \to M$ two A-linear morphisms. Then :*
  (i) $\mathrm{tr}_{M/A}(\psi \circ \phi) = \mathrm{tr}_{N/A}(\phi \circ \psi).$
  (ii) *If $\psi \circ \phi = a \cdot 1_M$ and $\phi \circ \psi = a \cdot 1_N$ for some $a \in A_*$, and if, furthermore, there exist $u \in \mathrm{End}_A(M)$, $v \in \mathrm{End}_A(N)$ such that $v \circ \phi = \phi \circ u$, then $a \cdot (\mathrm{tr}_{M/A}(u) - \mathrm{tr}_{N/A}(v)) = 0.$*

Proof. (i): By lemma 2.4.29(i), the natural morphism $N \otimes_A \mathrm{alHom}_A(M, A) \to \mathrm{alHom}_A(M, N)$ is an isomorphism (and similarly when we exchange the roles of

M and N). By A-linearity, we can therefore assume that ϕ (resp. ψ) is of the form $x \mapsto n \cdot \alpha(x)$ for some $n \in N_*$, $\alpha : M \to A$ (resp. of the form $x \mapsto m \cdot \beta(x)$ for some $m \in M_*$, $\beta : N \to A$). Then a simple computation yields:

$$\phi \circ \psi = \omega_{N/A}(n \cdot \alpha(m) \otimes \beta) \qquad \psi \circ \psi = \omega_{M/A}(m \cdot \beta(n) \otimes \alpha)$$

and the claim follows directly from the definition of the trace morphism. For (ii) we compute using (i): $a \cdot \mathrm{tr}_{M/A}(u) = \mathrm{tr}_{M/A}(\psi \circ \phi \circ u) = \mathrm{tr}_{M/A}(\psi \circ v \circ \phi) = \mathrm{tr}_{N/A}(v \circ \phi \circ \psi) = a \cdot \mathrm{tr}_{N/A}(v)$. $\qquad\square$

Lemma 4.1.3. *Let M, N be two almost finitely generated projective A-modules, $\phi \in \mathrm{End}_A(M)$ and $\psi \in \mathrm{End}_A(N)$. Then*

$$\mathrm{tr}_{M \otimes_A N/A}(\phi \otimes \psi) = \mathrm{tr}_{M/A}(\phi) \cdot \mathrm{tr}_{N/A}(\psi).$$

Proof. As usual we can suppose that $\phi = \omega_{M/A}(m \otimes \alpha)$, $\psi = \omega_{N/A}(n \otimes \beta)$ for some $\alpha \in M^*_*$, $\beta \in N^*_*$. Then $\phi \otimes \psi = \omega_{M \otimes_A N/A}((m \otimes n) \otimes (\alpha \otimes \beta))$ and the sought identity follows by explicit calculation. $\qquad\square$

Proposition 4.1.4. *Let $\underline{M} = (0 \to M_1 \xrightarrow{i} M_2 \xrightarrow{p} M_3 \to 0)$ be an exact sequence of almost finitely generated projective A-modules, and let $\underline{u} = (u_1, u_2, u_3) : \underline{M} \to \underline{M}$ be an endomorphism of $\underline{M}$, given by endomorphisms $u_i : M_i \to M_i$ ($i = 1, 2, 3$). Then we have $\mathrm{tr}_{M_2/A}(u_2) = \mathrm{tr}_{M_1/A}(u_1) + \mathrm{tr}_{M_3/A}(u_3)$.*

Proof. Suppose first that there exists a splitting $s : M_3 \to M_2$ for p, so that we can view u_2 as a matrix $\begin{pmatrix} u_1 & v \\ 0 & u_3 \end{pmatrix}$, where $v \in \mathrm{Hom}_A(M_3, M_1)$. By additivity of the trace, we are then reduced to show that $\mathrm{tr}_{M_2/A}(i \circ v \circ p) = 0$. By lemma 4.1.2(i), this is the same as $\mathrm{tr}_{M_3/A}(p \circ i \circ v)$, which obviously vanishes. In general, for any $a \in \mathfrak{m}$ we consider the morphism $\mu_a = a \cdot 1_{M_3}$ and the pull back morphism $\underline{M} * \mu_a \to \underline{M}$

$$
\begin{array}{ccccccccc}
0 & \longrightarrow & M_1 & \xrightarrow{\ i\ } & M_2 & \xrightarrow{\ p\ } & M_3 & \longrightarrow & 0 \\
 & & \| & & \uparrow{\scriptstyle \phi} & & \uparrow{\scriptstyle \mu_a} & & \\
0 & \longrightarrow & M_1 & \longrightarrow & P & \xrightarrow{\ p'\ } & M_3 & \longrightarrow & 0.
\end{array}
$$

Pick a morphism $j : M_3 \to M_2$ such that $p \circ j = a \cdot 1_{M_3}$; the pair $(j, 1_{M_3})$ determines a morphism $\sigma : M_3 \to P$ such that $\sigma \circ p' = 1_{M_3}$, i.e. the sequence $\underline{M} * \mu_a$ is split exact; this sequence also inherits the endomorphism $\underline{u} * \mu_a = (u_1, v, u_3)$, for a certain $v \in \mathrm{End}_A(P)$. The pair of morphisms $(a \cdot 1_{M_2}, p)$ determines a morphism $\psi : M_2 \to P$, and it is easy to check that $\phi \circ \psi = a \cdot 1_{M_2}$ and $\psi \circ \phi = a \cdot 1_P$. We can therefore apply lemma 4.1.2 to deduce that $a \cdot (\mathrm{tr}_{P/A}(v) - \mathrm{tr}_{M/A}(u_2)) = 0$. By the foregoing we know that $\mathrm{tr}_{P/A}(v) = \mathrm{tr}_{M_1/A}(u_1) + \mathrm{tr}_{M_3/A}(u_3)$, so the claim follows. $\qquad\square$

Lemma 4.1.5. *Let A be a V^a-algebra.*

(i) *If $P := M \otimes_A N$ is an almost projective and faithfully flat (resp. and almost finitely generated) A-module, then so are M and N.*

(ii) *If $M \otimes_A N \simeq A$, then the evaluation map* $\mathrm{ev}_M : M \otimes_A M^* \to A$ *is an isomorphism.*

(iii) *An invertible A-module is faithfully flat and almost finitely generated projective.*

(iv) *An epimorphism $\phi : M \to N$ of invertible A-modules is an isomorphism.*

Proof. Clearly (iii) is just a special case of (i). We show (i): By proposition 2.4.28(iv) we know that $\mathscr{E}_{P/A} = A$; however, one checks easily that $\mathscr{E}_{P/A} \subset \mathscr{E}_{N/A}$, whence

$$(4.1.6) \qquad\qquad \mathscr{E}_{N/A} = A.$$

Therefore N will be faithfully flat, as soon as it is shown to be almost projective, again by virtue of proposition 2.4.28(iv). In any case, (4.1.6) means that, for every $\varepsilon \in \mathfrak{m}$, we can find an almost element of the form $\sum_{i=1}^{n} x_i \otimes \phi_i \in N \otimes_A N^*$, such that $\sum_{i=1}^{n} \phi_i(x_i) = \varepsilon$. We use such an element to define morphisms $A \to N^n \to A$ whose composition equals $\varepsilon \cdot \mathbf{1}_N$. After tensoring by M, we obtain morphisms $M \to P \to M$ whose composition is $\varepsilon \cdot \mathbf{1}_M$. Then, since P is almost projective, it follows easily that so must be M; similarly, if P is almost finitely generated, the same follows for M. By symmetry, the same holds for N.

(ii): Notice that, by (i), we know already that M and N are almost finitely generated projective. By lemma 4.1.3 we deduce that $\mathrm{tr}_{M/A}(\mathbf{1}_M) \cdot \mathrm{tr}_{N/A}(\mathbf{1}_N) = 1$, so both factors are invertible in A_*. It follows that the morphism $A \to \mathrm{End}_A(M)$ given by $a \mapsto a \cdot \mathbf{1}_M$ provides a splitting for the trace morphism (and similarly for N in place of M). Thus we can write $\mathrm{End}_A(M) \simeq A \oplus X$, $\mathrm{End}_A(N) \simeq A \oplus Y$ for some A-modules X, Y. However, on one hand we have a natural isomorphism $\mathrm{End}_A(M) \otimes_A \mathrm{End}_A(N) \simeq A$; on the other hand, we have a decomposition $\mathrm{End}_A(M) \otimes_A \mathrm{End}_A(N) \simeq A \oplus X \oplus Y \oplus (X \otimes_A Y)$; working out the identifications, one sees that the induced isomorphism $A \oplus X \oplus Y \oplus (X \otimes_A Y) \simeq A$ restricts to the identity morphism on the direct summand A; it follows that $X = Y = 0$, which readily implies the claim.

(iv): In view of (ii) we can replace ϕ by $\phi \otimes_A \mathbf{1}_{M^*}$, and thereby assume that $M = A$. Then $N \simeq A/\mathrm{Ker}\,\phi$; it is clear that such a module is faithfully flat if and only if $\mathrm{Ker}\,\phi = 0$. By (iii), the claim follows. $\square$

Lemma 4.1.5 explains why we do not insist, in the definition of an invertible A-module, that it should be almost projective or almost finitely generated: both conditions can be deduced.

4.1.7.   Suppose now that B is an almost finite projective A-algebra. For any $b \in B_*$, denote by $\mu_b : B \to B$ the B-linear morphism $b' \mapsto b \cdot b'$. The map $b \mapsto \mu_b$ defines a B-linear monomorphism $\mu : B \to \mathrm{End}_A(B)^a$. The composition

$$\mathrm{Tr}_{B/A} := \mathrm{tr}_{B/A} \circ \mu : B \to A$$

will also be called the almost trace morphism of the A-algebra B.

Proposition 4.1.8. *Let A and B be as in (4.1.7).*

(i) *If $\phi : A \to B$ is an isomorphism, then:* $\mathrm{Tr}_{B/A} = \phi^{-1}$.

(ii) *If C any other A-algebra, then:* $\mathrm{Tr}_{C\otimes_A B/C} = \mathbf{1}_C \otimes_A \mathrm{Tr}_{B/A}$.

(iii) *If C is an almost finite projective B-algebra, then:*

$$\mathrm{Tr}_{C/A} = \mathrm{Tr}_{B/A} \circ \mathrm{Tr}_{C/B}.$$

Proof. (i) and (ii) are left as exercises for the reader. We verify (iii). It comes down to checking that the following diagram commutes:

$$
\begin{array}{ccccc}
C \otimes_B \mathrm{alHom}_B(C, B) & \longrightarrow & C \otimes_B \mathrm{alHom}_A(C, B) & \xrightarrow{\ \sim\ } & C \otimes_A \mathrm{alHom}_A(C, A) \\
\Big\downarrow{\scriptstyle \mathrm{ev}_{C/B}} & & & & \Big\downarrow{\scriptstyle \mathrm{ev}_{C/A}} \\
B & & \xrightarrow{\qquad\qquad \mathrm{Tr}_{B/A} \qquad\qquad} & & A.
\end{array}
$$

Therefore, pick $c \in C_*$ and $\phi \in \mathrm{Hom}_B(C, B)$. For every $\varepsilon \in \mathfrak{m}$ we can find elements $b_1, ..., b_k \in B_*$ and $\phi_1, ..., \phi_k \in \mathrm{Hom}_A(C, A)$ such that $\varepsilon \cdot \phi(x) = \sum_i b_i \cdot \phi_i(x)$ for every $x \in C_*$. The B-linearity of ϕ translates into the identity:

$$(4.1.9) \qquad \sum_i b_i \cdot \phi_i(b \cdot x) = \sum_i b \cdot b_i \cdot \phi_i(x) \qquad \text{for all } b \in B_*, x \in C_*.$$

Then $\varepsilon \cdot \mathrm{ev}_{C/B}(c \otimes \phi) = \sum_i b_i \cdot \phi_i(c)$ and we need to show that

$$(4.1.10) \qquad \mathrm{Tr}_{B/A}\Big(\sum_i b_i \cdot \phi_i(c)\Big) = \sum_i \phi_i(c \cdot b_i).$$

For every $i \leq k$, let $\mu_i : A \to B$ be the morphism $a \mapsto b_i \cdot a$ (for all $a \in A_*$); furthermore, let $\mu_c : B \to C$ be the morphism $b \mapsto c \cdot b$ (for all $b \in B_*$). In view of (4.1.9), the left-hand side of (4.1.10) is equal to $\mathrm{tr}_{B/A}(\sum_i \mu_i \circ \phi_i \circ \mu_c)$. By lemma 4.1.2(i), we have $\mathrm{tr}_{B/A}(\mu_i \circ \phi_i \circ \mu_c) = \mathrm{tr}_{A/A}(\phi_i \circ \mu_c \circ \mu_i) = \phi_i(c \cdot b_i)$ for every $i \leq k$. The claim follows. $\qquad\square$

Corollary 4.1.11. *Let B be a faithfully flat almost finite projective and étale A-algebra. Then $\mathrm{Tr}_{B/A} : B \to A$ is an epimorphism.*

Proof. Let $C = \mathrm{Coker}(\mathrm{Tr}_{B/A})$ and $\mathrm{Tr}_{B/B\otimes_A B}$ the trace morphism for the morphism of almost V-algebras $\mu_{B/A}$. By faithful flatness, the natural morphism $C \hookrightarrow C \otimes_A B = \mathrm{Coker}(\mathrm{Tr}_{B\otimes_A B/B})$ is a monomorphism, hence it suffices to show that $\mathrm{Tr}_{B\otimes_A B/B}$ is an epimorphism (here $B \otimes_A B$ is viewed as a B-algebra via the second factor). However, from proposition 4.1.8(i) and (iii) we see that $\mathrm{Tr}_{B/B\otimes_A B}$ is a right inverse for $\mathrm{Tr}_{B\otimes_A B/B}$. The claim follows. $\qquad\square$

4.1.12. It is useful to introduce the A-linear morphism

$$t_{B/A} := \mathrm{Tr}_{B/A} \circ \mu_{B/A} : B \otimes_A B \to A.$$

We can view $t_{B/A}$ as a bilinear form; it induces an A-linear morphism

$$\tau_{B/A} : B \to B^* := \mathrm{alHom}_A(B, A)$$

characterized by the equality $t_{B/A}(b_1 \otimes b_2) = \tau_{B/A}(b_1)(b_2)$ for all $b_1, b_2 \in B_*$. We say that $t_{B/A}$ is *a perfect pairing* if $\tau_{B/A}$ is an isomorphism.

Lemma 4.1.13. *Let $A \to B$ be an almost finite projective morphism of V^a-algebras and C any A-algebra. Denote by*

$$\eta_{B,C} : C \otimes_A \text{alHom}_A(B, A) \to \text{alHom}_C(C \otimes_A B, C)$$

the natural isomorphism provided by lemma 2.4.31(i). Then:

(i) *$\tau_{B/A}$ is B-linear (for the B-module structure of B^* defined in remark 2.4.25(i));*

(ii) *$\eta_{B,C}$ is $C \otimes_A B$-linear;*

(iii) *$\eta_{B,C} \circ (1_C \otimes \tau_{B/A}) = \tau_{C \otimes_A B/C}$.*

Proof. For any $b \in B_*$, let $\xi_b : B \to A$ the A-linear morphism defined by the rule $b' \mapsto \text{Tr}_{B/A}(b' \cdot b)$ for all $b' \in B_*$. Then, directly from the definition we can compute: $(\eta_{B,C} \circ (1_C \otimes \tau_{B/A}))(c \otimes b)(c' \otimes b') = \eta_{B,C}(c \otimes \xi_b)(c' \otimes b') = c \cdot c' \cdot \text{Tr}_{B/A}(b' \cdot b)$ for all $b, b' \in B_*$, $c, c' \in C_*$. But by proposition 4.1.8(ii), the latter expression can be rewritten as $\tau_{C \otimes_A B/C}(c \otimes b)(c' \otimes b')$, which shows (iii). The proofs of (i) and (ii) are similar direct verifications: we show (i) and leave (ii) to the reader. Let us pick any $b, b', b'' \in B_*$; then $(b \cdot \tau_{B/A}(b'))(b'') = \tau_{B/A}(b')(bb'') = \text{Tr}_{B/A}(bb'b'') = (\tau_{B/A}(bb'))(b'')$. $\square$

Theorem 4.1.14. *An almost finite projective morphism $\phi : A \to B$ of almost V-algebras is étale if and only if the trace form $t_{B/A}$ is a perfect pairing.*

Proof. By lemma 4.1.13, we have a commutative diagram:

$$
(4.1.15) \quad
\begin{array}{ccc}
(B \otimes_A B) \otimes_B B & \xrightarrow{\;\sim\;} & B \otimes_A B \\
{\scriptstyle 1_{B \otimes_A B \otimes_B \tau_B}} \downarrow & & \downarrow {\scriptstyle 1_B \otimes_A \tau_B} \quad \searrow {\scriptstyle \tau_{B \otimes_A B/B}} \\
(B \otimes_A B) \otimes_B B^* & \xrightarrow{\;\sim\;} & B \otimes_A B^* \xrightarrow{\;\eta_{B,B}\;} \text{alHom}_B(B \otimes_A B, B)
\end{array}
$$

in which all the morphisms are $B \otimes_A B$-linear (in the left column we take the B-module structure on $B \otimes_A B$ given by multiplication on the right factor). Suppose now that ϕ is étale; then, by corollary 3.1.9, there is an isomorphism of B-algebras: $B \otimes_A B \simeq I_{B/A} \oplus B$. It follows that $\tau_{B \otimes_A B/B} = \tau_{B/B} \oplus \tau_{I_{B/A}/B}$. Especially, $1_B \otimes_{B \otimes_A B} \tau_{B \otimes_A B/B}$ is the identity morphism of B (by proposition 4.1.8(i)). This means that in the diagram $B \otimes_{B \otimes_A B}$ (4.1.15) all the arrows are isomorphisms. In particular, $\tau_{B/A}$ is an isomorphism, as claimed.

To prove the converse, we consider the almost element ζ_B of the $B \otimes_A B$-module $B \otimes_A B^*$. Viewing B^* as a B-module in the natural way (cp. remark 2.4.25(i)), we also get a scalar multiplication morphism $\sigma_{B^*/B} : B \otimes_A B^* \to B^*$ (see (2.2.6)).

Claim 4.1.16. With the above notation we have:

$$I_{B/A} \cdot \zeta_B = 0 \quad \text{and} \quad \sigma_{B^*/B}(\zeta_B) = \text{Tr}_{B/A}.$$

Proof of the claim. Notice that $\omega_{B/A}$ is also $B \otimes_A B$-linear for the $B \otimes_A B$-module structure on $\text{End}_A(B)$ such that $((b \otimes b') \cdot \phi)(b'') = b' \cdot \phi(b \cdot b'')$ for every $b, b', b'' \in B_*$ and every $\phi \in \text{End}_A(B)$. We compute:

$$\omega_{B/A}((b \otimes b') \cdot \zeta_B)(b'') = ((b \otimes b') \cdot \omega_{B/A}(\zeta_B))(b'') = b \cdot b' \cdot b''.$$

Whence $x \cdot \zeta_B = (\mu_{B/A}(x) \otimes 1) \cdot \zeta_B$ for every $x \in B \otimes_A B_*$ which implies the first claimed identity. Next, for every $b \in B_*$ we compute:

$$\sigma_{B^*/B}(\zeta_B)(b) = \mathrm{ev}_B((1 \otimes b) \cdot \zeta_B) = (\mathrm{tr}_{B/A} \circ \omega_{B/A})((1 \otimes b) \cdot \zeta_B)$$
$$= \mathrm{tr}_{B/A}((1 \otimes b) \cdot \omega_{B/A}(\zeta_B)) = \mathrm{tr}_{B/A}((1 \otimes b) \cdot 1_B) = \mathrm{Tr}_{B/A}(b).$$

The claim follows. $\diamond$

Now, if $\tau_{B/A}$ is an isomorphism, we can define $e := (1_B \otimes \tau_{B/A}^{-1})(\zeta_B)$. From claim 4.1.16 and lemma 4.1.13(i) we derive that $I_{B/A} \cdot e = 0$ and $\tau_{B/A}(\sigma_{B/B}(e)) = \mathrm{Tr}_{B/A}$. The latter equality implies that $\sigma_{B/B}(e) = 1$, in other words $\mu_{B/A}(e) = 1$. We see therefore that e satisfies conditions (ii) and (iii) of proposition 3.1.4 and therefore also condition (i), since the latter is an easy consequence of the other two. Thus $A \to B$ is an étale morphism, as claimed. $\square$

Remark 4.1.17. By inspection of the proof of theorem 4.1.14, we see that the following has been shown. Let $A \to B$ be an almost finite projective étale morphism of V^a-algebras. Then $(1_B \otimes \tau_{B/A})(e_{B/A}) = \zeta_B$.

Definition 4.1.18. The *nilradical* of an almost algebra A is the ideal $\mathrm{nil}(A) = \mathrm{nil}(A_*)^a$ (where, for a ring R, we denote by $\mathrm{nil}(R)$ the ideal of nilpotent elements in R). We say that A is *reduced* if $\mathrm{nil}(A) \simeq 0$.

4.1.19. Notice that, if R is a V-algebra, then every nilpotent ideal in R^a is of the form I^a, where I is a nilpotent ideal in R (indeed, it is of the form I^a where I is an ideal, and $\mathrm{m} \cdot I$ is seen to be nilpotent). It follows easily that $\mathrm{nil}(A)$ is the colimit of the nilpotent ideals in A; moreover $\mathrm{nil}(R)^a = \mathrm{nil}(R^a)$. Using this one sees that $A/\mathrm{nil}(A)$ is reduced.

Proposition 4.1.20. *Let $A \to B$ be an étale almost finite projective morphism of almost algebras. If A is reduced then B is reduced as well.*

Proof. For given $\varepsilon \in \mathrm{m}$, pick a sequence of morphisms $B \overset{u_\varepsilon}{\to} A^n \overset{v_\varepsilon}{\to} B$ such that $v_\varepsilon \circ u_\varepsilon = \varepsilon \cdot 1_B$; let $\mu_b : B \to B$ be multiplication by $b \in B_*$ and define $\nu_b : A^n \to A^n$ by $\nu_b = v_\varepsilon \circ \mu_b \circ u_\varepsilon$. One verifies easily that $\nu_b^m = \varepsilon^{m-1} \cdot \nu_{b^m}$ for all integers $m > 0$. Now, suppose that $b \in \mathrm{nil}(B_*)$. It follows that $b^m = 0$ for m sufficiently large, hence $\nu_b^m = 0$ for m sufficiently large. Let $\mathfrak{p}$ be any prime ideal of A_*; let $\pi : A_* \to A_*/\mathfrak{p}$ be the natural projection and F the fraction field of $A_*/\mathfrak{p}$. The F-linear morphism $\nu_{b*} \otimes_{A_*} 1_F$ is nilpotent on the vector space F^n, hence $\pi \circ \mathrm{tr}_{A_*^n/A_*}(\nu_{b*}) = \mathrm{tr}_{F^n/F}(\nu_{b*} \otimes_{A_*} 1_F) = 0$. This shows that $\mathrm{tr}_{A_*^n/A_*}(\nu_{b*})$ lies in the intersection of all prime ideals of A_*, hence it is nilpotent. Since by hypothesis A is reduced, we get $\mathrm{tr}_{A_*^n/A_*}(\nu_{b*}) = 0$, whence $\mathrm{tr}_{A^n/A}(\nu_b) = 0$. Using lemma 4.1.2(i) we deduce $\varepsilon \cdot \mathrm{tr}_{B/A}(\mu_b) = 0$, and finally, $\mathrm{tr}_{B/A}(b) = 0$. Now, for any $b' \in B_*$, the almost element bb' will be nilpotent as well, so the same conclusion applies to it. This shows that $\tau_{B/A}(b) = 0$. But by hypothesis B is étale over A, hence theorem 4.1.14 yields $b = 0$, as required. $\square$

Remark 4.1.21. Let M be an A-module. We say that an almost element a of A is *M-regular* if the multiplication morphism $m \mapsto am : M \to M$ is a monomorphism. Assume **(A)** (see (2.1.6)) and suppose furthermore that m is generated by a

multiplicative system $\mathscr{S}$ which is a cofiltered semigroup under the preorder structure $(\mathscr{S}, \succ)$ induced by the divisibility relation in V. We say that $\mathscr{S}$ is *archimedean*
if, for all $s, t \in \mathscr{S}$ there exists $n > 0$ such that $s^n \succ t$. Suppose that $\mathscr{S}$ is
archimedean and that A is a reduced almost algebra. Then $\mathscr{S}$ consists of A-regular
elements. Indeed, by hypothesis $\mathrm{nil}(A_*)^a = 0$; since the annihilator of $\mathscr{S}$ in A_* is
0 we get $\mathrm{nil}(A_*) = 0$. Suppose that $s \cdot a = 0$ for some $s \in \mathscr{S}$ and $a \in A_*$. Let
$t \in \mathscr{S}$ be arbitrary and pick $n > 0$ such that $t^n \succ s$. Then $(ta)^n = 0$ hence $ta = 0$
for all $t \in \mathscr{S}$, hence $a = 0$.

Definition 4.1.22. Let $\phi : A \to B$ be an almost finite projective morphism of V^a-
algebras. By (4.1.12), we can assign to ϕ a B-linear trace morphism $\tau_{B/A} : B \to$
B^*. The *different ideal* of the morphism ϕ is the ideal

$$\mathscr{D}_{B/A} := \mathrm{Ann}_B(\mathrm{Coker}\,\tau_{B/A}) \subset B.$$

Lemma 4.1.23. (i) *Let M be an invertible A-module. The map*

$$\mathscr{I}_A(A) \to \mathscr{I}_A(M) \quad : \quad I \mapsto IM$$

is an isomorphism of uniform spaces. Its inverse is the map : $N \mapsto \mathrm{Ann}_A(M/N)$.

(ii) *Let $M_1 \xrightarrow{\phi} M_2 \xrightarrow{\psi} M_3$ be two A-linear morphisms of invertible A-modules*
  M_i $(i \leq 3)$ and C an A-algebra. Then:
  (a) $\mathrm{Ann}_A(\mathrm{Coker}(\psi \circ \phi)) = \mathrm{Ann}_A(\mathrm{Coker}\,\phi) \cdot \mathrm{Ann}_A(\mathrm{Coker}\,\psi)$.
  (b) $\mathrm{Ann}_C(\mathrm{Coker}(1_C \otimes_A \psi)) = C \cdot \mathrm{Ann}_A(\mathrm{Coker}\,\psi)$.
  (c) $\mathrm{Ann}_A(\mathrm{Coker}\,\phi) = \mathrm{Ann}_A(\mathrm{Coker}\,\phi^* : M_2^* \to M_1^*)$.

Proof. (i): It is easy to check that the given map is uniformly continuous (see definition 2.3.1(i)), and it admits a uniformly continuous left inverse given by the stated
rule. Hence it remains only to verify the surjectivity of the map. Pick an isomor-
phism $\alpha : M \otimes_A M^* \xrightarrow{\sim} A$, and let $N \subset M$; then $\alpha(N \otimes_A M^*)$ is an ideal I.
However, $I = \alpha(IM \otimes_A M^*)$; since M^* is faithfully flat (lemma 4.1.5(iii)), it
follows that $N = IM$.

(ii.a): According to (i) we can write $\mathrm{Im}(\phi) = IM_2$ and $\mathrm{Im}(\psi) = JM_3$ for
ideals $I, J \subset A$; therefore $\mathrm{Im}(\psi \circ \phi) = IJM_3$, and the claim follows easily.

(ii.b): Let again $\mathrm{Im}(\psi) = JM_3$; then $\mathrm{Im}(1_C \otimes_A \psi) = JC \otimes_A M_3$, and the
claim follows easily.

(ii.c): Indeed, let $J := \mathrm{Ann}_A(\mathrm{Coker}\,\phi)$; by (i) we have $JM_2 = \mathrm{Im}\,\phi$, hence
$\phi \otimes_A 1_{A/J} = 0$ and therefore $0 = (\phi \otimes_A 1_{A/J})^* = \phi^* \otimes_A 1_{A/J}$, so that
$\mathrm{Ann}_A(\mathrm{Coker}\,\phi^*) \subset J$. By symmetry, the converse inclusion holds as well, which
is the claim. $\qquad\qquad\square$

Lemma 4.1.24. *Let $\phi : A \to B$ be a morphism of V^a-algebras as in definition*
4.1.22. Let C be an A-algebra. Suppose that either C is flat over A, or B^* is an
invertible B-module for its natural B-module structure. Then:

$$\mathscr{D}_{C \otimes_A B/C} = \mathscr{D}_{B/A} \cdot (C \otimes_A B).$$

Proof. Under the stated assumptions, $\mathrm{alHom}_A(B, A)$ is an almost finitely generated
projective A-module. In particular, $\mathrm{Coker}\,\tau_{B/A}$ is almost finitely generated; If C

is flat over A, it follows by lemma 2.4.6 that $\operatorname{Ann}_{C \otimes_A B}(C \otimes_A \operatorname{Coker} \tau_{B/A}) = \mathscr{D}_{B/A} \cdot (C \otimes_A B)$; if B^* is an invertible B-module, the same holds by virtue of lemma 4.1.23(ii.b). However, by lemma 4.1.13(iii), the trace pairing is preserved under arbitrary base changes, so: $C \otimes_A \operatorname{Coker} \tau_{B/A} \simeq \operatorname{Coker}(\mathbf{1}_C \otimes_A \tau_{B/A}) \simeq \operatorname{Coker} \tau_{C \otimes_A B/C}$, which shows the claim. $\square$

Proposition 4.1.25. *Let $B \to C$ be a morphism of A-algebras. Suppose that B (resp. C) is an almost finite projective A-algebra (resp. B-algebra) and that $B^* := \operatorname{alHom}_A(B, A)$ (resp. $C^* := \operatorname{alHom}_B(C, B)$) is an invertible B-module (resp. C-module) for its natural B-module (resp. C-module) structure. Then*

$$\mathscr{D}_{C/A} = \mathscr{D}_{C/B} \cdot \mathscr{D}_{B/A}.$$

Proof. Let $C^*_{/A} := \operatorname{alHom}_A(C, A)$; following remark 2.4.25(ii) we obtain a C-linear isomorphism

$$\xi : \operatorname{alHom}_B(C, B^*) \xrightarrow{\sim} C^*_{/A}$$

given by the rule: $\phi \mapsto (c \mapsto \phi(c)(1))$ for every $\phi \in \operatorname{Hom}_B(C, B^*)$ and $c \in C_*$.

Claim 4.1.26. $C^*_{/A}$ is an invertible C-module.

Proof of the claim. By lemma 2.4.29(i), the natural morphism $\lambda : C^* \otimes_B B^* \to \operatorname{alHom}_B(C, B^*)$ is a C-linear isomorphism. After composing with ξ we deduce a C-linear isomorphism: $C^* \otimes_B B^* \xrightarrow{\sim} C^*_{/A}$, whence the claim. $\diamond$

Unwinding the definitions, one verifies that the following diagram commutes:

$$
\begin{array}{ccc}
C & \xrightarrow{\;\tau_{C/A}\;} & C^*_{/A} \\
{\scriptstyle \tau_{C/B}}\downarrow & & \uparrow{\scriptstyle \xi} \\
C^* & \xrightarrow{\operatorname{alHom}_B(C, \tau_{B/A})} & \operatorname{alHom}_B(C, B^*).
\end{array}
$$

Thus, taking into account claim 4.1.26, and lemma 4.1.23(ii.a), we have

$$\operatorname{Ann}_C(\operatorname{Coker} \tau_{C/A}) = \operatorname{Ann}_C(\operatorname{Coker} \tau_{C/B}) \cdot \operatorname{Ann}_C(\operatorname{Coker}(\operatorname{alHom}_B(C, \tau_{B/A}))).$$

However:

$$\operatorname{Coker}(\operatorname{alHom}_B(C, \tau_{B/A})) \simeq C^* \otimes_B \operatorname{Coker} \tau_{B/A}$$
$$\simeq C^* \otimes_C (C \otimes_B \operatorname{Coker} \tau_{B/A})$$

by lemma 2.4.29(i.b). By lemma 4.1.5(iii), C^* is a faithfully flat C-module; hence:

$$\operatorname{Ann}_C(\operatorname{Coker}(\operatorname{alHom}_B(C, \tau_{B/A}))) = \operatorname{Ann}_C(C \otimes_B \operatorname{Coker} \tau_{B/A})$$
$$= C \cdot \operatorname{Ann}_B(\operatorname{Coker} \tau_{B/A})$$

by lemma 4.1.23(ii.b). The assertion follows. $\square$

Lemma 4.1.27. *Let $\phi : A \to B$ be an almost finite projective morphism of V^a-algebras. Then ϕ is étale if and only if $\mathscr{D}_{B/A} = B$.*

Proof. By theorem 4.1.14 it follows easily that $\mathscr{D}_{B/A} = B$ whenever ϕ is étale. Conversely, suppose that $\mathscr{D}_{B/A} = B$; it then follows that $\tau_{B/A}$ is an epimorphism, and one derives an isomorphism of B-modules: $B^* \simeq B/\mathrm{Ker}\,\tau_{B/A}$. However, the assumption on B implies that the natural morphism $B \to (B^*)^*$ is an isomorphism of B-modules; especially, $\mathrm{Ann}_B B^* = 0$, therefore $\mathrm{Ker}\,\tau_{B/A} = 0$, *i.e.* $\tau_{B/A}$ is an isomorphism, and we conclude again by theorem 4.1.14. $\qquad\square$

The following lemma will be useful when we will compute the different ideal in situations such as those contemplated in proposition 6.3.13.

Lemma 4.1.28. *Let A be a V^a-algebra, B an almost finite almost projective A-algebra, and let $\{B_\alpha \mid \alpha \in J\}$ be a net of A-subalgebras of B, with B_α almost finite projective over A for every $\alpha \in J$, such that $\lim_{\alpha \in J} B_\alpha = B$ in $\mathscr{I}_A(B)$. Then*

$$\lim_{\alpha \in J} \mathscr{D}_{B_\alpha/A} = \mathscr{D}_{B/A}.$$

Proof. For given $\alpha \in J$, let $\varepsilon \in V$ such that $\varepsilon B \subset B_\alpha$; lemma 4.1.2(ii) implies that $\varepsilon \cdot \mathrm{Tr}_{B_\alpha/A}(b) = \varepsilon \cdot \mathrm{Tr}_{B/A}(b)$ for every $b \in B_{\alpha*}$. Hence the diagrams:

$$
\begin{array}{ccc}
B & \xrightarrow{\ \mu_\varepsilon\ } & B_\alpha \\
{\scriptstyle \varepsilon \cdot \tau_{B/A}}\big\downarrow & & \big\downarrow{\scriptstyle \tau_{B_\alpha/A}} \\
B^* & \longrightarrow & B_\alpha^*
\end{array}
\qquad
\begin{array}{ccc}
B_\alpha & \longrightarrow & B \\
{\scriptstyle \tau_{B_\alpha/A}}\big\downarrow & & \big\downarrow{\scriptstyle \varepsilon \cdot \tau_{B/A}} \\
B_\alpha^* & \xrightarrow{\ \mu_\varepsilon^*\ } & B^*
\end{array}
$$

commute. The rightmost diagram implies that $\mathscr{D}_{B_\alpha/A} \cdot \mathrm{Im}\,\mu_\varepsilon^* \subset \mathrm{Im}(\varepsilon \cdot \tau_{B/A}) \subset \mathrm{Im}\,\tau_{B/A}$. Hence $\varepsilon \cdot \mathscr{D}_{B_\alpha/A} \subset \mathscr{D}_{B/A}$, so finally $\mathrm{Ann}_V(B/B_\alpha) \cdot \mathscr{D}_{B_\alpha/A} \subset \mathscr{D}_{B/A}$. From the leftmost diagram we deduce that $\varepsilon \cdot \mathscr{D}_{B/A}$ (which is an ideal in B_α) annihilates $\mathrm{Coker}(\varepsilon \cdot \tau_{B/A} : B \to B^*)$ and on the other hand $\mathrm{Ann}_V(B/B_\alpha)$ obviously annihilates $\mathrm{Coker}(B^* \to B_\alpha^*)$; we deduce that $\mathrm{Ann}_V(B/B_\alpha)^2 \cdot \mathscr{D}_{B/A} \subset \mathscr{D}_{B_\alpha/A}$, whence the claim. $\qquad\square$

4.2. Endomorphisms of $\widehat{\mathbb{G}}_m$.

This section is dedicated to a discussion of the universal ring that classifies endomorphisms of the formal group $\widehat{\mathbb{G}}_m$. The results of this section will be used in sections 4.3 and 4.4.

4.2.1.   For every ring R and every integer $n \geq 0$ we introduce the "n-truncated" version of $\widehat{\mathbb{G}}_{m,R}$. This is the scheme $\mathbb{G}_{m,R}(n) := \mathrm{Spec}\, R[T]/(T^{n+1})$, endowed with a multiplication morphism μ which makes it a group object in the category of R-schemes equipped with a quasi-coherent ideal whose $(n+1)$-th power is 0; namely, μ is the morphism of affine schemes associated to the co-multiplication map

$$R[T]/(T^{n+1}) \to R[T,S]/(T,S)^{n+1} \qquad T \mapsto T + S + T \cdot S.$$

Then in the category of formal schemes we have a natural identification

$$\widehat{\mathbb{G}}_{m,R} \simeq \operatorname*{colim}_{n \in \mathbb{N}} \mathbb{G}_{m,R}(n).$$

4.2.2. In the terminology of [56, §II.4], $\mathbb{G}_m(n)$ is the n-bud of $\widehat{\mathbb{G}}_m$. We will be mainly interested in the endomorphisms of $\mathbb{G}_m(n)$, but before we can get to that, we will need some complements on buds over artinian rings. Therefore, suppose we have a cartesian diagram of artinian rings

(4.2.3)
$$\begin{array}{ccc} R_3 & \longrightarrow & R_1 \\ \downarrow & & \downarrow \\ R_2 & \longrightarrow & R_0 \end{array}$$

such that one of the two maps $R_i \to R_0$ ($i = 1, 2$) is surjective. For any ring R, we define the category $\mathbf{Bud}(n, d, R)$ of n-buds over R whose underlying R-algebra is isomorphic to $R[T_1, ..., T_d]/(T_1, ..., T_d)^{n+1}$.

Lemma 4.2.4. *Let S be a finite flat augmented algebra over a local noetherian ring R; let I be the augmentation ideal, and suppose that $I^{n+1} = 0$. Let κ be the residue field of R, and suppose that $S \otimes_R \kappa \simeq \kappa[t_1, ..., t_d]/(t_1, ..., t_d)^{n+1}$. Then $S \simeq R[T_1, ..., T_d]/(T_1, ..., T_d)^{n+1}$.*

Proof. Let $\varepsilon : S \to R$ be the augmentation map. For every $i = 1, ..., d$, pick a lifting $T_i' \in S$ of t_i; set $T_i = T_i' - \varepsilon(T_i')$. By Nakayama's lemma, the monomials $T_1^{a_1} \cdot ... \cdot T_d^{a_d}$ with $\sum_{i=1}^d a_i \leq n$ generate the R-module S. Furthermore, under the stated hypothesis, S is a free R-module, and its rank is equal to $\dim_\kappa S \otimes_R \kappa$; hence the above monomials form an R-basis of S. Clearly the elements T_i lie in the augmentation ideal of S, therefore every product of $n + 1$ of them equals zero; in other words, the natural morphism $R[X_1, ..., X_d] \to S$ given by $X_i \mapsto T_i$ is surjective, with kernel containing $J := (X_1, ..., X_d)^{n+1}$; but by comparing the ranks over R we see that this kernel cannot be larger than J. The assertion follows. $\square$

Proposition 4.2.5. *In the situation of (4.2.3), the natural functor*

$$\mathbf{Bud}(n, d, R_3) \to \mathbf{Bud}(n, d, R_1) \times_{\mathbf{Bud}(n,d,R_0)} \mathbf{Bud}(n, d, R_2)$$

is an equivalence of categories.

Proof. Let $\mathcal{M}_{i,\mathrm{proj}}$ ($i = 0, ..., 3$) be the category of projective R_i-modules. By our previous discussion on descent, we already know that (4.2.3) induces a natural equivalence between $\mathcal{M}_{3,\mathrm{proj}}$ and the 2-fibered product $\mathcal{M}_{1,\mathrm{proj}} \times_{\mathcal{M}_{0,\mathrm{proj}}} \mathcal{M}_{2,\mathrm{proj}}$. It is easy to see that this equivalence respects the rank of R_i-modules, hence induces a similar equivalence for the categories $\mathcal{M}_{i,r}$ of free R_i-modules of finite rank r, for every $r \in \mathbb{N}$. Given two objects $M := (M_1, M_2, \alpha : M_1 \otimes_{R_1} R_0 \xrightarrow{\sim} M_2 \otimes_{R_2} R_0)$ and $N := (N_1, N_2, \beta : N_1 \otimes_{R_1} R_0 \xrightarrow{\sim} N_2 \otimes_{R_2} R_0)$, define the tensor product $M \otimes N := (M_1 \otimes_{R_1} N_1, M_2 \otimes_{R_2} N_2, \alpha \otimes_{R_0} \beta)$. Then one checks easily that the above equivalences respect tensor products. It follows formally that one has analogous equivalences for the categories of finite flat R_i-algebras. From there, one further obtains equivalences on the categories of such R_i-algebras that are augmented over R_i, and even on the subcategories $R_i\text{-}\mathbf{Alg}_{\mathrm{aug.fl.}}^{(n)}$ of those augmented

R_i-algebras such that the $(n+1)$-th power of the augmentation ideal vanishes. These categories admit finite coproducts, that are constructed as follows. For augmented R_i-algebras $\varepsilon_A : A \to R_i$ and $\varepsilon_B : B \to R_i$, set

$$(A \to R_i) \otimes (B \to R_i) := (A \otimes_{R_i} B/\mathrm{Ker}(\varepsilon_A \otimes_{R_i} \varepsilon_B)^{n+1} \to R_i).$$

This is a coproduct of A and B. By formal reasons, the foregoing equivalences of categories respect these coproducts. Finally, an object of $\mathbf{Bud}(n, d, R_i)$ can be defined as a commutative group object in $(R_i\text{-}\mathbf{Alg}^{(n)}_{\mathrm{aug.fl.}})^o$, such that its underlying R_i-algebra is isomorphic to $R_i[T_1, ..., T_d]/(T_1, ..., T_d)^{n+1}$. By formal categorical considerations we see that the foregoing equivalence induces equivalences on the commutative group objects in the respective categories. It remains to check that an R_3-algebra S such that $S \otimes_{R_3} R_i \simeq R_i[T_1, ..., T_d]/(T_1, ..., T_d)^{n+1}$, (for $i = 1, 2$) is itself of the form $R_i[T_1, ..., T_d]/(T_1, ..., T_d)^{n+1}$. However, this follows readily from lemma 4.2.4 and the fact that one of the maps $R_3 \to R_i$ $(i = 1, 2)$ is surjective. $\square$

4.2.6.   For a given ring R, the endomorphisms of $\mathbb{G}_{m,R}(n)$ (as a group object in a category as in (4.2.1)) are all the polynomials $f(T) := a_1 \cdot T + ... + a_n \cdot T^n$ such that $f(T) + f(S) + f(T) \cdot f(S) \equiv f(T + S + T \cdot S) \pmod{(T, S)^{n+1}}$. This relationship translates into a finite set of polynomial identities for the coefficients $a_1, ..., a_n$, and using these identities we can therefore define a quotient $\mathscr{G}_n$ of the ring in n indeterminates $\mathbb{Z}[X_1, ..., X_n]$ which will be the "universal ring of endomorphisms" of $\mathbb{G}_m(n)$, i.e., such that $X_1 \cdot T + X_2 \cdot T^2 + ... + X_n \cdot T^n$ is an endomorphism of $\mathbb{G}_{m,\mathscr{G}_n}(n)$ and such that, for every ring R, and every $f(T)$ as above, the map $\mathbb{Z}[X_1, ..., X_n] \to R$ given by $X_i \mapsto a_i$ $(i = 1, ..., n)$ factors through a (necessarily unique) map $\mathscr{G}_n \to R$. One of the main results of this section will be a simple and explicit description of the ring $\mathscr{G}_n$.

Proposition 4.2.7. *$\mathscr{G}_n$ is a smooth $\mathbb{Z}$-algebra.*

Proof. We know already that $\mathscr{G}_n$ is of finite type over $\mathbb{Z}$, therefore it suffices to show that, for every prime ideal $\mathfrak{p}$ of $\mathscr{G}_n$, the local ring $\mathscr{G}_{n,\mathfrak{p}}$ is formally smooth over $\mathbb{Z}$ for the $\mathfrak{p}$-adic topology (see [30, Ch.IV, Prop.17.5.3]). Therefore, let $R_1 \to R_0$ be a surjective homomorphism of local artinian rings; we need to show that the natural map $\mathrm{End}(\mathbb{G}_{m,R_1}(n)) \to \mathrm{End}(\mathbb{G}_{m,R_0}(n))$ is surjective. Let $f \in \mathrm{End}(\mathbb{G}_{m,R_0}(n))$; we define an automorphism χ of

$$\mathbb{G}_{m,R_0}(n) \times \mathbb{G}_{m,R_0}(n) := \mathrm{Spec}\, R_0[T, S]/(T, S)^{n+1}$$

the n-bud of $\widehat{\mathbb{G}}_m \times \widehat{\mathbb{G}}_m$, by setting $(T, S) \mapsto (T, f(T) + S + f(T) \cdot S)$. Then, thanks to proposition 4.2.5, we obtain an n-bud X_n over $R_2 := R_1 \times_{R_0} R_1$, by gluing two copies of $\mathbb{G}_{m,R_1}(n) \times \mathbb{G}_{m,R_1}(n)$ along the automorphism χ.

Claim 4.2.8.   The n-bud X_n is isomorphic to $\mathbb{G}_{m,R_2}(n) \times \mathbb{G}_{m,R_2}(n)$ if and only if χ lifts to an automorphism of $\mathbb{G}_{m,R_1}(n) \times \mathbb{G}_{m,R_1}(n)$.

Proof of the claim. Taking into account the decription of $B(n, d, R_2)$ as 2-fibered product of categories, the proof amounts to a simple formal verification, which is best left to the reader. $\Diamond$

Claim 4.2.9. Fix one of the two natural quotient maps $\phi : R_2 \to R_1$. There exists a compatible system of k-buds X_k over R_2 for every $k > n$, such that X_k reduces to X_{k-1} over R_2, and specializes to $\mathbb{G}^2_{m,R_1}(k)$ under the base change ϕ.

Proof of the claim. We can assume that $n > 0$. In case R_2 is a torsion-free $\mathbb{Z}$-algebra, this follows from [56, Ch.II, §4.10] and an easy induction. If R_2 is a general artinian ring, choose a torsion-free $\mathbb{Z}$-algebra R_3 with a surjective homomorphism $R_3 \to R_2$. By *loc. cit.* (and an easy induction) we can find an n-bud Y_n over R_3 such that Y_n specializes to X_n on the quotient R_2, and Y_n reduces to $\mathbb{G}^2_{m,R_3}(1)$ over R_3. Then, again by *loc. cit*, we can find a compatible system of k-buds Y_k on R_3 for every $k > n$, such that Y_k reduces to Y_{k-1} over R_3 and specializes to $\mathbb{G}^2_{m,R_1}(k)$ over the quotient R_1 of R_3. The claim holds if we take X_k equal to the specialization of Y_k over R_2. $\Diamond$

The direct limit (in the category of formal schemes) of the system $(X_k)_{k \geq n}$ is a formal group $\widehat{X}$ over R_2, such that $\widehat{X} \otimes_{R_2} R_1 \simeq \widehat{\mathbb{G}}_{m,R_1} \times \widehat{\mathbb{G}}_{m,R_1}$. This formal group gives rise to a p-divisible group $(\widehat{X}(n))_{n \geq 0}$, where $\widehat{X}(n)$ is the kernel of multiplication by p^n in $\widehat{X}$. For every $m \in \mathbb{N}$, $\widehat{X}(m)$ is a finite flat group scheme over R_2, such that $\widehat{X}(m) \times_{R_2} R_1 \simeq \mu_{p^m,R_1} \times \mu_{p^m,R_1}$. Denote by $\widehat{X}(m)^*$ the Cartier dual of $\widehat{X}(m)$ (cp. [61, §III.14]). Then $\widehat{X}(m)^* \times_{R_2} R_1 \simeq (\mathbb{Z}/p^m\mathbb{Z})^2_{R_1}$, in particular it has p^{2m} connected components. Since the pair (R_2, R_1) is henselian, it follows that $\widehat{X}(m)^*$ must have p^{2m} connected components as well, and consequently $\widehat{X}(m)^* \simeq (\mathbb{Z}/p^m\mathbb{Z})^2_{R_2}$. Finally, this shows that $\widehat{X}(m) \simeq \mu_{p^m,R_2} \times \mu_{p^m,R_2}$, whence $\widehat{X} \simeq \widehat{\mathbb{G}}_{m,R_2} \times \widehat{\mathbb{G}}_{m,R_2}$. From claim 4.2.8, we deduce that χ lifts to an automorphism χ' of $\mathbb{G}_{m,R_1}(n) \times \mathbb{G}_{m,R_1}(n)$. Let $i : \mathbb{G}_{m,R_1}(n) \to \mathbb{G}_{m,R_1}(n) \times \mathbb{G}_{m,R_1}(n)$ and $\pi : \mathbb{G}_{m,R_1}(n) \times \mathbb{G}_{m,R_1}(n) \to \mathbb{G}_{m,R_1}(n)$ be respectively the imbedding of the first factor, and the projection onto the second factor; clearly $\pi \circ \chi' \circ i$ yields a lifting of $f(T)$, as required. $\square$

4.2.10.   Next, let us remark that, for every $n \in \mathbb{N}$, the polynomial $(1+T)^X - 1 :=$ $X \cdot T + \binom{X}{2} \cdot T^2 + \ldots + \binom{X}{n} \cdot T^n \in \mathbb{Q}[X,T]$ is an endomorphism of $\mathbb{G}_{m,\mathbb{Q}[X]}(n)$. As a consequence, there is a unique ring homomorphism $\mathscr{G}_n \to \mathbb{Z}[X, \binom{X}{2}, \ldots, \binom{X}{n}]$ representing this endomorphism. The following theorem will show that this homomorphism is an isomorphism.

Theorem 4.2.11. *The functor*

$$\mathbb{Z}\text{-}\mathbf{Alg} \to \mathbf{Set} \qquad R \mapsto \mathrm{End}_R(\mathbb{G}_{m,R}(n))$$

is represented by the ring $\mathbb{Z}[X, \binom{X}{2}, \ldots, \binom{X}{n}]$.

Proof. The above discussion has already furnished us with a natural surjective map $\rho : \mathscr{G}_n \to \mathbb{Z}[X, \binom{X}{2}, \ldots, \binom{X}{n}]$. Thus, it suffices to show that this map is injective.

Claim 4.2.12. $\rho \otimes_{\mathbb{Z}} \mathbf{1}_{\mathbb{Q}}$ is an isomorphism.

Proof of the claim. First of all, the map ρ can be characterized in the following way. The identity map $\mathscr{G}_n \to \mathscr{G}_n$ determines an endomorphism $f(T) := a_1 \cdot T +$

... $+ a_n \cdot T^n$ of $\mathbb{G}_{m,\mathscr{G}_n}(n)$; then ρ is the unique ring homomorphism such that $\rho(f) := f(a_1) \cdot T + ... + f(a_n) \cdot T^n = (1 + T)^X - 1$. On the other hand, the ring $\mathscr{G}_n \otimes_\mathbb{Z} \mathbb{Q}$ represents endomorphisms of $\mathbb{G}_m(n)$ in the category of $\mathbb{Q}$-algebras. However, for every $n \in \mathbb{N}$ and for every $\mathbb{Q}$-algebra R, there is an isomorphism

$$\log : \mathbb{G}_{m,R}(n) \xrightarrow{\sim} \mathbb{G}_{a,R}(n)$$

to the n-bud of the additive formal group $\widehat{\mathbb{G}}_{a,R}$. We can assume that $n > 0$, and in that case the endomorphism group of $\mathbb{G}_{a,R}(n)$ is easily computed, and found to be isomorphic to R. In other words, the universal ring representing endomorphisms of $\mathbb{G}_a(n)$ over $\mathbb{Q}$-algebras is just $\mathbb{Q}[X]$ when $n > 0$, and the bijection $\mathrm{Hom}_{\mathbb{Q}\text{-}\mathbf{Alg}}(\mathbb{Q}[X], R) \simeq \mathrm{End}(\mathbb{G}_{a,R}(n))$ assigns to a homomorphism $\phi : \mathbb{Q}[X] \to R$, the endomorphism $g_\phi(T) := \phi(X) \cdot T$. It follows that, for any $\mathbb{Q}$-algebra R there is a natural bijection $\mathrm{Hom}_{\mathbb{Q}\text{-}\mathbf{Alg}}(\mathbb{Q}[X], R) \simeq \mathrm{End}(\mathbb{G}_{m,R}(n))$ given by:

$$(\phi : \mathbb{Q}[X] \to R) \mapsto \exp(\phi(X) \cdot \log(1 + T)) - 1 = (1 + T)^{\phi(X)} - 1.$$

Especially, $f(T)$ can be written in the form $(1 + T)^{\psi(X)} - 1$ for a unique ring homomorphism $\psi : \mathbb{Q}[X] \to \mathscr{G}_n \otimes_\mathbb{Z} \mathbb{Q}$. Clearly ψ is inverse to $\rho \otimes_\mathbb{Z} 1_\mathbb{Q}$. $\diamond$

In view of claim 4.2.12, we are thus reduced to showing that $\mathscr{G}_n$ is a flat $\mathbb{Z}$-algebra, which follows readily from proposition 4.2.7. $\square$

4.2.13. Furthermore, $\mathscr{G}_n$ is endowed with a co-addition, *i.e.* a ring homomomorphism $\mathscr{G}_n \to \mathscr{G}_n \otimes_\mathbb{Z} \mathscr{G}_n$ satisfying the usual co-associativity and co-commutativity conditions. The co-addition is given by the rule:

$$\mathrm{coadd} : \mathscr{G}_n \to \mathscr{G}_n \otimes_\mathbb{Z} \mathscr{G}_n \qquad \binom{X}{k} \mapsto \sum_{i+j=k} \binom{X}{i} \otimes \binom{X}{j}.$$

Moreover, for every $k \in \mathbb{Z}$, we have a ring homomorphism $\pi_k : \mathscr{G}_n \to \mathbb{Z}$, which corresponds to the endomorphism of $\mathbb{G}_{m,\mathbb{Z}}(n)$ given by the rule: $T \mapsto (1+T)^k - 1$ (raising to the k-th power in $\mathbb{G}_{m,\mathbb{Z}}(n)$). Hence we derive, for every $k \in \mathbb{Z}$, a ring homomorphism

$$(4.2.14) \qquad \mathscr{G}_n \xrightarrow{\mathrm{coadd}} \mathscr{G}_n \otimes_\mathbb{Z} \mathscr{G}_n \xrightarrow{1_{\mathscr{G}_n} \otimes \pi_k} \mathscr{G}_n.$$

Remark 4.2.15. (i) Suppose $n > 0$; on $\mathscr{G}_n \otimes_\mathbb{Z} \mathbb{Q} = \mathbb{Q}[X]$, (4.2.14) is the unique map such that $\binom{X}{i} \mapsto \binom{X+k}{i}$ for all $i \le n$, therefore we see that $\binom{X+k}{i} \in \mathscr{G}_n$ for all $k \in \mathbb{Z}$, $n > 0$ and $0 \le i \le n$. Moreover, (4.2.14) is clearly an automorphism for every $k \in \mathbb{Z}$.

(ii) It is also interesting (though it will not be needed in this work) to remark that $\mathscr{G}_n$ is endowed additionally with a co-composition structure, so that $\mathscr{G}_n$ is actually a co-ring, and it represents the functor $R \mapsto \mathrm{End}(\mathbb{G}_{m,R}(n))$ from $\mathbb{Z}$-algebras to unitary rings. One can check that the co-composition map is given by the rule:

$$\binom{X}{k} \mapsto \sum_\phi \binom{X}{\phi} \otimes \prod_{j \in \mathbb{N}^*} \binom{Y}{j}^{\phi(j)}$$

where ϕ ranges over all the functions $\phi : \mathbb{N}^* := \mathbb{N} \setminus \{0\} \to \mathbb{N}$ subject to the condition that $\sum_{j \in \mathbb{N}^*} j \cdot \phi(j) = k$, and $\binom{X}{\phi} := \dfrac{X(X-1) \cdot \ldots \cdot (X - \sum_{j \in \mathbb{N}^*} \phi(j) + 1)}{\prod_{j \in \mathbb{N}^*} \phi(j)!}$.

To show that $\binom{X}{\phi} \in \mathscr{G}_n$, one notices that $\binom{X}{\phi} = \prod_{j \in \mathbb{N}^*} \binom{X - \sum_{i=1}^{j-1} \phi(i)}{\phi(j)}$ and then uses (i).

4.2.16.   For the rest of this section we fix a prime number p and we let $v_p : \mathbb{Q} \to \mathbb{Z} \cup \{\infty\}$ be the p-adic valuation.

Lemma 4.2.17. *For every $n > 0$, the ring $\mathscr{G}_{n,(p)} := \mathscr{G}_n \otimes_{\mathbb{Z}} \mathbb{Z}_{(p)}$ is the $\mathbb{Z}_{(p)}$-algebra generated by the polynomials X, $\binom{X}{p}, \binom{X}{p^2}, \ldots, \binom{X}{p^k}$, where k is the unique integer such that $p^k \leq n < p^{k+1}$.*

Proof. We proceed by induction on n. It suffices to prove that $\binom{X}{n}$ is contained in the $\mathbb{Z}_{(p)}$-algebra $R := \mathbb{Z}_{(p)}[\binom{X}{p}, \binom{X}{p^2}, \ldots, \binom{X}{p^k}]$. We will use the following (easily verified) identity which holds in $\mathbb{Q}[X]$ for every $i, j \in \mathbb{N}$:

$$(4.2.18) \qquad \binom{X}{i+j} = \binom{X}{i} \cdot \binom{X-i}{j} \cdot \binom{i+j}{j}^{-1}.$$

Suppose first that n is a multiple of p^k, and write $n = (b+1)p^k$ for some $b < p - 1$. If $b = 0$, there is nothing to prove, so we can even assume that $b > 0$. We apply (4.2.18) with $i = b \cdot p^k$ and $j = p^k$. By remark 4.2.15(i), $\binom{X - b \cdot p^k}{p^k}$ is in R, and so is $\binom{X}{b \cdot p^k}$, by induction. The claim will therefore follow in this case, if we show that $\binom{(b+1)p^k}{p^k}$ is invertible in $\mathbb{Z}_{(p)}$. However, this is clear, since $v_p(i) = v_p(i + b \cdot p^k)$ for every $i = 1, \ldots, p^k$. Finally, it remains consider the case where $n = b \cdot p^k + a$ for some $b > 0$ and $0 < a < p^k$. This is dealt with in the same way: apply (4.2.18) with $i = b \cdot p^k$ and $j = a$ and use the previous case. $\square$

Lemma 4.2.19. *Let $k \in \mathbb{N}$. If R is a flat $\mathbb{Z}_{(p)}$-algebra and $f \in R$, then the following two conditions are equivalent:*

(i) $\binom{f}{p^i} \in R$ *for every $i = 1, \ldots, k$;*

(ii) *locally on $\operatorname{Spec} R$ there exists $j \in \mathbb{Z}$ such that $f \equiv j \pmod{p^k}$.*

Proof. We may assume that R is local. For $k = 0$ there is nothing to prove. For $k = 1$ we have $\binom{f}{p} = u \cdot p^{-1} \cdot \prod_{i=0}^{p-1}(f - i)$ for a unit u of R. Then the assertion holds since all but possibly one of the $f - i$ are invertible. For $k > 1$, by induction we can write $f = i + p \cdot g$ for some $g \in R$ and $0 \leq i < p$. Since $v_p(p^k!) = 1 + p + p^2 + \ldots + p^{k-1}$, we have

$$\binom{f}{p^k} = u \cdot p^{-1-p-p^2-\ldots-p^{k-1}} \cdot \prod_{\substack{j \equiv i \pmod{p} \\ 0 \leq j < p^k}} (f - j) = u' \cdot \binom{g}{p^{k-1}}$$

for some units $u, u' \in R$. The claim follows. $\square$

4.2.20. For every integer $k \geq 0$, we construct a scheme X_k by gluing the affine schemes $\operatorname{Spec}\mathbb{Z}_{(p)}[\frac{X-i}{p^k}]$ $(0 \leq i < p^k)$ along their general fibres. For every $k \in \mathbb{N}$ and every $i \in \mathbb{N}$ with $0 \leq i < p^{k+1}$ there is an obvious imbedding $\mathbb{Z}_{(p)}[\frac{X-i}{p^k}] \subset \mathbb{Z}_{(p)}[\frac{X-i}{p^{k+1}}]$. By gluing the duals of these imbeddings, we obtain, for every $k \in \mathbb{N}$, a morphism of schemes $\rho_k : X_{k+1} \to X_k$. Let also $\xi_k : \operatorname{Spec}\mathcal{G}_{p^{k+1}} \to \operatorname{Spec}\mathcal{G}_{p^k}$ be the morphism which is dual to the imbedding $\mathcal{G}_{p^k} \subset \mathcal{G}_{p^{k+1}}$.

Proposition 4.2.21. *With the notation of (4.2.20) we have:*

(i) *For given $n > 0$, let k be the unique integer such that $p^k \leq n < p^{k+1}$. There is a natural isomorphism of schemes: $\pi_k : X_k \xrightarrow{\sim} \operatorname{Spec}\mathcal{G}_n \otimes_{\mathbb{Z}} \mathbb{Z}_{(p)}$.*

(ii) *For every $k \in \mathbb{N}$ the diagram of schemes:*

$$
\begin{array}{ccc}
X_{k+1} & \xrightarrow{\pi_{k+1}} & \operatorname{Spec}\mathcal{G}_{p^{k+1}} \otimes_{\mathbb{Z}} \mathbb{Z}_{(p)} \\
\rho_k \downarrow & & \downarrow \xi_k \otimes_{\mathbb{Z}} \mathbb{Z}_{(p)} \\
X_k & \xrightarrow{\pi_k} & \operatorname{Spec}\mathcal{G}_{p^k} \otimes_{\mathbb{Z}} \mathbb{Z}_{(p)}
\end{array}
$$

commutes.

Proof. By lemma 4.2.17 we may assume that $n = p^k$. By lemma 4.2.19, we see that both X_k and $\operatorname{Spec}\mathcal{G}_{p^k} \otimes_{\mathbb{Z}} \mathbb{Z}_{(p)}$ represent the same functor from the category of flat $\mathbb{Z}_{(p)}$-schemes to the category of sets. Since both schemes are flat over $\operatorname{Spec}\mathbb{Z}_{(p)}$, (i) follows. It is similarly clear that $\xi_k \otimes_{\mathbb{Z}} \mathbb{Z}_{(p)}$ and ρ_k represent the same natural transformation of functors, so (ii) follows. $\square$

Corollary 4.2.22. (i) *For given $n > 0$, let k be the unique integer such that $p^{k-1} \leq n < p^k$. Then there is a natural ring isomorphism*

$$\mathbb{Z}/p^k\mathbb{Z} \xrightarrow{\sim} \operatorname{End}(\mathbb{G}_{m,\mathbb{F}_p}(n)) \quad i \mapsto (1+T)^i - 1$$

(ii) *Let R be a ring such that $\mathbb{F}_p \subset R$. Then there is a natural ring isomorphism*

$$\mathscr{C}^0(\operatorname{Spec}R, \mathbb{Z}_p) \xrightarrow{\sim} \operatorname{End}_R(\widehat{\mathbb{G}}_{m,R}) \quad \beta \mapsto (1+T)^\beta - 1.$$

Proof. (i): By lemma 4.2.17 we can assume $n = p^k - 1$. In this case, it is clear that the polynomials $(1+T)^i - 1$ are all distinct in $\mathbb{F}_p[T]/(T^{n+1})$ for $i = 0, ..., p^k - 1$ and they form a subring of $\operatorname{End}(\mathbb{G}_{m,\mathbb{F}_p}(n))$. However, an endomorphism of $\mathbb{G}_{m,\mathbb{F}_p}(n)$ corresponds to a unique point in $\operatorname{Spec}\mathcal{G}_n(\mathbb{F}_p)$. From proposition 4.2.21(i) we derive that $\operatorname{Spec}\mathcal{G}_n \otimes_{\mathbb{Z}} \mathbb{F}_p$ is the union of the special fibres of the affine schemes $\operatorname{Spec}\mathbb{Z}[\frac{X-i}{p^{k-1}}]$, for $i = 0, ..., p^{k-1} - 1$. Each of those contribute an affine line $\mathbb{A}^1_{\mathbb{F}_p}$, so $\operatorname{Spec}\mathcal{G}_n \otimes_{\mathbb{Z}} \mathbb{F}_p$ consists of exactly p^{k-1} connected components. In total, we have therefore exactly p^k points in $\operatorname{Spec}\mathcal{G}_n(\mathbb{F}_p)$, so (i) follows.

(ii): To give an endomorphism of $\widehat{\mathbb{G}}_m$ is the same as giving a compatible system of endomorphisms of $\mathbb{G}_m(n)$, one for each $n \in \mathbb{N}$. In case $\mathbb{F}_p \subset R$, lemma 4.2.17 shows that this is also equivalent to the datum of a compatible system of morphisms $\phi_k : \operatorname{Spec}R \to \operatorname{Spec}\mathcal{G}_{p^k} \otimes_{\mathbb{Z}} \mathbb{F}_p$, for every $k \geq 0$. From proposition 4.2.21(ii) we can further deduce that, under the morphism ξ_k, each of the p^{k+1} connected components of $\operatorname{Spec}\mathcal{G}_{p^{k+1}} \otimes_{\mathbb{Z}} \mathbb{F}_p$ gets mapped onto one of the p^{k+1} rational points of $\operatorname{Spec} \mathscr{G}_{p^k} \otimes_{\mathbb{Z}} \mathbb{F}_p$. Since $\phi_{k-1} = \xi_k \circ \phi_k$, we see that the scheme-theoretic image of ϕ_{k-1} is contained in $\operatorname{Spec} \mathscr{G}_{p^{k-1}}(\mathbb{F}_p)$ (with the reduced subscheme structure), for every $k > 0$. Taking (i) into account, we see that an endomorphism of $\widehat{\mathbb{G}}_{m,R}$ is the same as the datum of a compatible system of continuous maps $\operatorname{Spec} R \to \mathbb{Z}/p^k\mathbb{Z}$. Since the p-adic topology of $\mathbb{Z}_p$ is the inverse limit of the discrete topologies on the $\mathbb{Z}/p^k\mathbb{Z}$, the claim follows.   □

4.3. Modules of almost finite rank.
Let A be a V^a-algebra, P an almost finitely generated projective A-module and $\phi \in \operatorname{End}_A(P)$.

4.3.1. We say that ϕ is Λ-*nilpotent* if there exists an integer $i > 0$ such that $\Lambda_A^i \phi = 0$. Notice that the Λ-nilpotent endomorphisms of P form a bilateral ideal of the unitary ring $\operatorname{End}_A(P)$. Notice also that $\Lambda_A^i P$ is an almost finitely generated projective A-module for every $i \geq 0$; indeed, this is easily shown by means of lemma 2.4.15. For a Λ-nilpotent endomorphism ϕ we introduce the notation

$$\det(\mathbf{1}_P + \phi) := \sum_{i \geq 0} \operatorname{tr}_{\Lambda_A^i P/A}(\Lambda_A^i \phi).$$

Notice that the above sum consists of only finitely many non-zero terms, so that $\det(\mathbf{1}_P + \phi)$ is a well defined element of A_*.

Lemma 4.3.2. *Let P be an almost finitely generated projective A-module.*

(i) *If ϕ is a Λ-nilpotent endomorphism of P and $\alpha : A \to A'$ is any morphism of V^a-algebras, set $P' := P \otimes_A A'$. Then:*

$$\det(\mathbf{1}_{P'} + \phi \otimes_A \mathbf{1}_{A'}) = \alpha(\det(\mathbf{1}_P + \phi)).$$

(ii) *Let $\phi, \psi \in \operatorname{End}_A(P)$ such that $\phi \circ \psi$ and $\psi \circ \phi$ are Λ-nilpotent. Then:*

$$\det(\mathbf{1}_P + \phi \circ \psi) = \det(\mathbf{1}_P + \psi \circ \phi).$$

Proof. (i) is a straightforward consequence of the definitions. (ii) follows directly from lemma 4.1.2(i).   □

4.3.3. Now, let $\phi, \psi \in \operatorname{End}_A(P)$ be two endomorphisms. Set

$$B := A[X, Y]/(X^n, Y^n) \quad \text{and} \quad P_B := P \otimes_A B.$$

ϕ and ψ induce endomorphisms of P_B that we denote again by the same letters. Clearly $X \cdot \phi$ and $Y \cdot \psi$ are Λ-nilpotent; hence we get elements $\det(\mathbf{1}_{P_B} + X \cdot \phi)$, $\det(\mathbf{1}_{P_B} + Y \cdot \psi)$ and $\det(\mathbf{1}_{P_B} + X \cdot \phi + Y \cdot \psi + XY \cdot \psi \circ \phi)$ in B_*. Notice that any element of B_* can be written uniquely as an A_*-linear combination of the monomials $X^i Y^j$ with $0 \leq i, j < n$. Moreover, it is clear that $\det(\mathbf{1}_{P_B} + X \cdot \phi) = \sum_{0 \leq i < n} \operatorname{tr}_{\Lambda_A^i P/A}(\Lambda_A^i \phi) \cdot X^i$, and similarly for ψ.

Proposition 4.3.4. *With the above notation, the following identity holds :*

(4.3.5) $\det(\mathbf{1}_{P_B} + X\phi) \cdot \det(\mathbf{1}_{P_B} + Y\psi) = \det(\mathbf{1}_{P_B} + X\phi + Y\psi + XY\psi\phi).$

Proof. First of all we remark that, when P is a free A-module of finite rank, the above identity is well-known, and easily verified by working with matrices with entries in A_*. Suppose next that P is arbitrary, but $\phi = \varepsilon \cdot \phi'$, $\psi = \varepsilon \cdot \psi'$ for some $\phi', \psi' \in \text{End}_A(P)$ and $\varepsilon \in \mathfrak{m}$. Pick a free A-module F of finite rank, and morphisms $u : P \to F$, $v : F \to P$ such that $v \circ u = \varepsilon \cdot 1_P$. Set $\phi_\varepsilon := u \circ \phi' \circ v : F \to F$ and define similarly ψ_ε. Clearly $\det(1_{P_B} + X \cdot \phi) = \det(1_{P_B} + X \cdot \varepsilon \cdot \phi') = \det(1_{P_B} + X \cdot v \circ u \circ \phi') = \det(1_{P_B} + X \cdot \phi_\varepsilon)$ (by lemma 4.3.2(ii)) and similarly for the other terms appearing in (4.3.5). Thus we have reduced this case to the case of a free A-module. Finally, we deal with the general case. The foregoing shows that the sought identity is known at least when ϕ and ψ are replaced by $\varepsilon \cdot \phi$, resp. $\varepsilon \cdot \psi$, for any $\varepsilon \in \mathfrak{m}$. Equivalently, consider the A-algebra endomorphism $\alpha : B \to B$ defined by $X \mapsto \varepsilon \cdot X$, $Y \mapsto \varepsilon \cdot Y$ and let C be the B-algebra structure on B determined by α; by lemma 4.3.2(i) we have

$$\det(1_{P_B} + X \cdot \varepsilon \cdot \phi) = \det(1_{P_C} + (X \cdot \phi) \otimes_B 1_C) = \alpha(\det(1_{P_B} + X \cdot \phi))$$

and similarly for the other terms appearing in (4.3.5). Thus, the images under α of the two members of (4.3.5) coincide. But applying α to a monomial of the form $a \cdot X^i Y^j$ has the effect of multiplying it by ε^{i+j}; by (**B**), the $(i + j)$-th powers of elements of $\mathfrak{m}$ generate $\mathfrak{m}$, hence the claim follows easily. $\quad\square$

Corollary 4.3.6. *If* $\phi, \psi \in \text{End}_A(P)$ *are two* Λ-*nilpotent endomorphisms, then*

$$\det(1_P + \phi) \cdot \det(1_P + \psi) = \det(1_P + \phi + \psi + \phi \circ \psi).$$

Proof. If $A[[X,Y]] := A_*[[X,Y]]^a$, then $A[X,Y]_* = A_*[[X,Y]]$. Set $P' := P \otimes_A A[[X,Y]]$ and for every endomorphism α of P' such that $\text{Im} \, \alpha \subset XP' + YP'$, let $\det(1 + \alpha) := \sum_{i \geq 0} \text{tr}_{\Lambda_A^i P'/A[[X,Y]]}(\Lambda_{A[[X,Y]]}^i \alpha)$. Then proposition 4.3.4 implies that the analogue of (4.3.5) holds in $A_*[[X,Y]]$; when ϕ and ψ are Λ-nilpotent, all terms are polynomials and the claim then follows by evaluating the polynomials $\det(1_{P'} + X \cdot \phi)$, $\det(1_{P'} + Y \cdot \psi)$ and $\det(1_{P'} + X \cdot \phi + Y \cdot \psi + XY \cdot \phi \circ \psi)$ for $X = Y = 1$. $\quad\square$

4.3.7. Next, for P as above, set

$$\chi_P(X) := \sum_{i \geq 0} \text{tr}_{\Lambda_A^i P/A}(\Lambda_A^i 1_P) \cdot X^i \in A_*[[X]]$$
$$\psi_P(X) := \sum_{i \geq 0} \text{tr}_{\text{Sym}_A^i P/A}(1_{\text{Sym}_A^i P}) \cdot X^i \in A_*[[X]].$$

Corollary 4.3.8. *Let* P *be an almost finitely generated projective A-module. Then:*

(i) *the power series* $\chi_P(X)$ *is an endomorphism of the formal group* $\widehat{\mathbb{G}}_{m,A_*}$.
(ii) $\chi_P(X) \cdot \psi_P(-X) = 1$.
(iii) $\chi_P(X) \in 1 + \mathscr{E}_{P/A_*}[[X]]$.

Proof. (i) follows from proposition 4.3.4. For (ii), recall that, for every $n > 0$ there is an acyclic Koszul complex (cp. [17, Ch.X, §9, n.3, Prop.3])

$$0 \to \Lambda_A^n Q \to ... \to (\Lambda_A^j Q) \otimes_A (\text{Sym}_A^{n-j} Q) \to ... \to \text{Sym}_A^n Q \to 0.$$

From proposition 4.1.4 we derive, by a standard argument, that the trace is an additive function on arbitrary bounded acyclic complexes. Then, taking into account lemma 4.1.3 we obtain: $\sum_{i=0}^{n}(-1)^i \cdot \mathrm{tr}_{\Lambda_A^{n-i}P/A}(1_{\Lambda_A^{n-i}P}) \cdot \mathrm{tr}_{\mathrm{Sym}_A^i P/A}(1_{\mathrm{Sym}_A^i P}) = 0$ for every $n > 0$. This is equivalent to the sought identity. To show (iii) we remark more precisely that $\mathscr{E}_{\Lambda_A^r P/A} \subset \mathscr{E}_{P/A}$ for every $r > 0$. Indeed, set $B := A/\mathscr{E}_{P/A}$. Then, by proposition 2.4.28(i),(iii): $\mathscr{E}_{\Lambda_A^r P/A} \cdot B = \mathscr{E}_{\Lambda_B^r(P \otimes_A B)/B} = 0$, whence the claim. □

Definition 4.3.9. Let P be an almost finitely generated projective A-module.

(i) The *formal rank* of P is the ring homomorphism

$$\mathrm{f.rk}_A(P) : \mathscr{G}_\infty := \mathbb{Z}[\alpha, \binom{\alpha}{2}, ...] \to A_*$$

associated to $\chi_P(X)$.

(ii) We say that P is *of almost finite rank* if, for every $\varepsilon \in \mathfrak{m}$, there exists an integer $i \geq 0$ such that $\varepsilon \cdot \Lambda_A^i P = 0$.

(iii) We say that P is *of finite rank* if there exists an integer $i \geq 0$ such that $\Lambda_A^i P = 0$.

(iv) Let $r \in \mathbb{N}$; we say that P has *constant rank equal to r* if $\Lambda_A^{r+1} P = 0$ and $\Lambda_A^r P$ is an invertible A-module.

Remark 4.3.10. (i) It follows easily from lemma 2.3.7(vi) that every uniformly almost finitely generated projective A-module is of finite rank.

(ii) Notice that if P is of finite rank, then $\chi_P(X)$ is a polynomial, whence it defines an endomorphism of the algebraic group $\mathbb{G}_{m,A_*}$. In this case, it follows that $\chi_P(X)$ is of the form $(1 + X)^\alpha$, where $\alpha : \mathrm{Spec}\, A_* \to \mathbb{Z}$ is a continuous function (where $\mathbb{Z}$ is seen as a discrete topological space). More precisely, there is an obvious injective ring homomorphism

$$(4.3.11) \qquad \mathscr{C}^0(\mathrm{Spec}\, A_*, \mathbb{Z}) \to \mathrm{End}_A(\mathbb{G}_{m,A_*}) \qquad \beta \mapsto (1+X)^\beta$$

which allows to identify the continuous function α with the formal rank of P. Moreover, if $\Lambda_A^i P = 0$, it is clear that $\alpha(\mathrm{Spec}\, A_*) \subset \{0, ..., i-1\}$.

The main result of this section is theorem 4.3.28, which describes general modules of almost finite rank as infinite products of modules of finite rank. The first step is lemma 4.3.12, concerned with the case of an A-module of rank one.

Lemma 4.3.12. *Let P be an almost finitely generated projective A-module such that $\Lambda_A^2 P = 0$. There exists V^a-algebras A_0, A_1 and an isomorphism of V^a-algebras $A \simeq A_0 \times A_1$ such that $P \otimes_A A_0 = 0$ and $P \otimes_A A_1$ is an invertible A_1-module.*

Proof. Since the natural map $P \times P \to \Lambda_A^2 P$ is universal for alternating A-bilinear maps on $P \times P$, we have

$$(4.3.13) \qquad f(p) \cdot q = f(q) \cdot p \qquad \text{for every } f \in (P^*)_* \text{ and } p, q \in P_*.$$

Using (4.3.13) we derive $\omega_{P/A}(p \otimes f)(q) = \mathrm{tr}_{P/A}(\omega_{P/A}(p \otimes f)) \cdot q$ for every $f \in (P^*)_*$ and $p, q \in P_*$. In other words, $\omega_{P/A}(p \otimes f) = \mathrm{tr}_{P/A}(\omega_{P/A}(p \otimes f)) \cdot 1_P$, for every $f \in (P^*)_*, p \in P_*$. By linearity we finally deduce

$$(4.3.14) \qquad \phi = \mathrm{tr}_{P/A}(\phi) \cdot 1_P \qquad \text{for all } \phi \in \mathrm{End}_A(P).$$

Now, by remark 4.3.10(ii), the hypothesis $\Lambda_A^2 P = 0$ also implies that $\chi_P(X) = (1 + X)^\alpha$, for a continuous function $\alpha : \operatorname{Spec} A_* \to \{0, 1\}$. We can decompose accordingly $A = A_0 \times A_1$, so that $\alpha(\operatorname{Spec} A_{i*}) = i$, which gives the sought decomposition. We can now treat separately the two cases $A = A_0$ and $A = A_1$. In case $\operatorname{f.rk}_A(P) = 0$, then $\operatorname{tr}_{P/A}(1_P) = 0$, and then (4.3.14) implies that $P = 0$. In case $\operatorname{f.rk}_A(P) = 1$, then $\operatorname{tr}_{P/A}(1_P) = 1$ and (4.3.14) implies that the natural map $A \to \operatorname{End}_A(P) : a \mapsto a \cdot 1_P$ is an inverse for $\operatorname{tr}_{P/A}$, thus $A \simeq P \otimes_A P^*$.   □

The next step consists in analyzing the structure of A-modules of finite rank. To this purpose we need some preliminaries of multi-linear algebra.

4.3.15.   For every $n \geq 0$ let $\mathbf{n} := \{1, ..., n\}$; for a subset $I \subset \mathbf{n}$ let $|I|$ be the cardinality of I; for a given partition $\mathbf{n} = I \cup J$, let $\prec$ denote the total ordering on $\mathbf{n}$ that restricts to the usual ordering on I and on J, and such that $i \prec j$ for every $i \in I$, $j \in J$. Finally let ε_{IJ} be the sign of the unique order-preserving bijection $(\mathbf{n}, <) \to (\mathbf{n}, \prec)$.

Let M be any A-module. Given elements $m_1, m_2, ..., m_n$ in M_*, and $I \subset \mathbf{n}$ a subset of elements $i_1 < i_2 < ... < i_{|I|}$, let $m_I := m_{i_1} \wedge ... \wedge m_{i_{|I|}} \in \Lambda_A^{|I|} M_*$ (with the convention that $m_\varnothing = 1 \in A_* = \Lambda_A^0 M_*$).

4.3.16.   Let M, N be any two A-modules. For every $i, j \geq 0$ there is a natural morphism

$$(4.3.17) \qquad \Lambda_A^i M \otimes_A \Lambda_A^j N \to \Lambda_A^{i+j}(M \oplus N)$$

determined by the rule:

$$m_1 \wedge ... \wedge m_i \otimes n_1 \wedge ... \wedge n_j \mapsto (m_1, 0) \wedge (m_2, 0) \wedge ... \wedge (0, n_1) \wedge ... \wedge (0, n_j)$$

for all $m_1, ..., m_i \in M_*$ and $n_1, ..., n_j \in N$. The morphisms (4.3.17) assemble to an isomorphism of A-modules

$$(4.3.18) \qquad \Lambda_A^\bullet M \otimes_A \Lambda_A^\bullet N \to \Lambda_A^\bullet(M \oplus N).$$

Clearly, there is a unique graded A-algebra structure on $\Lambda_A^\bullet M \otimes_A \Lambda_A^\bullet N$ such that (4.3.18) is an isomorphism of (graded-commutative) A-algebras. Explicitly, given $x_i \in \Lambda_A^{a_i} M_*, y_i \in \Lambda_A^{b_i} N_*$ $(i = 1, 2)$ one verifies easily that the product on $\Lambda_A^\bullet M \otimes_A \Lambda_A^\bullet N$ is fixed by the rule

$$(4.3.19) \qquad (x_1 \otimes y_1) \cdot (x_2 \otimes y_2) = (-1)^{a_2 b_1} \cdot (x_1 \wedge x_2) \otimes (y_1 \wedge y_2).$$

Then $\Lambda_A^\bullet M \otimes_A \Lambda_A^\bullet N$ is even a bigraded A-algebra, if we let $\Lambda_A^i M \otimes_A \Lambda_A^j N$ be the graded component of bidegree (i, j).

4.3.20.   Next, let $\delta : M \to M \oplus M$ be the diagonal morphism $m \mapsto (m, m)$ (for all $m \in M_*$). It induces a morphism $\Lambda_A^\bullet \delta : \Lambda_A^\bullet M \to \Lambda_A^\bullet(M \oplus M)$ of A-algebras. We let $\Delta : \Lambda_A^\bullet M \to \Lambda_A^\bullet M \otimes_A \Lambda_A^\bullet M$ be the composition of the morphism $\Lambda_A^\bullet \delta$ and the inverse of the isomorphism (4.3.18), taken with $M = N$. For every $a, b \geq 0$ we also let $\Delta_{a,b} : \Lambda_A^\bullet M \to \Lambda_A^a M \otimes_A \Lambda_A^b M$ be the composition of Δ and the projection onto the graded component of bidegree (a, b). The morphisms $\Delta_{a,b}$ are usually called "co-multiplication morphisms". An easy calculation shows that:

(4.3.21) $$\Delta_{a,b}(x_1 \wedge x_2 \wedge \ldots \wedge x_{a+b}) = \sum_{I,J} \varepsilon_{IJ} \cdot x_I \otimes x_J$$

where the sum ranges over all the partitions $a + b = I \cup J$ such that $|I| = a$. Let now $x_1, \ldots, x_a, y_1, \ldots, y_b \in M_*$. Since Δ is a morphism of A-algebras, we have $\Delta(x_{\mathbf{a}} \wedge y_{\mathbf{b}}) = \Delta(x_{\mathbf{a}}) \cdot \Delta(y_{\mathbf{b}})$. Hence, using (4.3.19) and (4.3.21) one deduces easily:

(4.3.22) $$\Delta_{a,b}(x_{\mathbf{a}} \wedge y_{\mathbf{b}}) = \sum_{I,J,K,L} \varepsilon_{IJ} \cdot \varepsilon_{KL} \cdot (-1)^{|J|} \cdot (x_I \wedge y_K) \otimes (x_J \wedge y_L).$$

where the sum runs over all partitions $I \cup J = \mathbf{a}$, $K \cup L = \mathbf{b}$ such that $|J| = |K|$. The following lemma is analogous to Sylvester's lemma (see [39, §8.1, Lemma 2, p.108]).

Lemma 4.3.23. *Suppose that* $\Lambda_A^{a+1} M = 0$ *for some integer* $a \geq 0$. *Let* $0 < b \leq a$ *and* $x_1, \ldots, x_a, y_1, \ldots, y_b \in M_*$. *The following identity holds in* $(\Lambda_A^a M \otimes_A \Lambda_A^b M)_*$:

$$x_{\mathbf{a}} \otimes y_{\mathbf{b}} = \sum_{I,J} \varepsilon_{JI} \cdot (x_J \wedge y_{\mathbf{b}}) \otimes x_I$$

where the sum ranges over all the partitions $\mathbf{a} = I \cup J$ *such that* $|I| = b$.

Proof. For a given subset $B \subset \mathbf{b}$ we let

$$\gamma(y_B) := \sum_{I,J} \varepsilon_{IJ} \cdot (x_I \wedge y_B) \otimes x_J - x_{\mathbf{a}} \otimes y_B$$

where the sum is taken over all the partitions $I \cup J = \mathbf{a}$ such that $|J| = |B|$. Notice that $\gamma(y_\varnothing) = 0$. We have to show that $\gamma(y_{\mathbf{b}}) = 0$. To this purpose we show the following:

Claim 4.3.24. If $|B| > 0$, then

(4.3.25) $$\Delta_{a,|B|}(x_{\mathbf{a}} \wedge y_B) = \sum_{K,L} \varepsilon_{KL} \cdot (-1)^{|K|} \cdot \gamma(y_K) \wedge y_L$$

where the sum ranges over all the partitions $K \cup L = B$.

Proof of the claim. (The term $\gamma(y_K) \wedge y_L$ means that we apply exterior multiplication by y_L to the second factor of the tensor products appearing in the expression for $\gamma(y_K)$.) Using (4.3.22), the difference between the two sides of (4.3.25) is seen to be equal to

$$\sum_{K,L} \varepsilon_{KL} \cdot (-1)^{|K|} \cdot x_{\mathbf{a}} \otimes (y_K \wedge y_L) = \sum_{K,L} (-1)^{|K|} \cdot x_{\mathbf{a}} \otimes y_B$$

where the sum runs over all partitions $K \cup L = B$. A standard combinatorial argument shows that this expression can be rewritten as $x_{\mathbf{a}} \otimes y_B \cdot \sum_{k=0}^{|B|} (-1)^k \cdot \binom{|B|}{k}$, which vanishes if $|B| > 0$. $\diamond$

To conclude the proof of the lemma, we remark that $\Delta_{a,b}$ vanishes if $b > 0$ because by assumption $\Lambda_A^{a+1} M = 0$; then the assertion follows by induction on $|B|$, using claim 4.3.24. $\square$

Lemma 4.3.26. *Let P be an A-module such that $\Lambda_A^{n+1} P = 0$ and assume that either P is flat or 2 is invertible in A_*. Then $\Lambda_A^2 (\Lambda_A^n P) = 0$.*

Proof. For any A-module M and $r \geq 0$ there exists an antisymmetrization operator (cp. [15, Ch. III, §7.4, Remarque])

$$a_r : M^{\otimes r} \to M^{\otimes r} \quad : \quad m_1 \otimes ... \otimes m_r \mapsto \sum_{\sigma \in S_r} \mathrm{sgn}(\sigma) \cdot m_{\sigma(1)} \otimes ... \otimes m_{\sigma(r)}.$$

Clearly a_r factors thorough $\Lambda_A^r M$, and in case M is free of finite rank, it is easy to check (just by arguing with basis elements) that the induced map $\bar{a}_r : \Lambda_A^r M \to \mathrm{Im}(a_r)$ is an isomorphism. This is still true also in case $r!$ is invertible in A_*, since in that case one checks that $a_r / r!$ is idempotent (see *loc. cit.*). More generally, if M is flat then, by [55, Ch.I, Th.1.2], $M_!$ is the filtered colimit of a direct system of free A_*-modules of finite rank, so also in this case $\bar{a}_r$ is an isomorphism. Notice that, again by [55, Ch.I, Th.1.2], if P is flat, then $\Lambda_A^k P$ is also flat, for every $k \geq 0$. Hence, to prove the lemma, it suffices to verify that $\mathrm{Im}(a_2 : (\Lambda_A^n P)^{\otimes 2} \to (\Lambda_A^n P)^{\otimes 2}) = 0$ when $\Lambda_A^{n+1} P = 0$. However, this follows easily from lemma 4.3.23. $\quad\square$

We are now ready to return to A-modules of finite rank.

Proposition 4.3.27. *Let P be an almost projective A-module of finite rank; say that $\Lambda_A^r P = 0$. There exists a natural decomposition $A \simeq A_0 \times A_1 \times ... \times A_{r-1}$ such that $P_i := P \otimes_A A_i$ is an A_i-module of constant rank equal to i (and of formal rank i) for every $i = 0, ..., r - 1$.*

Proof. We proceed by induction on r; the case $r = 2$ is covered by lemma 4.3.12. By lemma 4.3.26 we have $\Lambda_A^2 (\Lambda_A^{r-1} P) = 0$, so by lemma 4.3.12, there is a decomposition $A \simeq A'_{r-1} \times A'_{r-2}$, such that for $P_i := P \otimes_A A'_i$ ($i = r - 2, r - 1$) the following holds. $\Lambda_{A'_{r-2}}^{r-1} (P_{r-2}) = 0$ and $\Lambda_{A'_{r-1}}^{r-1} (P_{r-1})$ is an invertible A_{r-1}-module. It follows in particular that $\chi_{P_{r-1}}(X)$ is a polynomial of degree $r - 1$, and its leading coefficient is invertible in A_{r-1}. Hence $\chi_{P_{r-1}}(X) = (1 + X)^{r-1}$. By induction, A'_{r-2} admits a decomposition $A'_{r-2} \simeq A_{r-2} \times ... \times A_0$ with the stated properties; it suffices then to take $A_{r-1} := A'_{r-1}$. $\quad\square$

Theorem 4.3.28. *Let P be an almost projective A-module of almost finite rank. Then there exists a natural decomposition $A \simeq \prod_{i=0}^{\infty} A_i$ such that:*

(i) $\lim_{i \to \infty} \mathrm{Ann}_{V^a}(A_i) = V^a$ *(for the uniform structure of definition 2.3.1);*

(ii) *for $i \in \mathbb{N}$, let $P_i := P \otimes_A A_i$; then $P \simeq \prod_{i=0}^{\infty} P_i$ and every P_i is an A_i-module of finite constant rank equal to i.*

Proof. Let $\{\mathfrak{m}_\lambda\}_{\lambda \in I}$ be the filtered family of finitely generated subideals of $\mathfrak{m}$. For every $\lambda \in I$, let $A_\lambda := A / \mathrm{Ann}_A(\mathfrak{m}_\lambda)$. By hypothesis, $P_\lambda := P \otimes_A A_\lambda$ is an A_λ-module of finite rank; say that the rank is $r(\lambda)$. By proposition 4.3.27 we have natural decompositions $A_\lambda \simeq A_{\lambda,0} \times ... \times A_{\lambda,r(\lambda)}$ such that $P \otimes_A A_{\lambda,i}$ is an $A_{\lambda,i}$-module of constant rank equal to i for every $i \leq r(\lambda)$. The naturality of the decomposition means that for every $\lambda, \mu \in I$ such that $\mathfrak{m}_\lambda \subset \mathfrak{m}_\mu$, we have $A_{\mu,i} \otimes_A A_\lambda \simeq A_{\lambda,i}$ for every $i \leq r(\mu)$. In particular, $\mathfrak{m}_\lambda A_{\mu,i} = 0$ for every $i > r(\lambda)$. By considering the short exact sequence of cofiltered systems of A-modules:

$$0 \to (\mathrm{Ann}_A(\mathfrak{m}_\lambda))_{\lambda \in I} \to (A)_{\lambda \in I} \to (A_\lambda)_{\lambda \in I} \to 0$$

we deduce easily that $A \simeq \lim_{\lambda \in I} A_\lambda$ and therefore we obtain a decomposition $A \simeq \prod_{i=0}^\infty A_i$, with $A_i := \lim_{\lambda \in I} A_{\lambda,i}$ for every $i \in \mathbb{N}$. Notice that, for every $i \in \mathbb{N}$ and every $\lambda \in I$, the natural morphism $A_i \to A_{\lambda,i}$ is surjective with kernel killed by $\mathfrak{m}_\lambda$. It follows easily that $\Lambda_{A_i}^{i+1}(P \otimes_A A_i) = 0$ and $\Lambda_{A_i}^i(P \otimes_A A_i)$ is an invertible A_i-module. Furthermore, for every $\lambda \in I$, $\mathfrak{m}_\lambda \cdot A_i = 0$ for all $i > r(\lambda)$, which implies (i). Finally, for every $\lambda \in I$, $\mathfrak{m}_\lambda$ kills the kernel of the projection $P \to \prod_{i=0}^{r(\lambda)} P_i$, so P is isomorphic to the infinite product of the P_i. $\square$

4.4. Localization in the flat site.
Throughout this section P denotes an almost finitely generated projective A-module. The following definition introduces the main tool used in this section.

Definition 4.4.1. The *splitting algebra* of P is the A-algebra:

$$\mathrm{Split}(A, P) := \mathrm{Sym}_A^\bullet(P \oplus P^*)/(1 - \zeta_P).$$

We endow $\mathrm{Sym}_A^\bullet(P \oplus P^*)$ with the structure of graded algebra such that P is placed in degree one and P^* in degree -1. Then ζ_P is a homogeneous element of degree zero, and consequently $\mathrm{Split}(A, P)$ is also a graded A-algebra.

4.4.2. We define a functor $S : A\text{-}\mathbf{Alg} \to \mathbf{Set}$ by assigning to every A-algebra B the set $S(B)$ of all pairs (x, ϕ) where $x \in (P \otimes_A B)_*$, $\phi : P \otimes_A B \to B$ such that $\phi(x) = 1$.

Lemma 4.4.3. (i) $\mathrm{Split}(A, P)$ *is a flat A-algebra.*
(ii) $\mathrm{Split}(A', P \otimes_A A') \simeq \mathrm{Split}(A, P) \otimes_A A'$ *for every A-algebra A'.*
(iii) $\mathrm{Split}(A, P)$ *represents the functor S.*

Proof. We have $\zeta_P^k \in (\mathrm{Sym}_A^k P) \otimes_A (\mathrm{Sym}_A^k P^*) \subset \mathrm{gr}^0(\mathrm{Sym}_A^{2k}(P \oplus P^*))$ for every $k \geq 0$. It is easy to verify the formula:

$$(4.4.4) \qquad \mathrm{gr}^k\mathrm{Split}(A, P) \simeq \mathop{\mathrm{colim}}_{j \in \mathbb{Z}} (\mathrm{Sym}_A^{k+j} P) \otimes_A (\mathrm{Sym}_A^j P^*)$$

where the transition maps in the direct system are given by multiplication by ζ_P. In particular, it is clear that $\mathrm{gr}^k\mathrm{Split}(A, P)$ is a flat A-module, so (i) holds. (ii) is immediate. To show (iii), let us introduce the functor $T : A\text{-}\mathbf{Alg} \to \mathbf{Set}$ that assigns to every A-algebra B the set of all pairs (x, ϕ) where $x \in (P \otimes_A B)_*$ and $\phi : P \otimes_A B \to B$. So S is a subfunctor of T.

Claim 4.4.5. The functor T is represented by the A-algebra $\mathrm{Sym}_A^\bullet(P \oplus P^*)$.

Proof of the claim. Indeed, there are natural bijections:

$$(P \otimes_A B)_* \xrightarrow{\sim} \mathrm{Hom}_B(P^* \otimes_A B, B) \xrightarrow{\sim} \mathrm{Hom}_A(P^*, B) \xrightarrow{\sim} \mathrm{Hom}_{A\text{-}\mathbf{Alg}}(\mathrm{Sym}_A^\bullet P^*, B)$$

which show that the functor $B \mapsto (P \otimes_A B)_*$ is represented by the A-algebra $\text{Sym}_A P^*$. Working out the definitions, one finds that the composition of these bijections assigns to an element $x \in (P \otimes_A B)_*$ the unique A-algebra morphism $f_x : \text{Sym}_A^\bullet P^* \to B$ such that $f_x(\psi) = \psi(x)$ for every $\psi \in \text{Hom}_A(P, A)$. Similarly, the functor $B \mapsto \text{Hom}_B(P \otimes_A B, B)$ is represented by $\text{Sym}_A^\bullet P$, and again, one checks that the bijection assigns to $\phi : P \otimes_A B \to B$ the unique A-algebra morphism $g_\phi : \text{Sym}_A^\bullet P \to B$ such that $g_\phi(p) = \phi(p)$ for every $p \in P_*$. It follows that T is represented by $(\text{Sym}_A^\bullet P) \otimes_A (\text{Sym}_A^\bullet P^*) \simeq \text{Sym}_A^\bullet (P \oplus P^*)$. $\Diamond$

For every A-algebra B we have a natural map: $\alpha_B : T(B) \to B_*$ given by $(x, \phi) \mapsto \phi(x)$. This defines a natural transformation of functors $\alpha : T \to (-)_*$. Moreover, let us consider the trivial map $\beta_B : T(B) \to B_*$ that sends everything onto the element $1 \in B_*$. β is another natural transformation from T to the almost elements functor. Clearly :

$$S(B) = \text{Equal}(\, T(B) \underset{\beta_B}{\overset{\alpha_B}{\rightrightarrows}} B_* \,).$$

We remark that the functor $B \mapsto B_*$ on A-algebras is represented by $A[X] := \text{Sym}_A^\bullet A$. Therefore there are morphisms $\alpha^*, \beta^* : A[X] \to \text{Sym}_A^\bullet (P \oplus P^*)$ that represent these natural transformations. It follows that S is represented by the A-algebra

$$\text{Coequal}(\, A[X] \underset{\beta^*}{\overset{\alpha^*}{\rightrightarrows}} \text{Sym}_A^\bullet (P \oplus P^*) \,).$$

To determine α^* and β^* it suffices to calculate them on the element $X \in A[X]_*$. It is easy to see that $\alpha^*(X) = 1$. To conclude the proof it suffices therefore to show:

Claim 4.4.6. $\beta^*(X) = \zeta_P$.

Proof of the claim. In view of the definitions, and using the notation of the proof of claim 4.4.5, the claim amounts to the identity: $(g_\phi \otimes f_x)(\zeta_P) = \phi(x)$ for every $(x, \phi) \in T(B)$. By naturality, it suffices to show this for $B = A$. Now, for every $\varepsilon \in \mathfrak{m}$, we can write $\varepsilon \cdot \zeta_P = \sum_i q_i \otimes \psi_i$ for some $q_i \in P$, $\psi_i \in P^*$ and we have $\sum_i q_i \cdot \psi_i(b) = \varepsilon \cdot b$ for all $b \in P_*$. Hence $(g_\phi \otimes f_x)(\sum_i q_i \otimes \psi_i) = \sum_i \phi(q_i) \cdot \psi_i(x) = \phi(\sum_i q_i \cdot \psi_i(x)) = \phi(\varepsilon x)$ and the claim follows. $\square$

Remark 4.4.7. The construction of the splitting algebra occurs already, in a tannakian context, in Deligne's paper [23] : see the proof of lemma 7.15 in *loc.cit.*

4.4.8.   We recall that for every $k \geq 0$ and any A-module P there are natural morphisms

$$\text{Sym}_A^k P \xrightarrow{\alpha_P} \Gamma_A^k P \xrightarrow{\beta_P} \text{Sym}_A^k P$$

such that $\beta_P \circ \alpha_P = k! \cdot 1_{\text{Sym}_A^k P}$ and $\alpha_P \circ \beta_P = k! \cdot 1_{\Gamma_A^k P}$. (To obtain the morphisms, one can consider the A_*-module $P_!$, thus one can assume that P is a module over a usual ring; then α_P is obtained by extending multiplicatively the identity morphism $\text{Sym}_A^1 P = P \to P = \Gamma_A^1 P$, and β_P is deduced from the homogeneous degree k polynomial law $P \otimes_A B \to (P \otimes_A B)^{\otimes k}$ defined by $x \mapsto x^{\otimes k}$.) Moreover an element of $\mathrm{Sym}_{A_*}^k P_*^*$ yields a homogeneous degree k polynomial law $P_* \to A_*$; hence one has a natural map $\mathrm{Sym}_A^k P^* \to (\Gamma_A^k P)^*$, which – by usual reductions – is an isomorphism when P is almost finitely generated projective.

Lemma 4.4.9. *With the notation of* (4.4.8) *we have:*

$$(\alpha_P \otimes_A 1_{\mathrm{Sym}_A^k P^*})(\zeta_P^k) = k! \cdot \zeta_{\Gamma_A^k P}.$$

Proof. Suppose first that P is a free A-module, let $e_1, ..., e_n$ be a base of P_* and $e_1^*, ..., e_n^*$ the dual base of P^*. Then $\Gamma_A^k P$ is the free A-module generated by the basis $e_1^{[n_1]} \cdot ... \cdot e_k^{[n_k]}$ where $0 \le n_i \le k$ for $i = 1, ..., k$ and $\sum_j n_j = k$. The dual of this basis is the basis of $\mathrm{Sym}_A^k P^*$ consisting of the elements $e_1^{*n_1} \cdot ... \cdot e_k^{*n_k}$. Furthermore, $\zeta_P = \sum_i e_i \otimes e_i^*$ and therefore $\zeta_P^k = \sum_{\underline{n}} \binom{k}{\underline{n}} (e_1^{n_1} \cdot ... \cdot e_k^{n_k}) \otimes (e_1^{*n_1} \cdot ... \cdot e_k^{*n_k})$, where $\underline{n} := (n_1, ..., n_k)$ ranges over the multi-indices submitted to the above conditions and $\binom{k}{\underline{n}} := \frac{k!}{n_1! \cdot ... \cdot n_k!}$. Then the claim follows straightforwardly from the identity: $\alpha_P(e_1^{n_1} \cdot ... \cdot e_k^{n_k}) = n_1! \cdot ... \cdot n_k! \cdot e_1^{[n_1]} \cdot ... \cdot e_k^{[n_k]}$. For the general case we shall use the following

Claim 4.4.10. Let M, F be almost finitely generated projective A-modules, $\varepsilon \in \mathfrak{m}$, $u : M \to F$ and $v : F \to M$ morphisms with $v \circ u = \varepsilon \cdot 1_M$. Then $v \otimes u^*(\zeta_F) = \varepsilon \cdot \zeta_M$.

Proof of the claim. We have a commutative diagram

$$
\begin{array}{ccc}
F \otimes_A F^* & \xrightarrow{\;v \otimes u^*\;} & M \otimes_A M^* \\
\downarrow & & \downarrow \\
\mathrm{End}_A(F) & \longrightarrow & \mathrm{End}_A(M)
\end{array}
$$

where the vertical morphisms are the natural ones, and where the bottom morphism is given by $\psi \mapsto v \circ \phi \circ u$. Then the claim follows by an easy diagram chase. ◊

Pick morphisms $u : P \to F$ and $v : F \to P$ with $v \circ u = \varepsilon \cdot 1_P$ and F free of finite rank. We consider the commutative diagram

$$
\begin{array}{ccc}
\mathrm{Sym}_A^k F \otimes_A \mathrm{Sym}_A^k F^* & \xrightarrow{\;\mathrm{Sym}_A^k v \otimes \mathrm{Sym}_A^k u^*\;} & \mathrm{Sym}_A^k P \otimes_A \mathrm{Sym}_A^k P^* \\
\downarrow{\scriptstyle \alpha_F \otimes 1_{\mathrm{Sym}_A^k F^*}} & & \downarrow{\scriptstyle \alpha_P \otimes 1_{\mathrm{Sym}_A^k P^*}} \\
\Gamma_A^k F \otimes_A \mathrm{Sym}_A^k F^* & \xrightarrow{\;\Gamma_A^k v \otimes \mathrm{Sym}_A^k u^*\;} & \Gamma_A^k P \otimes_A \mathrm{Sym}_A^k P^*.
\end{array}
$$

By claim 4.4.10, we have $v \otimes u^*(\zeta_F) = \varepsilon \cdot \zeta_P$; whence $\mathrm{Sym}_A^k v \otimes \mathrm{Sym}_A^k u^*(\zeta_F^k) = \varepsilon^k \cdot \zeta_P^k$. Moreover, we remark that $(\Gamma_A^k v) \circ (\mathrm{Sym}_A^k u^*)^* = (\Gamma_A^k v) \circ (\Gamma_A^k u) = \varepsilon^k \cdot 1_{\Gamma_A^k P}$, therefore claim 4.4.10 (applied for $M = \Gamma_A^k P$) yields $\Gamma_A^k v \otimes \mathrm{Sym}_A^k u^*(\zeta_{\Gamma_A^k F}) = \varepsilon^k \cdot \zeta_{\Gamma_A^k P}$. Since we already know the lemma for F, a simple diagram chase shows that $(\alpha \otimes_A 1_{\mathrm{Sym}_A^k P^*})(\varepsilon^k \cdot \zeta_P^k) = \varepsilon^k \cdot k! \cdot \zeta_{\Gamma_A^k P}$. Since the k-th powers of elements of $\mathfrak{m}$ generate $\mathfrak{m}$, the claim follows. □

Lemma 4.4.11. *For P as above, suppose that ζ_P is nilpotent in $\mathrm{Sym}_A^\bullet(P \oplus P^*)_*$.*

   (i) *If $\mathbb{Q} \subset A_*$, then $\chi_P(X) = (1 + X)^{-\alpha}$ for some continuous function $\alpha : \mathrm{Spec}\, A_* \to \mathbb{N}$.*

   (ii) *If $\mathbb{F}_p \subset A_*$, then $\mathscr{E}_{P/A}$ is a Frobenius-nilpotent ideal.*

Proof. (i): Since $\mathbb{Q} \subset A_*$, then $k!$ is invertible in A_* for every $k \geq 0$. Hence, say that $\zeta_P^k = 0$; by lemma 4.4.9 it follows that $\mathrm{Sym}_A^k P = 0$, thus $\psi_P(X)$ is a polynomial. By corollary 4.3.8, $\psi_P(-X)$ defines an endomorphism of $\mathbb{G}_{m,A_*}$, therefore $\psi_P(X) = (1 + X)^\alpha$ for some continuous function $\alpha : \mathrm{Spec}\, A_* \to \mathbb{N}$; then the claim follows by corollary 4.3.8(ii).

   (ii): Let $(p, f) \in (P \oplus P^*)_*$; in the notation of the proof of lemma 4.4.3, we can write $(p, f) \in T(A)$. It follows that (p, f) corresponds to a morphism of A-algebras $\mathrm{Sym}_A^\bullet(p, f)^* : \mathrm{Sym}_A^\bullet(P \oplus P^*) \to A$. In particular $\mathrm{Sym}_A^\bullet(p, f)^*(\zeta_P^k) = \mathrm{Sym}_A^\bullet(p, f)^*(\zeta_P)^k$ for every $k \geq 0$. By inspecting the proof of claim 4.4.6, we deduce $\mathrm{Sym}_A^\bullet(p, f)^*(\zeta_P) = f(p)$ for every $f \in P^*$ and $p \in P$. By hypothesis, $\zeta_P^{p^n} = 0$ for every sufficiently large n. It follows that $\mathscr{E}_{P/A}$ is Frobenius-nilpotent. $\square$

Lemma 4.4.12. *Let R_0 be a noetherian commutative ring, R an R_0-algebra and M a flat R-module. Then $M = 0$ if and only if $M \otimes_{R_0} \kappa = 0$ for every residue field κ of R_0 (i.e., for every field $\kappa := \mathrm{Frac}(R_0/\mathfrak{p})$, where $\mathfrak{p}$ is some prime ideal of R_0).*

Proof. Clearly we have only to show the direction $\Leftarrow$. It suffices to show that $M_\mathfrak{p} = 0$ for every prime ideal of R_0. Hence we can assume that R_0 is local, in particular of finite Krull dimension. We proceed by induction on the dimension of R_0. If $\dim R_0 = 0$, then R_0 is a local artinian ring, hence a power of its maximal ideal $\mathfrak{m}$ is equal to 0. By assumption, $M/\mathfrak{m}M = 0$, i.e. $M = \mathfrak{m}M$. Then $M = \mathfrak{m}^k M$ for every $k \geq 0$, so $M = 0$. Next, suppose that $\dim R_0 = d > 0$ and the lemma already known for all rings of dimension strictly less than d. Assume first that R_0 is an integral domain and pick $f \in S := R_0 \setminus \{0\}$. Then R_0/fR_0 has dimension strictly less than d, so by induction we have $M/fM = 0$, i.e. $M = fM$. Due to the flatness of M, we have: $\mathrm{Ann}_M(f) = \mathrm{Ann}_R(f) \cdot M = \mathrm{Ann}_R(f) \cdot fM = 0$. This implies that the kernel of the natural map $M \to S^{-1}M$ is trivial. On the other hand, by hypothesis $S^{-1}M = 0$, whence $M = 0$ in this case. For a general R_0 of dimension d, notice that the above argument implies that, for every minimal prime ideal $\mathfrak{p}$ of R_0, we have $\mathfrak{p}M = M$. But the product of all (finitely many) minimal prime ideals is contained in the nilpotent radical $\mathfrak{R}$ of R_0, whence $\mathfrak{R}M = M$, and finally $M = 0$ as claimed. $\square$

4.4.13. Let now $\mathfrak{p} \in \mathrm{Spec}\, A_*$. By composing $\mathrm{f.rk}_A(P)$ with the map $A_* \to A_*/\mathfrak{p}$, we obtain a ring homomorphism

$$\mathrm{f.rk}_A(P, \mathfrak{p}) : \mathscr{G}_\infty \to A_*/\mathfrak{p}.$$

In case $(A_*/\mathfrak{p})^a \neq 0$, we can interpret $\mathrm{f.rk}_A(P, \mathfrak{p})$ as the formal rank of $P \otimes_A (A/\mathfrak{p}^a)$. More precisely, let $\pi : A \to A/\mathfrak{p}^a$ be the natural projection; then π_* factors through a map $\pi' : A_*/\mathfrak{p} \to (A/\mathfrak{p}^a)_*$ and we have: $\mathrm{f.rk}_{A/\mathfrak{p}^a}(P \otimes_A (A/\mathfrak{p}^a)) = \pi' \circ \mathrm{f.rk}_A(P, \mathfrak{p})$.

Even if $(A_*/\mathfrak{p})^a = 0$, the morphism $\mathrm{f.rk}_A(P,\mathfrak{p})$ can still be interpreted as the map associated to an endomorphism of $\widehat{\mathbb{G}}_{m,A_*/\mathfrak{p}}$, so it still makes sense to ask whether $\mathrm{f.rk}_A(P,\mathfrak{p})$ is an integer, as indicated in remark 4.3.10(ii).

Lemma 4.4.14. *Let P be as in (4.4) and $\mathfrak{p} \in \mathrm{Spec}\, A_*$. If B is any $A_{*\mathfrak{p}}^a$-algebra and $\mathfrak{q} \in \mathrm{Spec}\, B_*$ such that $r(\mathfrak{q}) := \mathrm{f.rk}_B(P \otimes_A B, \mathfrak{q})$ is an integer, then $r(\mathfrak{p}) := \mathrm{f.rk}_A(P,\mathfrak{p})$ is also an integer and $r(\mathfrak{p}) = r(\mathfrak{q})$.*

Proof. Indeed, let us consider the natural maps $\mathscr{G}_\infty \to A_* \to B_*$; under the assumptions, the contraction of $\mathfrak{q}$ in A_* is contained in $\mathfrak{p}$. Since the image of $\binom{\alpha}{i}$ in $B_*/\mathfrak{q}$ is $\binom{r(\mathfrak{q})}{i}$, it follows that the same holds in $A_*/\mathfrak{p}$.   □

Definition 4.4.15. We say that an A-module P *admits infinite splittings* if there is an infinite chain of decompositions of the form: $P \simeq A \oplus P_1$, $P_1 \simeq A \oplus P_2$, $P_2 \simeq A \oplus P_3$, ...

Theorem 4.4.16. *Let P be as in (4.4). The following conditions are equivalent:*

(i) *P is of almost finite rank.*
(ii) *For all A-algebras $B \neq 0$, we have: $\bigcap_{r>0} \mathscr{E}_{\Lambda_B^r(P \otimes_A B)/B} = 0$.*
(iii) *For all A-algebras $B \neq 0$, $P_B := P \otimes_A B$ does not admit infinite splittings, and moreover if $P_B \simeq B^n \oplus Q$ for some B-module Q such that $\chi_Q(X) = (1 + X)^{-\alpha}$ for some continuous function $\alpha : \mathrm{Spec}\, B_* \to \mathbb{N}$, then $Q = 0$.*
(iv) *For all A-algebras $B \neq 0$, P_B does not admit infinite splittings, and moreover if $P_B \simeq B^n \oplus Q$ for some B-module Q, then :*
   (a) *If $\mathbb{F}_p \subset B_*$ and $Q = IQ$ for a Frobenius-nilpotent ideal $I \subset B$, then $Q = 0$;*
   (b) *If $\mathbb{Q} \subset B_*$ and $\mathrm{Sym}_B^r Q = 0$ for some $r \geq 1$, then $Q = 0$.*

Proof. (i) $\Rightarrow$ (ii): Indeed, from proposition 2.4.28(iii) one sees that, for every A-module of almost finite rank, every A-algebra B and every $\varepsilon \in \mathfrak{m}$, there exists $r \geq 0$ such that $\varepsilon \cdot \mathscr{E}_{\Lambda_B^r(P \otimes_A B)/B} = 0$.

(ii) $\Rightarrow$ (iii): Let $B \neq 0$ be an A-algebra; by hypothesis, there exists $r \geq 0$ such that $J_r := \mathscr{E}_{\Lambda_B^r(P \otimes_A B)/B} \neq B$. Suppose that $P \otimes_A B$ admits infinite splittings. The B/J_r-module $P_B/J_r P_B$ has rank $< r$, and at the same time it admits infinite splittings, a contradiction.

Suppose next, that there is a decomposition $P \otimes_A B \simeq B^n \oplus Q$; then Q is obviously of almost finite rank. Suppose that $\chi_Q(X) = (1 + X)^\alpha$ has the shape described in (ii). We reduce easily to the case where α is a constant function. However, $\chi_{Q/J_r Q}(X)$ is a polynomial of degree $< r$, thus $\alpha = 0$, and then $Q = 0$ by theorem 4.3.28(ii).

(iii) $\Rightarrow$ (iv): Suppose that $\mathbb{F}_p \subset B_*$ and $Q = IQ$ for some Frobenius-nilpotent ideal I. Then $\chi_Q(X) \in 1 + I_*[[X]]$, which means that the image of $\chi_Q(X)$ in $\mathrm{End}(\widehat{\mathbb{G}}_{m,B_*/I_*})$ is the trivial endomorphism. But we have a commutative diagram

$$\begin{array}{ccc}
\mathscr{C}^0(\operatorname{Spec} B_*, \mathbb{Z}_p) & \longrightarrow & \operatorname{End}(\widehat{\mathbb{G}}_{m,B_*}) \\
\downarrow & & \downarrow \\
\mathscr{C}^0(\operatorname{Spec} B_*/I_*, \mathbb{Z}_p) & \longrightarrow & \operatorname{End}(\widehat{\mathbb{G}}_{m,B_*/I_*})
\end{array}$$

where the horizontal maps are those defined in remark 4.3.10(ii), and are bijective by corollary 4.2.22. The left vertical map is induced by restriction to the closed subset $\operatorname{Spec} B_*/I_*$, and since I is Frobenius-nilpotent, it is a bijection as well. It follows that the right vertical map is bijective, whence $\mathrm{f.rk}_B(Q) = 0$, and finally $Q = 0$ by (iii).

Next, consider the case when $\mathbb{Q} \subset B_*$ and $\operatorname{Sym}_B^r Q = 0$ for some $r \geq 1$. It follows that ζ_Q is nilpotent in $\operatorname{Sym}_B^\bullet(Q \oplus Q^*)$. By lemma 4.4.11(i), $\chi_Q(X) = (1 + X)^{-\alpha}$ for some continuous $\alpha : \operatorname{Spec} B_* \to \mathbb{N}$. Then (iii) implies that $Q = 0$.

To show that (iv) $\Rightarrow$ (i), we will use the following:

Claim 4.4.17. Assume (iv). Then: $\operatorname{Split}(B, Q) = 0 \Rightarrow Q = 0$.

Proof of the claim. Suppose $Q \neq 0$ and $\operatorname{Split}(B, Q) = 0$; then (4.4.4) implies that for every $\varepsilon \in \mathfrak{m}$ there exists $j \geq 0$ such that $\varepsilon \cdot \zeta_Q^j = 0$. We have $\varepsilon Q \neq 0$ for some $\varepsilon \in \mathfrak{m}$. From the flatness of Q, we derive $\operatorname{Ann}_{\operatorname{Sym}_B^\bullet(Q \oplus Q^*)}(\varepsilon) = \operatorname{Ann}_B(\varepsilon) \cdot \operatorname{Sym}_B^\bullet(Q \oplus Q^*)$, hence we can replace B by $B/\operatorname{Ann}_B(\varepsilon)$, Q by $Q/\operatorname{Ann}_B(\varepsilon) \cdot Q$, thereby achieving that ζ_Q is nilpotent in $\operatorname{Sym}_B^\bullet(Q \oplus Q^*)$ and still $Q \neq 0$. Using lemma 4.4.12 (and the functoriality of $\operatorname{Split}(B, Q)$ for base extensions $B \to B'$) we can further assume that B_* contains either $\mathbb{Q}$ or one of the finite fields $\mathbb{F}_p$. If $\mathbb{Q} \subset B_*$, then $k!$ is invertible in B_* for every $k \geq 0$; by lemma 4.4.9 it follows that $\operatorname{Sym}_B^k Q = 0$, whence $Q = 0$ by (iv), a contradiction. If $\mathbb{F}_p \subset B_*$, then by lemma 4.4.11(ii), $\mathscr{E}_{Q/B}$ is Frobenius-nilpotent. However, from proposition 2.4.28(i),(iii) it follows easily that $Q = \mathscr{E}_{Q/B} \cdot Q$, whence $Q = 0$, again by (iv), and again a contradiction. In either case, this shows that $\operatorname{Split}(B, Q) \neq 0$, as claimed. ◊

Claim 4.4.18. Assume (iv). Then $A/\mathscr{E}_{P/A}$ is a flat A-algebra.

Proof of the claim. It suffices to show that $A_\mathfrak{p}/(\mathscr{E}_{P/A})_\mathfrak{p}$ is a flat $A_\mathfrak{p}$-algebra for every prime ideal $\mathfrak{p} \subset A_*$. If $\mathscr{E}_{P/A_*}$ is not contained in $\mathfrak{p}$, then $(\mathscr{E}_{P/A})_\mathfrak{p} = A_\mathfrak{p}$, so there is nothing to prove in this case. We assume therefore that

(4.4.19) $$\mathscr{E}_{P/A_*} \subset \mathfrak{p}.$$

We will show that $P_\mathfrak{p} = 0$ in such case, whence $A_\mathfrak{p}/(\mathscr{E}_{P/A})_\mathfrak{p} = A_\mathfrak{p}$, so the claim will follow. From (4.4.19) and corollary 4.3.8(iii) we know already that

(4.4.20) $$\mathrm{f.rk}_A(P, \mathfrak{p}) = 0.$$

Suppose that $P_\mathfrak{p} \neq 0$; then there exists $\varepsilon \in \mathfrak{m}$ such that $\varepsilon P_\mathfrak{p} \neq 0$. Define inductively $A_0 := A_\mathfrak{p}$, $Q_0 := P_\mathfrak{p}$, $A_{i+1} := \operatorname{Split}(A_i, Q_i)$ and Q_{i+1} as an A_{i+1}-module such that $Q_i \otimes_{A_i} A_{i+1} \simeq A_{i+1} \oplus Q_{i+1}$, for every $i \geq 0$ (the existence of Q_{i+1} is assured by lemma 4.4.3(iii)). Then $\operatorname{colim}_{n \in \mathbb{N}} A_n \simeq 0$, since, after base change to this V^a-algebra, P admits infinite splittings. This implies that there exists $n \in \mathbb{N}$

such that $\varepsilon A_{n+1} = 0$ and $\varepsilon A_n \neq 0$. However, since A_{n+1} is flat over A_n, we have: $A_{n+1} = \mathrm{Ann}_{A_{n+1}}(\varepsilon) = \mathrm{Ann}_{A_n}(\varepsilon) \cdot A_{n+1}$. Set $A' := A_n/\mathrm{Ann}_{A_n}(\varepsilon)$; then $\mathrm{Split}(A', Q_n \otimes_{A_n} A') = 0$, so $Q_n \otimes_{A_n} A' = 0$ by claim 4.4.17. By flatness of Q_n, this means that $\varepsilon Q_n = 0$; in particular, $n > 0$. By definition, $Q_{n-1} \otimes_{A_{n-1}} A_n \simeq A_n \oplus Q_n$; it follows that $Q' := Q_{n-1} \otimes_{A_{n-1}} A' \simeq A'$, in particular $\mathrm{f.rk}_{A'}(Q') = 1$ and consequently $\mathrm{f.rk}_{A'}(P \otimes_A A') = n > 0$; in view of lemma 4.4.14, this contradicts (4.4.20), therefore $P_\mathfrak{p} = 0$, as required. $\Diamond$

Claim 4.4.21. Assuming (iv), the natural morphism

$$\phi : A \to (A/\mathscr{E}_{P/A}) \times \mathrm{Split}(A, P)$$

is faithfully flat.

Proof of the claim. The flatness is clear from claim 4.4.18. Hence, to prove the claim, it suffices to show that $((A/\mathscr{E}_{P/A}) \times \mathrm{Split}(A, P)) \otimes_A (A/J) \neq 0$ for every proper ideal $J \subset A,$. But the construction of ϕ commutes with arbitrary base changes $A \to A'$, therefore we are reduced to verify that $(A/\mathscr{E}_{P/A}) \times \mathrm{Split}(A, P) \neq 0$ when $A \neq 0$. By claim 4.4.17, this can fail only when $P = 0$; but in this case $\mathscr{E}_{P/A} = 0$, so the claim follows. $\Diamond$

We can now conclude the proof of the theorem: define inductively as in the proof of claim 4.4.18: $A_0 := A$, $Q_0 := P$, $A_{i+1} := \mathrm{Split}(A_i, Q_i)$ and Q_{i+1} as an A_{i+1}-module such that $Q_i \otimes_{A_i} A_{i+1} = A_{i+1} \oplus Q_{i+1}$, for every $i \geq 0$. The same argument as in *loc. cit.* shows that, for every $\varepsilon \in \mathfrak{m}$, there exists $n \in \mathbb{N}$ such that $\varepsilon A_n = 0$. Moreover, by claim 4.4.21 (and an easy induction), $B := A_0/\mathscr{E}_{Q_0/A_0} \times A_1/\mathscr{E}_{Q_1/A_1} \times \dots \times A_{n-1}/\mathscr{E}_{Q_{n-1}/A_{n-1}} \times A_n$ is a faithfully flat A-algebra. However, one checks easily by induction that $P \otimes_A (A_i/\mathscr{E}_{Q_i/A_i})$ is a free $A_i/\mathscr{E}_{Q_i/A_i}$-module of rank i, for every $i < n$. Hence, $\Lambda^n_B (P \otimes_A B) \simeq \Lambda^n_{A_n} (P \otimes_A A_n)$, which is therefore killed by ε. By faithful flatness, so is $\Lambda^n_A P$. The proof is concluded. $\square$

Proposition 4.4.22. *If P is a faithfully flat almost projective A-module of almost finite rank, then $\mathrm{Split}(A, P)$ is faithfully flat over A.*

Proof. If $A = 0$ there is nothing to prove, so we assume that $A \neq 0$. In this case, it suffices to show that $\mathrm{Split}(A, P) \otimes_A A/I \neq 0$ for every proper ideal I of A. However, $\mathrm{Split}(A, P) \otimes_A A/I \simeq \mathrm{Split}(A/I, P/IP)$, and since P is faithfully flat, $P/IP \neq 0$; hence we are reduced to showing that $\mathrm{Split}(A, P) \neq 0$ when P is faithfully flat. Suppose that $\mathrm{Split}(A, P) = 0$; then (4.4.4) implies that for every $\varepsilon \in \mathfrak{m}$ there exists $j \geq 0$ such that $\varepsilon \cdot \zeta_P^j = 0$. Since $P \neq 0$, we have $\varepsilon P \neq 0$ for some $\varepsilon \in \mathfrak{m}$. From the flatness of P, we derive $\mathrm{Ann}_{\mathrm{Sym}^\bullet_A(P \oplus P^*)}(\varepsilon) = \mathrm{Ann}_A(\varepsilon) \cdot \mathrm{Sym}^\bullet_A(P \oplus P^*)$, hence we can replace A by $A/\mathrm{Ann}_A(\varepsilon)$, P by $P/\mathrm{Ann}_A(\varepsilon) \cdot P$, which allows us to assume that ζ_P is nilpotent in $\mathrm{Split}(A, P)$. Using lemmata 4.4.12 and 4.4.3(ii), we can further assume that A_* contains either $\mathbb{Q}$ or one of the finite fields $\mathbb{F}_p$. If $\mathbb{Q} \subset A_*$, then $k!$ is invertible in A_* for every $k \geq 0$; by lemma 4.4.9 it follows that $\mathrm{Sym}^k_A P = 0$, whence $P = 0$ by the implication (i)$\Rightarrow$(iv) of theorem 4.4.16; this contradicts our assumptions, so the proposition is proved in this case. Finally, suppose that $\mathbb{F}_p \subset A_*$, then by lemma 4.4.11(ii), $\mathscr{E}_{P/A}$ is Frobenius-nilpotent.

However, since P is faithfully flat, proposition 2.4.28(iv) says that $\mathcal{E}_{P/A} = A$, so $A = 0$, which again contradicts our assumptions.   □

4.4.23.   For any V^a-algebra A we have a (large) fpqc site on the category $(A\text{-}\mathbf{Alg})^o$ (in some fixed universe!); as usual, this site is defined by the pretopology whose covering families are the finite families $\{\operatorname{Spec} C_i \to \operatorname{Spec} B \mid i = 1, ..., n\}$ such that the induced morphism $B \to C_1 \times ... \times C_n$ is faithfully flat (notation of (3.3.3)).

Theorem 4.4.24. *Every almost projective A-module of finite rank is locally free of finite rank in the fpqc topology of $(A\text{-}\mathbf{Alg})^o$.*

Proof. We may assume (by proposition 4.3.27) that the given module is of constant rank r. We iterate the construction of $\operatorname{Split}(A, P)$ to split off successive free sub-modules of rank one. We use the previous characterization of modules of finite rank (proposition 4.3.27) to show that this procedure stops after r iterations. By proposition 4.4.22, the output of this procedure is a faithfully flat A-algebra.   □

Theorem 4.4.24 allows to prove easily results on almost projective modules of finite rank, by reduction to the case of free modules. Here are a few examples of this method.

Lemma 4.4.25. *Let P be an almost projective A-module of constant rank equal to $r \in \mathbb{N}$. Then, for every integer $0 \leq k \leq r$, the natural morphism*

(4.4.26) $$\Lambda_A^k P \otimes_A \Lambda_A^{r-k} P \to \Lambda_A^r P   :   x \otimes y \mapsto x \wedge y$$

is a perfect pairing.

Proof. By theorem 4.4.24, there exists a faithfully flat A-algebra B such that $P_B := P \otimes_A B$ is a free B-module of rank r. It suffices to prove the assertion for the B-module P_B, in which case the claim is well known.   □

4.4.27.   Keep the assumptions of lemma 4.4.25. Taking $k = 1$ in (4.4.26), we derive a natural isomorphism

$$\beta_P : (\Lambda_A^{r-1} P)^* \xrightarrow{\sim} P \otimes_A (\Lambda_A^r P)^*.$$

Now, let us consider an A-linear morphism $\phi : P \to Q$ of A-modules of constant rank equal to r. We set

$$\psi := \beta_P \circ (\Lambda_A^{r-1}\phi)^* \circ \beta_Q^{-1} : Q \otimes_A (\Lambda_A^r Q)^* \to P \otimes_A (\Lambda_A^r P)^*.$$

Proposition 4.4.28. *With the notation of (4.4.27), we have:*

$$\psi \circ (\phi \otimes_A \mathbf{1}_{(\Lambda_A^r Q)^*}) = \mathbf{1}_P \otimes_A (\Lambda_A^r \phi)^*   \text{ and }   (\phi \otimes_A \mathbf{1}_{(\Lambda_A^r P)^*}) \circ \psi = \mathbf{1}_Q \otimes_A (\Lambda_A^r \phi)^*.$$

Especially, ϕ is an isomorphism if and only if the same holds for $\Lambda_A^r \phi$.

Proof. After faithfully flat base change, we can assume that P and Q are free modules of rank r. Then we recognize Cramer's rule in the above identities.   □

4.4.29. Keep the assumption of lemma 4.4.25 and let ϕ be an A-linear endomorphism of P. Set $B := A[X, X^{-1}]$ and $P_B := P \otimes_A B$. Obviously ϕ induces a Λ-nilpotent endomorphism of P_B, which we denote by the same letter. Hence we can define

$$\chi_\phi(X) := X^r \cdot \det(1_{P_B} - X^{-1} \cdot \phi) \in A_*[X]$$

(notation of (4.3.1)).

Proposition 4.4.30. *With the notation of (4.4.29), we have* $\chi_\phi(\phi) = 0$ *in* $\operatorname{End}_A(P)$.

Proof. Again, we can reduce to the case of a free A-module of rank r, in which case we conclude by Cayley-Hamilton. □

Corollary 4.4.31. *Keep the assumptions of (4.4.29), and suppose that ϕ is integral over a subring $S \subset A_*$. Then the coefficients of χ_ϕ are integral over S.*

Proof. The assumption means that ϕ satisfies an equation of the kind

$$\phi^n + \sum_{i=0}^{n-1} a_i \phi^i = 0$$

where $a_i \in S$ for every $i = 0, ..., n-1$. We can assume that P is free, in which case we reduce to the case of an endomorphism of a free R-module of finite rank, where R is a usual commutative ring containing S; we can further suppose that R is of finite type over $\mathbb{Z}$, and it is easily seen that we can even replace R by its associated reduced ring $R/\operatorname{nil}(R)$. Then R injects into a finite product of fields $\prod_i K_i$, and we can replace S by its integral closure in $\prod_i K_i$, which allows us to reduce to the case where R is a field. In this case the coefficients of χ_ϕ are elementary symmetric polynomials in the eigenvalues of ϕ, so it suffices to show that these eigenvalues $e_1, ..., e_r$ are integral over S. But this is clear, since we have more precisely $e_j^n + \sum_{i=0}^{n-1} a_i e_j^i = 0$ for every $j \leq r$. □

To conclude this section, we want to apply the previous results to analyze in some detail the structure of invertible modules : it turns out that the notion of invertibility is rather more subtle than for usual modules over rings.

Definition 4.4.32. Let M be an invertible A-module. Clearly $M \otimes_A M$ is invertible as well, consequently the map $A \to \operatorname{End}_A(M \otimes_A M) : a \mapsto a \cdot 1_{M \otimes_A M}$ is an isomorphism (by the proof of lemma 4.1.5(ii)). Especially, for the transposition endomorphism $\theta_{M|M}$ of $M \otimes_A M : x \otimes y \mapsto y \otimes x$, there exists a unique element $u_M \in A_*$ such that $\theta_{M|M} = u_M \cdot 1_{M \otimes_A M}$. Clearly $u_M^2 = 1$. We say that M is *strictly invertible* if $u_M = 1$.

Lemma 4.4.33. *For an invertible A-module M the following are equivalent:*

   (i) *M is strictly invertible;*
   (ii) *$\Lambda_A^2 M = 0$;*
   (iii) *M is of almost finite rank;*
   (iv) *there exists a faithfully flat A-algebra B such that $M \otimes_A B \simeq B$.*

Proof. (i) $\Rightarrow$ (ii): Indeed, the condition $u_M = 1$ says that the antisymmetrizer operator $a_2 : M^{\otimes 2} \to M^{\otimes 2}$ vanishes (cp. the proof of lemma 4.3.26); since M is flat, (ii) follows.

(ii) $\Rightarrow$ (iii) and (iv) $\Rightarrow$ (i) are obvious. To show that (iii) $\Rightarrow$ (iv) let us set $B :=$ Split(A, M); by proposition 4.4.22 B is faithfully flat over A, and $B \otimes_A M \simeq B \oplus X$ for some B-module X. Clearly $B \otimes_A M$ is an invertible B-module, therefore, by lemma 4.1.5(ii), the evaluation morphism gives an isomorphism

$$(B \oplus X) \otimes_A (B \oplus X^*) \simeq B \oplus X \oplus X^* \oplus (X \otimes_A X^*) \simeq B.$$

By inspection, the restriction of the latter morphism to the direct summand B equals the identity of B; hence $X = 0$ and (iv) follows. □

Lemma 4.4.34. *If M is invertible, then* $\operatorname{tr}_{M/A}(1_M) = u_M$.

Proof. Pick arbitrary $f \in M_*^*$, $m, n \in M_*$. Then, directly from the definition of u_M we deduce that $f(m) \cdot n = u_M \cdot f(n) \cdot m$. In other words, $\omega_{M/A}(n \otimes f) = u_M \cdot \operatorname{ev}_{M/A}(n \otimes f) \cdot 1_M$. By linearity we deduce that $\phi = u_M \cdot \operatorname{tr}_{M/A}(\phi) \cdot 1_M$ for every $\phi \in \operatorname{End}_A(M)$. By letting $\phi := 1_M$, and taking traces on both sides, we obtain: $\operatorname{tr}_{M/A}(1_M) = u_M \cdot \operatorname{tr}_{M/A}(1_M)^2$. But since M is invertible, $\operatorname{tr}_{M/A}(1_M)$ is invertible in A_*, whence $u_M \cdot \operatorname{tr}_{M/A}(1_M) = 1$, which is equivalent to the sought identity. □

Proposition 4.4.35. *Let M be an invertible A-module. Then:*

(i) *$M \otimes_A M$ is strictly invertible.*

(ii) *There exists a natural decomposition $A \simeq A_1 \times A_{-1}$ where $M \otimes_A A_1$ is strictly invertible, A_{-1*} is a $\mathbb{Q}$-algebra and $\operatorname{Sym}^2_{A_{-1}}(M \otimes_A A_{-1}) = 0$.*

Proof. (i): It is clear that $M^{\otimes n}$ is invertible for every n. Let $\sigma \in S_n$ be any permutation; it is easy to verify that the morphism

$$\sigma_M : M^{\otimes n} \to M^{\otimes n} \; : \; x_1 \otimes x_2 \otimes \ldots \otimes x_n \mapsto x_{\sigma(1)} \otimes x_{\sigma(2)} \otimes \ldots \otimes x_{\sigma(n)}$$

equals $u_M^{\operatorname{sgn}(\sigma)} \cdot 1_{M^{\otimes n}}$. Especially, the transposition operator on $(M \otimes_A M)^{\otimes 2}$ acts via the permutation: $x \otimes y \otimes z \otimes w \mapsto z \otimes w \otimes x \otimes y$ whose sign is even. Therefore $u_{M \otimes_A M} = 1$, which is (i).

(ii): It follows from (i) that the antisymmetrizer operator a_2 on $(M \otimes_A M)^{\otimes 2}$ vanishes; *a fortiori* it vanishes on the quotient $(\Lambda^2_A M)^{\otimes 2}$, therefore $\Lambda^2_A(\Lambda^2_A M) \simeq \operatorname{Im}(a_2 : (\Lambda^2_A M)^{\otimes 2} \to (\Lambda^2_A M)^{\otimes 2}) = 0$. Then lemma 4.3.12 says that there exists a natural decomposition $A \simeq A_1 \times A_{-1}$ such that $(\Lambda^2_A M) \otimes_A A_1 = 0$ and $(\Lambda^2_A M) \otimes_A A_{-1}$ is invertible. To show that A_{-1*} is a $\mathbb{Q}$-algebra, it is enough to show that $A_{-1}/pA_{-1} = 0$ for every prime p. Up to replacing A by A/pA, we reduce to verifying that, if $\mathbb{F}_p \subset A_*$ and M is invertible, then M is of almost finite rank. To this aim, it suffices to verify that the equivalent condition (iv) of theorem 4.4.16 is satisfied. If $B \neq 0$ and $M_B := M \otimes_A B \simeq B \oplus X$, then the argument in the proof of lemma 4.4.33 shows that $X = 0$ and therefore M_B does not admit infinite splittings. Finally, it remains only to verify condition (a) of *loc. cit.* So suppose that $M_B \simeq B^n \oplus Q$. If $n > 0$, we have just seen that $Q = 0$; if $n = 0$, and $Q/IQ = 0$ for some ideal I, then by the faithfulness of M (lemma 4.1.5(iii)) we must have $I = B$;

if I is Frobenius nilpotent it follows that $B = 0$. Finally, set $M_{-1} := M \otimes_A A_{-1}$; notice that, since A_{-1*} is a $\mathbb{Q}$-algebra, the endomorphism group of $\widehat{\mathbb{G}}_{m,A_{-1*}}$ is isomorphic to A_{-1*}, and therefore $\chi_{M_{-1}}(X) = (1+X)^\alpha$, where α is an element of A_{-1*} which can be determined by looking at the coefficient of $\chi_{M_{-1}}(X)$ in degree 1. One finds $\alpha = \mathrm{tr}_{M_{-1}/A_{-1}}(1_{M_{-1}})$. In view of lemma 4.4.34, we can rewrite $\alpha = u_{M_{-1}}$; therefore

$$(4.4.36) \qquad \mathrm{tr}_{\Lambda^2_{A_{-1}} M_{-1}/A_{-1}}(1_{\Lambda^2_{A_{-1}} M_{-1}}) = \binom{u_{M_{-1}}}{2}.$$

On the other hand, since $\Lambda^2_{A_{-1}} M_{-1}$ is an invertible A_{-1}-module of finite rank, we know that the left-hand side of (4.4.36) equals 1; consequently $u_{M_{-1}} = -1$. This means that, in $(M_{-1}^{\otimes 2})_*$, the identity $x \otimes y = -y \otimes x$ holds for every $x, y \in M_{-1*}$; therefore, the kernel of the projection $M_{-1}^{\otimes 2} \to \mathrm{Sym}^2_{A_{-1}} M_{-1}$ contains all the elements of the form $2 \cdot x \otimes y$; in other words, multiplication by 2 is the zero morphism in $\mathrm{Sym}^2_{A_{-1}} M_{-1}$; since A_{-1*} is a $\mathbb{Q}$-algebra, this at last shows that $\mathrm{Sym}^2_{A_{-1}} M_{-1}$ vanishes, and concludes the proof of the proposition. □

4.5. Construction of quotients by flat equivalence relations. We will need to recall some generalities on groupoids, which we borrow from [25, Exp. V].

4.5.1. If $\mathscr{C}$ is any category admitting fibred products and a final object, a $\mathscr{C}$-*groupoid* is the datum of two objects X_0, X_1 of $\mathscr{C}$, together with "source" and "target" morphisms $s, t : X_1 \to X_0$, an "identity" morphism $\iota : X_0 \to X_1$ and a further "composition" morphism $c : X_2 \to X_1$, where X_2 is the fibre product in the cartesian diagram:

$$
\begin{array}{ccc}
X_2 & \xrightarrow{\;t'\;} & X_1 \\
{\scriptstyle s'}\downarrow & & \downarrow{\scriptstyle s} \\
X_1 & \xrightarrow{\;t\;} & X_0.
\end{array}
$$

The datum $(X_0, X_1, s, t, c, \iota)$ is subject to the following condition. For every object S of $\mathscr{C}$, the set $X_0(S) := \mathrm{Hom}_{\mathscr{C}}(S, X_0)$ is the set of objects of a groupoid, with set of morphisms given by $X_1(S)$, and for every $\phi \in X_1(S)$, the source and target of ϕ are respectively $s(\phi) := \phi \circ s$ and $t(\phi) := \phi \circ t$; furthermore the composition law in $X_1(S)$ is given by $c(S) : X_1(S) \times_{X_0(S)} X_1(S) \to X_1(S)$ and $\iota(S)$ sends every object to the corresponding identity morphism. The above conditions amount to saying that $s \circ \iota = t \circ \iota = 1_{X_0}$ and that the diagrams

$$(4.5.2) \qquad
\begin{array}{ccc}
X_2 & \overset{c}{\underset{t'}{\rightrightarrows}} & X_1 \\
{\scriptstyle s'}\downarrow & & \downarrow{\scriptstyle s} \\
X_1 & \overset{s}{\underset{t}{\rightrightarrows}} & X_0
\end{array}
\qquad
\begin{array}{ccc}
X_2 & \xrightarrow{\;c\;} & X_1 \\
{\scriptstyle t'}\downarrow & & \downarrow{\scriptstyle t} \\
X_1 & \xrightarrow{\;t\;} & X_0
\end{array}
$$

are commutative and cartesian in $\mathscr{C}$ both for the square made up from the upper arrows and for the square made up from the lower arrows, and moreover that commutativity holds for diagrams that translate the associativity of c and the identity property of ι: cp. [25, Exp. V §1].

One says that the groupoid $G := (X_0, X_1, s, t, c, \iota)$ *has trivial automorphisms*, if the morphism $(s,t) : X_1 \to X_0 \times X_0$ is a (categorical) monomorphism. (This translates in categorical terms the requirement that for every object S of $\mathscr{C}$, and every $x \in X_0(S)$, the automorphism group of x in $G(S)$ is trivial.)

It is sometimes convenient to denote by $X \times_{(\alpha, \beta)} Z$ the fibre product of two morphisms $\alpha : X \to Y$ and $\beta : Z \to Y$.

4.5.3. Given a groupoid G, and a morphism $X_0' \to X_0$, we obtain a new groupoid $G_{X_0'} := G \times_{X_0} X_0'$; the easiest way to define it is by describing its S-points. Indeed, for a set A let A_i be the category with object set A and having exactly one morphism between any two objects. Then for any object S of $\mathscr{C}$

$$G_{X_0'}(S) := G(S) \times_{X_0(S)_i} X_0'(S)_i$$

(strict fibred product of categories). Moreover, suppose that $\mathscr{C}$ admits finite coproducts and that all such coproducts are disjoint universal (cp. [4, Exp.II, Def.4.5]). Denote by $Y \amalg Z$ the coproduct of two objects Y and Z of $\mathscr{C}$. Let $G' := (X_0', X_1', s', t', c', \iota')$ be another groupoid; one can define a groupoid $G \amalg G'$ by taking the datum

$$(X_0 \amalg X_0', X_1 \amalg X_1', s \amalg s', t \amalg t', c \amalg c', \iota \amalg \iota').$$

4.5.4. We will be concerned with groupoids in the category $A\text{-}\mathbf{Alg}^o$ of affine A-schemes, where A is any V^a-algebra. We will use the general terminology for almost schemes introduced in (3.3.3), and complemented by the following:

Definition 4.5.5. Let $G := (X_0, X_1, s, t, c, \iota)$ be a groupoid in the category of A-schemes, $\phi : X \to Y$ a morphism of affine A-schemes.

(i) We say that ϕ is a *closed imbedding* (resp. is *almost finite*, resp. is *étale*, resp. is *flat*, resp. is *almost projective*) if the corresponding morphism $\phi^\sharp : \mathscr{O}_Y \to \mathscr{O}_X$ is an epimorphism of $\mathscr{O}_Y$-modules (resp. enjoys the same property). We say that ϕ is an *open and closed imbedding* if it induces an isomorphism $X \xrightarrow{\sim} Y_1$ onto one of the factors of a decomposition $Y = Y_1 \amalg Y_2$.

(ii) We say that G is a *closed equivalence relation* if the morphism $(s,t) : X_1 \to X_0 \times X_0$ is a a closed imbedding. We say that G is *flat* (resp. *étale*, resp. *almost finite*, resp. *almost projective*) if the morphism $s : X_1 \to X_0$ enjoys the same property. We say that G is *of finite rank* if $\mathscr{O}_{X_1}$ is an almost projective $\mathscr{O}_{X_0}$-module of finite rank, when the structure of $\mathscr{O}_{X_0}$-module on $\mathscr{O}_{X_1}$ is deduced from $s^\sharp$. Furthermore, we set $X_0/G := \operatorname{Spec} \mathscr{O}_{X_0}^G$, where $\mathscr{O}_{X_0}^G \subset \mathscr{O}_{X_0}$ is the equalizer of the morphisms $s^\sharp$ and $t^\sharp$.

4.5.6. Let $G := (X_0, X_1, s, t, c, \iota)$ be a groupoid of finite rank in A-$\mathbf{Alg}^o$. By assumption $\mathcal{O}_{X_1}$ is an almost projective $\mathcal{O}_{X_0}$-module of finite rank, hence, by proposition 4.3.27, there is a decomposition $\mathcal{O}_{X_0} \simeq \prod_{i=0}^{r} B_i$ such that $C_i := \mathcal{O}_{X_1} \otimes_{\mathcal{O}_{X_0}} B_i$ is of constant rank equal to i, for $i = 0, ..., r$. Since s has a section, we have $B_0 = 0$. Set $X_{0,i} := \operatorname{Spec} B_i$.

Lemma 4.5.7. *In the situation* (4.5.6), *there is a natural isomorphism of groupoids:*

$$G \simeq (G \times_{X_0} X_{0,1}) \amalg ... \amalg (G \times_{X_0} X_{0,r}).$$

Proof. For every $i \le r$, let $\alpha_i : X_{0,i} \to X_0$ be the open and closed imbedding defined by (4.5.6). Set $X_{1,i} := X_{0,i} \times_{(\alpha_i,s)} X_1$ (so $X_{1,i} = \operatorname{Spec} C_i$). Moreover, let $X'_{1,i} := X_{0,i} \times_{(\alpha_i,t)} X_1$, $\beta_i : X'_{1,i} \to X_1$ the open and closed imbedding (obtained by pulling back α_i), $X_{2,i} := X_{1,i} \times_{(\beta_i,s')} X_2$ and $X'_{2,i} := X'_{1,i} \times_{(\beta_i,s')} X_2$. There follow natural decompositions $X_1 \simeq X'_{1,1} \amalg ... \amalg X'_{1,r}$ and $X_2 \simeq X'_{2,1} \amalg ... \amalg X'_{2,r}$, such that s' decomposes as a coproduct of morphisms $X'_{2,i} \to X'_{1,i}$. By the construction of $X_{0,i}$, it is clear that $\mathcal{O}_{X'_{2,i}}$ has constant rank equal to i as an $\mathcal{O}_{X'_{1,i}}$-module, for every $i \le r$. In other words, the above decompositions fulfill the conditions of proposition 4.3.27. Similarly, we obtain decompositions $X_1 \simeq X_{1,1} \amalg ... \amalg X_{1,r}$ and $X_2 \simeq X_{2,1} \amalg ... \amalg X_{2,r}$ which fulfill the same conditions. However, these conditions characterize uniquely the factors occuring in it, thus $X_{1,i} = X'_{1,i}$ for $i \le r$. The claim follows easily. $\quad\square$

Lemma 4.5.8. *Let* $G := (X_0, X_1, s, t, c, \iota)$ *be a groupoid of finite rank in* A-$\mathbf{Alg}^o$. *If G has trivial automorphisms, then it is a closed equivalence relation.*

Proof. Using lemma 4.5.7 we reduce easily to the case where $\mathcal{O}_{X_1}$ is of constant rank, say equal to $r \in \mathbb{N}$. Let $Y := X_0 \times X_0$; since G has trivial automorphisms, the morphism $(s,t) : X_1 \to Y$ is a monomorphism; equivalently, the natural projections $\operatorname{pr}_1, \operatorname{pr}_2 : X_1 \times_Y X_1 \to X_1$ are isomorphisms. Let

$$D := \operatorname{Im}((s,t)^\sharp : \mathcal{O}_Y \to \mathcal{O}_{X_1}).$$

It follows that the natural morphisms $\operatorname{pr}_1^\sharp, \operatorname{pr}_2^\sharp : \mathcal{O}_{X_1} \to \mathcal{O}_{X_1} \otimes_D \mathcal{O}_{X_1}$ are isomorphisms and consequently,

(4.5.9) $(\mathcal{O}_{X_1}/D) \otimes_D \mathcal{O}_{X_1} = 0.$

We need to show that $\mathcal{O}_{X_1} = D$, or equivalently, that $\mathcal{O}_{X_1}/D = 0$. However, by theorem 4.4.24, we can find a faithfully flat $\mathcal{O}_{X_0}$-algebra B such that $C := B \otimes_{\mathcal{O}_{X_0}} \mathcal{O}_{X_1} \simeq B^r$. Let $D' := B \otimes_{\mathcal{O}_{X_0}} D$ (here the $\mathcal{O}_{X_0}$-algebra structure of D is deduced from $s^\sharp$); it follows that C is a faithful finitely generated D'-module. It suffices to show that $C/D' = 0$, and we know already from (4.5.9) that $(C/D') \otimes_{D'} C = 0$. By proposition 3.4.4 it follows that C/D' is a flat D'-module; consequently $C/D' \subset (C/D') \otimes_{D'} C$, and the claim follows. $\quad\square$

4.5.10.   Let B be an A-algebra, P an almost finitely generated projective B-module. For every integer $i \geq 0$, we define a B-linear morphism

(4.5.11) $$\Gamma_B^i(\mathrm{End}_B(P)^a) \to B$$

as follows (see (8.1.14) for the definition of the functor $\Gamma_B^i : B\text{-}\mathbf{Mod} \to B\text{-}\mathbf{Mod}$).
Proposition 8.1.17(i), applied to the B_*-modules $L := \mathrm{End}_B(P)$, $N := P_*$ and the functor $\mathscr{F}_i := \Lambda_{B_*}^i$, yields a natural morphism

(4.5.12) $$\Gamma_{B_*}^i(\mathrm{End}_B(P)) \otimes_{B_*} \Lambda_{B_*}^i P_* \to \Lambda_{B_*}^i(\mathrm{End}_B(P) \otimes_{B_*} P_*)$$

On the other hand, the natural morphism $\mathrm{End}_B(P) \otimes_{B_*} P_* \to P_*$ induces

(4.5.13) $$\Lambda_{B_*}^i(\mathrm{End}_B(P) \otimes_{B_*} P_*) \to \Lambda_{B_*}^i P_*.$$

By composing (4.5.12) and (4.5.13) and passing to almost modules we obtain the morphism

$$\Gamma_B^i(\mathrm{End}_B(P)^a) \otimes_B \Lambda_B^i P \to \Lambda_B^i P$$

or equivalently:

(4.5.14) $$\Gamma_B^i(\mathrm{End}_B(P)^a) \to \mathrm{End}_B(\Lambda_B^i P)^a.$$

Then (4.5.11) is defined as the composition of (4.5.14) and the morphism $\mathrm{tr}_{\Lambda_B^i P/B}$.

4.5.15.   Let C be an almost finite projective B-algebra. Define $\mu : C \to \mathrm{End}_B(C)^a$ as in (4.1.7). By composition of $\Gamma_B^i \mu$ and (4.5.11) we obtain a B-linear morphism

$$\Gamma_B^i C \to B$$

characterized by the condition: $c^{[i]} \mapsto \sigma_i(c) := \mathrm{tr}_{\Lambda_B^i C/B}(\Lambda_B^i \mu(c))$.

4.5.16.   The construction of (4.5.15) applies especially to an almost finite projective groupoid $G := (X_0, X_1, s, t, c, \iota)$. In such case one verifies, using the cartesian diagrams (4.5.2), that $\sigma_i(t^\sharp(f)) \in \mathscr{O}_{X_0}^G$ for every $f \in \mathscr{O}_{X_{0*}}$ and every $i \leq r$: the argument is the same as in the proof of [25, Exp.V, Th.4.1]. In this way one obtains $\mathscr{O}_{X_0}^G$-linear morphisms

(4.5.17) $$T_{G,i} : \Gamma_{\mathscr{O}_{X_0}^G}^i \mathscr{O}_{X_0} \xrightarrow{\Gamma^i t^\sharp} \Gamma_{\mathscr{O}_{X_0}}^i \mathscr{O}_{X_1} \to \mathscr{O}_{X_0}^G \qquad f^{[i]} \mapsto \sigma_i(t^\sharp(f)).$$

Theorem 4.5.18. *Let* $G := (X_0, X_1, s, t, c, \iota)$ *be an étale almost finite and closed equivalence relation in* $A\text{-}\mathbf{Alg}^o$. *Then* G *is effective and the natural morphism* $X_0 \to X_0/G$ *is étale and almost finite projective.*

Proof. See [25, Exp.IV, §3.3] for the definition of effective equivalence relation. By (4.5.2), we have an identification $X_2 \simeq X_1 \times_{(s,s)} X_1$; therefore, the natural diagonal morphism $X_1 \to X_1 \times_{(s,s)} X_1$ gives a section $\delta : X_1 \to X_2$ of the morphism $s' : X_2 \to X_1$. Furthermore, since $X_2 = X_1 \times_{(s,t)} X_1$, the pair of morphisms $(1_{X_1}, \iota \circ s) : X_1 \to X_1$ induces another morphism $\psi_0 : X_1 \to X_2$; similarly, let $\psi_1 : X_1 \to X_2$ be the morphism induced by the pair $(\iota \circ t, 1_{X_1})$ (these are the degeneracy maps of the simplicial complex associated to G: cp. [25, Exp. V, §1]). By arguing with T-points (and exploiting the interpretation (4.5.1) of

$X_0(T)$, $X_1(T)$, etc.) one checks easily, first that $\psi_1 = \delta$, and second, that the two commutative diagrams

(4.5.19)

$$
\begin{array}{ccc}
X_0 & \xrightarrow{\iota} & X_1 \\
\downarrow{\scriptstyle \iota} & & \downarrow{\scriptstyle \psi_1} \\
X_1 & \xrightarrow{\psi_0} & X_2
\end{array}
\qquad
\begin{array}{ccc}
X_1 & \xrightarrow{t} & X_0 \\
\downarrow{\scriptstyle \psi_1} & & \downarrow{\scriptstyle \iota} \\
X_2 & \xrightarrow{t'} & X_1
\end{array}
$$

are cartesian. Since by assumption s is étale, corollary 3.1.9 implies that δ is an open and closed imbedding; consequently the same holds for ι. Let $e_0 \in \mathscr{O}_{X_{1*}}$ (resp. $e_1 \in \mathscr{O}_{X_{2*}}$) be the idempotent corresponding to the open and closed imbedding ι (resp. δ); since G is a closed equivalence relation, for every $\varepsilon \in \mathfrak{m}$ we can write $\varepsilon \cdot e_0 = \sum_{i=1}^{n} s^{\sharp}(b_i) \cdot t^{\sharp}(b_i')$ for some $b_i, b_i' \in \mathscr{O}_{X_{1*}}$. In view of (4.5.19) we deduce that $\varepsilon \cdot e_1 = \sum_{i=1}^{n} (t'^{\sharp} \circ s^{\sharp}(b_i)) \cdot (t'^{\sharp} \circ t^{\sharp}(b_i'))$. However, $s \circ t' = t \circ s'$ and $t \circ t' = t \circ c$, consequently

$$
\varepsilon \cdot e_1 = \sum_{i=1}^{n} (s'^{\sharp} \circ t^{\sharp}(b_i)) \cdot (c^{\sharp} \circ t^{\sharp}(b_i')).
$$

Finally, thanks to remark 4.1.17, and again (4.5.2), we can write:

(4.5.20) $\displaystyle \varepsilon \cdot f = \sum_{i=1}^{n} s^{\sharp} \circ \mathrm{Tr}_{X_1/X_0}(f \cdot t^{\sharp}(b_i)) \cdot t^{\sharp}(b_i') \quad$ for every $f \in \mathscr{O}_{X_{1*}}$.

If we now let $f := t^{\sharp}(g)$ in (4.5.20) we deduce:

$$
\varepsilon \cdot t^{\sharp}(g) = \sum_{i=1}^{n} s^{\sharp}(T_{G,1}(g \cdot b_i)) \cdot t^{\sharp}(b_i') = \sum_{i=1}^{n} t^{\sharp}(T_{G,1}(g \cdot b_i)) \cdot t^{\sharp}(b_i')
$$

for every $g \in \mathscr{O}_{X_{0*}}$. Since $t^{\sharp}$ is injective, this means that:

(4.5.21) $\displaystyle \varepsilon \cdot g = \sum_{i=1}^{n} T_{G,1}(g \cdot b_i) \cdot b_i' \quad$ for every $g \in \mathscr{O}_{X_{0*}}$.

It follows easily that $\mathscr{O}_{X_0}$ is an almost finitely generated projective $\mathscr{O}_{X_0}^G$-module. Furthermore, let us introduce the bilinear pairing

$$
t_G := T_{G,1} \circ \mu_{\mathscr{O}_{X_0}/\mathscr{O}_{X_0}^G} : \mathscr{O}_{X_0} \otimes_{\mathscr{O}_{X_0}^G} \mathscr{O}_{X_0} \to \mathscr{O}_{X_0}^G.
$$

Claim 4.5.22. t_G is a perfect pairing.

Proof of the claim. We have to show that the associated $\mathscr{O}_{X_0}$-linear morphism

$$
\tau_G : \mathscr{O}_{X_0} \to \mathscr{O}_{X_0}^* := \mathrm{alHom}_{\mathscr{O}_{X_0}^G}(\mathscr{O}_{X_0}, \mathscr{O}_{X_0}^G)
$$

is an isomorphism. From (4.5.21) it follows easily that τ_G is a monomorphism. Let $\phi : \mathscr{O}_{X_0} \to \mathscr{O}_{X_0}^G$ be a $\mathscr{O}_{X_0}^G$-linear morphism; it remains only to show that, for every $\varepsilon \in \mathfrak{m}$, there exists $b \in \mathscr{O}_{X_{0*}}$ such that $\tau_G(b) = \varepsilon \cdot \phi$. However, by applying ϕ to the two sides of (4.5.21) we get

$$
\varepsilon \cdot \phi(g) = \sum_{i=1}^{n} t_G(g \otimes b_i) \cdot \phi(b_i').
$$

In other words, $\varepsilon \cdot \phi = \tau_G(\sum_{i=1}^n b_i \cdot \phi(b_i'))$, as required.   ◊

By assumption, the morphism $\pi : C := \mathscr{O}_{X_0} \otimes_{\mathscr{O}_{X_0}^G} \mathscr{O}_{X_0} \to \mathscr{O}_{X_1}$ induced by the pair $(s^\sharp, t^\sharp)$ is an epimorphism. Moreover, by construction, we have the identity:

$$\mathrm{Tr}_{X_1/X_0} \circ \pi = 1_{\mathscr{O}_{X_0}} \otimes_{\mathscr{O}_{X_0}^G} T_{G,1}.$$

By claim 4.5.22 we see that $1_{\mathscr{O}_{X_0}} \otimes_{\mathscr{O}_{X_0}^G} T_{G,1}$ induces a perfect pairing $C \otimes_{\mathscr{O}_{X_0}} C \to \mathscr{O}_{X_0}$; on the other hand, Tr_{X_1/X_0} is already a perfect pairing, by theorem 4.1.14. It then follows that π must be a monomorphism, hence $C \simeq \mathscr{O}_{X_1}$, which shows that G is effective; then it is easy to verify that $T_{G,1}$ is actually the trace of the $\mathscr{O}_{X_0}^G$-algebra $\mathscr{O}_{X_0}$, which is consequently étale over $\mathscr{O}_{X_0}^G$.   □

Proposition 4.5.23. *Let $G = (X_0, X_1, s, t, c, \iota)$ be a groupoid of finite rank. Then $\mathscr{O}_{X_{0*}}$ is integral over $\mathscr{O}_{X_{0*}}^G$.*

Proof. By lemma 4.5.7 we can reduce to the case where the rank of $\mathscr{O}_{X_1}$ is constant, say equal to r. The assertion is then a direct consequence of the following:

Claim 4.5.24. Let $f \in \mathscr{O}_{X_{0*}}$. With the notation of (4.5.16) we have:

$$(t^\sharp(f))^r - T_{G,1}(f) \cdot (t^\sharp(f))^{r-1} + T_{G,2}(f) \cdot (t^\sharp(f))^{r-2} + \dots + (-1)^r T_{G,r}(f) = 0.$$

Proof of the claim. We apply proposition 4.4.30 (*i.e.* Cayley-Hamilton's theorem) to the endomorphism $t^\sharp(f) \cdot 1_{\mathscr{O}_{X_1}} : \mathscr{O}_{X_1} \to \mathscr{O}_{X_1}$.   □

Proposition 4.5.25. *Let G be an étale closed equivalence relation of finite rank. Then G is universally effective and the morphism $X_0 \to X_0/G$ is étale, faithfully flat and almost finite projective.*

Proof. Everything is known by theorem 4.5.18, except for the faithfulness, which follows from the following:

Claim 4.5.26. Under the assumptions of proposition 4.5.23, let $I \subset \mathscr{O}_{X_0}^G$ be an ideal such that $I \cdot \mathscr{O}_{X_0} = \mathscr{O}_{X_0}$. Then $I = \mathscr{O}_{X_0}^G$.

Proof of the claim. First of all, let $\mathscr{O}_{X_0} \simeq \prod_{i=1}^r B_i$ be the decomposition as in (4.5.6); one derives easily a corresponding decomposition $\mathscr{O}_{X_0}^G \simeq \prod_{i=1}^r B_i^G$, so we can assume that the rank of $\mathscr{O}_{X_1}$ is constant, equal to r. Let $J \subset \mathscr{O}_{X_0}^G$ be any ideal, and set $C := \mathscr{O}_{X_0}^G/J$. We have a natural isomorphism

$$\Gamma_C^i(\mathscr{O}_{X_0} \otimes_{\mathscr{O}_{X_0}^G} C) \simeq \Gamma_{\mathscr{O}_{X_0}^G}^i(\mathscr{O}_{X_0}) \otimes_{\mathscr{O}_{X_0}^G} C.$$

Composing with (4.5.17) $\otimes_{\mathscr{O}_{X_0}^G} C$, we derive a C-linear morphism:

$$\psi_i : \Gamma_C^i(\mathscr{O}_{X_0} \otimes_{\mathscr{O}_{X_0}^G} C) \to C.$$

By inspecting the construction, one shows easily that $\psi_i(1^{[i]}) = \binom{r}{i}$ (indeed, by flat base change one reduces to the case where $\mathscr{O}_{X_1}$ is a free B-module of rank r, in which case the result is obvious). Let us now take $J = I$. Then $\Gamma_C^i(\mathscr{O}_{X_0} \otimes_{\mathscr{O}_{X_0}^G} C) = 0$ for every $i > 0$, whence $1 = \psi_r(1^{[r]}) = 0$ in C, and the claim follows.   □

Proposition 4.5.27. *Keep the assumptions of proposition* 4.5.25.

(i) *If $\mathscr{O}_{X_0}$ is an almost finite (resp. almost finitely presented, resp. flat, resp. almost projective) A-module, then the same holds for the A-module $\mathscr{O}_X^G$.*

(ii) *If X_0 is a weakly unramified (resp. unramified, resp. weakly étale, resp. étale) Spec A-scheme, then the same holds for the Spec A-scheme X_0/G.*

Proof. By proposition 4.5.25, $\mathscr{O}_{X_0}$ is a faithfully flat almost finitely generated projective $\mathscr{O}_{X_0}^G$-module, hence $\mathscr{E}_{\mathscr{O}_{X_0}/\mathscr{O}_{X_0}^G} = \mathscr{O}_{X_0}^G$ by proposition 2.4.28(iv). It follows easily that, for every $\varepsilon \in \mathfrak{m}$ there exists $n \in \mathbb{N}$ such that $\varepsilon \cdot 1_{\mathscr{O}_{X_0}^G}$ factors as a composition of $\mathscr{O}_{X_0}^G$-linear morphisms:

$$(4.5.28) \qquad \mathscr{O}_{X_0}^G \to \mathscr{O}_{X_0}^n \to \mathscr{O}_{X_0}^G.$$

The assertions for "almost finite" and for "almost projective" are immediate consequences. To prove the assertion for "almost finitely presented" we use the criterion of proposition 2.3.16(ii). Indeed, let $(N_\lambda, \phi_{\lambda\mu} \mid \lambda, \mu \in \Lambda)$ be a filtered system of A-modules; we apply the natural transformation (2.3.17) to the sequence of morphisms (4.5.28) : since $\mathscr{O}_{X_0}$ is almost finitely presented, so is $\mathscr{O}_{X_0}^n$, hence the claim follows by a little diagram chase. The assertions for "flat" and "weakly unramified" are easy and shall be left to the reader. To conclude, it suffices to consider the assertion for "unramified". Now, by proposition 4.5.25 it follows that $\mathscr{O}_{X_0} \otimes_A \mathscr{O}_{X_0}$ is an almost finitely generated projective $\mathscr{O}_{X_0}^G \otimes_A \mathscr{O}_{X_0}^G$-module; since by assumption $\mathscr{O}_{X_0}$ is an almost projective $\mathscr{O}_{X_0} \otimes_A \mathscr{O}_{X_0}$-module, we deduce from lemma 2.4.7 that $\mathscr{O}_{X_0}$ is an almost projective $\mathscr{O}_{X_0}^G \otimes_A \mathscr{O}_{X_0}^G$-module. Using (4.5.28) we deduce that $\mathscr{O}_{X_0}^G$ is almost projective over $\mathscr{O}_{X_0}^G \otimes_A \mathscr{O}_{X_0}^G$ as well. $\qquad\square$

5. HENSELIZATION AND COMPLETION OF ALMOST ALGEBRAS

This chapter deals with more advanced aspects of almost commutative algebra: we begin with the definitions of *Jacobson radical* $\mathrm{rad}(A)$ of an almost algebra A, of *henselian pair* and *henselization* of a pair (A, I), where $I \subset A$ is an ideal contained in $\mathrm{rad}(A)$. These notions are especially well behaved when I is a *tight* ideal (definition 5.1.5), in which case we can also prove a version of Nakayama's lemma (lemma 5.1.7).

In section 5.3 we explain what is a *linear topology* on an A-module and an A-algebra; as usual, one is most interested in the case of I-adic topologies. In case A is I-adically complete and I is tight, we show that the functor $B \mapsto B/IB$ from almost finitely presented étale A-algebras to almost finitely presented étale A/I-algebras is an equivalence (theorem 5.3.27). For the proof we need some criteria to ensure that an A-algebra is unramified under various conditions : such results are collected in section 5.2, especially in theorem 5.2.12 and its corollary 5.2.15.

In section 5.5, theorem 5.3.27 is further generalized to the case where the pair (A, I) is tight henselian (see theorem 5.5.7, that also contains an analogous statement concerning almost finitely generated projective A-modules). The proof is a formal patching argument, which can be outlined as follows. First one reduces to the case where I is principal, say generated by f, and since I is tight, one can assume that $f \in \mathfrak{m}$; hence, we want to show that a given étale almost finitely presented A/fA-algebra B_0 lifts uniquely to an A-algebra B of the same type; in view of section 5.3 one can lift B_0 to an étale algebra $B^\wedge$ over the f-adic completion $A^\wedge$ of A; on the other hand, the almost spectrum $\mathrm{Spec}\, A$ is a usual scheme away from the closed subscheme defined by I, so we can use standard algebraic geometry to lift $B^\wedge[f^{-1}]$ to an étale algebra B' over $A[f^{-1}]$. Finally we need to show that $B^\wedge$ and B' can be patched in a unique way over $\mathrm{Spec}\, A$; this amounts to showing that certain commutative diagrams of functors are 2-cartesian (proposition 5.5.6).

The techniques needed to construct B' are borrowed from Elkik's paper [31]; for our purpose we need to extend and refine slightly Elkik's results, to deal with non-noetherian rings. This material is presented in section 5.4; its usefulness transcends the modest applications to almost ring theory presented here.

The second main thread of the chapter is the study of the *smooth locus* of an almost scheme; first we consider the affine case: as usual, an affine scheme X over $S := \mathrm{Spec}\, A$ can be identified with the fpqc sheaf that it represents; then the smooth locus X_{sm} of X is a certain natural subsheaf, defined in terms of the cotangent complex $\mathbb{L}_{X/S}$. To proceed beyond simple generalities one needs to impose some finiteness conditions on X, whence the definition of *almost finitely presented* scheme over S. For such affine S-schemes we can show that the smooth locus enjoys a property which we could call *almost formal smoothness*. Namely, suppose that $I \subset A$ is a tight ideal such that the pair (A, I) is henselian; suppose furthermore that $\sigma_0 : S_0 := \mathrm{Spec}\, A/I \to X$ is a section lying in the smooth locus of X; in this situation it does not necessarily follow that σ_0 extends to a full section $\sigma : S \to X$, however σ always exists if σ_0 extends to a section over some closed subscheme of the form $\mathrm{Spec}\, A/\mathfrak{m}_0 I$ (for a finitely generated subideal $\mathfrak{m}_0 \subset \mathfrak{m}$).

Next we consider quasi-projective almost schemes; if X is such a scheme, the invertible sheaf $\mathcal{O}_X(1)$ defines a quasi-affine $\mathbb{G}_m$-torsor $Y \to X$, and we define the smooth locus X_{sm} just as the projection of the smooth locus of Y. This is presumably not the best way to define X_{sm}, but anyway it suffices for the applications of section 5.8. In the latter we consider again a tight henselian pair (A, I), and we study some deformation problems for G-torsors, where G is a closed subgroup scheme of GL_n defined over Spec A and fulfilling certain general assumptions (see (5.8.4)). For instance, theorem 5.8.21 says that every G-torsor over the closed subscheme Spec A/I extends to a G-torsor over the whole of Spec A; the extension is however not unique, but any two such extensions are close in a precise sense (theorem 5.8.19) : here the almost formal smoothness of the quasi-projective almost scheme $(\mathrm{GL}_n/G)^a$ comes into play and accounts for the special quirks of the situation.

5.1. Henselian pairs.

Definition 5.1.1. Let A be a V^a-algebra. The *Jacobson radical* of A is the ideal $\mathrm{rad}(A) := \mathrm{rad}(A_*)^a$ (where, for a ring R, we have denoted by $\mathrm{rad}(R)$ the usual Jacobson ideal of R).

Lemma 5.1.2. *Let R be a V-algebra, $I \subset R$ an ideal. Then $I^a \subset \mathrm{rad}(R^a)$ if and only if $\mathfrak{m}I \subset \mathrm{rad}(R)$.*

Proof. Let us remark the following:

Claim 5.1.3. If S is any ring, $J \subset S$ an ideal, then $J \subset \mathrm{rad}(S)$ if and only if, for every $x \in J$ there exists $y \in J$ such that $(1 + x) \cdot (1 + y) = 1$.

Proof of the claim. Suppose that $J \subset \mathrm{rad}(S)$ and let $x \in J$; then $1 + x$ is not contained in any maximal ideal of S, so it is invertible. Find some $u \in S$ with $u \cdot (1+x) = 1$; setting $y := u - 1$, we derive $y = -x - xy \in J$. Conversely, suppose that the condition of the claim holds for all $x \in J$. Let $x \in J$; we have to show that $x \in \mathrm{rad}(S)$. If this were not the case, there would be a maximal ideal $\mathfrak{q} \subset S$ such that $x \notin \mathfrak{q}$; then we could find $a \in S$ such that $x \cdot a \equiv -1 \pmod{\mathfrak{q}}$, so $1 + x \cdot a \in \mathfrak{q}$, especially $1 + a \cdot x$ is not invertible, which contradicts the assumption. $\Diamond$

Let $\phi : \mathfrak{m}I \to R \to R_*^a$ be the natural composed map.

Claim 5.1.4. $\mathrm{Im}\,\phi$ is an ideal of R_*^a and $\mathfrak{m}I \subset \mathrm{rad}(R)$ if and only if $\mathrm{Im}\,\phi \subset \mathrm{rad}(R_*^a)$.

Proof of the claim. The first assertion is easy to check, and clearly we only have to verify the "if" direction of the second assertion, so suppose that $\mathrm{Im}\,\phi \subset \mathrm{rad}(R_*^a)$. Notice that $\mathrm{Ker}\,\phi$ is a square-zero ideal of R. Then, using claim 5.1.3, we deduce that for every $x \in \mathfrak{m} \cdot I$ there exists $z \in \mathfrak{m} \cdot I$ such that $(1 + x) \cdot (1 + z) = 1 + a$, where $a \in \mathrm{Ker}\,\phi$, so $a^2 = 0$. Consequently $(1 + x) \cdot (1 + z) \cdot (1 - a) = 1$, so the element $y := z - a - z \cdot a$ fulfills the condition of claim 5.1.3. $\Diamond$

It is clear that $I^a \subset \mathrm{rad}(R^a) := \mathrm{rad}(R_*^a)^a$ if and only if $\mathrm{Im}\,\phi \subset \mathrm{rad}(R_*^a)$, so the lemma follows from claim 5.1.4. $\square$

Given a V^a-algebra A and an ideal $I \subset \mathrm{rad}(A)$, one can ask whether the obvious analogue of Nakayama's lemma holds for almost finitely generated A-modules. As stated in lemma 5.1.7, this is indeed the case, at least if the ideal I has the property singled out by the following definition, which will play a constant role in the sequel.

Definition 5.1.5. Let I be an ideal of a V^a-algebra A. We say that I is *tight* if there exists a finitely generated subideal $\mathfrak{m}_0 \subset \mathfrak{m}$ and an integer $n \in \mathbb{N}$ such that $I^n \subset \mathfrak{m}_0 A$.

5.1.6.   In the following arguments we will use the following notational convention. Given an ideal I of a V^a-algebra A, we denote by I_*^n the ideal of A_* of all the almost elements of I^n; when we wish to denote the n-th power of I_* we shall write $(I_*)^n$.

Lemma 5.1.7. *Let A be a V^a-algebra, $I \subset \mathrm{rad}(A)$ a tight ideal. If M is an almost finitely generated A-module with $IM = M$, we have $M = 0$.*

Proof. Under the assumptions of the lemma we can find a finitely generated A-module Q such that $\mathfrak{m}_0 M \subset Q \subset M$. It follows that $M = I^n M \subset \mathfrak{m}_0 M \subset Q$, so $M = Q$ and actually M is finitely generated. Let $M_0 \subset M_*$ be a finitely generated A_*-submodule with $M_0^a = M$; clearly $\mathfrak{m} M_0 \subset I_* M_0$ and $\mathfrak{m}(I_*)^n \subset \mathfrak{m}_0 A_*$. Therefore:

$$\mathfrak{m}_0 M_0 \subset \mathfrak{m}^{n+2} M_0 \subset I_* \mathfrak{m}^{n+1} M_0 \subset ... \subset (I_*)^{n+1} \mathfrak{m} M_0 \subset I_* \mathfrak{m} \cdot \mathfrak{m}_0 M_0 \subset \mathfrak{m}_0 M_0$$

and consequently $\mathfrak{m} M_0 = \mathfrak{m}_0 M_0 = \mathfrak{m} I_* \cdot \mathfrak{m}_0 M_0$. By lemma 5.1.2 we have $\mathfrak{m} I_* \subset \mathrm{rad}(A_*)$, hence $\mathfrak{m}_0 M_0 = 0$ by Nakayama's lemma, thus finally $\mathfrak{m} M_0 = 0$, *i.e.* $M = 0$, as claimed. $\square$

Corollary 5.1.8. *Let A, I be as in lemma 5.1.7; suppose that $\phi : N \to M$ is an A-linear morphism of almost finitely generated projective A-modules such that $\phi \otimes_A 1_{A/I}$ is an isomorphism. Then ϕ is an isomorphism.*

Proof. From the assumptions we derive that $\mathrm{Coker}(\phi) \otimes_A (A/I) = 0$, hence $\mathrm{Coker}\,\phi = 0$, in view of lemma 5.1.7. By lemma 2.3.18(ii) it follows that $\mathrm{Ker}\,\phi$ is almost finitely generated; moreover, since M is flat, $\mathrm{Ker}(\phi) \otimes_A (A/I) = \mathrm{Ker}(\phi \otimes_A 1_{A/I}) = 0$, whence $\mathrm{Ker}\,\phi = 0$, again by lemma 5.1.7. $\square$

Definition 5.1.9. Let A be a V^a-algebra, $I \subset \mathrm{rad}(A)$ an ideal. We say that (A, I) is a *henselian pair* if $(A_*, \mathfrak{m} \cdot I_*)$ is a henselian pair. If in addition, I is tight, we say that (A, I) is a *tight henselian pair*.

Remark 5.1.10. For the convenience of the reader, we recall without proofs a few facts about henselian pairs.

(i) For a ring R, let $R_{\mathrm{red}} := R/\mathrm{nil}(R)$ and denote by $\sqrt{I}$ the radical of the ideal I. Then the pair (R, I) is henselian if and only if the same holds for the pair $(R_{\mathrm{red}}, \sqrt{I \cdot R_{\mathrm{red}}})$.

(ii) Suppose that $I \subset \mathrm{rad}(R)$; then (R, I) is a henselian pair if and only if the same holds for the pair $(\mathbb{Z} \oplus I, I)$, where $\mathbb{Z} \oplus I$ is endowed with the ring structure such that $(a, x) \cdot (b, y) := (ab, ay + bx + xy)$ for every $a, b \in \mathbb{Z}$, $x, y \in I$. Indeed, this follows easily from the following criterion (iii), which is shown in [66, Ch.XI, §2, Prop.1].

(iii) Assume $I \subset \mathrm{rad}(R)$; then the pair (R, I) is henselian if and only if every monic polynomial $p(X) \in R[X]$ such that $p(X) \equiv (X^2 - X)^m \pmod{I[X]}$ decomposes as a product $p(X) = q(X) \cdot r(X)$ where $q(X), r(X)$ are monic polynomials in $R[X]$ and $q(X) \equiv X^m \pmod{I[X]}$, $r(X) \equiv (X - 1)^m \pmod{I[X]}$. Moreover, under the assumption $I \subset \mathrm{rad}(R)$ the foregoing properties uniquely determine $q(X)$ and $r(X)$, whenever such a decomposition exists.

(iv) Let $J \subset I$ be a subideal. If the pair (R, I) is henselian, the same holds for (R, J).

(v) If (R, I) is a henselian pair, and $R \to S$ is an integral ring homomorphism, then the pair (S, IS) is henselian ([66, Ch.XI, §2, Prop. 2]).

Lemma 5.1.11. *Let R be a V-algebra, $I \subset \mathrm{rad}(R)$ an ideal. Then the pair (R^a, I^a) is henselian if and only if the same holds for the pair $(R, \mathfrak{m}I)$.*

Proof. It comes down to checking that $(R, \mathfrak{m}I)$ is henselian if and only if $(R_*^a, \mathfrak{m}I_*^a)$ is. To this aim, let $\phi : R \to R_*^a$ be the natural map. Let $S := \mathrm{Im}\,\phi$ and $J := \phi(\mathfrak{m}I) \subset S$. Since $\mathrm{Ker}\,\phi$ is a square-zero ideal in R, it follows from remark 5.1.10(i) that $(R, \mathfrak{m}I)$ is henselian if and only if (S, J) is. However, it is clear that the induced map $J \to \mathfrak{m}I_*^a$ is bijective, hence remark 5.1.10(ii) and lemma 5.1.2 say that (S, J) is henselian if and only if $(R_*^a, \mathfrak{m}I_*^a)$ is. $\square$

5.1.12.   Suppose that (R, I) is a henselian pair, with R a V-algebra. Then, in view of lemma 5.1.11 and remark 5.1.10(iv) we see that (R^a, I^a) is also henselian. This gives a way of producing plenty of henselian pairs.

Lemma 5.1.13. *Let $\phi : A \to B$ be an almost finite morphism, $I \subset A$ an ideal.*

(i) *The induced ring homomorphism $A_* \to \phi(A_*) + \mathfrak{m}B_*$ is integral.*

(ii) *If $I \subset \mathrm{rad}(A)$, then $IB \subset \mathrm{rad}(B)$.*

(iii) *If (A, I) is a henselian pair, the same holds for the pair (B, IB).*

Proof. (i): For a given finitely generated subideal $\mathfrak{m}_0 \subset \mathfrak{m}$, pick a finitely generated submodule $Q \subset B_*$ with $\mathfrak{m}_0 B_* \subset Q$; notice that $(\mathfrak{m}_0 Q)^2 \subset \mathfrak{m}_0 Q$, hence $\phi(A_*) + \mathfrak{m}_0 Q$ is a subring of B_*, finite over A_*. As $\phi(A_*) + \mathfrak{m}B_*$ is a filtered union of such subrings, the assertion follows. (ii) follows from (i) and from lemma 5.1.2. (iii) is a direct consequence of (i), of remark 5.1.10(v) and of (5.1.12). $\square$

5.1.14.   Given a V^a-algebra A and an ideal $I \subset A$, we define the *henselization* of the pair (A, I) as the unique pair (up to unique isomorphism) $(A^{\mathrm{h}}, I^{\mathrm{h}})$ which satisfies the (almost version of the) usual universal property (cp. [66, Ch.XI, §2, Déf.4]). Suppose $A = R^a$ and $I = J^a$ for a V-algebra R and an ideal $J \subset R$, and let (R', J') be a henselization of the pair $(R, \mathfrak{m}J)$; one can easily check that (R'^a, J'^a) is a henselization of (A, I). It follows in particular that, if $I \subset \mathrm{rad}(A)$ and $(A^{\mathrm{h}}, I^{\mathrm{h}})$ is a henselization of (A, I), then the morphism $A \to A^{\mathrm{h}}$ is faithfully flat ([66, Ch.XI, §2]).

Lemma 5.1.15. *Let* $\mathfrak{m}_0 \subset \mathfrak{m}$ *be a finitely generated subideal. Then there exists an integer* $n := n(\mathfrak{m}_0) > 0$ *such that* $((\mathfrak{m}_0 A)_*)^n \subset \mathfrak{m}_0 A_*$ *for any* V^a-*algebra* A.

Proof. We proceed by induction on the number k of generators of $\mathfrak{m}_0$. To start out, let B be any V^a-algebra and $f \in B_*$ any almost element. The endomorphism $B \to B \; : \; x \mapsto f \cdot x$ induces an isomorphism $\alpha : \overline{B} := B/\mathrm{Ann}_B(f) \overset{\sim}{\to} fB$, and a commutative diagram:

$$
\begin{array}{ccc}
\overline{B}_* \otimes_V \overline{B}_* & \xrightarrow{\;\mu_{\overline{B}_*}\;} & \overline{B}_* \\
{\scriptstyle \alpha_* \otimes \alpha_*}\Big\downarrow & & \Big\downarrow{\scriptstyle f\cdot\alpha_*} \\
(fB)_* \otimes_V (fB)_* & \xrightarrow{\;\mu_{B_*}\;} & (fB)_*.
\end{array}
$$

It follows that

$$(5.1.16) \qquad\qquad f \cdot (fB)_* = ((fB)_*)^2$$

as ideals of B_*, which takes care of the case $k = 1$. Suppose now that $k > 1$; let us write $\mathfrak{m}_0 = x_1 \cdot V + \mathfrak{m}_1$, where $\mathfrak{m}_1$ is generated by $k - 1$ elements. We apply (5.1.16) with $B := A/\mathfrak{m}_1 A$ and $f := x_1$ to deduce:

$$(\mathfrak{m}_0 A/\mathfrak{m}_1 A)_* \cdot (\mathfrak{m}_0 A/\mathfrak{m}_1 A)_* = x_1 \cdot (\mathfrak{m}_0 A/\mathfrak{m}_1 A)_* \subset \mathrm{Im}((x_1 A)_* \to (A/\mathfrak{m}_1 A)_*).$$

Therefore $((\mathfrak{m}_0 A)_*)^2 \subset (x_1 A)_* + (\mathfrak{m}_1 A)_*$. After raising the latter inclusion to some high power, the inductive assumption on $\mathfrak{m}_1$ allows to conclude. $\qquad\square$

Corollary 5.1.17. *Let* A *be a* V^a-*algebra,* $I \subset \mathrm{rad}(A)$ *a tight ideal. Then:*

    (i) $I_* \subset \mathrm{rad}(A_*)$.

    (ii) *If* (A, I) *is a henselian pair, then the same holds for the pair* (A_*, I_*).

Proof. For integers n, m large enough we can write (notation as in (5.1.6)) :

$$(I_*)^{nm} \subset (I_*^n)^m \subset ((\mathfrak{m}_0 A)_*)^m \subset \mathfrak{m}_0 A_*$$

thanks to lemma 5.1.15. Then by lemma 5.1.2 we deduce $(I_*)^{nm+1} \subset \mathrm{rad}(A_*)$, which implies (i). Similarly, we deduce from lemma 5.1.15 that $\sqrt{\mathfrak{m} I_*} = \sqrt{I_*}$, so (ii) follows from remark 5.1.10(i). $\qquad\square$

Proposition 5.1.18. *Let* (A, I) *be a tight henselian pair. The natural morphism* $A \to A/I$ *induces a bijection from the set of idempotents of* A_* *to the set of idempotents of* $(A/I)_*$.

Proof. Pick an integer $m > 0$ and a finitely generated subideal $\mathfrak{m}_0 \subset \mathfrak{m}$ such that $I^m \subset \mathfrak{m}_0 \cdot A$. We suppose first that $\widetilde{\mathfrak{m}}$ has homological dimension ≤ 1. By corollary 5.1.17(ii), the quotient map $A_* \to A_*/I_*$ induces a bijection on idempotents. So we are reduced to showing that the natural injective map $A_*/I_* \to (A/I)_*$ induces a surjection on idempotents. To this aim, it suffices to show that an idempotent almost element $\overline{e} : V^a \to A/I$ always lifts to an almost element $e : V^a \to A$; indeed, the image of e inside A_*/I_* will then necessarily agree with $\overline{e}$. Now, the obstruction to the existence of e is a class $\omega_1 \in \mathrm{Ext}^1_{V^a}(V^a, I)$. On the other hand, proposition 2.5.13(i) ensures that, for every integer $n > 0$, $\overline{e}$ admits a unique idempotent lifting

$\bar{e}_n : V^a \to A/I^n$. Let more generally $\omega_n \in \mathrm{Ext}^1_{V^a}(V^a, I^n)$ be the obstruction to the existence of a lifting of $\bar{e}_n$ to an almost element of A; the imbedding $I^n \subset I$ induces a map

$$(5.1.19) \qquad \mathrm{Ext}^1_{V^a}(V^a, I^n) \to \mathrm{Ext}^1_{V^a}(V^a, I)$$

and clearly the image of ω_n under (5.1.19) agrees with ω_1. Thus, the proposition will follow in this case from the following:

Claim 5.1.20. For n sufficiently large, the map (5.1.19) vanishes identically.

Proof of the claim. We will prove more precisely that the natural map

$$(5.1.21) \qquad \mathrm{Ext}^1_{V^a}(V^a, \mathfrak{m}_0 I) \to \mathrm{Ext}^1_{V^a}(V^a, I)$$

vanishes; the claim will follow for $n := m + 1$. A choice of generators $\varepsilon_1, ..., \varepsilon_k$ for $\mathfrak{m}_0$ determines an epimorphism $\phi : I^{\oplus k} \to \mathfrak{m}_0 I$; notice that $\mathrm{Ext}^2_{V^a}(V^a, \mathrm{Ker}\,\phi) = 0$ due to lemma 2.4.14(i),(ii) and the assumption on the homological dimension of $\tilde{\mathfrak{m}}$. Thus, the induced map

$$(5.1.22) \qquad \mathrm{Ext}^1_{V^a}(V^a, I^{\oplus k}) \to \mathrm{Ext}^1_{V^a}(V^a, \mathfrak{m}_0 I)$$

is surjective. To prove the claim, it suffices therefore to show that the composition of (5.1.22) and (5.1.21) vanishes, which is obvious, since the modules in question are almost zero. ◇

Finally suppose that $\mathfrak{m}$ is arbitrary; by theorem 2.1.12(i.b),(ii.b), $\mathfrak{m}$ is the colimit of a filtered family of subideals $(\mathfrak{m}_\lambda \subset \mathfrak{m} \mid \lambda \in \Lambda)$, such that $\mathfrak{m}_0 \subset \mathfrak{m}_\lambda = \mathfrak{m}_\lambda^2$ and $\tilde{\mathfrak{m}}_\lambda := \mathfrak{m}_\lambda \otimes_V \mathfrak{m}_\lambda$ is V-flat for every $\lambda \in \Lambda$, and moreover each $\tilde{\mathfrak{m}}_\lambda$ has homological dimension ≤ 1. We may suppose that $(A, I) = (R, J)^a$ for a henselian pair (R, I), where R is a V-algebra and $J^m \subset \mathfrak{m}_0 R$. Each pair $(V, \mathfrak{m}_\lambda)$ is a basic setup, and we denote by (A_λ, I_λ) the tight henselian pair corresponding to (R, J) in the almost category associated to $(V, \mathfrak{m}_\lambda)$ (so A_λ is the image of R under the localization functor $V\text{-}\mathbf{Alg} \to (V, \mathfrak{m}_\lambda)^a\text{-}\mathbf{Alg}$). Let $\bar{e}$ be an idempotent almost element of A/I (which is an object of $(V, \mathfrak{m})^a\text{-}\mathbf{Alg}$); $\bar{e}$ is represented by a unique V-linear map $f : \tilde{\mathfrak{m}} \to R/J$ and by the foregoing, for every $\lambda \in \Lambda$ the restriction $f_\lambda : \tilde{\mathfrak{m}}_\lambda \to R/J$ lifts to a unique idempotent map $g_\lambda : \tilde{\mathfrak{m}}_\lambda \to R$. By uniqueness, the maps g_λ glue to a map $\mathrm{colim}_{\lambda \in \Lambda}\, g_\lambda : \tilde{\mathfrak{m}} \to R$ which is the sought lifting of $\bar{e}$. □

5.2. Criteria for unramified morphisms.

The following lemma generalizes a case of [46, Partie II, lemma 1.4.2.1].

Lemma 5.2.1. *Let $A \to C$ be a morphism of V^a-algebras, $f \in A_*$ any almost element and M a C-module. Suppose that $M[f^{-1}]$ is a flat $C[f^{-1}]$-module, M/fM is a flat C/fC-module, $\mathrm{Tor}^A_1(C, A/fA) = 0$ and $\mathrm{Tor}^A_i(M, A/fA) = 0$ for $i = 1, 2$. Then M is a flat C-module.*

Proof. Let N be any C-module; we need to show that $\mathrm{Tor}^C_1(M, N) = 0$. To this aim we consider the short exact sequence $0 \to K \to N \xrightarrow{j} N[f^{-1}] \to L \to 0$ where j is the natural morphism. We have

$$\mathrm{Tor}^C_1(M, N[f^{-1}]) = \mathrm{Tor}^{C[f^{-1}]}_1(M[f^{-1}], N[f^{-1}]) = 0.$$

Therefore we are reduced to showing :

Claim 5.2.2. $\operatorname{Tor}_1^C(M, K) = \operatorname{Tor}_2^C(M, L) = 0$.

Proof of the claim. Notice that $K = \bigcup_{k>0} \operatorname{Ann}_K(f^k)$, and similarly for L, so it suffices to show that $\operatorname{Tor}_i^C(M, Q) = 0$ for $i = 1, 2$ and every C-module Q such that $f^k Q = 0$ for some integer $k > 0$. By considering the short exact sequence $0 \to \operatorname{Ann}_Q(f^{k-1}) \to Q \to Q' \to 0$, an easy induction on k further reduces to showing that $\operatorname{Tor}_i^C(M, Q) = 0$ for $i = 1, 2$, in case $fQ = 0$. However, the morphisms $C \to C/fC$ and $A \to C$ determine base change spectral sequences (cp. [77, Th.5.6.6])

$$E_{pq}^2 := \operatorname{Tor}_p^{C/fC}(\operatorname{Tor}_q^C(M, C/fC), Q) \Rightarrow \operatorname{Tor}_{p+q}^C(M, Q)$$
$$F_{pq}^2 := \operatorname{Tor}_p^C(\operatorname{Tor}_q^A(C, A/fA), M) \Rightarrow \operatorname{Tor}_{p+q}^A(M, A/fA).$$

The spectral sequence F yields an exact sequence:

$$\operatorname{Tor}_2^A(M, A/fA) \longrightarrow \operatorname{Tor}_2^C(C/fC, M) \longrightarrow \operatorname{Tor}_1^A(C, A/fA) \otimes_C M$$
$$\longrightarrow \operatorname{Tor}_1^A(M, A/fA) \longrightarrow \operatorname{Tor}_1^C(C/fC, M) \longrightarrow 0$$

which implies $\operatorname{Tor}_i^C(M, C/fC) = 0$ for $i = 1, 2$. Thus $E_{pq}^2 = 0$ for $q = 1, 2$ and we get

$$\operatorname{Tor}_i^C(M, Q) \simeq \operatorname{Tor}_i^{C/fC}(M/fM, Q) \qquad \text{for } i = 1, 2.$$

Since M/fM is a flat C/fC-module, the claim follows. $\qquad\square$

Lemma 5.2.3. *Let M be an A-module, $f \in A_*$ an almost element, and denote by $A^\wedge := \lim_{n \in \mathbb{N}} (A/f^n A)$ the f-adic completion of A. Then we have:*

  (i) $\operatorname{Ann}_A(M[f^{-1}]) \cdot \operatorname{Ann}_A(M \otimes_A A^\wedge) \subset \operatorname{Ann}_A(M)$.
  (ii) *If M is almost finitely generated and $M/fM = M[f^{-1}] = 0$, then $M = 0$.*

Proof. (i): Let $a \in \operatorname{Ann}_A(M[f^{-1}])_*$, $b \in \operatorname{Ann}_A(M \otimes_A A^\wedge)_*$ and denote by $\mu_a, \mu_b : M \to M$ the scalar multiplication morphisms. From $a \cdot M[f^{-1}] = 0$ we deduce that $aM \subset \bigcup_{n>0} \operatorname{Ann}_M(f^n)$; it follows that the natural morphism $aM \to (aM) \otimes_A A^\wedge$ is an isomorphism. Now the claim follows by inspecting the commutative diagram:

$$
\begin{array}{ccccc}
M & \longrightarrow & M \otimes_A A^\wedge & \xrightarrow{\ 0 = \mu_b \otimes 1_{A^\wedge}\ } & M \otimes_A A^\wedge \\
{\scriptstyle \mu_a}\downarrow & & & & \downarrow{\scriptstyle \mu_a \otimes 1_{A^\wedge}} \\
aM & \xrightarrow{\ \mu_b\ } & aM & \xrightarrow{\ \sim\ } & (aM) \otimes_A A^\wedge.
\end{array}
$$

(ii): For a given finitely generated subideal $\mathfrak{m}_0 \subset \mathfrak{m}$, pick a finitely generated A-module $Q \subset M$ such that $\mathfrak{m}_0 M \subset Q$. By assumption, $Q[f^{-1}] = 0$; hence there exists an integer $n \geq 0$ such that $\mathfrak{m}_0 \cdot f^n Q = 0$, whence $\mathfrak{m}_0^2 \cdot f^n M = 0$.

However $M = fM$ by assumption, thus $\mathfrak{m}_0^2 M = 0$, and finally $M = 0$, since $\mathfrak{m}_0$ is arbitrary. □

Theorem 5.2.4. *Let A be a V^a-algebra, $f \in A_*$ any almost element and $I \subset A$ an almost finitely generated ideal with $I^2 = I$. Suppose that both $I[f^{-1}] \subset A[f^{-1}]$ and $(I + fA)/fA \subset A/fA$ are generated by idempotents in $A[f^{-1}]_*$ and respectively $(A/fA)_*$. Then I is generated by an idempotent of A_*.*

Proof. We start out with the following:

Claim 5.2.5. Let $I, J \subset A$ be two ideals such that J is nilpotent and $I^2 = I$. If $\overline{I} := \mathrm{Im}(I \to A/J)$ is generated by an idempotent of $(A/J)_*$, then I is generated by an idempotent of A_*.

Proof of the claim. One reduces easily to the case where $J^2 = 0$. We apply proposition 2.5.13 to the non-unitary extension $0 \to J \cap I \to I \to \overline{I} \to 0$ to derive that the idempotent $\overline{e}$ that generates $\overline{I}$ lifts uniquely to an idempotent $e \in I_*$. Then e induces decompositions $A \simeq A_0 \times A_1$ and $I = (IA_0) \times (IA_1)$ such that $e = 0$ (resp. $e = 1$) on A_0 (resp. on A_1). We can therefore reduce to the cases $A = A_0$ or $A = A_1$. If $A = A_0$, then $I \subset J$, hence $I = I^2 = 0$. If $A = A_1$ then $1 \in I_*$, so $I = A$; in either case, the claim holds. ◇

Claim 5.2.6. In the situation of (3.4.7), suppose that $I \subset A_0$ is an ideal such that IA_1 and IA_2 are generated by idempotents. Then I is generated by an idempotent of A_{0*}.

Proof of the claim. By applying termwise the functor $M \mapsto M_*$ to (3.4.8), we obtain a commutative diagram (3.4.8)$_*$ which is still cartesian. By assumption, we can find idempotents $e_1 \in A_{1*}$, $e_2 \in A_{2*}$ with $e_i A_i = IA_i$ $(i = 1, 2)$. It is then easy to see that the images of e_1 and e_2 agree in A_{3*}; consequently there exists a unique element $e_0 \in A_{0*}$ whose image in A_i agrees with e_i for $i = 1, 2$. Such element is necessarily an idempotent; moreover, since the natural A_0-linear morphism $I \to (IA_1) \times (IA_2)$ is a monomorphism, we deduce easily that $(1 - e_0) \cdot I = 0$, whence $I \subset e_0 A_0$. Let $M := (e_0 A_0)/I$; clearly $M \otimes_{A_0} A_i = 0$ for $i = 1, 2$. Using lemma 3.4.18(i) we deduce that $M = 0$, whence $e_0 \in I$, as stated. ◇

Suppose next that f is a regular element of A_*. Denote by $A^\wedge$ the f-adic completion of A. One verifies easily that the natural commutative diagram

(5.2.7)
$$\begin{array}{ccc} A & \longrightarrow & A[f^{-1}] \\ \downarrow & & \downarrow \\ A^\wedge & \longrightarrow & A^\wedge[f^{-1}] \end{array}$$

is cartesian.

Claim 5.2.8. In the situation of (5.2.7), let $I \subset A$ be an ideal such that both $I[f^{-1}] \subset A[f^{-1}]$ and $I \cdot A^\wedge \subset A^\wedge$ are generated by idempotents. Then I is generated by an idempotent.

Proof of the claim. As in the proof of claim 5.2.6, we find in A_* an idempotent e such that $eA^\wedge = IA^\wedge$ and $eA[f^{-1}] = I[f^{-1}]$, and deduce that $I \subset eA$. Let $M := (eA)/I$; on one hand we have $M \otimes_A A^\wedge = 0$. On the other hand, $M[f^{-1}] = 0$, *i.e.* $M = \bigcup_{n \in \mathbb{N}} \mathrm{Ann}_M(f^n)$, which implies that $M \otimes_A A^\wedge = M$. Hence $M = 0$, and the claim follows. $\Diamond$

Claim 5.2.9. The theorem holds in case A is f-adically complete.

Proof of the claim. Say that $\mathrm{Im}(I \to A/fA) = \bar{e}_1 \cdot (A/fA)$, for an idempotent $\bar{e}_1 \in (A/fA)_*$. For every $n > 0$ set $\bar{I}_n := \mathrm{Im}(I \to A/f^n A)$; applying proposition 2.5.13 to the non-unitary extensions

$$0 \to (f^n A/f^{n+1} A) \cap \bar{I}_{n+1} \to \bar{I}_{n+1} \to \bar{I}_n \to 0$$

we construct recursively a compatible system of idempotents $\bar{e}_n \in \bar{I}_{n*}$, and by claim 5.2.5, $\bar{e}_n$ generates $\bar{I}_n$ for every $n > 0$; whence an element $e \in \lim_{n \in \mathbb{N}} (A/f^n A)_* \simeq (\lim_{n \in \mathbb{N}} A/f^n A)_* \simeq A_*$. Clearly e is again an idempotent, and it induces decompositions $A \simeq A_0 \times A_1$, $I \simeq (IA_0) \times (IA_1)$, such that $e = 0$ (resp. $e = 1$) in A_0 (resp. in A_1). We can thus assume that either $A = A_0$ or $A = A_1$. If $A = A_0$, then $\bar{e}_n = 0$ for every $n > 0$, so $I \subset \bigcap_{n>0} f^n A = 0$. If $A = A_1$, then $\bar{e}_n = 1$ for every $n > 0$, *i.e.* $\bar{I}_n = A/f^n A$ for every $n > 0$, that is, I is dense in A. It follows easily that $\mathrm{Ann}_A(I) = 0$. Using lemma 2.4.6 we deduce $\mathrm{Ann}_{A[f^{-1}]}(I[f^{-1}]) = 0$; since $I[f^{-1}]$ is generated by an idempotent, this means that $I[f^{-1}] = A[f^{-1}]$. Set $M := A/I$; the foregoing shows that M fulfills the hypotheses of lemma 5.2.3(ii), whence $M = 0$, which implies the claim. $\Diamond$

Claim 5.2.10. The theorem holds in case f is a non-zero-divisor of A_*.

Proof of the claim. According to claim 5.2.8, it suffices to prove that the ideal $IA^\wedge \subset A^\wedge$ is generated by an idempotent. However, since $A^\wedge/fA^\wedge \simeq A/fA$, this follows directly from our assumptions and from claim 5.2.9. $\Diamond$

After these preparations, we are ready to prove the theorem. Let

$$T := \bigcup_{n>0} \mathrm{Ann}_A(f^n).$$

By claim 5.2.10, we know already that the ideal $I \cdot (A/T)$ is generated by an idempotent, and the same holds for the ideal $IA^\wedge$. Since $T \otimes_A A^\wedge \simeq T$, it follows that the natural morphism $\phi : T \to TA^\wedge$ is an epimorphism; one verifies easily that $J := \mathrm{Ker}\, \phi = T \cap (\bigcap_{n>0} f^n A)$, and then it is clear that $J^2 = 0$. Furthermore, it follows that the natural commutative diagram

$$\begin{array}{ccc} A/J & \longrightarrow & A/T \\ \downarrow & & \downarrow \\ A^\wedge & \longrightarrow & A^\wedge/TA^\wedge \end{array}$$

is cartesian. Taking claim 5.2.6 into account, we deduce that the image of I in A/J is generated by an idempotent. Lastly, we invoke claim 5.2.5 to show that I is indeed generated by an idempotent.   □

Remark 5.2.11. One may wonder whether every almost finitely generated ideal $I \subset A$ such that $I = I^2$, is generated by an idempotent. We do not know the answer to this question.

Theorem 5.2.12. *Let $\phi : A \to B$ be a morphism of V^a-algebras, $I, J \subset A$ any two ideals and $f \in A_*$ any almost element.*

  (i) *If $\phi \otimes_A 1_{A/I}$ and $\phi \otimes_A 1_{A/J}$ are weakly unramified, then the same holds for $\phi \otimes_A 1_{A/IJ}$.*
  (ii) *If $\phi \otimes_A 1_{A/I}$ and $\phi \otimes_A 1_{A/J}$ are unramified, the same holds for $\phi \otimes_A 1_{A/IJ}$.*
  (iii) *If ϕ is flat and both $\phi \otimes_A 1_{A/fA}$ and $\phi \otimes_A 1_{A[f^{-1}]}$ are weakly unramified, then ϕ is weakly unramified.*
  (iv) *If ϕ is almost finite and both $\phi \otimes_A 1_{A/fA}$ and $\phi \otimes_A 1_{A[f^{-1}]}$ are unramified, then ϕ is unramified.*

Proof. To start out, we remark the following:

Claim 5.2.13. Let A' be any A-algebra and set $B' := B \otimes_A A'$; then for every $B' \otimes_{A'} B'$-module M we have a natural isomorphism:
$$\mathrm{Tor}_1^{B \otimes_A B}(B, M) \simeq \mathrm{Tor}_1^{B' \otimes_{A'} B'}(B', M).$$

Proof of the claim. Notice that the short exact sequence of $B \otimes_A B$-modules

(5.2.14) $0 \to I_{B/A} \to B \otimes_A B \to B \to 0.$

is split exact as a sequence of B-modules (and *a fortiori* as a sequence of A-modules); hence $(5.2.14) \otimes_A N$ is again exact for every A-module N. We deduce easily that $\mathrm{Tor}_1^{B \otimes_A B}(B, N \otimes_A B \otimes_A B) = 0$ for every A-module N. On the other hand, the morphism $B \otimes_A B \to B' \otimes_{A'} B'$ determines a base change spectral sequence (cp. [77, Th.5.6.6])
$$E_{pq}^2 := \mathrm{Tor}_p^{B' \otimes_{A'} B'}(\mathrm{Tor}_q^{B \otimes_A B}(B, B' \otimes_{A'} B'), M) \Rightarrow \mathrm{Tor}_{p+q}^{B \otimes_A B}(B, M)$$
for every $B' \otimes_A B'$-module M. The foregoing yields $E_{p1}^2 = 0$ for every $p \in \mathbb{N}$, so the claim follows.   ◊

In order to show (i) we may replace A by A/IJ and therefore assume that $IJ = 0$; then we have to prove that $\mathrm{Tor}_1^{B \otimes_A B}(B, N) = 0$ for every $B \otimes_A B$-module N. By a simple devissage we can further reduce to the case where either $IN = 0$ or $JN = 0$. Say $IN = 0$; we apply claim 5.2.13 with $A' := A/I$, $B' := B/IB$ to deduce $\mathrm{Tor}_1^{B \otimes_A B}(B, N) \simeq \mathrm{Tor}_1^{B' \otimes_{A'} B'}(B', N)$; however the latter module vanishes by our assumption on $\phi \otimes_A 1_{B'}$.

For (ii) we can again assume that $IJ = 0$. We need to show that B is an almost projective $B \otimes_A B$-module, and we know already that it is flat by (i). Set $I' := I(B \otimes_A B)$, $J' := J(B \otimes_A B)$; by assumption B/IB (resp. B/JB) is almost projective over $B \otimes_A B/I'$ (resp. $B \otimes_A B/J'$); in light of proposition 3.4.21 we deduce that $B/(IB \cap JB)$ is almost projective over $B \otimes_A B/(I' \cap J')$. Then, since $(I' \cap J')^2 \subset I'J' = 0$, the assertion follows from lemma 3.2.25(i).

Next we remark that, under the assumptions of (iii), the hypotheses of lemma 5.2.1 are fulfilled by $C := B \otimes_A B$ and $M := B$, so assertion (iii) holds. Finally, suppose that ϕ is almost finite; hence $\Omega_{B/A}$ is an almost finitely generated A-module, and by assumption we have $\Omega_{B/A} \otimes_A (A/fA) = \Omega_{B/A}[f^{-1}] = 0$, therefore $\Omega_{B/A} = 0$, in view of lemma 5.2.3(ii). Set $I := \text{Ker}(B \otimes_A B \to B)$; it follows that $I^2 = I$, and I is almost finitely generated as an A-module, so a fortiori as a $B \otimes_A B$-module. Moreover, from proposition 3.1.4 and our assumption on $\phi \otimes_A 1_{A/fA}$ and $\phi[f^{-1}]$, it follows that both $I[f^{-1}] \subset B \otimes_A B[f^{-1}]$ and $\text{Im}(I \to B \otimes_A (B/fB))$ are generated by idempotents. Then theorem 5.2.4 says that also I is generated by an idempotent of $(B \otimes_A B)_*$, so that B is almost projective as a $B \otimes_A B$-module, by remark 3.1.8, which proves (iv). $\square$

Corollary 5.2.15. *Let $I \subset \text{rad}(A)$ be a tight ideal. If B is an almost finite A-algebra and B/IB is unramified over A/I, then B is unramified over A.*

Proof. Under the stated assumptions, $\Omega_{B/A}$ is an almost finitely generated A-module such that $\Omega_{B/A} \otimes_A A/I = 0$; by lemma 5.1.7 it follows that

$$(5.2.16) \qquad\qquad \Omega_{B/A} = 0.$$

Set $I_{B/A} := \text{Ker}(\mu_{B/A} : B \otimes_A B \to B)$; we derive that $I_{B/A} = I_{B/A}^2$. Pick $n > 0$ and a finitely generated subideal $\mathfrak{m}_0 \subset \mathfrak{m}$ with $I^n \subset \mathfrak{m}_0 A$; let $\varepsilon_1, ..., \varepsilon_k$ be a set of generators for $\mathfrak{m}_0$.

Claim 5.2.17. $I_{B/A}[\varepsilon_i^{-1}]$ is generated by an idempotent of $B \otimes_A B[\varepsilon_i^{-1}]$, for every $i \leq k$.

Proof of the claim. Notice that $A[\varepsilon_i^{-1}]$ is a (usual) V-algebra (that is, the localization functor $A[\varepsilon_i^{-1}]_*\text{-}\mathbf{Mod} \to A[\varepsilon_i^{-1}]\text{-}\mathbf{Mod}$ is an equivalence) and $B[\varepsilon_i^{-1}]$ is a finite $A[\varepsilon_i^{-1}]$-algebra; from (5.2.16) and [66, Ch.III, Prop.9] we deduce that $B[\varepsilon_i^{-1}]$ is unramified over $A[\varepsilon_i^{-1}]$, which implies the claim. $\diamond$

Claim 5.2.18. $I_{B/A}/\mathfrak{m}_0 I_{B/A}$ is generated by an idempotent as well.

Proof of the claim. We apply theorem 5.2.12(ii) with $J = I^m$ to deduce, by induction on m, that $B/I^{m+1}B$ is an unramified A/I^{m+1}-algebra for every $m \in \mathbb{N}$; it follows that $B/\mathfrak{m}_0 B$ is unramified over $A/\mathfrak{m}_0 A$, whence the claim. $\diamond$

Using theorem 5.2.12(iv) and claims 5.2.17, 5.2.18, the assertion is now easily verified by induction on the number k of generators of $\mathfrak{m}_0$, $\square$

The following lemma shows that morphisms of unramified A-algebras can be read off from their "graphs". This result will be useful in section 5.5.

5.2.19. Let $A \to B$ be an unramified morphism of V^a-algebras, and C any A-algebra. Suppose that $e \in (B \otimes_A C)_*$ is an idempotent; to e we can attach a morphism of A-algebras:

$$\Gamma(e) : C \to B \otimes_A C \to e \cdot (B \otimes_A C) \qquad c \mapsto 1 \otimes c \mapsto e \cdot (1 \otimes c) \quad (c \in C_*)$$

and, in case $\Gamma(e)$ is an isomorphism, we obtain a morphism $\phi_e : B \to C$ after composing $\Gamma(e)^{-1}$ and the natural morphism

$$\Delta(e) : B \to B \otimes_A C \to e \cdot (B \otimes_A C) \qquad b \mapsto b \otimes 1 \mapsto e \cdot (b \otimes 1) \quad (b \in B_*)$$

Conversely, to a given a morphism $\phi : B \to C$ of A-algebras, we can associate an idempotent $e_\phi \in (B \otimes_A C)_*$, by setting $e_\phi := (1_B \otimes_A \phi)(e_{B/A})$, where $e_{B/A}$ is the idempotent provided by proposition 3.1.4. Then, since the commutative diagram

$$
\begin{array}{ccc}
B \otimes_A B & \xrightarrow{\ 1_B \otimes \phi\ } & B \otimes_A C \\
{\scriptstyle \mu_{B/A}} \downarrow & & \downarrow {\scriptstyle \mu_{C/A} \circ (\phi \otimes 1_C)} \\
B & \xrightarrow{\quad \phi \quad} & C
\end{array}
$$

is cocartesian, we deduce from proposition 3.1.4 that the corresponding morphism $\Gamma(e_\phi)$ is an isomorphism. In fact, the following holds:

Lemma 5.2.20. *The rule* $\phi \mapsto e_\phi$ *of (5.2.19) induces a natural bijection from* $\mathrm{Hom}_{A\text{-}\mathbf{Alg}}(B, C)$ *onto the set of idempotents* $e \in (B \otimes_A C)_*$ *such that* $\Gamma(e)$ *is an isomorphism. Its inverse is the rule* $e \mapsto \phi_e$.

Proof. Let $\phi : B \to C$ be given and set $e := e_\phi$. We need to show that $\phi_e = \phi$, or equivalently $\Delta(e) = \Gamma(e) \circ \phi$. This translates into the equality:

$$(5.2.21) \qquad e \cdot (b \otimes 1) = e \cdot (1 \otimes \phi(b)) \quad \text{for every } b \in B_*.$$

The latter can be rewritten as $(1 \otimes \phi)(e_{B/A} \cdot (1 \otimes b - b \otimes 1)) = 0$ which follows from identity (iii) in proposition 3.1.4. Conversely, for e given such that $\Gamma(e)$ is an isomorphism, set $\phi := \phi_e$; we have to check that $e_\phi = e$. By construction of ϕ we have $\Delta(e) = \Gamma(e) \circ \phi$, which means that (5.2.21) holds for the pair (e, ϕ). Let $J \subset B \otimes_A C$ be the ideal generated by the elements of the form $b \otimes 1 - 1 \otimes \phi(b)$, for all $b \in B_*$; we know that $eJ = 0$, which means that the natural morphism $B \otimes_A C \to e(B \otimes_A C)$ factors through the natural morphism $B \otimes_A C \to e_\phi(B \otimes_A C)$. Since both $\Gamma(e)$ and $\Gamma(e_\phi)$ are isomorphisms, it follows that $e(B \otimes_A C) = e_\phi(B \otimes_A C)$, so $e = e_\phi$. $\qquad\square$

5.3. Topological algebras and modules.

Definition 5.3.1. Let A be a V^a-algebra, M an A-module.

  (i) A *linear topology* on M is a non-empty family $\mathscr{L}$ of submodules of M that satisfies the following conditions. If $I, J \in \mathscr{L}$, then $I \cap J \in \mathscr{L}$; if $I \in \mathscr{L}$ and $I \subset J$, then $J \in \mathscr{L}$. We say that a submodule I is *open* if $I \in \mathscr{L}$.

  (ii) Let $\mathscr{F}$ be a family of submodules of M. The *topology generated by* $\mathscr{F}$ is the smallest linear topology containing $\mathscr{F}$.

(iii) Let $\mathscr{L}$ be a linear topology on M and $I \subset M$ a submodule. The *closure* of I is the submodule $\overline{I} := \bigcap_{J \in \mathscr{L}} (I + J)$.

(iv) We say that the topology $\mathscr{L}$ is *complete* if the natural morphism $M \to \lim_{I \in \mathscr{L}} M/I$ is an isomorphism.

(v) Let $\mathscr{L}_M$, $\mathscr{L}_A$ be topologies on M, resp. on A, and I an open ideal of A. We say that the topology $\mathscr{L}_M$ is *I-preadic* (resp. *I-c-preadic*) if the family of submodules $(I^n M \mid n \in \mathbb{N})$ (resp. $(\overline{I^n} \cdot M \mid n \in \mathbb{N})$) is a cofinal subfamily in $\mathscr{L}_M$. In either case, we say that I is an *ideal of definition* for $\mathscr{L}_M$. Furthermore, we say that the topology $\mathscr{L}_M$ is *I-adic* if it is I-preadic and complete. We similarly define an *I-c-adic* topology.

5.3.2.   One defines as usual the notions of *continuous*, resp. *open* morphism of A-modules, of *induced topology* on submodules of a linearly topologized module and of *completion* of a topological module. Notice that, if A is an I-c-preadic topological V^a-algebra and M an I-c-preadic A-module (for some open ideal $I \subset A$) then we have $\overline{I^n M} = \overline{I^n} \cdot M$ for every $n \in \mathbb{N}$.

Lemma 5.3.3. *Let M, N be two A-modules endowed with descending filtrations $\mathscr{F}_M := (M_n \mid n \in \mathbb{N})$, $\mathscr{F}_N := (N_n \mid n \in \mathbb{N})$ so that $M_0 = M$, $N_0 = N$ and M (resp. N) is complete for the topology generated by $\mathscr{F}_M$ (resp. by $\mathscr{F}_N$). Suppose furthermore that $\phi : M \to N$ is a morphism of A-modules which respects the filtrations, and such that the induced morphisms $\mathrm{gr}^i \phi : \mathrm{gr}^i M \to \mathrm{gr}^i N$ are epimorphisms for every $i \in \mathbb{N}$. Then ϕ is an epimorphism.*

Proof. Under the assumptions, the induced morphism $\phi_n : M/M_n \to N/N_n$ is an epimorphism for every $n \in \mathbb{N}$. Clearly it suffices to show then that $\lim_{n \in \mathbb{N}}^1 \mathrm{Ker}\, \phi_n$ vanishes. However, from $\mathrm{Coker}(\mathrm{gr}^i \phi) = 0$ one deduces by the snake lemma that the induced morphism $\mathrm{Ker}\, \phi_{i+1} \to \mathrm{Ker}\, \phi_i$ is an epimorphism, so the claim follows. $\square$

5.3.4.   Let $(A_n; \phi_n : A_{n+1} \to A_n \mid n \in \mathbb{N})$ be a projective system of V^a-algebras. Then $A_\infty := \lim_{n \in \mathbb{N}} A_n$ is naturally endowed with a linear topology, namely the topology $\mathscr{L}$ generated by the family $(\mathrm{Ker}(A_\infty \to A_n) \mid n \in \mathbb{N})$. We call $\mathscr{L}$ the *projective topology* on A_∞.

Lemma 5.3.5. *Let $(A_n; \phi_n : A_{n+1} \to A_n \mid n \in \mathbb{N})$ be a projective system of V^a-algebras, A_∞ its projective limit, and suppose that:*

(a) *ϕ_n is an epimorphism for every $n \in \mathbb{N}$.*

(b) *$\mathrm{Ker}\, \phi_n = \mathrm{Ker}(\phi_0 \circ \ldots \circ \phi_n : A_{n+1} \to A_0)^{n+1}$ for every $n \in \mathbb{N}$.*

Then we have:

(i) *Let $I := \mathrm{Ker}(A_\infty \to A_0)$. The projective topology of A_∞ is I-c-preadic and is complete, and moreover $\overline{I^{n+1}} = \mathrm{Ker}(A_\infty \to A_n)$ for every $n \in \mathbb{N}$.*

(ii) *Conversely, suppose that A is any complete linearly topologized V^a-algebra and $I \subset A$ an open ideal such that the topology of A is I-c-preadic. Set $A_n := A/\overline{I^{n+1}}$ for every $n \in \mathbb{N}$. Then the projective system $(A_n \mid n \in \mathbb{N})$ satisfies conditions (a) and (b).*

Proof. To show (i) it suffices to show that $\overline{I^{n+1}} = \mathrm{Ker}(A_\infty \to A_n)$ for every $n \in \mathbb{N}$. We endow $\overline{I^{n+1}}$ (resp. $\mathrm{Ker}(A_\infty \to A_n)$) with the descending filtration $F_i := \overline{I^{n+1+i}}$ (resp. $G_i := \mathrm{Ker}(A_\infty \to A_{n+i})$) ($i \in \mathbb{N}$). Notice first that the natural morphism $A_\infty \to A_{k+1}$ induces an isomorphism $I/G_{k+1} \xrightarrow{\sim} \mathrm{Ker}(A_{k+1} \to A_0)$. By (b) we deduce isomorphisms

$$(5.3.6) \qquad (I^{k+1} + G_{k+1})/G_{k+1} \xrightarrow{\sim} \mathrm{Ker}\,\phi_k \qquad \text{for every } k \in \mathbb{N}.$$

Especially, $I^{k+1} \subset \mathrm{Ker}(A_\infty \to A_k)$. Since the latter is a closed ideal, we deduce $\overline{I^{k+1}} \subset \mathrm{Ker}(A_\infty \to A_k)$ for every $k \in \mathbb{N}$. In other words, we have $F_i \subset G_i$ for every $i \in \mathbb{N}$. Since $\mathrm{gr}^i G_\bullet \simeq \mathrm{Ker}\,\phi_{i+n}$, we derive easily from (5.3.6) that the induced morphism $\mathrm{gr}^i F_\bullet \to \mathrm{gr}^i G_\bullet$ is an epimorphism for every $i \in \mathbb{N}$, so the natural morphism $F_0 \to G_0/G_i$ is an epimorphism; *i.e.* $\overline{I^{n+1}}$ is both closed and dense in $\mathrm{Ker}(A_\infty \to A_n)$ and (i) follows. Under the assumptions of (ii) it is obvious that condition (a) holds. To show (b) comes down to verifying the identity $I^{n+1} + \overline{I^{n+2}} = \overline{I^{n+1}}$, which is obvious since $\overline{I^{n+2}}$ is an open ideal. $\qquad\square$

Remark 5.3.7. Notice that, in the situation of lemma 5.3.5, the pair (A_∞, I) is henselian. One sees this as follows, using the criterion of remark 5.1.10(iii). Let $R_n := A_*/\overline{I^{n+1}}_*$, $J_n := I_*/\overline{I^{n+1}}_*$ for every $n \in \mathbb{N}$; let also $R := \lim_{n \in \mathbb{N}} R_n$ and denote by J the image of $\lim_{n \in \mathbb{N}} J_n$ in R. Using the criterion of claim 5.1.3 one verifies easily that $J \subset \mathrm{rad}(R)$. Moreover, since J_n is a nilpotent ideal of R_n, the pair (R_n, J_n) is henselian. Let now $p(X) \in R[X]$ be as in remark 5.1.10(iii), and denote by $p_n(X)$ the image of $p(X)$ in $R_n(X)$, for every $n \in \mathbb{N}$, especially $p_0(X) = X^m(X-1)^m$; by *loc.cit.* we derive a compatible family of decompositions $p_n(X) = q_n(X) \cdot r_n(X)$ in $R_n[X]$ for every $n \in \mathbb{N}$; hence $p(X)$ admits a similar decomposition in $R[X]$, so (R, J) is a henselian pair, which implies the contention, in view of lemma 5.1.11.

Lemma 5.3.8. *Let A_∞ be the projective limit of a system of V^a-algebras satisfying conditions* (a) *and* (b) *of lemma 5.3.5. Let $J \subset A_\infty$ be a finitely generated ideal such that $\overline{J}$ is open. Then:*

(i) *$J = \overline{J}$.*

(ii) *Every finite system of generators $x_1, ..., x_r$ of J defines an open morphism $A_\infty^r \to J$ (for the topologies on A_∞^r and J induced from A_∞).*

(iii) *J^k is open for every $k \in \mathbb{N}$.*

Proof. By lemma 5.3.5 the topology on A_∞ is c-preadic for some defining ideal I. The assumption means that $\overline{I^n} \subset \overline{J}$ for a sufficiently large integer n.

Claim 5.3.9. The natural morphism: $(J \cap \overline{I^m})/(J \cap \overline{I^{m+1}}) \to \overline{I^m}/\overline{I^{m+1}}$ is an isomorphism for every $m \geq n$.

Proof of the claim. It is an easy verification that shall be left to the reader.   ◇

Let $x_1, ..., x_r \in J_*$ be a system of generators for J, and $f : A_\infty^r \to J$ the corresponding epimorphism. Set $M_k := f^{-1}(J \cap \overline{I^k})$. To start out, M_k contains $(\overline{I^k})^{\oplus r}$, whence M_k is open in A_∞^r for every $k \in \mathbb{N}$. This already shows that f

is continuous. We define a descending filtration on M_n by setting $F_k := \overline{I^k M_n}$ for every $k \in \mathbb{N}$. In view of claim 5.3.9, the morphism f induces an epimorphism $M_n \to \overline{I^n}/\overline{I^{n+1}}$, hence for every $k \in \mathbb{N}$, the morphisms:

(5.3.10) $\qquad \overline{I^k} \cdot M_n \to (\overline{I^k} \cdot \overline{I^n})/(\overline{I^k} \cdot \overline{I^{n+1}}) \to \overline{I^{n+k}}/\overline{I^{n+k+1}}$

are epimorphisms as well. The composition of the morphisms (5.3.10) extends to an epimorphism $F_k \to \overline{I^{n+k}}/\overline{I^{n+k+1}}$, for every $k \in \mathbb{N}$. In other words, if we endow $\overline{I^n}$ with its I-c-preadic filtration, then f induces a morphism $\phi : M_n \to \overline{I^n}$ of filtered modules, such that $\mathrm{gr}^\bullet \phi$ is an epimorphism of graded modules. By assumption $\overline{I^n}$ is complete for its filtration. Similarly, M_n is complete for its filtration $F_\bullet$; indeed, this follows by remarking that $(\overline{I^{n+k}})^{\oplus r} \subset F_k \subset (\overline{I^k})^{\oplus r}$ for every $k \in \mathbb{N}$. Then lemma 5.3.3 says that ϕ is an epimorphism, i.e. $f(M_n) = \overline{I^n}$; it follows that J is an open ideal, hence equal to its closure. More generally, the same argument proves that for every $k \in \mathbb{N}$, $f(F_k) = \overline{I^{n+k}}$. Since by the foregoing, $(F_k \mid k \in \mathbb{N})$ is a cofinal system of open submodules of A_∞^r, we deduce that f is an open morphism. Finally the identity

$$f((J^n)^{\oplus r}) = J^{n+1} \qquad \text{for every } n \in \mathbb{N}$$

together with (ii) and an easy induction, yields (iii). $\square$

5.3.11. Let I be an ideal of definition for the projective topology on a ring A_∞ as in lemma 5.3.5(i). We wish now to give a criterion to ensure that the projective topology on A_∞ is actually I-adic. This can be achieved in case I is tight, as shown by proposition 5.3.12, which generalizes [26, Ch.0, Prop.7.2.7].

Proposition 5.3.12. *Let A_∞ be as in lemma 5.3.8 and I an ideal of definition for the projective topology of A_∞. Assume that I is tight and moreover that $I/\overline{I^2}$ is an almost finitely generated A_∞-module. Then :*

(i) *There exists $n \in \mathbb{N}$ and a finitely generated ideal $J \subset I$ such that $I^n \subset J$.*
(ii) *The topology of A_∞ is I-adic.*

Proof. Using the natural epimorphism $(I/\overline{I^2})^{\otimes k} \to \overline{I^k}/\overline{I^{k+1}}$ we deduce that $\overline{I^k}/\overline{I^{k+1}}$ is almost finitely generated for every $k \geq 0$; then the same holds for $I/\overline{I^{k+1}}$. Let $\mathfrak{m}_0 \subset \mathfrak{m}$ be a finitely generated subideal and $n \in \mathbb{N}$ such that $I^n \subset \mathfrak{m}_0 A_\infty$; pick a finitely generated ideal $\mathfrak{m}_1 \subset \mathfrak{m}$ with $\mathfrak{m}_0 \subset \mathfrak{m}_1^2$; we can find a finitely generated A_∞-module $Q \subset I/\overline{I^{n+2}}$ such that $\mathfrak{m}_1 \cdot (I/\overline{I^{n+2}}) \subset Q$; up to replacing Q by $\mathfrak{m}_1 Q$, we can then achieve that Q is generated by the images of finitely many almost elements $x_1, ..., x_t$ of I and moreover $\mathfrak{m}_0 \cdot (I/\overline{I^{n+2}}) \subset Q$. Then, since $\overline{I^{n+2}}$ is open we deduce

(5.3.13) $\qquad \overline{I^{n+1}}/\overline{I^{n+2}} \subset Q.$

Let $J \subset A_\infty$ be the ideal generated by $x_1, ..., x_t$. From (5.3.13) we deduce that the natural morphism $(J \cap \overline{I^{n+1}})/(J \cap \overline{I^{n+2}}) \to \overline{I^{n+1}}/\overline{I^{n+2}}$ is an isomorphism. Since $I^k(J \cap \overline{I^{n+1}}) \subset J \cap \overline{I^{n+1+k}}$, the same holds more generally for the morphisms $(J \cap \overline{I^{n+1+k}})/(J \cap \overline{I^{n+2+k}}) \to \overline{I^{n+1+k}}/\overline{I^{n+2+k}}$, for every $k \in \mathbb{N}$. This easily implies that $J' := J \cap \overline{I^{n+1}}$ is *dense* in $\overline{I^{n+1}}$, *i.e.* $\overline{J'} = \overline{I^{n+1}}$; especially, $\overline{J}$ is open in A_∞ and therefore $J = \overline{J}$ by lemma 5.3.8(i), which proves assertion (i). Assertion (ii) follows easily from (i) and lemma 5.3.8(iii).   □

5.3.14. Let now $(A_n \mid n \in \mathbb{N})$ be an inverse system of V^a-algebras satisfying conditions (a) and (b) of lemma 5.3.5 and let $I := \mathrm{Ker}(A_\infty \to A_0)$. The induced functors $A_{n+1}\text{-}\mathbf{Mod} \to A_n\text{-}\mathbf{Mod} : M \mapsto A_n \otimes_{A_{n+1}} M$ define an inverse system of categories $(A_n\text{-}\mathbf{Mod} \mid n \in \mathbb{N})$. In such situation one can define a natural functor

$$(5.3.15) \qquad A_\infty\text{-}\mathbf{Mod}_{\mathrm{top}} \to 2\text{-}\lim_{n \in \mathbb{N}} A_n\text{-}\mathbf{Mod}$$

from the category of topological A_∞-modules whose topology is I-c-adic (and of continuous A_∞-linear morphisms) to the 2-limit of the foregoing inverse system of categories (see [47, Ch.I] for generalities on 2-categories). Namely, let $I := \mathrm{Ker}(A_\infty \to A_0)$; then to an A_∞-module M one associates the compatible system $(M/\overline{I^{n+1}M} \mid n \in \mathbb{N})$.

Lemma 5.3.16. *The functor* (5.3.15) *is an equivalence of categories.*

Proof. We claim that a quasi-inverse to (5.3.15) can be given by associating to any compatible system $(M_n \mid n \in \mathbb{N})$ the A_∞-module $M_\infty := \lim_{n \in \mathbb{N}} M_n$, with the linear topology generated by the submodules $K_n := \mathrm{Ker}(M_\infty \to M_n)$, for all $n \in \mathbb{N}$. In order to show that this topology on M_∞ is I-c-adic (where $I := \mathrm{Ker}(A_\infty \to A_0)$), it suffices to verify that $K_n = \overline{I^{n+1}M_\infty}$ for every $n \in \mathbb{N}$. Clearly $I^{n+1}M_\infty \subset K_n$, hence we are reduced to showing that $K_n \subset K_m + I^{n+1}M_\infty$ for every $m > n$ or equivalently, that K_n/K_m equals the image of $I^{n+1}M_\infty$ inside M_m, which is obvious.   □

In the following we will seek conditions under which (5.3.15) can be refined to equivalences between interesting subcategories, for instance to almost finitely generated, or almost projective modules.

Lemma 5.3.17. *Let M_∞ be a topological A_∞-module whose topology is I-c-adic (for some ideal of definition $I \subset A_\infty$). Suppose that $N \subset M_\infty$ is a finitely generated submodule such that $\overline{N}$ is open in M_∞. Then :*

    (i) $N = \overline{N}$.

    (ii) *Every finite set of generators $x_1, ..., x_r$ of N determines an open morphism* $A_\infty^r \to N$.

    (iii) *For every open ideal $J \subset A_\infty$, the submodule JN is open in M_∞.*

Proof. *Mutatis mutandis*, this is the same as the proof of lemma 5.3.8; the details can thus be safely entrusted to the reader.   □

Lemma 5.3.18. *Keep the notation of* (5.3.14) *and assume that I is a tight ideal. Let M be a topological A_∞-module whose topology is I-c-adic and let $(M_n \mid n \in \mathbb{N})$ be the image of M under* (5.3.15). *Suppose also that M_0 is an almost finitely generated A_0-module. Then:*

    (i) *M admits an open finitely generated submodule.*

(ii) M is almost finitely generated.

(iii) $M \otimes_{A_\infty} A_n \simeq M_n$ for every $n \in \mathbb{N}$.

Proof. Using lemma 3.2.25(ii) and the hypothesis on M_0 we deduce that M_n is almost finitely generated for every $n \in \mathbb{N}$. Choose finitely generated $\mathfrak{m}_0 \subset \mathfrak{m}$ and $n \geq 0$ such that $I^n \subset \mathfrak{m}_0 A_\infty$; it follows easily that there are almost elements $x_1, ..., x_t$ of M whose images in M_{n+1} generate a submodule Q such that $\mathfrak{m}_0 M_{n+1} \subset Q$. This implies that $\overline{I^{n+1} M}/\overline{I^{n+2} M} \subset Q$. Let $N \subset M$ be the submodule generated by $x_1, ..., x_t$. Then, arguing as in the proof of proposition 5.3.12 we see that $\overline{I^{n+1} M} \subset \overline{N}$. Hence $N = \overline{N}$ in view of lemma 5.3.17(i), so (i) holds. Furthermore, by lemma 5.3.17(iii), $\overline{I^n} \cdot M$ is open for every $n \in \mathbb{N}$ (since it contains $\overline{I^n} \cdot N$), so (iii) follows easily. Finally, M is almost finitely generated because the natural morphism $M_n \to M/N$ is an epimorphism. $\qquad\square$

5.3.19. In the situation of (5.3.14), let M be an A_∞-module with an I-c-adic topology, and let $(M_n \mid n \in \mathbb{N})$ be its image under (5.3.15). We suppose now that M_n is an almost projective A_n-module for every $n \in \mathbb{N}$. Let us choose an epimorphism $f : A_\infty^{(S)} \to M$. We endow the free A_∞-module $A_\infty^{(S)}$ with the topology generated by the family of submodules $(\overline{I^n} \cdot A_\infty^{(S)} \mid n \in \mathbb{N})$. Notice that f is continuous for this topology, hence it extends to a continuous morphism $f^\wedge : F_\infty \to M$ on the completion F_∞ of $A_\infty^{(S)}$.

Lemma 5.3.20. *With the notation of (5.3.19), we have:*

(i) *$f^\wedge$ is topologically almost split, i.e., for every $\varepsilon \in \mathfrak{m}$ there exists a continuous morphism $g : M \to F_\infty$ such that $f^\wedge \circ g = \varepsilon \cdot 1_M$.*

(ii) *If M is almost finitely generated, then M is almost projective.*

Proof. By lemma 5.3.16 the set of continuous morphisms $M \to F_\infty$ is in natural bijection with $\varprojlim_{n \in \mathbb{N}} \mathrm{Hom}_{A_n}(M_n, F_n)$, where $F_n := F_\infty/\overline{I^{n+1} F_\infty}$ for every $n \in \mathbb{N}$. On the other hand, under the assumptions of the lemma, f induces an epimorphism:

$$\varprojlim_{n \in \mathbb{N}} \mathrm{alHom}_{A_n}(M_n, F_n) \to \varprojlim_{n \in \mathbb{N}} \mathrm{alHom}_{A_n}(M_n, M_n).$$

Assertion (i) follows easily. Suppose now that M is almost finitely generated; then for every finitely generated subideal $\mathfrak{m}_0 \subset \mathfrak{m}$ we can find $m \geq 0$ and a morphism $h : A_\infty^m \to M$ such that $\mathfrak{m}_0 \cdot \mathrm{Coker}\, h = 0$. Let $\varepsilon_1, ..., \varepsilon_t$ be a system of generators for $\mathfrak{m}_0$; then for every $i, j \leq t$ we can pick a morphism $\phi : A_\infty^{(S)} \to A_\infty^m$ such that $h \circ \phi = \varepsilon_i \cdot \varepsilon_j \cdot f$; after taking completions, this relation becomes $h \circ \phi^\wedge = \varepsilon_i \cdot \varepsilon_j \cdot f^\wedge$. Choose $g : M \to F_\infty$ as in (i); we deduce $\varepsilon_i \cdot \varepsilon_j \cdot \varepsilon \cdot 1_M = h \circ \phi^\wedge \circ g$, which as usual shows that $\varepsilon_i \cdot \varepsilon_j \cdot \varepsilon$ annihilates $\mathrm{alExt}^1_{A_\infty}(M, N)$ for every A_∞-module N, so (ii) holds. $\qquad\square$

5.3.21. For a V^a-algebra A, let us denote by $A\text{-}\mathbf{Mod}_{\mathrm{afpr}}$ the full subcategory of $A\text{-}\mathbf{Mod}$ consisting of all almost finitely generated projective A-modules. Let now $(A_n \mid n \in \mathbb{N})$ and $I \subset A_\infty$ be as in (5.3.14). We define natural functors

$$(5.3.22) \qquad A_\infty\text{-}\mathbf{Mod}_{\mathrm{afpr}} \to 2\text{-}\varprojlim_{n \in \mathbb{N}} A_n\text{-}\mathbf{Mod}_{\mathrm{afpr}}$$

(respectively

$$(5.3.23) \qquad A_\infty\text{-}\acute{\mathbf{E}}\mathbf{t}_{\text{afp}} \to A_0\text{-}\acute{\mathbf{E}}\mathbf{t}_{\text{afp}}$$

in the notation of (3.2.27)) by assigning to a given object M of A_∞-$\mathbf{Mod}_{\text{afpr}}$, the compatible system $(M \otimes_{A_\infty} A_n \mid n \in \mathbb{N})$ (resp. to an object B of A_∞-$\acute{\mathbf{E}}\mathbf{t}_{\text{afp}}$, the A_0-algebra $B \otimes_{A_\infty} A_0$).

Theorem 5.3.24. *In the situation of* (5.3.21), *suppose that I is tight. Then* (5.3.22) *is an equivalence of categories.*

Proof. We claim that a quasi-inverse to (5.3.22) is obtained by assigning to any compatible system $(M_n \mid n \in \mathbb{N})$ the A_∞-module $M := \lim_{n \in \mathbb{N}} M_n$. Indeed, by the proof of lemma 5.3.16 it results that M is endowed with a natural I-c-adic topology; then by lemma 5.3.18(ii) we deduce that M is almost finitely generated, and finally lemma 5.3.20(ii) says that M is almost projective. From lemma 5.3.18(iii) it follows that the functor thus defined is a right quasi-inverse to (5.3.22), so the latter is essentially surjective. Full faithfulness is a consequence of the following:

Claim 5.3.25. Let P be an almost finitely generated projective A_∞-module. Then the natural morphism

$$(5.3.26) \qquad P \to \lim_{n \in \mathbb{N}} (P \otimes_{A_\infty} A_n)$$

is an isomorphism.

Proof of the claim. This is clear if $P = A_\infty^r$ for some $r \geq 0$. For a general P one chooses, for every $\varepsilon \in \mathfrak{m}$, a sequence $P \to A_\infty^n \to P$ as in lemma 2.4.15 and applies to it the natural transformation (5.3.26); the claim follows by the usual diagram chase. $\qquad\square$

Theorem 5.3.27. *In the situation of* (5.3.21), *suppose that I is tight. Then* (5.3.23) *is an equivalence of categories.*

Proof. It follows easily from theorem 3.2.28(ii) that the natural functor

$$2\text{-}\lim_{n \in \mathbb{N}} A_n\text{-}\acute{\mathbf{E}}\mathbf{t}_{\text{afp}} \to A_0\text{-}\acute{\mathbf{E}}\mathbf{t}_{\text{afp}}$$

is an equivalence. Hence we are reduced to showing that the natural functor

$$(5.3.28) \qquad A_\infty\text{-}\acute{\mathbf{E}}\mathbf{t}_{\text{afp}} \to 2\text{-}\lim_{n \in \mathbb{N}} A_n\text{-}\acute{\mathbf{E}}\mathbf{t}_{\text{afp}}$$

is an equivalence. To this aim, we claim that the rule $(B_n \mid n \in \mathbb{N}) \mapsto \lim_{n \in \mathbb{N}} B_n$ defines a quasi-inverse to (5.3.28). Taking into account theorem 5.3.24, this will follow once we have shown:

Claim 5.3.29. Let B be an almost finitely generated projective A_∞-algebra such that $B_0 := B \otimes_{A_\infty} A_0$ is unramified. Then B is unramified.

Proof of the claim. To start out, we apply theorem 5.2.12(ii) with $J = I^n$ to deduce, by induction on n, that $B/I^{n+1}B$ is unramified over A_∞/I^{n+1} for every $n \in \mathbb{N}$. Set $B_n := B \otimes_{A_\infty} A_n$; it follows that B_n is unramified over A_n for every $n \in \mathbb{N}$. By proposition 3.1.4 we deduce that there exists a compatible system of idempotent almost elements e_n of $B_n \otimes_{A_n} B_n$, such that $e_n \cdot I_{B_n/A_n} = 0$ and $\mu_{B_n/A_n}(e_n) = 1$ for every $n \in \mathbb{N}$. Hence we obtain an idempotent in $(\lim_{n \in \mathbb{N}} B_n \otimes_{A_n} B_n)_* \simeq \lim_{n \in \mathbb{N}} (B_n \otimes_{A_n} B_n)_*$; however, the latter is isomorphic to $(B \otimes_{A_\infty} B)_*$, in view of claim 5.3.25. Then conditions (ii) and (iii) of proposition 3.1.4 follow easily by remarking that $\mu_{B/A_\infty} = \lim_{n \in \mathbb{N}} \mu_{B_n/A_n}$ and $I_{B/A_\infty} = \lim_{n \in \mathbb{N}} I_{B_n/A_n}$. $\square$

5.4. Henselian approximation of structures over adically complete rings. This section reviews and complements some results of Elkik's article [31]. Especially, we wish to show how the main theorems of *loc.cit.* generalize to the case of not necessarily noetherian rings. In principle, this is known ([31, Ch.III, §4, Rem.2, p.587] explains briefly how to adapt the proofs to make them work in some non-noetherian situations), but we feel that it is worthwhile to give more details.

Definition 5.4.1. Let R be a ring, $F := R[X_1, ..., X_N]$ a free R-algebra of finite type, $J \subset F$ a finitely generated ideal, and let $S := F/J$. We define an ideal of F by setting:
$$\mathsf{H}_R(F, J) := \mathrm{Ann}_F \, \mathrm{Ext}^1_S(\mathbb{L}_{S/R}, J/J^2).$$

Lemma 5.4.2. *In the situation of definition 5.4.1, we have:*
(i) *For any R-algebra R' let $F' := R' \otimes_R F$. Then:*
    (a) $\mathsf{H}_R(F, J) \cdot F' \subset \mathsf{H}_{R'}(F', JF')$, *and*
    (b) *the functor*
$$R\text{-}\mathbf{Alg} \to R\text{-}\mathbf{Mod} \quad : \quad R' \mapsto \mathsf{H}_{R'}(F', JF')$$
    commutes with filtered colimits of R-algebras.
(ii) *The subset $\mathrm{Spec}\, S \setminus V(\mathsf{H}_R(F, J) \cdot S)$ is the smooth locus of S over R.*
(iii) $\mathsf{H}_R(F, J)$ *annihilates $\mathrm{Ext}^1_S(\mathbb{L}_{S/R}, N)$ for every S-module N.*

Proof. According to [50, Ch.III, Cor.1.2.9.1] there is a natural isomorphism in $D(S\text{-}\mathbf{Mod})$:

(5.4.3) $$\tau_{[-1]}\mathbb{L}_{S/R} \simeq (0 \to J/J^2 \xrightarrow{\partial} S \otimes_F \Omega_{F/R} \to 0)$$

where ∂ is induced by the universal derivation $d : F \to \Omega_{F/R}$. Since $\Omega_{F/R}$ is a free F-module, we derive a natural isomorphism:

(5.4.4) $$\mathrm{Ext}^1_S(\mathbb{L}_{S/R}, J/J^2) \simeq \mathrm{End}_S(J/J^2)/\partial^* \mathrm{Hom}_F(\Omega_{F/R}, J/J^2)$$

Let now R' be an R-algebra and $h \in F$; in view of (5.4.4), the condition $h \in \mathsf{H}_R(F, J)$ means precisely that the scalar multiplication $h : J/J^2 \to J/J^2$ factors through ∂. It follows that $h : (J/J^2) \otimes_R R' \to (J/J^2) \otimes_R R'$ factors through $\partial \otimes_S 1_{S'}$. However, the latter map admits a factorization:
$$\partial \otimes_S 1_{S'} : (J/J^2) \otimes_R R' \xrightarrow{\alpha} J'/J'^2 \xrightarrow{\partial'} S' \otimes_{F'} \Omega_{F'/R'}$$

where α is a surjective map. It follows easily that the scalar multiplication h : $J'/J'^2 \to J'/J'^2$ factors through ∂', which shows (i.a).

Next, suppose that $(R_\lambda \mid \lambda \in \Lambda)$ is a filtered family of R-algebras whose colimit is R'; set $F_\lambda := R_\lambda \otimes_R F$, $S_\lambda := F_\lambda/JF_\lambda$. Suppose that $h \in F'$ is an element of $\mathsf{H}_{R'}(F', JF')$; then multiplication by h in JF'/J^2F' factors through $\partial' = \operatorname*{colim}_{\lambda \in \Lambda} (\partial_\lambda : JF_\lambda/J^2F_\lambda \to S_\lambda \otimes_F \Omega_{F/R})$. This factorization is given by an F-module map $\Omega_{F/R} \to JF/J^2F$, which in turns comes from a map $\beta : S_\lambda \otimes_F \Omega_{F/R} \to JF_\lambda/J^2F_\lambda$ for some $\lambda \in \Lambda$, and we can assume that h is the image of $h_\lambda \in F_\lambda$; as J is finitely generated, we increase λ, so that $\beta \circ \partial_\lambda$ is multiplication by h_λ. From this, (i.b) follows easily.

To show (ii), let us first pick any $h \in \mathsf{H}_R(F, J)$; we have to prove that S_h is smooth over R. However, the foregoing shows that multiplication by h on J/J^2 factors through $S \otimes_F \Omega_{F/R}$; if now h is invertible, this means that ∂ is a split imbedding, therefore S_h is formally smooth over R by the Jacobian criterion [28, Ch.0, Th.22.6.1]. Since S is of finite presentation over R, the assertion follows. Conversely, let $h \in F$ be an element such that S_h is smooth over R; we wish to show that $h^n \in \mathsf{H}_R(F, J)$ for n large enough. One can either prove this directly using the definition of $\mathsf{H}_R(F, J)$, or else by using the following lemma 5.4.6 and the well known fact that $h^n \in H_S$ for n large enough.

Finally we recall the natural isomorphism:

$$\operatorname{Hom}_{\mathsf{D}(S\text{-}\mathbf{Mod})}(\mathbb{L}_{S/R}, N[1]) \simeq \operatorname{Hom}_{\mathsf{D}(S\text{-}\mathbf{Mod})}(\tau_{[-1}\mathbb{L}_{S/R}, N[1])$$

which, in view of (5.4.3), shows that every morphism $\tau_{[-1}\mathbb{L}_{S/R} \to N[1]$ factors through a map $\tau_{[-1}\mathbb{L}_{S/R} \to J/J^2[1]$, whence (iii). $\qquad\square$

5.4.5.   In the situation of definition 5.4.1, choose a finite system of generators $f_1, ..., f_q$ for J; it is shown in [31, Ch.0, §2] how to construct an ideal, called H_S in *loc.cit.* with the same properties as in lemma 5.4.2(i.a),(ii). However, the definition of H_S depends explicitly on the choice of the generators $f_1, ..., f_q$. This goes as follows. For every integer p and every multi-index $(\alpha) = (\alpha_1, ..., \alpha_p) \in \mathbb{N}^p$ with $1 \le \alpha_1 < \alpha_2... < \alpha_p \le q$, set $|\alpha| := p$ and let $J_\alpha \subset J$ be the subideal generated by $f_{\alpha_1}, ..., f_{\alpha_p}$ and $\Delta_\alpha \subset F$ the ideal generated by the minors of order p of the Jacobian matrix $(\partial f_{\alpha_i}/\partial X_j \mid 1 \le i \le p, 1 \le j \le N)$. We set

$$H_S := \sum_{p \ge 0} \sum_{|\alpha|=p} \Delta_\alpha \cdot (J_\alpha : J).$$

Notice that we have always $J \subset \mathsf{H}_R(F, J)$; this does not necessarily hold for the H_S, and it is one of the reasons why we favor the former ideal. Though we won't be using the ideal H_S, we want to explain how it relates to the more intrinsic $\mathsf{H}_R(F, J)$. This is the purpose of the following:

Lemma 5.4.6. *With the notation of (5.4.5), we have: $H_S \subset \mathsf{H}_R(F, J)$.*

Proof. Let $p \in \mathbb{N}$, α a multi-index with $|\alpha| = p$, $\delta \in \Delta_p$ and $x \in F$ such that $xJ \subset J_\alpha$. We can suppose that $\delta = \det(M)$, where $M = (\partial f_{\alpha_j}/\partial X_{\beta_i} \mid 1 \le i, j \le p)$ for a certain multi-index β with $|\beta| = p$. We consider the maps

$$\phi_\alpha : S^p \to J/J^2 \quad : \quad e_i \mapsto f_{\alpha_i} \quad (\text{mod } J^2)$$

and $\pi_\beta : S \otimes_F \Omega_{F/R} \to S^p : dX_{\beta_j} \mapsto e_j$ for $j = 1, ..., p$ and $dX_k \mapsto 0$ if $k \notin \{\beta_1, ..., \beta_p\}$. Then the matrix of the composed S-linear map $\pi_\beta \circ \partial \circ \phi_\alpha : S^p \to S^p$ is none else than M. Let M' be the adjoint matrix of M, so that $M \cdot M' = \delta \cdot 1_{S^p}$, and we can compute:

$$\phi_\alpha \circ M' \circ \pi_\beta \circ \partial \circ \phi_\alpha = \phi_\alpha \circ (\delta \cdot 1_{S^p}) = (\delta \cdot 1_{J/J^2}) \circ \phi_\alpha.$$

In other words, the maps $\phi_\alpha \circ M' \circ \pi_\beta \circ \partial$ and $\delta \cdot 1_{J/J^2}$ agree on $\operatorname{Im} \phi_\alpha = (J_\alpha + J^2)/J^2$. Therefore: $(x \cdot \phi_\alpha) \circ M' \circ \pi_\beta \circ \partial = x \cdot \delta \cdot 1_{J/J^2}$, i.e. the scalar multiplication by $x \cdot \delta$ on J/J^2 factors through ∂; as in the proof of lemma 5.4.2 this implies that $x \cdot \delta \in \mathsf{H}_R(F, J)$, as claimed. $\qquad\square$

5.4.7.   Let R be a (not necessarily noetherian) ring, $t \in R$ a non-zero-divisor, $I \subset R$ an ideal, S an R-algebra of finite presentation, say of the form $S = F/J$ for some finitely generated ideal $J \subset F := R[X_1, ..., X_N]$. For any $a := (a_1, ..., a_N) \in R^N$, let $\mathfrak{p}_a \subset F$ be the ideal generated by $(X_1 - a_1, ..., X_N - a_N)$.

Lemma 5.4.8. *In the situation of (5.4.7), let* $n, h \in \mathbb{N}$ *and* $a \in R^N$ *such that*

$$(5.4.9) \qquad t^h \in \mathsf{H}_R(F, J) + \mathfrak{p}_a \qquad J \subset \mathfrak{p}_a + t^n I F \qquad n > 2h.$$

Then there exists $b \in R^N$ *such that*

$$b - a \in t^{n-h} I R^N \quad \text{and} \quad J \subset \mathfrak{p}_b + (t^{n-h} I)^2 F.$$

Proof. We are given a morphism $\sigma : \operatorname{Spec} R/t^n I \to \operatorname{Spec} S$ whose restriction to the closed subscheme $\operatorname{Spec} R/t^{n-h} I$ we denote by σ_0. The claim amounts to saying that there exists a lifting $\widetilde{\sigma} : \operatorname{Spec} R/(t^{n-h} I)^2 \to \operatorname{Spec} S$ of σ_0. By [50, Ch.III, Prop.2.2.2], the obstruction to the existence of an extension of σ to a morphism $\operatorname{Spec} R/(t^n I)^2 \to \operatorname{Spec} S$ is a class $\omega \in \mathbb{Ext}^1_S(\mathbb{L}_{S/R}, t^n I/(t^n I)^2)$. We have a commutative diagram:

$$
\begin{array}{ccc}
t^n I/(t^n I)^2 & \xrightarrow{\ \alpha\ } & t^{n-h} I/(t^{n-h} I)^2 \\
{\scriptstyle\beta}\downarrow & & \downarrow{\scriptstyle\gamma} \\
t^n I/(t^n I)^2 & \xrightarrow{\ \delta\ } & t^n I/t^{2n-h} I^2
\end{array}
$$

where α is induced by the inclusion $t^n I \subset t^{n-h} I$, β is the scalar multiplication by t^h, δ is the restriction to $t^n I$ of the natural projection $R/(t^n I)^2 \to R/t^{2n-h} I^2$, and γ is an isomorphism induced by scalar multiplication by $t^h : t^{n-h} I \xrightarrow{\sim} t^n I$. Since the S-module structure on $t^n I/(t^n I)^2$ is induced by extension of scalars via σ, it is clear that $\mathfrak{p}_a \cdot \omega = 0$; on the other hand, by lemma 5.4.2(iii) we know that $\mathsf{H}_R(F, J) \cdot \omega = 0$, hence $t^h \cdot \omega = \mathbb{Ext}^1_S(\mathbb{L}_{S/R}, \beta)(\omega) = 0$, and consequently $\mathbb{Ext}^1_S(\mathbb{L}_{S/R}, \alpha)(\omega) = 0$. Since the latter class is the obstruction to the existence of $\widetilde{\sigma}$, the assertion follows. $\qquad\square$

Lemma 5.4.8 is the basis of an inductive procedure that allows to construct actual sections of $X := \operatorname{Spec} S$, starting from approximate solutions of the system of equations defined by the ideal J. The sections thus obtained live in $X(R^\wedge)$, where

$R^\wedge$ is the tI-adic completion of R. In case I is finitely generated, $R^\wedge$ is (separated) complete for the $(tI)^\wedge$-adic topology, where $(tI)^\wedge$ is the topological closure of tI in $R^\wedge$; however, in later sections we will find situations where the relevant ideal I is not finitely generated; in such case, $R^\wedge$ is complete only for the $(tI)^\wedge$-c-adic topology (see definition 5.3.1(v)). In this section we carry out a preliminary study of some topologies on R-modules and on sets of sections of R-schemes; one of the main themes is to compare the topologies of, say $Z(R)$ where Z is an R-scheme, and of $Z(R^\wedge)$, where $R^\wedge$ is an adic completion of R. For this reason, it is somewhat annoying that, in passing from R to its completion, one is forced to replace a preadic topology by a c-preadic one. That is why we prefer to use a slightly coarser topology on R, described in the following:

Definition 5.4.10. Let R be any ring, $t \in R$ a non-zero-divisor, $I \subset R$ an ideal, M_0 (resp. N) a finitely generated R-module (resp. $R[t^{-1}]$-module).

  (i) The (t, I)-*preadic topology* of M_0 is the linear topology that admits the family of submodules $(t^n I M_0 \mid n \in \mathbb{N}_0)$ as a cofinal system of open neighborhoods of 0.
  (ii) The (t, I)-*preadic topology* of N is the linear topology $\mathscr{L}_N$ defined as follows. Choose a finitely generated R-module N_0 with an R-linear map $\phi : N_0 \to N$ such that $\phi[t^{-1}] : N_0[t^{-1}] \to N$ is onto; then $\mathscr{L}_N$ is the finest topology such that ϕ becomes an open map when we endow N_0 with the (t, I)-preadic topology. Therefore, the family of R-submodules $(\phi(t^k I N_0) \mid k \in \mathbb{N})$ is a cofinal system of open neighborhoods of $0 \in N$ for the topology $\mathscr{L}_N$.

The advantage of the (t, I)-adic topology is that the (t, I)-adic completion $R^\wedge$ of R is complete for a topology of the same type, namely for the $(t, I^\wedge)$-topology (where $I^\wedge \subset R^\wedge$ is the topological closure of I). Hence, for every R-scheme Z, the topologies on $Z(R)$ and $Z(R^\wedge)$ admit a uniform description. After this caveat, let us stress nevertheless that all of the results of this section hold as well for the c-adic topologies.

5.4.11. In the situation of definition 5.4.10, it is easy to verify that the (t, I)-preadic topology on N does not depend on the choice of N_0 and ϕ. Indeed, if $\phi' : N_1 \to N$ is another choice, then one checks that $\phi(t^k N_0) \subset \phi'(N_1)$ for $k \in \mathbb{N}$ large enough, and symmetrically $\phi'(t^k N_1) \subset \phi(N_0)$, which implies the assertion.

Moreover, it is easy to check that every $R[t^{-1}]$-linear map $M \to N$ of finitely generated $R[t^{-1}]$-modules is continuous for the respective (t, I)-preadic topologies.

Lemma 5.4.12. *Let R be a noetherian ring, S a finitely presented R-algebra, $\mathscr{I} \subset R$ an ideal, and suppose that the pair $(R, \mathscr{I})$ is henselian. Let $\overline{R}$ be the $\mathscr{I}$-adic completion of R and set $\overline{S} := \overline{R} \otimes_R S$. Let $U \subset \operatorname{Spec} S$ be an open subset smooth over $\operatorname{Spec} R$, and $\overline{U} \subset \operatorname{Spec} \overline{S}$ the preimage of U. Then, for every integer n and every $\overline{R}$-section $\overline{\sigma} : \operatorname{Spec} \overline{R} \to \operatorname{Spec} \overline{S}$ whose restriction to $\operatorname{Spec} \overline{R} \setminus V(\mathscr{I}\overline{R})$ factors through $\overline{U}$, there exists an R-section $\sigma : \operatorname{Spec} R \to \operatorname{Spec} S$ congruent to $\overline{\sigma}$ modulo $\mathscr{I}^n$, and whose restriction to $\operatorname{Spec} R \setminus V(\mathscr{I})$ factors through U.*

Proof. It is [31, Ch.II, Th.2 bis].   □

Proposition 5.4.13. *Keep the notation of* (5.4.7) *and let* $I \subset R$ *be an ideal such that* (R, tI) *is a henselian pair. Let* $a \in R^N$ *and* $n, h \geq 0$ *such that* (5.4.9) *holds. Then there exists* $b \in R^N$ *such that* $b - a \in t^{n-h} I R^N$ *and* $J \subset \mathfrak{p}_b$.

Proof. We consider first the following special case:

Claim 5.4.14. The proposition holds if R is complete for the tI-adic topology.

Proof of the claim. We apply repeatedly lemma 5.4.8 to obtain a tI-adically convergent sequence of elements $(a_m \in R^N \mid m \in \mathbb{N})$, with $a_0 := a$ and such that $a_m \equiv a$ $(\mathrm{mod}\ t^{n-h} I R^N)$ for every $m \in \mathbb{N}$. The limit b of the sequence $(a_m \mid m \in \mathbb{N})$ will do.   ◇

Let next R be a general ring; let $H \subset \mathsf{H}_R(F, J)$ be a finitely generated subideal such that $t^h \in H + \mathfrak{p}_a$. Up to replacing I by a subideal, we can assume that I is finitely generated, say by a finite subset $T \subset R$; furthermore, we can write $R = \bigcup_{\lambda \in \Lambda} R_\lambda$ for a filtered family of noetherian subrings $(R_\lambda \mid \lambda \in \Lambda)$ such that $\{t\} \cup T \subset R_\lambda$ and $a \in R_\lambda^N$ for every $\lambda \in \Lambda$, and we set $I_\lambda := T R_\lambda$ for every $\lambda \in \Lambda$. We can also assume that the pair (R_λ, tI_λ) is henselian for every $\lambda \in \Lambda$. Let $f_1, ..., f_q \in F$ be a finite set of generators for J and $g_1, ..., g_r \in F$ a finite set of generators for H; up to restricting to a cofinal family, we can then assume that $f_i, g_j \in F_\lambda := R_\lambda[X_1, ..., X_N]$ for every $\lambda \in \Lambda$ and every $i \leq q, j \leq r$. Let $J_\lambda \subset F_\lambda$ be the ideal generated by $f_1, ..., f_q$ and set $S_\lambda := F_\lambda/J_\lambda$; let also $H_\lambda \subset F_\lambda$ be the ideal generated by $g_1, ..., g_r$. Again after replacing Λ by a cofinal subfamily, we can achieve that $H_\lambda \subset \mathsf{H}_{R_\lambda}(F_\lambda, J_\lambda)$ for every $\lambda \in \Lambda$ (by lemma 5.4.2(i.b)). Finally, let $\mathfrak{p}_{\lambda,a} \subset F_\lambda$ be the ideal generated by $X_1 - a_1, ..., X_N - a_N$; we can assume that $t^h \in H_\lambda + \mathfrak{p}_{\lambda,a}$ and $J_\lambda \subset \mathfrak{p}_{\lambda,a} + t^n I_\lambda F_\lambda$ for every $\lambda \in \Lambda$. With this setup, let $\overline{R}_\lambda$ be the tI_λ-adic completion of R_λ; we can apply claim 5.4.14 to deduce that there exists $c \in \overline{R}_\lambda^N$ such that $c - a \in t^{n-h} I_\lambda \overline{R}_\lambda^N$ and such that $J_\lambda \subset \overline{\mathfrak{p}}_{\lambda,c}$ (where $\overline{\mathfrak{p}}_{\lambda,c}$ denotes the ideal generated by $X_1 - c_1, ..., X - c_n$ in $\overline{R}_\lambda[X_1, ..., X_N]$). Let $U_\lambda := \mathrm{Spec}\, S_\lambda \setminus V(H_\lambda \cdot S_\lambda)$ and $\overline{S}_\lambda := \overline{R}_\lambda \otimes_{R_\lambda} S_\lambda$; from lemma 5.4.2(ii) we know that U_λ is smooth over $\mathrm{Spec}\, R_\lambda$, and by construction c determines a $\overline{R}_\lambda$-section of $\mathrm{Spec}\, \overline{S}_\lambda$ whose restriction to $\mathrm{Spec}\, \overline{R}_\lambda \setminus V(tI_\lambda)$ factors through the preimage of U_λ. Therefore lemma 5.4.12 ensures that there exists an R_λ-section $\sigma : \mathrm{Spec}\, R_\lambda \to \mathrm{Spec}\, S_\lambda$ that agrees with c modulo $(tI_\lambda)^{n-h}$ and whose restriction to $\mathrm{Spec}\, R_\lambda \setminus V(tI_\lambda)$ factors through U. Let $\pi_\lambda : F_\lambda \to S_\lambda$ be the natural projection; there is a unique $b \in R_\lambda^N$ such that $\mathfrak{p}_{\lambda,b} = \pi_\lambda^{-1}(\mathrm{Ker}\, \sigma^\sharp : S_\lambda \to R_\lambda)$; this point b has the sought properties.   □

5.4.15.   Let now X be an affine scheme of finite type over $\mathrm{Spec}\, R[t^{-1}]$; for every $n \in \mathbb{N}$ let $\mathbb{A}^n_{R[t^{-1}]}$ be the n-dimensional affine space over $\mathrm{Spec}\, R[t^{-1}]$; for n large enough we can find a closed imbedding $j : X \to \mathbb{A}^n_{R[t^{-1}]}$ of $\mathrm{Spec}\, R[t^{-1}]$-schemes. The choice of coordinates on $\mathbb{A}^n_{R[t^{-1}]}$ yields a bijection $\mathbb{A}^n_{R[t^{-1}]}(R[t^{-1}]) \simeq R[t^{-1}]^n$, and then j induces an injective map $j_* : X(R[t^{-1}]) \subset R[t^{-1}]^n$. The (t, I)-*adic* *topology* of $X(R[t^{-1}])$ is defined as the subspace topology induced by j_* (where $R[t^{-1}]^n$ is endowed with its (t, I)-preadic topology).

Lemma 5.4.16. *Let X be as in (5.4.15); we have:*

(i) *The (t, I)-adic topology of $X(R[t^{-1}])$ is independent of the choice of closed imbedding j and of coordinates on $\mathbb{A}^n_{R[t^{-1}]}$.*

(ii) *If Y is another affine $R[t^{-1}]$-scheme, then the natural map*

$$(X \times_{R[t^{-1}]} Y)(R[t^{-1}]) \to X(R[t^{-1}]) \times Y(R[t^{-1}])$$

is a homeomorphism for the (t, I)-adic topologies.

(iii) *If $U \subset X$ is any open subset, and $\sigma \in U(R[t^{-1}])$, there exists $f \in \mathscr{O}_X(X)$ such that $D(f) := X \backslash V(f) \subset U$ and such that σ factors through $D(f)$.*

(iv) *Suppose that $tI \subset \mathrm{rad}(R)$. If $U \subset X$ is an affine open subset, then the (t, I)-adic topology on $U(R[t^{-1}])$ agrees with the topology induced from the (t, I)-adic topology on $X(R[t^{-1}])$, and $U(R[t^{-1}])$ is an open subset of $X(R[t^{-1}])$.*

(v) *If $X = \mathrm{Spec}\,(\mathrm{Sym}^{\bullet}_{R[t^{-1}]} P)$ for a finitely generated projective $R[t^{-1}]$-module P, then the natural bijection $X(R[t^{-1}]) \xrightarrow{\sim} P^*$ identifies the (t, I)-adic topology of $X(R[t^{-1}])$ with the (t, I)-preadic topology of P.*

Proof. (i): Suppose $j_1 : X \to \mathbb{A}^n_{R[t^{-1}]}$ and $j_2 : X \to \mathbb{A}^m_{R[t^{-1}]}$ are two closed imbeddings, τ_1 and τ_2 the respective topologies on $X(R[t^{-1}])$; by symmetry it suffices to show that the identity map $(X(R[t^{-1}]), \tau_1) \to (X(R[t^{-1}]), \tau_2)$ is continuous. However, j_2 can be extended to some morphism $\phi : \mathbb{A}^n_{R[t^{-1}]} \to \mathbb{A}^m_{R[t^{-1}]}$, and we come down to showing that the induced map $\phi_* : R[t^{-1}]^n \to R[t^{-1}]^m$ is continuous for the (t, I)-adic topology. We can further reduce to the case of $m = 1$, in which case ϕ is given by a single polynomial $f \in R[t^{-1}, T_1, ..., T_n]$, and ϕ_* is the map $(a_1, ..., a_n) \mapsto f(a_1, ..., a_n)$. Let $x_0 \in R[t^{-1}]^n$ and set $y_0 := f(x_0)$; we have to show that, for every $k \in \mathbb{N}$ there exists $h \in \mathbb{N}$ such that $f(x_0 + t^h I) \subset y_0 + t^k I$. The Taylor formula gives an identity of the form: $f(T) = y_0 + \sum_{r \in \mathbb{N}^n \backslash \{0\}} a_r \cdot (T - x_0)^r$, where $a_r \in R[t^{-1}]$ for every $r \in \mathbb{N}^n$ and $a_r = 0$ for all but finitely many r. Let $s \in \mathbb{N}$ be an integer large enough so that $t^s a_r \in R$ for every $r \in \mathbb{N}^n$; obviously $h := k + s$ will do. (ii) is easily reduced to the corresponding statement for (t, I)-adic topologies on direct sums of $R[t^{-1}]$-modules; we leave the details to the reader.

(iii): We can write $U = X \backslash V(J)$, where $J \subset S := \mathscr{O}_X(X)$, and a section $\sigma \in U(R[t^{-1}])$ induces a map $\phi : S \to R[t^{-1}]$ such that $\phi(J)R = R$. Take $a_i \in R$ and $f_i \in J$, $i = 1, ..., n$ with $\sum_i a_i \cdot \phi(f_i) = 1$ and set $f := \sum_i f_i a_i$; it follows that the image of σ is contained in the affine open subset $\mathrm{Spec}\, S[f^{-1}] \subset U$. To show (iv), we can suppose, thanks to (iii), that $U = X \backslash V(f)$ for some $f \in \mathscr{O}_X(X)$. Let $\phi_f : X \to \mathbb{A}^1_{R[t^{-1}]}$ be the morphism defined by f, and $\Gamma_f : X \to X \times_{R[t^{-1}]} \mathbb{A}^1_{R[t^{-1}]}$ its graph; choose a closed imbedding $X \subset \mathbb{A}^n_{R[t^{-1}]}$. The composition $j : X \to X \times_{R[t^{-1}]} \mathbb{A}^1_{R[t^{-1}]} \to \mathbb{A}^{n+1}_{R[t^{-1}]}$ is another closed imbedding, which induces the same topology on $X(R[t^{-1}])$ in view of (i). We have $j^{-1}(\mathbb{A}^n_{R[t^{-1}]} \times_{R[t^{-1}]} \mathbb{G}_{m, R[t^{-1}]}) =$

U, consequently we are reduced to showing the assertion for the open imbedding $\mathbb{A}^n_{R[t^{-1}]} \times_{R[t^{-1}]} \mathbb{G}_{m,R[t^{-1}]} \subset \mathbb{A}^{n+1}_{R[t^{-1}]}$. Using (ii) we further reduce to considering the imbedding $\mathbb{G}_{m,R[t^{-1}]} \subset \mathbb{A}^1_{R[t^{-1}]}$. The $R[t^{-1}]$-sections of $\mathbb{G}_m$ and of $\mathbb{A}^1$ are naturally identified with $R[t^{-1}]^\times$ and respectively $R[t^{-1}]$; if $a \in R[t^{-1}]^\times$, then the coset $a + atI$ is open in the (t,I)-adic topology of $R[t^{-1}]$ and it is contained in $R[t^{-1}]^\times$ if $tI \subset \mathrm{rad}(I)$. This shows the second assertion of (iv). To show the first assertion of (iv) we use the closed imbedding $\mathbb{G}_{m,R[t^{-1}]} \subset \mathbb{A}^2_{R[t^{-1}]}$ given by the rule: $a \mapsto (a, a^{-1})$. So, the set of $R[t^{-1}]$-sections of $\mathbb{G}_{m,R[t^{-1}]}$ is identified with the diagonal $\Delta := \{(a, a^{-1}) \mid a \in R[t^{-1}]^\times\}$. In terms of this imbedding, the (t,I)-adic topology of $\mathbb{G}_m(R[t^{-1}])$ admits the basis $U_{n,a} := \Delta \cap \{(a+x, a^{-1}+y) \mid x, y \in t^n I\}$, for n ranging over the positive integers and $a \in R[t^{-1}]^\times$. We have to prove that the image of $U_{n,a}$ is an open subset in the (t,I)-topology of $\mathbb{A}^1(R[t^{-1}])$. This comes down to showing that $(a+x)^{-1} \in a^{-1} + t^n I$ whenever $x \in atI \cap a^2 t^n I$, which we leave as an exercise for the reader.

(v) is easily reduced to the case where P is a free $R[t^{-1}]$-module of finite rank, in which case the assertion follows by inspecting the definitions. □

Lemma 5.4.17. *Let R be any ring, X an open subscheme of a projective R-scheme. Then every R-section $\mathrm{Spec}\, R \to X$ of X factors through an open imbedding $U \subset X$, where U is an affine R-scheme.*

Proof. By assumption X is a locally closed subset of $\mathbb{P}^n_R$, for some $n \in N$.

Claim 5.4.18. The lemma holds if X is a closed subscheme of $\mathbb{P}^n_R$.

Proof of the claim. Indeed, in this case we can assume $X = \mathbb{P}^n_R$. Therefore, let $\sigma \in \mathbb{P}^n_R(R)$; by [27, Ch.II, Th.4.2.4] σ corresponds to a rank one locally free quotient L of R^{n+1}. We choose a section of the projection $\pi_1 : R^{n+1} \to L$, hence a decomposition $R^{n+1} \simeq L \oplus \mathrm{Ker}\, \pi_1$; the induced projection $\pi_2 : R^{n+1} \to \mathrm{Ker}\, \pi_1$ determines a closed imbedding $\mathbb{P}(\mathrm{Ker}\, \pi_1) \to \mathbb{P}^n_R$ representing the transformation of functors that assigns to every rank one locally free quotient of $\mathrm{Ker}\, \pi_1$ the same module, seen as a quotient of R^{n+1} via the projection π_2. It is clear that the image of σ does not intersect the image of $\mathbb{P}(\mathrm{Ker}\, \pi_1)$, and the complement $\mathbb{P}^n_R \setminus \mathbb{P}(\mathrm{Ker}\, \pi_1)$ is affine (since this is the same as $\mathrm{Spec}\, (\mathrm{Sym}^\bullet(L^* \otimes_R \mathrm{Ker}\, \pi_1)))$. ◊

Thanks to claim 5.4.18 we can assume that X is an open subscheme of an affine scheme Y of finite type over $\mathrm{Spec}\, R$. In this case, the claim reduces to lemma 5.4.16(iii) (applied with $t = 1$). □

5.4.19.   Assume now that $tI \subset \mathrm{rad}(I)$ and let X be an open subscheme of a projective $R[t^{-1}]$-scheme; each affine open subset of X comes with a natural (t,I)-adic topology. By lemma 5.4.16(iv) these topologies agree on the intersections of any two such affine open subsets, and according to lemma 5.4.17 we have $X(R[t^{-1}]) = \bigcup_U U(R[t^{-1}])$, where U ranges on the family of all affine open subsets $U \subset X$, so $X(R[t^{-1}])$ can be endowed with a well defined (t,I)-adic topology, independent of all choices, for which each $U(R[t^{-1}])$ is an open subspace.

Lemma 5.4.20. *The closed subscheme U_n of $\operatorname{Spec}\mathbb{Z}[x_{11}, ..., x_{nn}]$ that classifies the $n \times n$ idempotent matrices is smooth over $\operatorname{Spec}\mathbb{Z}$.*

Proof. Clearly U_n is of finite type over $\operatorname{Spec}\mathbb{Z}$, hence it suffices to show that U_n is formally smooth. Therefore, let R_0 be a ring and $I \subset R_0$ an ideal with $I^2 = 0$; we need to show that the induced map $U_n(R_0) \to U_n(R_0/I)$ is surjective, *i.e.* that every $n \times n$ idempotent matrix $\overline{M}$ with entries in R_0/I lifts to an idempotent matrix with entries in R_0. Pick an arbitrary matrix $M \in M_n(R_0)$ that lifts $\overline{M}$; let $E := R_0[M] \subset M_n(R_0)$ be the commutative R_0-algebra generated by M, $\overline{E} \subset M_n(R_0/I)$ be the image of E, $J \subset E$ the kernel of the induced map $E \to \overline{E}$. We have $J^2 = 0$, so we can apply proposition 2.5.13(i) to lift $\overline{M}$ to some idempotent matrix in E. $\qquad\square$

Proposition 5.4.21. *Let $t \in R$ be a non-zero-divisor, $I \subset R$ an ideal, $R^\wedge := \lim_{n \in \mathbb{N}} R/t^n I$ the (t, I)-adic completion of R, $I^\wedge$ the topological closure of I in $R^\wedge$, and suppose that the pair (R, tI) is henselian. Let X be a smooth open subscheme of a projective $R[t^{-1}]$-scheme, and endow $X(R[t^{-1}])$ (resp. $X(R^\wedge[t^{-1}])$) with its (t, I)-adic (resp. $(t, I^\wedge)$-adic) topology. Then the natural map $X(R[t^{-1}]) \to X(R^\wedge[t^{-1}])$ has dense image.*

Proof. We begin with the following special case:

Claim 5.4.22. The proposition holds if X is affine.

Proof of the claim. Say that $X = \operatorname{Spec} S$, where S is some finitely presented smooth $R[t^{-1}]$-algebra, and let $\sigma : S \to R^\wedge[t^{-1}]$ be any element of $X(R^\wedge[t^{-1}])$. We have to show that there are elements of $X(R[t^{-1}])$ in every (t, I)-adic neighborhood of σ. To this aim, we pick a finitely presented R-algebra S_0 such that $S_0[t^{-1}] \simeq S$; after clearing some denominators we can assume that σ extends to a map $S_0 \to R^\wedge$. Let

$$(5.4.23) \qquad\qquad S_0 = R[X_1, ..., X_N]/J$$

be a finite presentation of S_0, and set $H := \mathsf{H}_R(R[X_1, ..., X_N], J)$ (notation of definition 5.4.1 : recall that $J \subset H$). By assumption we have

$$(5.4.24) \qquad\qquad t^h \in H \qquad \text{for } h \in \mathbb{N} \text{ large enough.}$$

The presentation (5.4.23) defines a closed imbedding $X \subset \mathbb{A}_R^N$, and we can then find a section $\sigma_0 : \operatorname{Spec} R \to \mathbb{A}_R^N$ that is (t, I)-adically close to σ, so that the restrictions of σ and σ_0 agree on $\operatorname{Spec} R/t^n I$. Let $\mathfrak{p} := \operatorname{Ker} \sigma_0^\sharp \subset R[X_1, ..., X_N]$ be the ideal corresponding to σ_0; it then follows that

$$(5.4.25) \qquad\qquad J \subset \mathfrak{p} + t^n I R[X_1, ..., X_N].$$

Up to enlarging n (which is harmless) we can assume that $n > 2h$. By assumption we have $\operatorname{Ann}_R(t) = 0$, hence from (5.4.24), (5.4.25) and proposition 5.4.13 we deduce the contention. $\qquad\diamond$

Claim 5.4.26. Suppose that $X = \mathbb{P}_{R[t^{-1}]}^r$. Then every section $\sigma : \operatorname{Spec} R^\wedge[t^{-1}] \to X$ factors through an open imbedding $U \subset X$, where U is an affine R-scheme.

Proof of the claim. By [27, Ch.II, Th.4.2.4], σ corresponds to a rank one projective quotient L of $(R^{\wedge}[t^{-1}])^{r+1}$; we can then find an idempotent $(r + 1) \times (r + 1)$ matrix e such that $\mathrm{Coker}(e) = L$. By lemma 5.4.20, the scheme U that represents the $(r + 1) \times (r + 1)$ idempotent matrices is smooth; clearly U decomposes as a disjoint union of open and closed subschemes $U = \bigcup_{n=0}^{r+1} U_n$, where U_n represents the subfunctor that classifies all $(r + 1) \times (r + 1)$ idempotent matrices of rank n, for every $n = 0, ..., r + 1$. Therefore each U_n is smooth and affine over $\mathrm{Spec}\,\mathbb{Z}$ and then by claim 5.4.22 it follows that e can be approximated closely by an idempotent matrix $e_0 \in M_{r+1}(R[t^{-1}])$ whose rank equals the rank of e. We have $e_0 \cdot e = e_0 \cdot (I_{r+1} - e_0 + e)$ and if e_0 is sufficiently (t, I)-adically close to e, the matrix $I_{r+1} - e_0 + e$ is invertible in $M_{r+1}(R)$, hence we can assume that:

$$(5.4.27) \qquad \mathrm{Im}(e_0 \cdot e) = \mathrm{Im}(e_0) \subset R^{\wedge}[t^{-1}]^{r+1}.$$

The projection

$$\pi : R[t^{-1}]^{r+1} \to \mathrm{Im}\,e_0 \quad : \quad x \mapsto e_0(x)$$

determines a closed imbedding $\mathbb{P}(\mathrm{Im}\,e_0) \to \mathbb{P}^r_{R[t^{-1}]}$ representing the transformation of functors that, to every $R[t^{-1}]$-algebra S and every rank one projective quotient of $S \otimes_{R[t^{-1}]} \mathrm{Im}(e_0)$ assigns the same module seen as a quotient of $S \otimes_{R[t^{-1}]} R[t^{-1}]^{r+1}$ via π. By (5.4.27), the restriction of $1_{R^{\wedge}} \otimes_R \pi$ to $R^{\wedge} \otimes_R \mathrm{Im}(e)$ is an isomorphism, hence it remains such after every base change $R^{\wedge}[t^{-1}] \to S$; this means that the image of σ lands in the complement of $\mathbb{P}(\mathrm{Im}\,e_0)$, which implies the contention. $\diamond$

Claim 5.4.28. Let X be any open subscheme of a projective $R[t^{-1}]$-scheme, $\sigma :$ $\mathrm{Spec}\,R^{\wedge}[t^{-1}] \to X$ any section. Then σ factors through an open imbedding $U \subset X$, where U is an affine R-scheme.

Proof of the claim. Due to claim 5.4.26 we can assume that X is an open subscheme of an affine $R[t^{-1}]$-scheme of finite type. Thus, we can write $X = Y \setminus V(J)$, where $J \subset S := \mathscr{O}_Y(Y)$, and a section $\sigma \in X(R^{\wedge}[t^{-1}])$ induces a map $\phi : S \to R^{\wedge}[t^{-1}]$ such that $\phi(J)R^{\wedge}[t^{-1}] = R^{\wedge}[t^{-1}]$. Pick $a_i \in R^{\wedge}[t^{-1}]$, $f_i \in J$ such that $\sum_i a_i \cdot \phi(f_i) = 1$; we choose $b_i \in R[t^{-1}]$, (t, I)-adically close to a_i $(i = 1, ..., n)$, so that $\sum_i \phi(f_i) \cdot (a_i - b_i) \in tIR^{\wedge}$. Since $tIR^{\wedge} \subset \mathrm{rad}(R^{\wedge})$, we deduce easily that $\sum_i \phi(f_i)b_i \in R^{\wedge}$ is invertible in $R^{\wedge}$. Set $f := \sum_i f_i b_i$; it follows that the image of σ is contained in the affine open subset $\mathrm{Spec}\,S[f^{-1}] \subset X$. $\diamond$

The proposition follows easily from claims (5.4.22) and (5.4.28). $\square$

Proposition 5.4.29. *Resume the assumptions of proposition 5.4.21 and let $\phi : X \to Y$ be a morphism of quasi-projective $R[t^{-1}]$-schemes; then we have:*

    (i) *The map $\phi_* : X(R[t^{-1}]) \to Y(R[t^{-1}])$ is continuous for the (t, I)-adic topologies.*

    (ii) *If ϕ is smooth, ϕ_* is an open map.*

Proof. (i): We can factor ϕ as a closed imbedding $X \to X \times Y$ followed by a projection $X \times Y \to Y$, so it suffices to prove the claim for the latter maps. The case of a projection is immediate, and the case of a closed imbedding follows straightforwardly from lemma 5.4.16(i).

(ii): Due to lemmata 5.4.16(iv) and 5.4.17 we can assume that both X and Y are affine schemes, say $X = \operatorname{Spec} S$, $Y = \operatorname{Spec} T$ where S and T are finitely generated $R[t^{-1}]$-algebras and S is smooth over T, especially S is a finitely presented T-algebra. Let $\sigma : \operatorname{Spec} R[t^{-1}] \to X$ be an element of $X(R[t^{-1}])$; after choosing presentations for S and T and clearing some denominators, we can find morphisms of finitely generated R-algebras $f : T_0 \to S_0$ and $g : S_0 \to R$ such that

$$f[t^{-1}] = \phi^{\sharp} : T \to S \qquad g[t^{-1}] = \sigma^{\sharp} : S \to R[t^{-1}]$$

and moreover f is of finite presentation. Let $X_0 := \operatorname{Spec} S_0$, $Y_0 := \operatorname{Spec} T_0$, $\phi_0 :=$ $\operatorname{Spec} (f)$, $\sigma_0 := \operatorname{Spec} (g)$.

Claim 5.4.30. There is an open neighborhood $U \subset Y(R[t^{-1}])$ of $\phi_*(\sigma)$ such that every $\xi \in U$ extends to a section $\xi_0 : \operatorname{Spec} R \to Y_0$.

Proof of the claim. By construction, $\phi_*(\sigma)$ is induced by the map $g \circ f : T_0 \to R$; any other section $\xi \in Y(R[t^{-1}])$ is determined by a map $\xi^{\sharp} : T \to R[t^{-1}]$; we have $T = T_0[t^{-1}]$ and T_0 is of finite type over R, say $T_0 = R[x_1, ..., x_N]/J$. Therefore $a_i := g \circ f(x_i) \in R$ for every $i \leq N$. The claim follows by observing that the set of sections $\xi \in Y(R[t^{-1}])$ such that $\xi^{\sharp}(x_i) - a_i \in R$ for every $i \leq N$ forms an open neighborhood of $\phi_*(\sigma)$. $\diamond$

Let $V \subset X(R[t^{-1}])$ be a given open neighborhood of σ; we have to show that, for every section $\xi : \operatorname{Spec} R[t^{-1}] \to Y$ sufficiently close to $\phi_*(\sigma)$, there is $\xi' \in V$ such that $\phi_*(\xi') = \xi$. To this aim, choose a presentation

$$(5.4.31) \qquad\qquad S_0 \simeq T_0[x_1, ..., x_N]/J$$

and let $H := \mathsf{H}_{T_0}(T_0[x_1, ..., x_N], J)$ (notation of definition 5.4.1). Since S is smooth over T, we have

$$(5.4.32) \qquad\qquad t^h \in H \qquad \text{for } h \in \mathbb{N} \text{ large enough.}$$

Let $\xi \in Y(R[t^{-1}])$; by claim 5.4.30 we can suppose that ξ extends to a section $\xi_0 : \operatorname{Spec} R \to Y_0$. Define $X_0(\xi_0)$ as the fibre product in the cartesian diagram

$$
\begin{array}{ccc}
X_0(\xi_0) & \longrightarrow & X_0 \\
\downarrow & & \downarrow{\scriptstyle \phi_0} \\
\operatorname{Spec} R & \xrightarrow{\ \xi_0\ } & Y_0.
\end{array}
$$

The presentation (5.4.31) induces a closed imbedding $X_0(\xi_0) \subset \mathbb{A}_R^N$ whose defining ideal $J(\xi_0) \subset R[x_1, ..., x_N]$ is the image of J; define similarly $H(\xi_0) \subset R[x_1, ..., x_N]$ as the image of H. From (5.4.32) we deduce that

$$(5.4.33) \qquad\qquad t^h \in H(\xi_0) \qquad \text{for } h \in \mathbb{N} \text{ large enough.}$$

Suppose now that ξ is sufficiently close to $\phi_*(\sigma)$; this means that the restrictions of ξ_0 and $\phi_{0*}(\sigma_0)$ agree on some closed subset $\operatorname{Spec} R/t^n I \subset \operatorname{Spec} R$. Hence, let $\overline{\sigma}_0 : \operatorname{Spec} R/t^n I \to X_0(\xi)$ be the restriction of σ_0, and choose any extension of $\overline{\sigma}_0$ to a morphism $\sigma_1 : \operatorname{Spec} R \to \mathbb{A}_R^N$. Finally, let $\mathfrak{p} := \operatorname{Ker} \sigma_1^{\sharp} \subset R[x_1, ..., x_N]$ be the ideal corresponding to σ_1. By construction we have

(5.4.34) $J(\xi_0) \subset \mathfrak{p} + t^n IR[x_1, ..., x_N].$

Finally, (5.4.33), (5.4.34) and proposition 5.4.13 show that, if $n > 2h$ – which can always be arranged – the restriction of $\overline{\sigma}_0$ to Spec $R/t^{n-h}I$ can be extended to a section $\xi' :$ Spec $R \to X_0(\xi_0)$, which will lie in the given open neighborhood V, provided n is large enough. $\square$

5.4.35.   Resume the assumptions of proposition 5.4.21 and let

$$G := (X_0, X_1, s, t, c, \iota)$$

be a groupoid of quasi-projective $R[t^{-1}]$-schemes (see (4.5.1) for our general notations concerning groupoids in a category). We have a natural map

(5.4.36) $\pi_0(G(R[t^{-1}])) \to \pi_0(G(R^{\wedge}[t^{-1}]))$

where, for any groupoid of sets $\mathscr{G} := (G_0, G_1, s_G, t_G, c_G, \iota_G)$, we denote by $\pi_0(\mathscr{G})$ the set of isomorphism classes of elements of G_0; this is the same as the set of connected components of the geometric realization of the simplicial set associated to the groupoid $\mathscr{G}$.

Theorem 5.4.37. *Keep the notation of (5.4.35), and suppose that X_0 is smooth over $R[t^{-1}]$ and that the morphism $(s, t) : X_1 \to X_0 \times_{R[t^{-1}]} X_0$ is smooth. Then (5.4.36) is a bijection.*

Proof. Let $\sigma \in X_0(R^{\wedge}[t^{-1}])$; since X_0 is smooth over $R[t^{-1}]$ we can find sections $\sigma_0 \in X_0(R[t^{-1}])$ with image $\sigma_0^{\wedge} \in X_0(R^{\wedge}[t^{-1}])$ arbitrarily (t, I)-adically close to σ (proposition 5.4.21). Then $(\sigma, \sigma_0^{\wedge})$ can be made arbitrarily close to $(\sigma, \sigma) \in X_0 \times_{R[t^{-1}]} X_0(R^{\wedge}[t^{-1}])$; since the latter lies in the image of $X_1(R^{\wedge}[t^{-1}])$ under the morphism (s, t), it follows that the same holds for $(\sigma, \sigma_0^{\wedge})$, provided $\sigma_0^{\wedge}$ is sufficiently close to σ (proposition 5.4.29(ii)). This shows that (5.4.36) is onto. Next, suppose that $\sigma, \tau \in X_0(R[t^{-1}])$ and that their images $\sigma^{\wedge}, \tau^{\wedge}$ in $X_0(R^{\wedge}[t^{-1}])$ lie in the same homotopy class. By definition, this means that $(\sigma^{\wedge}, \tau^{\wedge}) = (s, t)(\alpha)$ for some $\alpha \in X_1(R^{\wedge}[t^{-1}])$; as (s, t) is smooth, the closed subscheme $Y := (s, t)^{-1}(\sigma, \tau) \subset X_1$ is smooth over Spec R. By construction, α is in $Y(R^{\wedge}[t^{-1}])$, hence it can be approximated by a point $\alpha_0 \in Y(R[t^{-1}])$; clearly $(s, t)(\alpha_0) = (\sigma, \tau)$. $\square$

5.4.38.   Let $n \in \mathbb{N}$ and define U_n as in lemma 5.4.20; sometimes we identify U_n to the functor which it represents. We define a functor from $F_n : \mathbb{Z}\text{-}\mathbf{Alg} \to \mathbf{Set}$ as follows. Given a ring R, we let $F_n(R)$ be the set of all data of the form (S, T, ϕ, ψ), where $S, T \in M_n(R)$ are two idempotent matrices and $\phi, \psi \in M_n(R)$ are two other matrices submitted to the following conditions:

    (a) $\phi \cdot S = 0 = T \cdot \phi.$
    (b) $\psi \cdot T = 0 = S \cdot \psi.$
    (c) $(I_n - S) \cdot (\psi \cdot \phi - I_n) = 0 = (I_n - T) \cdot (\phi \cdot \psi - I_n).$

The meaning of (a) is that ϕ induces a map $\overline{\phi} :$ Coker$(S) \to$ Coker(T); likewise, (b) means that ψ induces a map $\overline{\psi} :$ Coker$(T) \to$ Coker(S). Finally (c) means that $\overline{\phi} \circ \overline{\psi}$ is the identity of Coker(T), and likewise for $\overline{\psi} \circ \overline{\phi}$. It is easy to see from this description that the functor F_n is representable by an affine $\mathbb{Z}$-scheme of finite type, which we shall denote by the same name. Moreover, the rule $(S, T, \phi, \psi) \mapsto (S, T)$ defines a natural transformation of functors $F_n \to U_n \times U_n$, whence a morphism of schemes:

$$(5.4.39) \qquad\qquad F_n \to U_n \times_{\mathbb{Z}} U_n.$$

Lemma 5.4.40. *The morphism* (5.4.39) *is smooth.*

Proof. Since (5.4.39) is clearly of finite presentation, it suffices to verify that it is formally smooth. Hence, let $R \to R_0$ be a surjective ring homomorphism with nilpotent kernel J, let $(S, T) \in U_n \times_{\mathbb{Z}} U_n(R)$ and $(S_0, T_0, \phi_0, \psi_0) \in F_n(R_0)$ such that (S_0, T_0) coincides with the image of (S, T) in $U_n \times_{\mathbb{Z}} U_n(R_0)$. We need to show that there exist $\phi, \psi \in M_n(R)$ lifting ϕ_0 and ψ_0, such that $(S, T, \phi, \psi) \in F_n(R)$. However, since S and T are idempotent, $P := \mathrm{Coker}(S)$ and $Q := \mathrm{Coker}(T)$ are finitely generated projective R-modules. According to (5.4.38), the induced map $\overline{\phi}_0 : P_0 =: P \otimes_R R_0 \to Q_0 := Q \otimes_R R_0$ is an isomorphism with inverse $\overline{\psi}_0$. Let $\pi_P : P \to P_0$, $\pi_Q : Q \to Q_0$ be the projections; since $\overline{\psi}_0 \circ \pi_Q : Q \to P_0$ is surjective, we can find a map $\overline{\phi} : P \to Q$ such that $\overline{\psi}_0 \circ \pi_Q \circ \overline{\phi} = \pi_P$. Using Nakayama's lemma one checks easily that $\overline{\phi}$ is an isomorphism that lifts $\overline{\phi}_0$. Let $\alpha : R^n \to Q$ and $\beta : R^n \to P$ be the natural projections. We set $\phi := (I_n - T) \circ \overline{\phi} \circ \beta$ and $\psi := (I_n - S) \circ \overline{\phi}^{-1} \circ \alpha$ and leave to the reader the verification of the identities (a)-(c) of (5.4.38). $\qquad\square$

Corollary 5.4.41. *Resume the assumptions of proposition* 5.4.21. *Then the base change functor* $R[t^{-1}]$-**Mod** $\to R^\wedge[t^{-1}]$-**Mod** $: M \mapsto M \otimes_R R^\wedge$ *induces a bijection from the set of isomorphism classes of finitely generated projective* $R[t^{-1}]$-*modules to the set of isomorphism classes of finitely generated projective* $R^\wedge[t^{-1}]$-*modules.*

Proof. Resume the notation of (5.4.38) and let $s, t : F_n \to U_n$ be the morphisms obtained by composing (5.4.39) with the two projections onto U_n. The datum (U_n, F_n, s, t) can be completed to a groupoid of schemes, by letting $\iota : U_n \to F_n$ be the morphism representing the transformation of functors: $S \mapsto (S, S, I_n, I_n)$, and $c : F_n \times_{U_n} F_n \to F_n$ the morphism representing the transformation: $((S, T, \phi_1, \psi_1), (T, T', \phi_2, \psi_2)) \mapsto (S, T', \phi_2 \cdot \phi_1, \psi_2 \cdot \psi_1)$. Since every finitely generated projective module can be realized as the cokernel of an idempotent endomorphism of a free module of finite rank, the assertion is a straightforward consequence of theorem 5.4.37 and lemmata 5.4.20, 5.4.40. $\qquad\square$

5.4.42. Let S be an $R[t^{-1}]$-scheme, $\mathscr{P}$ a quasi-coherent $\mathscr{O}_S$-module. An S-*algebra structure* on $\mathscr{P}$ is a datum $(\mu, \underline{1})$ consisting of a map $\mu : \mathscr{P} \otimes_{\mathscr{O}_S} \mathscr{P} \to \mathscr{P}$ and a global section $\underline{1} \in \mathscr{P}(S)$, such that $(\mathscr{P}, \mu, \underline{1})$ is an $\mathscr{O}_S$-algebra, *i.e.* such that μ and $\underline{1}$ satisfy the following conditions for every open subset $U \subset S$ and local sections $x, y, z \in \mathscr{P}(U)$:

    (a) $\mu(\mu(x \otimes y) \otimes z) = \mu(x \otimes \mu(y \otimes z))$.
    (b) $\mu(x \otimes y) = \mu(y \otimes x)$.

(c) $\mu(\underline{1} \otimes x) = x$.

We say that an S-algebra structure on $\mathscr{P}$ is étale if the datum $(\mathscr{P}, \mu, \underline{1})$ is an étale $\mathscr{O}_S$-algebra. We denote by $\mathrm{Alg}_S(\mathscr{P})$ (resp. $\mathrm{Et}_S(\mathscr{P})$) the set of all S-algebra structures (resp. étale S-algebra structures) on $\mathscr{P}$. If now P is a finitely presented $R[t^{-1}]$-module, we obtain a functor:

(5.4.43) $R[t^{-1}]$-**Scheme** $\to$ **Set** $S \mapsto \mathrm{Alg}_S(\mathscr{O}_S \otimes_{R[t^{-1}]} P)$.

Lemma 5.4.44. *Suppose that P is a finitely generated projective $R[t^{-1}]$-module. Then the functor* (5.4.43) *is representable by a finitely presented $R[t^{-1}]$-algebra.*

Proof. Clearly (5.4.43) is a sheaf on the fppf topology of $\mathrm{Spec}\, R[t^{-1}]$, hence it suffices to show that we can cover $\mathrm{Spec}\, R[t^{-1}]$ by finitely many affine Zariski open subsets U_i, such that the restriction of (5.4.43) to U_i is representable and finitely presented. However, P is locally free of finite rank on the Zariski topology of $\mathrm{Spec}\, R[t^{-1}]$, hence we can assume that P is a free $R[t^{-1}]$-module. Let $e_1, ..., e_n$ be a basis of P; then a multiplication law μ is determined by its values $a_{ij} \in P$ on $e_i \otimes e_j$; by writing $a_{ij} = \sum_k a_{ijk} e_k$ we obtain n^3 elements of $R[t^{-1}]$; likewise, $\underline{1}$ is represented by elements $b_1, ..., b_n \in R[t^{-1}]$, and conditions (a),(b) and (c) of (5.4.42) translate as a finite system of polynomial identities for the a_{ijk} and the b_l; in other words, our functor is represented by a quotient of the free polynomial algebra $R[t^{-1}, X_{ijk}, Y_l \mid i, j, k, l = 1, ..., n]$ by a finitely generated ideal, which is the contention. □

Lemma 5.4.45. *Keep the assumptions of lemma* 5.4.44. *Let X_P be an affine $R[t^{-1}]$-scheme representing the functor* (5.4.43). *Then the functor*

(5.4.46) $R[t^{-1}]$-**Scheme** $\to$ **Set** $S \mapsto \mathrm{Et}_S(\mathscr{O}_S \otimes_{R[t^{-1}]} P)$

is represented by an affine open subset $U_P \subset X_P$. Moreover, U_P is smooth over $\mathrm{Spec}\, R[t^{-1}]$.

Proof. Let us show first that the functor (5.4.46) is formally smooth. Indeed, suppose that Z is an affine $R[t^{-1}]$-scheme and $Z_0 \subset Z$ is a closed subscheme defined by a nilpotent ideal. Let $(\mathscr{O}_Z \otimes_{R[t^{-1}]} P, \mu_0)$ be an étale Z_0-algebra structure; we can lift it to some étale $\mathscr{O}_Z$-algebra $(\mathscr{Q}, \mu)$ and then $\mathscr{Q}$ is necessarily a locally free sheaf of finite rank (for instance by lemma 3.2.25(i),(iii)); from proposition 3.2.30 we deduce that $\mathscr{Q} \simeq \mathscr{O}_Z \otimes_{R[t^{-1}]} P$, whence the claim. Next, let $(\mathscr{P}, \mu, \underline{1})$ be the universal $\mathscr{O}_{X_P}$-algebra structure on $P \otimes_{R[t^{-1}]} \mathscr{O}_{X_P}$; let also $\delta \in (\Lambda^r_{\mathscr{O}_{X_P}} \mathscr{P}^*)^{\otimes 2}(X_P)$ be the discriminant of the trace form of $\mathscr{P}$ (where r is the rank of P). By theorem 4.1.14, a point $x \in X_P$ is in the zero locus of δ if and only if $\mathscr{P}_x$ is not étale over $\mathscr{O}_{X_P,x}$; therefore, the subset U_P over which $\mathscr{P}$ is étale is indeed open and affine. To conclude, it suffices to show that U_P represents the functor (5.4.46). This amounts to showing that, for every morphism $f : S \to X_P$ of $R[t^{-1}]$-schemes, the algebra $f^* \mathscr{P}$ is étale over $\mathscr{O}_S$ if and only if the image of f lands in U_P. However, the latter statement follows easily from [30, Ch.IV, Cor.17.6.2]. □

5.4.47. Let S be an $R[t^{-1}]$-scheme, $\mathscr{P}$ a quasi-coherent $\mathscr{O}_S$-module. We denote by $\mathrm{Aut}_{\mathscr{O}_S}(\mathscr{P})$ the group of $\mathscr{O}_S$-linear automorphisms of $\mathscr{P}$. Then, for a given finitely presented $R[t^{-1}]$-module P we obtain a group-valued functor

$$(5.4.48) \qquad R[t^{-1}]\text{-}\mathbf{Scheme} \to \mathbf{Grp} \qquad S \mapsto \mathrm{Aut}_{\mathscr{O}_S}(\mathscr{O}_S \otimes_{R[t^{-1}]} P).$$

Lemma 5.4.49. *Keep the assumptions of lemma 5.4.44. Then the functor* (5.4.48) *is representable by a finitely presented $R[t^{-1}]$-group scheme.*

Proof. It is analogous to the proof of lemma 5.4.44 : up to restricting to a Zariski open subset, we can assume that P is free of some rank n. Then the group scheme representing our functor is just $\mathrm{GL}_{n,R[t^{-1}]}$. $\square$

5.4.50. Let now P be a finitely generated projective $R[t^{-1}]$-module. For any $R[t^{-1}]$-scheme S, let $\mathscr{P}_S := \mathscr{O}_S \otimes_{R[t^{-1}]} P$; we wish to define an action of $\mathrm{Aut}_{\mathscr{O}_S}(\mathscr{P}_S)$ on the set $\mathrm{Et}_S(\mathscr{P}_S)$. Indeed, if $\mu : \mathscr{P}_S \otimes_{\mathscr{O}_S} \mathscr{P}_S \to \mathscr{P}_S$ is any étale S-algebra structure and $g \in \mathrm{Aut}_{\mathscr{O}_S}(\mathscr{P}_S)$, let μ^g be the unique S-algebra structure on $\mathscr{P}_S$ such that g is an isomorphism of étale $\mathscr{O}_S$-algebras: $g : (\mathscr{P}_S, \mu) \xrightarrow{\sim} (\mathscr{P}_S, \mu^g)$. It is obvious that the rule $(g, \mu) \mapsto \mu^g$ is a functorial group action. Let U_P be as in lemma 5.4.45, and let Aut_P be a $R[t^{-1}]$-scheme representing the functor 5.4.48; the functorial map $(g, \mu) \mapsto (\mu^g, \mu)$ is represented by a morphism of schemes:

$$(5.4.51) \qquad \mathrm{Aut}_P \times_{R[t^{-1}]} U_P \to U_P \times_{R[t^{-1}]} U_P.$$

Lemma 5.4.52. *The morphism* (5.4.51) *is étale.*

Proof. The map is clearly of finite presentation, hence it suffices to show that it is formally étale. Therefore, let $\phi : Z \to U_P \times_{R[t^{-1}]} U_P$ be a morphism of affine $R[t^{-1}]$-schemes, $Z_0 \subset Z$ a closed subscheme defined by a nilpotent ideal, and suppose that the restriction of ϕ to Z_0 lifts to a morphism $\psi_0 : Z_0 \to \mathrm{Aut}_P \times_{R[t^{-1}]} U_P$. We need to show that ϕ lifts uniquely to a morphism ψ that extends ψ_0. However, the datum of ϕ is equivalent to the datum consisting of a pair of Z-algebra structures $(\mathscr{P}_Z, \mu_1)$ and $(\mathscr{P}_Z, \mu_2)$. The datum of ψ_0 is equivalent to the datum of a Z_0-algebra structure μ_0 on $\mathscr{P}_{Z_0}$, and of an automorphism g_0 of $\mathscr{P}_{Z_0}$. Finally, the fact that ψ_0 lifts the restriction of ϕ means that $\mu_0 = \mu_2 \otimes_Z 1_{Z_0}$, and $g_0 : (\mathscr{P}_{Z_0}, \mu_1 \otimes_Z 1_{Z_0}) \xrightarrow{\sim} (\mathscr{P}_{Z_0}, \mu_2 \otimes_Z 1_{Z_0})$ is an isomorphism of étale $\mathscr{O}_{Z_0}$-algebras. By theorem 3.2.18(iii), such an isomorphism extends uniquely to an isomorphism of étale $\mathscr{O}_Z$-algebras $g : (\mathscr{P}_Z, \mu_1) \xrightarrow{\sim} (\mathscr{P}_Z, \mu_2)$. The datum (g, μ_2) is equivalent to the sought map ψ. $\square$

Proposition 5.4.53. *Resume the assumptions of proposition 5.4.21. Then the base change functor $R[t^{-1}]$-$\mathbf{Alg} \to R^\wedge[t^{-1}]$-$\mathbf{Alg}$ induces an equivalence of categories from the category of finite étale $R[t^{-1}]$-algebras to the category of finite étale $R^\wedge[t^{-1}]$-algebras.*

Proof. Let $(P^\wedge, \mu^\wedge)$ be a finite étale $R^\wedge[t^{-1}]$-algebra; in particular, $P^\wedge$ is a finitely generated projective $R^\wedge[t^{-1}]$-module, hence by corollary 5.4.41 we can find a finitely generated projective $R[t^{-1}]$-module P such that $R^\wedge[t^{-1}] \otimes_{R[t^{-1}]} P \simeq P^\wedge$. The functorial action (5.4.51) of Aut_P on the set of étale algebra structures on

P defines a groupoid of quasi-projective schemes. It then follows from theorem 5.4.37 and lemmata 5.4.45, 5.4.52 that the base change map from the set of isomorphism classes of $R[t^{-1}]$-algebra structures on P to the set of isomorphism classes of $R^\wedge[t^{-1}]$-algebra structures on $P^\wedge$ is bijective. This shows that the base change functor is essentially surjective on finite étale $R^\wedge[t^{-1}]$-algebras. To prove full faithfulness, let Y_1, Y_2 be two finite étale schemes over $X := \operatorname{Spec} R[t^{-1}]$; we let H be the functor that assigns to every $R[t^{-1}]$-scheme Z the set $\operatorname{Hom}_Z(Z \times_X Y_1, Z \times_X Y_2)$. Since the functors represented by Y_1 and Y_2 are locally constant bounded sheaves on the fpqc site of X, the same holds for the functor H, hence the latter is represented by a finite étale X-scheme (proposition 8.2.23), which we denote by the same name. We can view H as a trivial groupoid (*i.e.* such that for every X-scheme Z, the only morphisms of the groupoid $H(Z)$ are the identity morphisms of its objects). In this case, the associated morphism $(s,t) : H \to H \times_X H$ is none other than the diagonal morphism; especially, the latter is an open imbedding, hence theorem 5.4.37 applies and yields the contention.   □

5.5. Lifting theorems for henselian pairs.

For the considerations that follow, we wish to work with sheaves of almost modules (or almost algebras) on a scheme : this is a special case of the relative situation considered in section 3.3, so we will proceed somewhat briskly.

5.5.1.   Let X be a scheme over $\operatorname{Spec} V$. The constant sheaves V_X and $\mathfrak{m}_X$ defined by V and $\mathfrak{m}$ give a basic setup on the Zariski topos of X (see section 3.3). The Zariski sheaf $\mathscr{O}_X$ is a V_X-algebra, hence we obtain a V_X^a-algebra $\mathscr{O}_X^a$, which – according to the general notation of section 3.3 – is an object of the category V_X^a-**Mod** obtained by localizing the category V_X-**Mod**. A different viewpoint on these constructions is provided by [26, Ch.0, §3.1]; namely, one can also consider the category $(V^a\text{-}\mathbf{Mod})_X^{\sim}$ of sheaves of V^a-modules on the Zariski topology of X. There is a natural equivalence of categories:

$$(5.5.2) \qquad\qquad V_X^a\text{-}\mathbf{Mod} \to (V^a\text{-}\mathbf{Mod})_X^{\sim}$$

that sends a V_X-module $\mathscr{F}$ to the sheaf $U \mapsto \mathscr{F}(U)^a$ (for every open subset $U \subset X$). A quasi-inverse to (5.5.2) is defined by sending a sheaf $\mathscr{G}$ of V^a-modules to the object of V_X^a-**Mod** represented by the sheaf of V_X-modules $U \mapsto \mathscr{G}(U)_*$.

The upshot is that one can define $\mathscr{O}_X^a$ in either category, and the two versions correspond under (5.5.2).

As in (3.3.3) we have the category $\mathscr{O}_X^a$-**Mod** of sheaves of $\mathscr{O}_X^a$-modules on X (briefly: of $\mathscr{O}_X^a$-modules); these can be viewed as sheaves $\mathscr{F}$ of almost modules endowed with a scalar multiplication $\mathscr{O}_X^a \otimes_{V^a} \mathscr{F} \to \mathscr{F}$ satisfying the usual conditions. There is a functor

$$(5.5.3) \qquad\qquad \Gamma(X, \mathscr{O}_X^a)\text{-}\mathbf{Mod} \to \mathscr{O}_X^a\text{-}\mathbf{Mod} \qquad M \mapsto M^{\sim}$$

defined as one expects. We say that $\mathscr{F}$ is *quasi-coherent* if we can cover X by affine open subsets U_i, such that $\mathscr{F}_{|U_i}$ is in the essential image of a functor (5.5.3). We denote by $\mathscr{O}_X^a$-**Mod**$_{\text{qcoh}}$ the full subcategory of quasi-coherent $\mathscr{O}_X^a$-modules.

Similarly, we denote by $\mathscr{O}_X^a$-**Alg** (resp. $\mathscr{O}_X^a$-**Alg**$_{\text{qcoh}}$) the category of $\mathscr{O}_X^a$-algebras (resp. quasi-coherent $\mathscr{O}_X^a$-algebras) defined as one expects.

5.5.4. By specializing the generalities of (3.3.2) we obtain functors

$$\mathscr{O}_X^a\text{-}\mathbf{Mod} \to \mathscr{O}_X\text{-}\mathbf{Mod} \qquad \mathscr{F} \mapsto \mathscr{F}_! \quad (\text{resp. } \mathscr{F} \mapsto \mathscr{F}_*)$$

which are left (resp. right) adjoint to the localization $\mathscr{O}_X$-**Mod** $\to \mathscr{O}_X^a$-**Mod**. Namely, $\mathscr{F}_*$ is the sheaf $U \mapsto \mathscr{F}(U)_*$, and $\mathscr{F}_!$ is the sheaf associated to the presheaf $U \mapsto \mathscr{F}(U)_!$; as $\widetilde{\mathfrak{m}}$ is flat, one can check that this presheaf is already a sheaf on the site of quasi-compact and quasi-separated open subsets. The functor $\mathscr{F} \mapsto \mathscr{F}_!$ is exact and preserves quasi-coherence, hence it provides a left adjoint to the localization functor $\mathscr{O}_X$-**Mod**$_{\text{qcoh}} \to \mathscr{O}_X^a$-**Mod**$_{\text{qcoh}}$. The functor $\mathscr{F} \mapsto \mathscr{F}_*$ does not preserve quasi-coherence, in general.

5.5.5. Let R be a V-algebra and set $X := \operatorname{Spec} R$. Using the full faithfulness of the functor $\mathscr{F} \mapsto \mathscr{F}_!$ one can easily verify that the functor $M \mapsto M^\sim$ from R^a-modules to quasi-coherent $\mathscr{O}_X^a$-modules is an equivalence, whose quasi-inverse is given by the global section functor.

After these preliminaries, we are ready to state the following descent result which will be crucial for the proof of theorem 5.5.7.

Proposition 5.5.6. (i) *Let R be a V-algebra, $J \subset R$ a finitely generated ideal and $R \to R'$ a flat morphism inducing an isomorphism $R/J \to R'/JR'$. Let $X :=$ $\operatorname{Spec}(R)$, $X' := \operatorname{Spec}(R')$, $U := X \setminus V(J)$ and $U' := U \times_X X'$. Then the natural commutative diagrams of functors:*

$$
\begin{array}{ccc}
\mathscr{O}_X^a\text{-}\mathbf{Mod}_{\text{qcoh}} & \longrightarrow & \mathscr{O}_{X'}^a\text{-}\mathbf{Mod}_{\text{qcoh}} \\
\downarrow & & \downarrow \\
\mathscr{O}_U^a\text{-}\mathbf{Mod}_{\text{qcoh}} & \longrightarrow & \mathscr{O}_{U'}^a\text{-}\mathbf{Mod}_{\text{qcoh}}
\end{array}
\qquad
\begin{array}{ccc}
\mathscr{O}_X^a\text{-}\mathbf{Alg}_{\text{qcoh}} & \longrightarrow & \mathscr{O}_{X'}^a\text{-}\mathbf{Alg}_{\text{qcoh}} \\
\downarrow & & \downarrow \\
\mathscr{O}_U^a\text{-}\mathbf{Alg}_{\text{qcoh}} & \longrightarrow & \mathscr{O}_{U'}^a\text{-}\mathbf{Alg}_{\text{qcoh}}
\end{array}
$$

are 2-cartesian (that is, cartesian in the 2-category of categories).

(ii) *Let A be a V^a-algebra, $f \in A_*$ a non-zero-divisor, $A^\wedge$ the f-adic completion of A. Denote by A-$\mathbf{Mod}_f$ (resp. A-$\mathbf{Alg}_f$) the full subcategory of f-torsion free A-modules (resp. A-algebras), and similarly define $A^\wedge$-$\mathbf{Mod}_f$ (resp. $A^\wedge$-$\mathbf{Alg}_f$). Then the natural commutative diagrams of functors*

$$
\begin{array}{ccc}
A\text{-}\mathbf{Mod}_f & \longrightarrow & A^\wedge\text{-}\mathbf{Mod}_f \\
\downarrow & & \downarrow \\
A[f^{-1}]\text{-}\mathbf{Mod} & \longrightarrow & A^\wedge[f^{-1}]\text{-}\mathbf{Mod}
\end{array}
\qquad
\begin{array}{ccc}
A\text{-}\mathbf{Alg}_f & \longrightarrow & A^\wedge\text{-}\mathbf{Alg}_f \\
\downarrow & & \downarrow \\
A[f^{-1}]\text{-}\mathbf{Alg} & \longrightarrow & A^\wedge[f^{-1}]\text{-}\mathbf{Alg}
\end{array}
$$

are 2-cartesian.

Proof. (i): For the functors on $\mathscr{O}_X^a$-modules, one applies the functor $\mathscr{F} \mapsto \mathscr{F}_!$, thereby reducing to the corresponding assertion for quasi-coherent $\mathscr{O}_X$-modules. The latter is proved in [36, Prop.4.2] (actually, in *loc.cit.* one assumes that X' is faithfully flat over X, but one can reduce to such case after replacing X' by X' II $(U_1 \amalg \dots \amalg U_n)$, where $(U_i \mid i \leq n)$ is a finite cover of U by affine open subsets; notice also that *loc.cit.* omits the assumption that J is finitely generated, but the proof works only under such assumption. See also [60] for more results along these lines). Since all the functors involved commute with tensor products, the assertion about $\mathscr{O}_X^a$-algebras follows formally.

(ii): For modules one argues as in the proof of (i), except that instead of invoking [36], one uses [9, Theorem]. For algebras, one has to proceed a little more carefully, since the tensor product of two f-torsion free modules may fail to be f-torsion free. However, the 2-cartesian diagram for modules gives an equivalence of categories

$$\pi : A\text{-}\mathbf{Mod}_f \to A^\wedge\text{-}\mathbf{Mod}_f \times_{A^\wedge[f^{-1}]\text{-}\mathbf{Mod}} A[f^{-1}]\text{-}\mathbf{Mod}$$

whose quasi-inverse is given by the fibre product functor T as in (3.4.7). Hence, let (B_1, B_2, β) be the datum consisting of an $A[f^{-1}]$-algebra B_1, an $A^\wedge$-algebra B_2 and an isomorphism $\beta : B_1 \otimes_A A^\wedge \xrightarrow{\sim} B_2[f^{-1}]$ of $A^\wedge[f^{-1}]$-algebras; then $T(B_1, B_2, \beta) = B_1 \times_{B_2[f^{-1}]} B_2$, which has an obvious algebra structure, so it defines a right adjoint for the algebra version of the functor π, and the adjunction unit and counit are isomorphism, since they are so for modules. $\square$

Theorem 5.5.7. *Let (A, I) be a tight henselian pair, $\overline{P}$ an almost finitely generated projective A/I-module, $\mathfrak{m}_1 \subset \mathfrak{m}$ a finitely generated subideal. We have:*

(i) *If $\widetilde{\mathfrak{m}}$ has homological dimension ≤ 1, then there exists an almost finitely generated projective A-module P such that $P \otimes_A (A/I) \simeq \overline{P}$.*

(ii) *If P_1 and P_2 are two liftings of $\overline{P}$ as in (i) and if there exists an isomorphism $\overline{\beta} : P_1 \otimes_A (A/\mathfrak{m}_1 I) \xrightarrow{\sim} P_2 \otimes_A (A/\mathfrak{m}_1 I)$, then there exists an isomorphism $\beta : P_1 \to P_2$ such that $\beta \otimes_A 1_{A/I} = \overline{\beta} \otimes_A 1_{A/I}$.*

(iii) *With the notation of (3.2.27), the natural functor $A\text{-}\acute{\mathbf{E}}\mathbf{t}_{\mathrm{afp}} \to (A/I)\text{-}\acute{\mathbf{E}}\mathbf{t}_{\mathrm{afp}}$ is an equivalence of categories.*

Proof. We begin by showing (ii): indeed, the obstruction to the existence of a morphism $\alpha : P_1 \to P_2$ such that $\alpha \otimes_A 1_{(A/\mathfrak{m}_1 I)} = \beta$ is a class $\omega \in \mathrm{Ext}_A^1(P_1, \mathfrak{m}_1 I P_2)$. The same argument used in the proof of claim 5.1.20 shows that the natural map $\mathrm{Ext}_A^1(P_1, \mathfrak{m}_1 I P_2) \to \mathrm{Ext}_A^1(P_1, I P_2)$ vanishes identically, and proves the assertion.

Next we wish to show that the functor of (iii) is fully faithful. Therefore, let B, C be two almost finitely presented étale A-algebras, and $\overline{\phi} : B/IB \to C/IC$ a morphism. According to lemma 5.2.20, $\overline{\phi}$ is characterized by its associated idempotent, call it $\overline{e} \in (B \otimes_A C/IC)_*$. Set $D := B \otimes_A C$; according to lemma 5.1.13(iii), the pair (D, ID) is tight henselian. Then proposition 5.1.18 says that $\overline{e}$ lifts uniquely to an idempotent $e \in D_*$.

Claim 5.5.8. The morphism $\Gamma(e)$ is an isomorphism (notation of (5.2.19)).

Proof of the claim. Indeed, by naturality of Γ, we have $\Gamma(e) \otimes_A 1_{A/I} = \Gamma(\overline{e})$, so the assertion follows from corollary 5.1.8. $\diamond$

By claim 5.5.8 and lemma 5.2.20, e corresponds to a unique morphism $\phi : B \to C$ which is the sought lifting of $\overline{\phi}$. The remaining steps to complete the proof of (iii) will apply as well to the proof of (i). Pick an integer $n > 0$ and a finitely generated subideal $\mathfrak{m}_0 \subset \mathfrak{m}$ such that $I^n \subset \mathfrak{m}_0 A$; we notice that assertions (i) and (iii) also hold when I is nilpotent, since in this case they reduce to theorem 3.2.28(i.b),(ii). It follows easily that it suffices to prove the assertions for the pair (A, I^n), hence we can and do assume throughout that $I \subset \mathfrak{m}_0 A$.

Claim 5.5.9. (i) and (iii) hold if I is generated by a non-zero-divisor of A_*.

Proof of the claim. Say that $I = fA$, for some non-zero-divisor $f \in (\mathfrak{m}_0 A)_*$, and let $A^\wedge$ be the f-adic completion of A. By theorem 3.2.28(i), $\overline{P}$ lifts to a compatible system $(P_n \mid n \in \mathbb{N})$ of almost finitely generated projective A/I^{n+1}-modules; by theorem 5.3.24, the latter compatible system gives rise to a unique almost finitely generated projective $A^\wedge$-module $P^\wedge$. Notice that f is regular on $A^\wedge$, hence also on $P^\wedge$. Since $A^\wedge[f^{-1}]$ is a (usual) V-algebra, the $A^\wedge[f^{-1}]$-module $P^\wedge[f^{-1}]$ is finitely generated projective; it follows from corollary 5.4.41 that there exists a finitely generated projective $A[f^{-1}]$-module Q with an isomorphism $\beta : Q \otimes_A A^\wedge \simeq P^\wedge[f^{-1}]$. By proposition 5.5.6(ii) the datum $(P^\wedge, Q, \beta)$ determines a unique f-torsion-free A-module P which lifts $\overline{P}$. Since f is regular on both P and A, we have $\operatorname{Tor}_i^A(P, A/fA) = 0$ for $i = 1, 2$, hence P is A-flat, by virtue of lemma 5.2.1. Next, set $C := A[f^{-1}] \times A^\wedge$; from lemma 5.2.3(i) we deduce that $\operatorname{Ann}_C(P \otimes_A C)^2 \subset \operatorname{Ann}_A(P)$, and since $\operatorname{Tor}_1^A(C, P) = 0$, remark 3.2.26(i) implies that P is almost finitely presented, therefore almost projective over A, which shows that (i) holds. Likewise, let $\overline{B}$ be an almost finitely presented étale A/I-algebra; by theorem 5.3.27, $\overline{B}$ admits a unique lifting to an almost finitely presented étale $A^\wedge$-algebra $B^\wedge$. Then $B^\wedge[f^{-1}]$ is a finite étale $A^\wedge[f^{-1}]$-algebra, hence by proposition 5.4.53 there exists a unique finite étale $A[f^{-1}]$-algebra B_0 with an isomorphism $\beta : B_0 \otimes_A A^\wedge \xrightarrow{\sim} B^\wedge[f^{-1}]$. By proposition 5.5.6(ii), the datum $(B^\wedge, B_0, \beta)$ determines a unique f-torsion-free A-algebra B; the foregoing proof of assertion (i) applies to the A-module underlying B and shows that B is an almost finitely generated projective A-algebra. By construction B/fB is unramified over A/fA, so theorem 5.2.12(iv) applies and shows that B is unramified over A. Thus, we have shown that the functor of (iii) is essentially surjective under the present assumptions; since it is already known in general that this functor is fully faithful, assertion (iii) is completely proved in this case. ◇

Claim 5.5.10. Assertions (i) and (iii) hold if I is a principal ideal.

Proof of the claim. Say that $I = fA$ for some $f \in (\mathfrak{m}_0 A)_*$. Let

$$J := \bigcup_{n>0} \operatorname{Ann}_A(f^n).$$

We have a cartesian diagram

$$
\begin{array}{ccc}
A/(J \cap I) & \longrightarrow & A/I \\
\downarrow & & \downarrow \\
A/J & \longrightarrow & A/(I+J).
\end{array}
$$

Let $\overline{P}$ be as in (i) and let $\overline{B}$ be a finitely presented étale A/I-algebra; by claim 5.5.9 we can find an almost finitely generated projective A/J-module $\overline{P}_1$ with an isomorphism $\beta : \overline{P}_1/I\overline{P}_1 \xrightarrow{\sim} \overline{P}/J\overline{P}$ and a finitely presented étale A/J-algebra $\overline{B}_1$ that lifts $\overline{B}/J\overline{B}$. By proposition 3.4.21, the datum $(\overline{P}, \overline{P}_1, \beta)$ determines a unique almost finitely generated projective $A/(I \cap J)$-module $\overline{P}_2$; likewise, using corollary 3.4.22 we obtain an étale almost finitely presented $A/(I \cap J)$-algebra $\overline{B}_2$ that lifts $\overline{B}$. Next, let $K := \bigcap_{n>0} f^n A$ and set $N := K \cap J \cap I$; we have a cartesian diagram

$$
\begin{array}{ccc}
A/N & \longrightarrow & A/(I \cap J) \\
\downarrow & & \downarrow \\
A^\wedge & \longrightarrow & A^\wedge/(I \cap J)A^\wedge.
\end{array}
$$

Due to theorem 3.2.28(i.b) we can lift $\overline{P}_2 \otimes_A A^\wedge$ to an almost finitely generated projective $A^\wedge/(I \cap J)^2 A^\wedge$-module $\overline{P}_2^\wedge$. For the same reason, $\overline{P}_2^\wedge \otimes_A A/f^2 A$ can be lifted to a compatible family $(\overline{Q}_n \mid n \in \mathbb{N})$, where $\overline{Q}_n$ is almost finitely generated projective over $A/f^{n+2}A$ for every $n \in \mathbb{N}$. Finally, by theorem 5.3.24, the projective limit Q of the system $(\overline{Q}_n \mid n \in \mathbb{N})$ is an almost finitely generated projective $A^\wedge$-module. By construction, there is an isomorphism $\overline{\beta} : Q/f^2 Q \xrightarrow{\sim} \overline{P}_2^\wedge \otimes_A A/f^2 A$; by assertion (ii) it follows that there exists an isomorphism $\beta : Q/(I \cap J)^2 Q \xrightarrow{\sim} \overline{P}_2^\wedge$ that lifts $\overline{\beta} \otimes_A 1_{A/fA}$. By proposition 3.4.21, the datum $(\overline{P}_2, Q, \beta \otimes_A 1_{A/(I \cap J)})$ determines a unique almost finitely generated projective A/N-module P_1; since $N^2 = 0$, P_1 can be further lifted to an almost finitely generated projective A-module P, so assertion (i) holds in this case. The proof of assertion (iii) is analogous, but easier : we need to show that $\overline{B}_2$ lifts to an almost finitely presented étale A-algebra B; to this aim, it suffices to show that $\overline{B}_2$ lifts to an almost finitely presented étale A/N-algebra B_1, since in that case B_1 can be lifted to an almost finitely presented étale A-algebra B, by theorem 3.2.28(ii). To obtain B_1 it suffices to find an almost finitely presented étale $A^\wedge$-algebra $B^\wedge$ with an isomorphism $\beta : B^\wedge \otimes_A A/(I \cap J) \xrightarrow{\sim} \overline{B}_2 \otimes_A A^\wedge$; indeed, in this case the datum $(B^\wedge, \overline{B}_2, \beta)$ determines a unique almost finitely presented étale A/N-algebra in view of corollary 3.4.22. Finally, we consider the natural functors

$$
A^\wedge\text{-}\acute{\text{E}}\text{t}_{\text{afp}} \to A^\wedge/(I \cap J)A^\wedge\text{-}\acute{\text{E}}\text{t}_{\text{afp}} \to A/I\text{-}\acute{\text{E}}\text{t}_{\text{afp}}.
$$

By theorem 5.3.27, the composition of these two functors is an equivalence of categories and the rightmost functor is fully faithful by (ii), so the leftmost functor is an equivalence, thus B_1 as sought can be found, which concludes the proof of (iii) in this case. $\diamond$

Claim 5.5.11. Assertions (i) and (iii) hold if I is a finitely generated ideal.

Proof of the claim. We proceed by induction on the number n of generators of I, the case $n = 1$ being covered by claim 5.5.9. So suppose $n > 1$ and let $f_1, ..., f_n \in I_*$ be a finite set of generators of I. Let $\overline{P}$ be as in (i) and $\overline{B}$ any almost finitely presented étale A/I-algebra. We let $A' := A/f_1 A$ and $J := \text{Im}(I \to A')$. By lemma

5.1.13(iii) the pair (A', J) is again tight henselian, so by inductive assumption we can find lift $\overline{P}$ (resp. $\overline{B}$) to an almost finitely generated projective A'-module (resp. to an almost finitely presented étale A'-algebra) P' (resp. B'). Thence we apply claim 5.5.9 to further lift P' to a module P (resp. to an algebra B) as stated.   ◇

Let now (A, I) be a general tight henselian pair; we can find a henselian pair (R, J) over V such that $R^a = A$, $J^a = I$ and $J \subset \mathfrak{m}_0 R$. Denote by $(R^h, \mathfrak{m}_0 R^h)$ the henselization of the pair $(R, \mathfrak{m}_0 R)$, and let $\overline{R}^h := R^h / J R^h$. Let us write $J = \bigcup_{\lambda \in \Lambda} J_\lambda$, where J_λ runs over the filtered family of all finitely generated subideals of J; set $R_\lambda := R / J_\lambda$ and $R_\lambda^h := R^h \otimes_R R_\lambda$ for every $\lambda \in \Lambda$. Furthermore, let

$$
\begin{array}{llll}
X := \operatorname{Spec} R & U := X \setminus V(\mathfrak{m}_0) & \overline{X} := \operatorname{Spec} R/J \\
X^h := \operatorname{Spec} R^h & X_\lambda := \operatorname{Spec} R_\lambda & X_\lambda^h := X^h \times_X X_\lambda
\end{array}
$$

and denote by

$$
\phi^h : U^h := U \times_X X^h \to U \quad \text{and} \quad \phi_\lambda^h : U_\lambda^h := U \times_X X_\lambda^h \to U_\lambda := U \times_X X_\lambda
$$

the natural morphisms of schemes. Now, let $\overline{P}$ be an almost finitely generated projective A/I-module and $\overline{B}$ an almost finitely presented étale A/I-algebra; $\overline{P}$ induces a quasi-coherent $\mathscr{O}_{\overline{X}}^a$-module $\overline{P}^\sim$, and by restriction we obtain a quasi-coherent module $\overline{P}^\sim_{|\overline{U}}$ on $\overline{U} := U \cap \overline{X}$. Furthermore, since $\overline{P}$ is almost finitely presented, we see that $\overline{P}^\sim_{|\overline{U}}$ is finitely presented; by [29, Ch.IV, Th.8.5.2(ii)] it follows that for some $\lambda_0 \in \Lambda$ there exists a quasi-coherent finitely presented module $\mathscr{P}$ on U_{λ_0} whose restriction to the closed subscheme $\overline{U}$ agrees with $\overline{P}^\sim_{|\overline{U}}$. By restricting further, we can even achieve that $\mathscr{P}$ be locally free of finite rank ([29, Ch.IV, Prop.8.5.5]). Similarly, we can find a locally free $\mathscr{O}_{U_{\lambda_0}}$-algebra $\mathscr{B}$ such that $\mathscr{B}_{|\overline{U}} \simeq \overline{B}^\sim_{|\overline{U}}$.

Claim 5.5.12. Suppose that $\hom.\dim_V \widetilde{\mathfrak{m}} \leq 1$. For every almost finitely generated projective $(\overline{R}^h)^a$-module $\overline{Q}$ there exists an almost finitely generated projective $(R^h)^a$-module Q such that

$$
Q \otimes_{(R^h)^a} (\overline{R}^h)^a \simeq \overline{Q}.
$$

Proof of the claim. By theorem 3.2.28(i.b) we know that $\overline{Q}$ lifts to an almost finitely generated projective module Q_1 over $(R^h/J^2 R^h)^a$. Then $Q_2 := Q_1 \otimes_A (A/\mathfrak{m}_0^2 A)$ is an almost finitely generated projective $(R^h/\mathfrak{m}_0^2 R^h)^a$-module, therefore by claim 5.5.11 we can lift Q_2 to an almost finitely generated projective R^h-module Q (notice that $(R^h, \mathfrak{m}_0^2 R^h)$ is still a henselian pair). It remains only to show that Q is a lifting of $\overline{Q}$; however, by construction we have

$$
Q_1 \otimes_{(R^h)^a} (R^h/\mathfrak{m}_0^2 R^h)^a \simeq Q \otimes_{(R^h)^a} (R^h/\mathfrak{m}_0^2 R^h)^a
$$

so it follows from (ii) that $Q_1 \simeq Q/I^2 Q$, whence the claim.   ◇

Claim 5.5.13. The natural functor $(R^h)^a\text{-}\acute{\text{E}}\text{t}_{\text{afp}} \to (\overline{R}^h)^a\text{-}\acute{\text{E}}\text{t}_{\text{afp}}$ is an equivalence.

Proof of the claim. By claim 5.5.11, the natural functors

$$(\overline{R}^{\mathrm{h}})^a\text{-}\acute{\mathbf{E}}\mathbf{t}_{\mathrm{afp}} \to (R^{\mathrm{h}}/\mathfrak{m}_0 R^{\mathrm{h}})^a\text{-}\acute{\mathbf{E}}\mathbf{t}_{\mathrm{afp}} \quad \text{and} \quad (R^{\mathrm{h}})^a\text{-}\acute{\mathbf{E}}\mathbf{t}_{\mathrm{afp}} \to (R^{\mathrm{h}}/\mathfrak{m}_0 R^{\mathrm{h}})^a\text{-}\acute{\mathbf{E}}\mathbf{t}_{\mathrm{afp}}$$

are equivalences of categories. The claim is a formal consequence.   ◇

We apply claim 5.5.12 with $\overline{Q} := \overline{P} \otimes_A (R^{\mathrm{h}})^a$; let $Q^{\sim}$ be the quasi-coherent $\mathscr{O}_{X^{\mathrm{h}}}^a$-module associated to Q. By construction, the restriction $Q_{|U^{\mathrm{h}}}^{\sim}$ is a quasi-coherent $\mathscr{O}_{U^{\mathrm{h}}}$-module of finite presentation, and we have an isomorphism

$$\overline{\beta} : Q^{\sim}_{|\phi^{\mathrm{h}-1}(\overline{U})} \xrightarrow{\sim} \phi^{\mathrm{h}*}(\overline{P}_{|\overline{U}}^{\sim}).$$

It then follows by [29, Ch.IV, Cor.8.5.2.5] that there exists some $\mu \in \Lambda$ with $X_\mu \subset X_{\lambda_0}$, such that the isomorphism $\overline{\beta}$ extends to an isomorphism $\beta_\mu : Q^{\sim}_{|U_\mu^{\mathrm{h}}} \xrightarrow{\sim} \phi_\mu^{\mathrm{h}*}(\mathscr{P}_{|U_\mu})$. Similarly, by claim 5.5.13, we can find an almost finitely presented étale $(R^{\mathrm{h}})^a$-algebra C with an isomorphism $\gamma_\mu : C^{\sim}_{|U_\mu^{\mathrm{h}}} \simeq \phi^*(\overline{B}_{|U_\mu^{\mathrm{h}}}^{\sim})$. According to (5.5.5), the global section functors:

$$\mathscr{O}_{X_\lambda}^a\text{-}\mathbf{Mod}_{\mathrm{qcoh}} \to R_\lambda^a\text{-}\mathbf{Mod} \qquad \mathscr{O}_{X_\lambda^{\mathrm{h}}}^a\text{-}\mathbf{Mod}_{\mathrm{qcoh}} \to (R_\lambda^{\mathrm{h}})^a\text{-}\mathbf{Mod}$$

are equivalences. Clearly the localization functors:

$$\mathscr{O}_{U_\lambda}\text{-}\mathbf{Mod}_{\mathrm{qcoh}} \to \mathscr{O}_{U_\lambda}^a\text{-}\mathbf{Mod}_{\mathrm{qcoh}} \qquad \mathscr{O}_{U_\lambda^{\mathrm{h}}}\text{-}\mathbf{Mod}_{\mathrm{qcoh}} \to \mathscr{O}_{U_\lambda^{\mathrm{h}}}^a\text{-}\mathbf{Mod}_{\mathrm{qcoh}}$$

are equivalences as well, and similarly for the corresponding categories of algebras. By proposition 5.5.6(i) it follows that the natural diagrams:

$$
\begin{array}{ccc}
R_\lambda^a\text{-}\mathbf{Mod} & \longrightarrow & (R_\lambda^{\mathrm{h}})^a\text{-}\mathbf{Mod} \\
\downarrow & & \downarrow \\
\mathscr{O}_{U_\lambda}\text{-}\mathbf{Mod}_{\mathrm{qcoh}} & \xrightarrow{\phi_\lambda^{\mathrm{h}*}} & \mathscr{O}_{U_\lambda^{\mathrm{h}}}\text{-}\mathbf{Mod}_{\mathrm{qcoh}}
\end{array}
\qquad
\begin{array}{ccc}
R_\lambda^a\text{-}\mathbf{Alg} & \longrightarrow & (R_\lambda^{\mathrm{h}})^a\text{-}\mathbf{Alg} \\
\downarrow & & \downarrow \\
\mathscr{O}_{U_\lambda}\text{-}\mathbf{Alg}_{\mathrm{qcoh}} & \xrightarrow{\phi_\lambda^{\mathrm{h}*}} & \mathscr{O}_{U_\lambda^{\mathrm{h}}}\text{-}\mathbf{Alg}_{\mathrm{qcoh}}
\end{array}
$$

are 2-cartesian for every $\lambda \in \Lambda$. Hence, the datum $(\mathscr{P}_{|U_\mu}, Q \otimes_{R^a} R_\mu^a, \beta_\mu)$ determines uniquely a R_μ^a-module P_μ that lifts $\overline{P}$, and the datum $(\mathscr{B}_{|U_\mu}, C \otimes_{R^a} R_\mu^a, \gamma_\mu)$ determines a R_μ^a-algebra B_μ that lifts $\overline{B}$. Furthermore, since the natural morphism $U_\mu \amalg X_\mu^{\mathrm{h}} \to X_\mu$ is faithfully flat, it follows from remark 3.2.26(ii) that P_μ and B_μ are almost finitely generated projective over R_μ^a. For the same reasons, B_μ is unramified, hence étale over R_μ^a. Finally, we apply claim 5.5.11 to the henselian pair (R^a, J_μ^a) to lift P_μ and B_μ all the way to A, thereby concluding the proof of the theorem.   □

Lemma 5.5.14. *Suppose that $\widetilde{\mathfrak{m}}$ has homological dimension ≤ 1, and let (A, I) be a tight henselian pair, $\overline{A} := A/I$, $\overline{Q}$ an almost finitely generated projective $\overline{A}$-module, M an A-module and $\overline{\phi} : \overline{Q} \to M/IM$ an $\overline{A}$-linear epimorphism. Then there exists an almost finitely generated projective A-module Q and a morphism $\phi : Q \to M$ such that $\phi \otimes_A 1_{\overline{A}} = \overline{\phi}$.*

Proof. We begin with the following special case:

Claim 5.5.15. The lemma holds if $I^2 = 0$. Furthermore, in this case ϕ is an epimorphism.

Proof of the claim. First of all, notice that M/IM is almost finitely generated, hence the same holds for M, in view of lemma 3.2.25(ii). If now Q is an almost finitely generated projective A-module and $\phi : Q \to M$ is a morphism that lifts $\overline{\phi}$, we have $\mathrm{Coker}(\phi) \otimes_A A/I \simeq \mathrm{Coker}\,\overline{\phi} = 0$, whence $I\mathrm{Coker}\,\phi = \mathrm{Coker}\,\phi$ and therefore $\mathrm{Coker}\,\phi = 0$. In other words, the second assertion follows from the first. Define the A-module N as the fibre product in the cartesian diagram of A-modules:

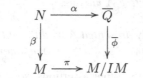

(where π is the natural projection). Notice that $IN = \mathrm{Ker}\,\alpha$; indeed, clearly $\alpha(IN) = 0$, and on the other hand $\beta(IN) = IM = \mathrm{Ker}\,\pi \simeq \mathrm{Ker}\,\alpha$. We derive an isomorphism $\overline{\psi} : \overline{Q} \xrightarrow{\sim} N/IN$, and clearly it suffices to find a morphism $Q \to N$ that lifts $\overline{\psi}$. Under our current assumptions, theorem 5.5.7(i) provides an almost finitely generated projective A-module Q_1 such that $Q_1 \otimes_A \overline{A} \simeq \overline{Q}$, which in turns determines an extension of A-modules $\underline{E} := (0 \to I \otimes_A \overline{Q} \to Q_1 \to \overline{Q} \to 0)$. Furthermore, $\overline{\psi}$ induces an epimorphism $\chi : I \otimes_A \overline{Q} \to IN$, whence an extension $\chi * \underline{E} := (0 \to IN \to Q_2 \to \overline{Q} \to 0)$. Another such extension is defined by $\underline{F} := (0 \to IN \to N \xrightarrow{\alpha} \overline{Q} \to 0)$. However, any extension $\underline{X}$ of $\overline{Q}$ by IN induces a morphism $u(\underline{X}) : I \otimes_{\overline{A}} \overline{Q} \to IN$, defined as in (3.2.19). Directly on the definition one can check that $u(\underline{X})$ depends only on the class of $\underline{X}$ in $\mathrm{Ext}^1_A(\overline{Q}, IN)$, and moreover, if $\underline{Y}$ is any such extension, then $u(\underline{X} + \underline{Y}) = u(\underline{X}) + u(\underline{Y})$ (where $\underline{X} + \underline{Y}$ denotes the Baer sum of the two extensions). We can therefore compute: $u(\chi * \underline{E} - \underline{F}) = \chi \circ u(\underline{E}) - u(\underline{F})$; but the definition of $\underline{E}$ is such that $u(\underline{E}) = 1_{I \otimes \overline{Q}}$ and by inspecting the construction of $\underline{F}$ we get $u(\underline{F}) = \chi$. So finally $u(\chi * \underline{E} - \underline{F}) = 0$; this means that $\chi * \underline{E} - \underline{F}$ is an extension of $\overline{A}$-modules, that is, its class is contained in the subgroup $\mathrm{Ext}^1_{\overline{A}}(\overline{Q}, IN) \subset \mathrm{Ext}^1_A(\overline{Q}, IN)$. Notice now that $\mathrm{Ext}^2_{\overline{A}}(\overline{Q}, \mathrm{Ker}\,\chi) = 0$, due to lemma 2.4.14(i),(ii); since χ is an epimorphism, it then follows that the induced map $\mathrm{Ext}^1_{\overline{A}}(\overline{Q}, \chi)$ is surjective. Hence there exists an extension $\underline{X} := (0 \to I \otimes_A \overline{Q} \to Q_3 \to \overline{Q} \to 0)$ of $\overline{A}$-modules, such that $\chi * \underline{X} = \chi * \underline{E} - \underline{F}$, i.e. $\underline{F} = \chi * (\underline{E} - \underline{X})$. Say $\underline{E}' := \underline{E} - \underline{X} = (0 \to I \otimes_A \overline{Q} \to Q \to \overline{Q} \to 0)$; by construction, $u(\underline{E}') = u(\underline{E})$, so Q is a flat A-module (see (3.2.19)) that lifts $\overline{Q}$. Finally, Q is almost finitely generated projective by lemma 3.2.25(i),(ii). The push-out $\underline{E}' \to \chi * \underline{E}'$ delivers the promised morphism $Q \to N$. $\diamond$

Next, since the pair (A, I^{n+1}) is still tight henselian for every $n \in \mathbb{N}$, an easy induction shows that the lemma holds when I is a nilpotent ideal. For the general case, pick $n > 0$ and a finitely generated subideal $\mathfrak{m}_0 \subset \mathfrak{m}$ with $I^n \subset \mathfrak{m}_0 A$; by the foregoing we can find an almost finitely generated projective A/I^{n+1}-module Q_{n+1} and an epimorphism $\phi_{n+1} : Q_{n+1} \to M/I^{n+1}M$. By theorem 5.5.7(i), we can lift Q_{n+1} to an almost finitely generated projective A-module Q; the obstruction to lifting the induced morphism $Q \to M/I^{n+1}M$ to a morphism $Q \to M$ is a class

$\omega \in \mathrm{Ext}^1_A(Q, I^{n+1}M)$; by the argument of claim 5.1.20 we see that the image of ω in $\mathrm{Ext}^1_A(Q, IM)$ vanishes, whence the claim. $\qquad\square$

Corollary 5.5.16. *Let A be a V^a-algebra, $I \subset \mathrm{rad}(A)$ a tight ideal and set $\overline{A} :=$ A/I; let P an almost finitely generated A-module, such that $\overline{P} := P \otimes_A \overline{A}$ is an almost projective $\overline{A}$-module. Then the following conditions are equivalent:*

    (i) *P is an almost projective A-module.*
    (ii) *P is a flat A-module.*
    (iii) *P is an almost finitely presented A-module and $\mathrm{Tor}^A_1(P, \overline{A}) = 0$.*
    (iv) *The natural morphism $P^* \to (\overline{P})^*$ is an epimorphism.*

Proof. Clearly (i) implies all the other assertions, so it suffices to show that each of the assertions (ii)-(iv) implies (i). Let us first remark that, in view of lemmata 2.3.14, 2.4.13 and theorem 2.1.12, we can assume that the homological dimension of $\widetilde{\mathfrak{m}}$ is ≤ 1. Furthermore, let (A^h, I^h) be the henselization of the pair (A, I); according to (5.1.14), the morphism $A \to A^h$ is faithfully flat. In view of remark 3.2.26(ii) and lemma 2.4.31(i) we deduce that each of the statements (i)-(iii) on P and $\overline{P}$ is equivalent to the corresponding statement (i)^h-(iii)^h made on the A^h-module $P \otimes_A A^h$ and the $(\overline{A} \otimes_A A^h)$-module $\overline{P} \otimes_A A^h$. Moreover, one checks easily that (iv)$\Rightarrow$(iv)^h. Thus, up to replacing (A, I) by (A^h, I^h) we can assume that (A, I) is a tight henselian pair (notice as well that $I^h = IA^h$). By lemma 5.5.14 we can find an almost finitely generated projective A-module Q and a morphism $\phi : Q \to P$ such that $\phi \otimes_A 1_{A/I}$ is an isomorphism. By lemma 5.1.7 we deduce easily that ϕ is an epimorphism. Suppose now that (ii) holds; then to deduce (i) it remains only to prove the following :

Claim 5.5.17. $\mathrm{Ker}\,\phi = 0$.

Proof of the claim. In view of proposition 2.4.28(v), it suffices to show that $\mathscr{E} := \mathscr{E}_{Q/A}(x) = 0$ for every $x \in \mathrm{Ker}\,\phi_*$ (see definition 2.4.23(iv)). However, since Q is A-flat, we can compute:

$$0 = \mathrm{Tor}^A_1(A/\mathscr{E}, P) \simeq \mathrm{Ker}(\mathrm{Ker}(\phi) \otimes_A (A/\mathscr{E}) \to Q/\mathscr{E}Q)$$
$$\simeq ((\mathrm{Ker}\,\phi) \cap \mathscr{E}Q)/(\mathscr{E} \cdot \mathrm{Ker}\,\phi).$$

That is, $(\mathrm{Ker}\,\phi) \cap \mathscr{E}Q = \mathscr{E} \cdot \mathrm{Ker}\,\phi$. By proposition 2.4.28(v) we have $x \in (\mathscr{E}Q)_*$; on the other hand we also know that $\mathrm{Ker}\,\phi \subset IQ$, whence $x \in (\mathscr{E}IQ)_*$. We apply once again proposition 2.4.28(v) to derive $\mathscr{E} = \mathscr{E}I$, so finally $\mathscr{E} = 0$ in view of lemma 5.1.7. $\qquad\Diamond$

Next, assume (iii); we compute: $0 = \mathrm{Tor}^A_1(P, \overline{A}) \simeq \mathrm{Ker}(\mathrm{Ker}(\phi) \otimes_A \overline{A} \to \mathrm{Ker}(\phi \otimes_A 1_{\overline{A}})) = \mathrm{Ker}(\phi) \otimes_A \overline{A}$. Since P is almost finitely presented, $\mathrm{Ker}\,\phi$ is almost finitely generated by lemma 2.3.18, whence $\mathrm{Ker}\,\phi = 0$ by lemma 5.1.7, so (i) holds. Finally, let $\phi^* : P^* \to Q^*$ be the transpose of the morphism ϕ; by lemma 2.4.31(i), the natural morphism $\psi : Q^* \to (Q/IQ)^*$ is an epimorphism. The composition $\psi \circ \phi^*$ factors through the transposed morphism $(\phi \otimes_A 1_{\overline{A}})^* : (\overline{P})^* \to (Q/IQ)^*$, so it is an epimorphism when (iv) holds; then lemma 5.1.7 implies easily that ϕ^* is an epimorphism; since it is obviously a monomorphism, we deduce that $P^* \overset{\sim}{\to} Q^*$. So $(\phi^*)^* : Q \simeq (Q^*)^* \to (P^*)^*$ is an isomorphism; since the latter factors through the natural morphism $P \to (P^*)^*$, we see that ϕ is a monomorphism and (i) follows. $\qquad\square$

5.6. Smooth locus of an affine almost scheme. Throughout this section we fix a V^a-algebra A and set $S := \operatorname{Spec} A$. Let X be an affine S-scheme. We often identify X with the functor it represents:

$$X : A\text{-}\mathbf{Alg} \to \mathbf{Set} \qquad T \mapsto X(T^o) := \operatorname{Hom}_{A\text{-}\mathbf{Alg}^o}(T^o, X).$$

The usual argument from faithfully flat descent shows that X is a sheaf for the fpqc topology of $A\text{-}\mathbf{Alg}^o$. In this section we aim to study, for every such X, the *smooth locus of X over S*, which will be a certain natural subsheaf of X. The starting point is the following:

Definition 5.6.1. Let S and X be as in (5.6). Given an affine S-scheme T and $\sigma \in X(T)$, we say that σ *lies in the smooth locus of X over S* if the following two conditions hold:

    (a) $H_1(L\sigma^* \mathbb{L}^a_{X/S}) = 0$ and

    (b) $H_0(L\sigma^* \mathbb{L}^a_{X/S})$ is an almost finitely generated projective $\mathscr{O}_T$-module.

We denote by $X_{\mathrm{sm}}(T) \subset X(T)$ the subset of all the T-sections of X that lie in the smooth locus of X over S.

5.6.2. Using remark 3.2.26(iii) one sees that X_{sm} is a subsheaf of X. Just as for usual schemes, in order to get a handle on the smooth locus X_{sm}, one often needs to assume that the almost scheme X satisfies some finiteness conditions. For our purposes, the following will do:

Definition 5.6.3. We say that the affine almost S-scheme X is *almost finitely presented* if there exists an almost finitely generated projective $\mathscr{O}_S$-module P, and an almost finitely generated ideal J of $S_P := \operatorname{Sym}^\bullet_{\mathscr{O}_S}(P)$, such that $X \simeq \operatorname{Spec} S_P/J$.

Lemma 5.6.4. *Let $X = \operatorname{Spec} S_F/J$, where $S_F := \operatorname{Sym}^\bullet_{\mathscr{O}_S}(F)$ for some flat $\mathscr{O}_S$-module F, and J is any ideal. Then there is a natural isomorphism in $\mathrm{D}(\mathscr{O}_X\text{-}\mathbf{Mod})$:*

$$\tau_{[-1]} \mathbb{L}^a_{X/S} \simeq (0 \to J/J^2 \to \mathscr{O}_X \otimes_{\mathscr{O}_S} F \to 0).$$

Proof. (The right-hand side is a complex placed in degrees -1 and 0.) Let us remark the following:

Claim 5.6.5. With the notation of the lemma, there is a natural isomorphism

$$\mathbb{L}^a_{\operatorname{Spec} S_F/S} \simeq S_F \otimes_{\mathscr{O}_S} F[0] \qquad \text{in } \mathrm{D}(S_F\text{-}\mathbf{Mod}).$$

Proof of the claim. Without the flatness assumption on F, the module $\Omega_{S_F/\mathscr{O}_S}$ is naturally isomorphic to $S_F \otimes_{\mathscr{O}_S} F$. Hence, in view of proposition 2.5.30, it remains to show that the homology of $\mathbb{L}^a_{S_F/\mathscr{O}_S}$ vanishes in non-zero degrees. But by proposition 8.1.7(ii) we know that

$$\mathbb{L}^a_{S_F/\mathscr{O}_S} \simeq \mathbb{L}^a_{\operatorname{Sym}^\bullet_{\mathscr{O}_{S*}}(F_!)/\mathscr{O}_{S*}}$$

and as $F_!$ is flat, the claim holds by [50, Ch.II, Prop.1.2.4.4]. $\qquad\diamond$

Using claim 5.6.5, the assertion can be shown as in the proof of [50, Ch.III, Cor.1.2.9.1]. □

5.6.6.   Let us consider a cartesian square of affine almost schemes:

$$
\begin{array}{ccc}
X' & \xrightarrow{\;f\;} & X \\
\downarrow & & \downarrow \\
S' & \longrightarrow & S.
\end{array}
$$

Just as for usual schemes, there is a natural morphism

$$\alpha : Lf^*\mathbb{L}^a_{X/S} \to \mathbb{L}^a_{X'/S'}.$$

Proposition 5.6.7. *In the situation of (5.6.6), we have:*

   (i) $H_0(\alpha)$ *is an isomorphism and* $H_1(\alpha)$ *is an epimorphism.*
   (ii) $f^{-1}(X_{\mathrm{sm}}) \subset X'_{\mathrm{sm}}$.

Proof. (i): It is enough to prove the analogous statement for the induced morphism $Lf^*\tau_{\lceil-1}\mathbb{L}^a_{X/S} \to \tau_{\lceil-1}\mathbb{L}^a_{X'/S'}$, which follows easily from the description of $\tau_{\lceil-1}\mathbb{L}^a_{X/S}$ in lemma 5.6.4, and its functoriality.

(ii): The assertion means that, for every affine S'-scheme T and every $\sigma \in X'(T)$ such that $f \circ \sigma$ lies in the smooth locus of X over S, we have already $\sigma \in X'_{\mathrm{sm}}(T)$, the set of T-sections of the smooth locus of X' over S'. However, (i) says that $\tau_{\lceil-1}\mathrm{Cone}(\alpha) \simeq 0$ in $\mathsf{D}(\mathcal{O}_{X'}\text{-}\mathbf{Mod})$; hence $\tau_{\lceil-1}\mathrm{Cone}(L\sigma^*(\alpha)) \simeq 0$ in $\mathsf{D}(\mathcal{O}_T\text{-}\mathbf{Mod})$, which implies the claim. □

5.6.8.   Let X be an almost finitely presented S-scheme and $t \in \mathfrak{m}$ any element. Then $A[t^{-1}]$ is a (usual) $V[t^{-1}]$-algebra, and we let $S_t := \mathrm{Spec}\, A[t^{-1}]$, $X_t := X \times_S S_t$. Both S_t and X_t are represented by (usual) affine schemes over $\mathrm{Spec}\, V[t^{-1}]$, and obviously X_t is finitely presented over S_t. Using lemma 5.6.4 it is also easy to see that the subfunctor $X_{\mathrm{sm},t} := X_{\mathrm{sm}} \cap X_t$ of the functor X is represented by the smooth locus of X_t over S_t, which is an open subscheme of X_t.

5.6.9.   In the situation of (5.6.8), suppose moreover that $I \subset A$ is a given ideal, and that t is regular in A. Let $R := A_*$; then t is a non-zero-divisor in R, and we have a well defined (t, I_*)-adic topology on $X_t(R[t^{-1}]) = X_t(S_t)$. Furthermore, it is clear that the restriction map:

(5.6.10) $X(S) \to X_t(S_t) \qquad \sigma \mapsto \sigma_*[t^{-1}]$

is injective. Consequently we can endow $X(S)$ with the (t, I)-*adic topology*, defined as the topology induced by the (t, I_*)-adic topology of $X_t(S_t)$. There is also another natural way of endowing $X(S)$ with a topology; namely, for any $f \in X(S)$ and $k \in \mathbb{N}$ one considers the subset $U_k(f) \subset X(S)$ consisting of all sections g whose restriction to the closed subscheme $S_k := \mathrm{Spec}\, A/t^k I$ agrees with f, and declares the family $U_k(f)$ to be a fundamental system of open neighborhoods of f. The topology thus defined shall be called the $S_\bullet$-*adic topology* on $X(S)$.

Lemma 5.6.11. *In the situation of (5.6.9), set* $A^\wedge := \lim_{n \in \mathbb{N}} A/t^n I$, *so that* $(A^\wedge)_* \simeq$ $\lim_{n \in \mathbb{N}} (A/t^n I)_*$ *has a filtration coming from the projective limit description. Let also* $(A_*)^\wedge$ *be the* (t, I_*)*-adic completion of* A_*. *Then the natural map*

$$(A_*)^\wedge \to (A^\wedge)_*$$

is an isomorphism of filtered rings.

Proof. The map is an inverse limit of the maps $\phi_n : A_*/t^n I_* \to (A/t^n I)_*$; moreover, since t^n is regular in A, we have $t^n I_* = (t^n I)_*$. Hence every ϕ_n is injective with cokernel injecting to $\mathrm{Ext}^1_{V^a}(V^a, t^n I)$, and the transition maps from the $(n+1)$-th cokernel to the n-th are 0 (cp. the proof of claim 5.1.20). □

Lemma 5.6.12. *In the situation of (5.6.9):*

(i) *The map (5.6.10) is an open imbedding for the respective* (t, I)*-adic and* (t, I_*)*-adic topologies.*

(ii) *The* (t, I)*-adic topology on* $X(S)$ *coincides with the* $S_\bullet$*-adic topology.*

Proof. Let us write $X = \mathrm{Spec}\, S_P/J$, where S_P is the symmetric algebra of an almost finitely generated projective A-module P, and set $\mathscr{P} := \mathrm{Spec}\, S_P$. Then X is a closed subscheme of $\mathscr{P}$ and X_t is a closed subscheme of $\mathscr{P}_t$, which is a vector bundle of finite rank over S_t. The (t, I_*)-adic topology of $X_t(S_t)$ is induced by the (t, I_*)-adic topology of $\mathscr{P}_t(S_t)$, and consequently the (t, I)-adic topology of $X(S)$ is induced by the (t, I)-adic topology of $\mathscr{P}(S)$. The commutative diagram of sets

(5.6.13)
$$\begin{array}{ccc} X(S) & \longrightarrow & X_t(S_t) \\ \downarrow & & \downarrow \\ \mathscr{P}(S) & \longrightarrow & \mathscr{P}_t(S_t) \end{array}$$

is cartesian, so we reduce to showing that the restriction map $\mathscr{P}(S) \to \mathscr{P}_t(S_t)$ is open. By assumption, there are finitely many almost elements $p_1, ..., p_n$ of P such that tP is contained in $\sum_{i=1}^n p_i A$; the topology on $X_t(S_t) = \mathrm{Hom}_A(P, A[t^{-1}])$ can be described in terms of a system of coordinates on X_t, in particular the one given by the p_i, so a fundamental system of open neighborhoods of a section $f \in X_t(S_t)$ is given by the family of subsets

$$V_k(f) := \{g \in X_t(S_t) \mid (f - g)(p_i) \in t^k I_* \text{ for } i = 1, ..., n\} \qquad (k \in \mathbb{N}).$$

Now, $f \in X(S)$ if and only if $f_*(P_*) \subset R$, and if $g \in V_1(f)$ then it is easily seen that $g \in X(S)$ if and only if $f \in X(S)$, *i.e.* the image of (5.6.10) is open, as stated in (i). Assertion (ii) follows by remarking that

$$U_{k+1}(f) \subset V_{k+1}(f) \subset U_k(f)$$

for every $k \in \mathbb{N}$. □

Proposition 5.6.14. *Let* X, T *be affine* S*-schemes and* $\sigma \in X(T)$ *a* T*-section,* $I \subset \mathrm{rad}(\mathscr{O}_T)$ *an ideal, and set* $T_0 := \mathrm{Spec}\, \mathscr{O}_T/I$; *suppose that the restriction* $\sigma_0 \in X(T_0)$ *of* σ *lies in the smooth locus of* X. *Suppose moreover that either:*

(a) I is nilpotent, or

(b) I is tight and X is almost finitely presented over S.

Then $\sigma \in X_{\mathrm{sm}}(T)$.

Proof. Suppose that (a) holds; for any quasi-coherent $\mathscr{O}_T$-module $\mathscr{F}$, let us denote by $\mathrm{Fil}_I^\bullet \mathscr{F}$ the I-adic filtration on $\mathscr{F}$. We can write $\tau_{[-1}L\sigma^*\mathbb{L}_{X/S}^a \simeq (0 \to N \overset{\phi}{\to} P \to 0)$ for two $\mathscr{O}_T$-modules N and P, and we can assume that P is almost projective over $\mathscr{O}_T$, so that the natural morphism

$$\tau_{[-1}(\mathscr{O}_{T_0} \overset{\mathbf{L}}{\otimes}_{\mathscr{O}_T} L\sigma^*\mathbb{L}_{X/S}^a) \to (0 \to N/IN \overset{\mathrm{gr}_I^0 \phi}{\to} P/IP \to 0)$$

is an isomorphism in $\mathsf{D}(\mathscr{O}_{T_0}\text{-}\mathbf{Mod})$. Hence, the assumption on σ means that $\mathrm{gr}_I^0 \phi$ is a monomorphism with almost finitely generated projective cokernel over $\mathscr{O}_{T_0}$. We consider, for every integer $i \in \mathbb{N}$, the commutative diagram:

$$\begin{array}{ccc}
\mathrm{gr}_I^i N & \overset{\mathrm{gr}_I^i \phi}{\longrightarrow} & \mathrm{gr}_I^i P \\
\alpha_i \uparrow & & \uparrow \beta_i \\
\mathrm{gr}_I^0 N \otimes_{\mathscr{O}_T} \mathrm{gr}_I^i \mathscr{O}_T & \overset{\mathrm{gr}_I^0 \phi \otimes_{\mathscr{O}_T} 1_{\mathrm{gr}_I^i \mathscr{O}_T}}{\longrightarrow} & \mathrm{gr}_I^0 P \otimes_{\mathscr{O}_T} \mathrm{gr}_I^i \mathscr{O}_T.
\end{array}$$

Since P is almost projective (especially, flat) β_i is an isomorphism. Moreover, since $\mathrm{gr}_I^0 \phi$ is a monomorphism with almost projective cokernel, the long exact Tor sequence shows that $\mathrm{gr}_I^0 \phi \otimes_{\mathscr{O}_T} 1_{\mathrm{gr}_I^i \mathscr{O}_T}$ is a monomorphism. It follows that α_i is a monomorphism for every $i \in \mathbb{N}$, and since it is obviously an epimorphism, we deduce that α_i is an isomorphism and $\mathrm{gr}_I^\bullet \phi$ is a monomorphism, therefore the same holds for ϕ. Let $C := \mathrm{Coker}(N \to P)$; we deduce easily that $\mathrm{Tor}_1^{\mathscr{O}_T}(\mathscr{O}_{T_0}, C) = 0$, and then it follows from the local flatness criterion (see [58, Ch.8,Th.22.3]) that C is a flat $\mathscr{O}_T$-module. Finally lemma 3.2.25(i),(ii) says that C is almost finitely generated projective, whence the claim, in case (a).

Next, suppose that (b) holds; by lemma 5.6.4 there is an isomorphism:

$$\tau_{[-1}L\sigma^*\mathbb{L}_{X/S}^a \simeq (0 \to N \overset{\phi}{\to} Q \to 0)$$

where N is almost finitely generated and Q is almost finitely generated projective over $\mathscr{O}_T$. Since Q is flat, we have

$$\tau_{[-1}(\mathscr{O}_{T_0} \overset{\mathbf{L}}{\otimes}_{\mathscr{O}_T} L\sigma^*\mathbb{L}_{X/S}^a) \simeq (0 \to N/IN \overset{\phi_0}{\to} Q/IQ \to 0)$$

and by assumption $\mathrm{Ker}\, \phi_0 = 0$ and $\mathrm{Coker}\, \phi_0$ is an almost finitely generated projective $\mathscr{O}_{T_0}$-module. Using the long exact Ext sequence we deduce that N/IN is almost projective. Thus, the bottom arrow of the natural commutative diagram

$$\begin{array}{ccc}
Q^* & \overset{\phi^*}{\longrightarrow} & N^* \\
\alpha \downarrow & & \downarrow \beta \\
(Q/IQ)^* & \overset{\phi_0^*}{\longrightarrow} & (N/IN)^*
\end{array}$$

is an epimorphism. Invoking twice corollary 5.5.16 we find first that α is an epimorphism (whence so is β), and then that N is almost projective. It then follows that $\phi^* \otimes_{\mathscr{O}_S} 1_{\mathscr{O}_{S_0}}$ is an epimorphism as well, hence the same holds for ϕ^*, by applying Nakayama's lemma 5.1.7 to Coker ϕ^*. The long exact Ext sequence then shows that Ker ϕ^* is almost projective, and the latter is almost finitely generated as well, by lemma 2.3.18(iii). Dualizing, we see that ϕ is a monomorphism and Coker $\phi \simeq (\text{Ker } \phi^*)^*$ is almost finitely generated projective, which is the claim. $\square$

Lemma 5.6.15. *Let X be an affine S-scheme, and suppose we are given a cartesian diagram of $\mathscr{O}_S$-algebras as (3.4.8). There follows a cartesian diagram of sets:*

$$
\begin{array}{ccc}
X_{\mathrm{sm}}(\operatorname{Spec} A_0) & \longrightarrow & X_{\mathrm{sm}}(\operatorname{Spec} A_2) \\
\downarrow & & \downarrow \\
X_{\mathrm{sm}}(\operatorname{Spec} A_1) & \longrightarrow & X_{\mathrm{sm}}(\operatorname{Spec} A_3).
\end{array}
$$

Proof. We have to check that every section $\sigma \in X(\operatorname{Spec} A_0)$ whose restrictions to $\operatorname{Spec} A_1$ and $\operatorname{Spec} A_2$ lie in the smooth locus of X over S, lies itself in the smooth locus. Hence, let $\tau_{[-1}L\sigma^* \mathbb{L}_{X/S} \simeq (0 \to N \xrightarrow{\phi} P \to 0)$, for some A_0-modules N and P, chosen so that P is almost projective. Hence $\tau_{[-1}(A_i \overset{\mathbf{L}}{\otimes}_{A_0} L\sigma^* \mathbb{L}_{X/S}) \simeq (0 \to N_i \xrightarrow{\phi_i} P_i \to 0)$ (where $N_i := A_i \otimes_{A_0} N$ and likewise for P_i, $i = 1, 2$). By assumption, Ker $\phi_i = 0$ and Coker ϕ_i is almost finitely generated projective over A_i ($i = 1, 2$). It follows that N_i is almost projective for $i = 1, 2$. Using lemma 3.4.18(iii) and proposition 3.4.21 we deduce that Coker ϕ is almost finitely generated projective over A_0 and N is almost projective. In particular, N is flat and consequently $N \subset N_1 \oplus N_2$, so Ker $\phi = 0$, and the assertion follows. $\square$

Lemma 5.6.16. *Let X and Y be two affine S-schemes such that $\operatorname{Tor}_i^{\mathscr{O}_S}(\mathscr{O}_X, \mathscr{O}_Y) = 0$ for every $i > 0$. Then we have a natural isomorphism of sheaves:*

$$(X \times_S Y)_{\mathrm{sm}} \simeq X_{\mathrm{sm}} \times Y_{\mathrm{sm}}.$$

Proof. Let $\pi_X : X \times_S Y \to X$ be the natural projection, and define likewise π_Y. By theorem 2.5.36, our assumptions imply that the natural morphism $\pi_X^* \mathbb{L}_{X/S}^a \oplus \pi_Y^* \mathbb{L}_{Y/S}^a \to \mathbb{L}_{X \times_S Y/S}^a$ is a quasi-isomorphism. Let now T be an affine S-scheme and $(\sigma, \tau) \in X \times_S Y(T) = X(T) \times Y(T)$; we derive a natural isomorphism:

$$L(\sigma, \tau)^* \mathbb{L}_{X \times_S Y/S} \simeq L\sigma^* \mathbb{L}_{X/S} \oplus L\tau^* \mathbb{L}_{Y/S}$$

from which the claim follows straightforwardly. $\square$

Theorem 5.6.17. *Assume that $\mathrm{hom.dim}_V \widetilde{\mathfrak{m}} \leq 1$. Let X be an almost finitely presented affine S-scheme, $I \subset \mathscr{O}_S$ an ideal such that the pair $(\mathscr{O}_S, I)$ is henselian, $\mathfrak{m}_0 \subset \mathfrak{m}$ a finitely generated subideal, and set $S_n := \operatorname{Spec} \mathscr{O}_S / \mathfrak{m}_0^n I$ for every $n \in \mathbb{N}$. Then we have:*

(i) $\operatorname{Im}(X_{\mathrm{sm}}(S_1) \to X_{\mathrm{sm}}(S_0)) = \operatorname{Im}(X_{\mathrm{sm}}(S) \to X_{\mathrm{sm}}(S_0))$.

(ii) *Set* $X_{\mathrm{sm}}^\wedge(S) := \lim_{n \in \mathbb{N}} X_{\mathrm{sm}}(S_n)$, *and endow* $X_{\mathrm{sm}}^\wedge(S)$ *with the pro-discrete topology. Then the natural map* $X_{\mathrm{sm}}(S) \to X_{\mathrm{sm}}^\wedge(S)$ *has dense image.*

Proof. We begin with an easy reduction:

Claim 5.6.18. In order to prove (i), we can assume that I is a tight ideal.

Proof of the claim. Indeed, let $\mathfrak{m}_0 \subset \mathfrak{m}$ be any finitely generated subideal, and choose $\mathfrak{m}_1 \subset \mathfrak{m}$ such that $\mathfrak{m}_0 \subset \mathfrak{m}_1^2$. Set $I' := \mathfrak{m}_1 I$, $S_0' := \mathrm{Spec}\, \mathscr{O}_S/I'$, $S_1' := \mathrm{Spec}\, \mathscr{O}_S/\mathfrak{m}_1 I'$. Notice that I' is tight, and suppose that the assertion is known for this ideal. By a simple chase on the commutative diagram:

$$
\begin{array}{ccccc}
X_{\mathrm{sm}}(S) & \longrightarrow & X_{\mathrm{sm}}(S_0) & \longleftarrow & X_{\mathrm{sm}}(S_1) \\
\| & & \uparrow & & \downarrow \\
X_{\mathrm{sm}}(S) & \longrightarrow & X_{\mathrm{sm}}(S_0') & \longleftarrow & X_{\mathrm{sm}}(S_1')
\end{array}
$$

the assertion can then be deduced for I as well. ◇

Let $\sigma_0 \in X_{\mathrm{sm}}(S_0)$ that admits an extension $\widetilde{\sigma} \in X_{\mathrm{sm}}(S_1)$, pick finitely many generators $\varepsilon_1, ..., \varepsilon_k$ for $\mathfrak{m}_0$, and define a map $\phi : I^{\oplus k} \to \mathfrak{m}_0 I$ by the rule: $(a_1, ..., a_k) \mapsto \sum_{i=1}^k \varepsilon_i \cdot a_i$.

Claim 5.6.19. Suppose that $I^2 = 0$ (so that I is an $\mathscr{O}_{S_0}$-module); then the map $\mathrm{Ext}^1_{\mathscr{O}_{S_0}}(L\sigma_0^* \mathbb{L}_{X/S}^a, \phi)$ is onto and $\mathrm{Ext}^1_{\mathscr{O}_{S_0}}(L\sigma_0^* \mathbb{L}_{X/S}^a, I)^a = 0$.

Proof of the claim. For the first assertion we use the spectral sequence

$$
E_2^{pq} := \mathrm{Ext}^p_{\mathscr{O}_{S_0}}(H_q(L\sigma_0^* \mathbb{L}_{X/S}^a), I^{\oplus k}) \Rightarrow \mathrm{Ext}^{p+q}_{\mathscr{O}_{S_0}}(L\sigma_0^* \mathbb{L}_{X/S}^a, I^{\oplus k})
$$

and a similar one for $\mathfrak{m}_0 I$. Since σ_0 lies in the smooth locus of X, the E_2^{01} terms vanish, hence we are reduced to verifying that the map $\mathrm{Ext}^1_{\mathscr{O}_{S_0}}(H_0(L\sigma_0^* \mathbb{L}_{X/S}^a), \phi)$ is surjective. However, the long exact Ext sequence shows that the cokernel of this map is a submodule of $\mathrm{Ext}^2_{\mathscr{O}_{S_0}}(H_0(L\sigma_0^* \mathbb{L}_{X/S}^a), \mathrm{Ker}\, \phi)$; again the assumption that σ_0 lies in the smooth locus of X, and $\mathrm{hom.dim}_V \widetilde{\mathfrak{m}} \le 1$, imply that the latter Ext group vanishes by lemma 2.4.14(ii). The same spectral sequence argument also proves the second assertion. ◇

Claim 5.6.20. Assertion (i) holds if $I^2 = 0$.

Proof of the claim. We need to show that there exists a morphism $\sigma : S \to X$ extending σ_0; then $\sigma \in X_{\mathrm{sm}}(S)$ in view of proposition 5.6.14. In other words, we have to find $\sigma^\sharp$ that fits into a morphism of extensions of $\mathscr{O}_S$-algebras:

$$
\begin{array}{ccccccccc}
0 & \longrightarrow & 0 & \longrightarrow & \mathscr{O}_X & = & \mathscr{O}_X & \longrightarrow & 0 \\
 & & \downarrow & & {\scriptstyle \sigma^\sharp}\downarrow & & \downarrow{\scriptstyle \sigma_0^\sharp} & & \\
0 & \longrightarrow & I & \longrightarrow & \mathscr{O}_S & \longrightarrow & \mathscr{O}_{S_0} & \longrightarrow & 0.
\end{array}
$$

By proposition 3.2.16, the obstruction to the existence of $\sigma^\sharp$ is a class

$$\omega \in \mathrm{Ext}^1_{\mathscr{O}_X}(\mathbb{L}^a_{X/S}, I) \simeq \mathrm{Ext}^1_{\mathscr{O}_{S_0}}(L\sigma_0^*\mathbb{L}^a_{X/S}, I).$$

Likewise, the obstruction to extending $\widetilde{\sigma}$ is a class $\widetilde{\omega} \in \mathrm{Ext}^1_{\mathscr{O}_{S_0}}(L\sigma_0^*\mathbb{L}^a_{X/S}, \mathfrak{m}_0 I)$, and ω is the image of $\widetilde{\omega}$ under the map $\mathrm{Ext}^1_{\mathscr{O}_{S_0}}(L\sigma_0^*\mathbb{L}^a_{X/S}, j)$ (where we have denoted by $j : \mathfrak{m}_0 I \to I$ the inclusion). From claim 5.6.19 one deduces easily that:

(a) $\mathrm{Ext}^1_{\mathscr{O}_{S_0}}(L\sigma_0^*\mathbb{L}^a_{X/S}, j \circ \phi)$ vanishes, and therefore,

(b) $\mathrm{Ext}^1_{\mathscr{O}_{S_0}}(L\sigma_0^*\mathbb{L}^a_{X/S}, j)$ must already vanish.

Thus $\omega = 0$, and the assertion holds. ◊

Claim 5.6.21. Choose a finitely generated subideal $\mathfrak{m}_1 \subset \mathfrak{m}$ such that $\mathfrak{m}_0 \subset \mathfrak{m}_1^2$. The section σ_0 can be lifted to an element of $\varprojlim_{n \in \mathbb{N}} X_{\mathrm{sm}}(\mathrm{Spec}\,\mathscr{O}_S/\mathfrak{m}_1 I^n)$.

Proof of the claim. For every $n > 0$, let $T_n := \mathrm{Spec}\,\mathscr{O}_S/\mathfrak{m}_1^2 I^n$ and $j_n :$ $\mathrm{Spec}\,\mathscr{O}_S/\mathfrak{m}_1 I^n \to T_n$ the natural morphism; we construct by induction on $n \in \mathbb{N}$ a sequence of section $\sigma_n \in X(T_n)$, such that the family $(\sigma_i \circ j_i \mid i > 0)$ defines an element of $\varprojlim_{n \in \mathbb{N}} X_{\mathrm{sm}}(\mathrm{Spec}\,\mathscr{O}_S/\mathfrak{m}_1 I^n)$. To this aim, we take σ_1 equal to the restriction of $\widetilde{\sigma}$; suppose then that $n > 1$ and that σ_{n-1} is already given. Notice that the image $J \subset \mathscr{O}_{T_n}$ of $\mathfrak{m}_1 I^{n-1}$ satisfies $J^2 = 0$. By the claim 5.6.20, it follows that $\sigma_{n-1} \circ j_{n-1}$ extends to a section in $X_{\mathrm{sm}}(T_n)$, and this we call σ_n. ◊

Let us now show how to deduce assertion (i) from (ii). By claim 5.6.18 we can suppose that $I^m \subset \mathfrak{m}_0 \mathscr{O}_S$ for some $m \geq 0$ and a finitely generated subideal $\mathfrak{m}_0 \subset \mathfrak{m}$; let $\mathfrak{m}_1 \subset \mathfrak{m}$ be as in claim 5.6.21; clearly for every $n \in \mathbb{N}$ there exists $m \geq 0$ such that $\mathfrak{m}_1 I^m \subset \mathfrak{m}_0^n I$, hence claim 5.6.21 shows that σ_0 can be lifted to an element of $X^\wedge_{\mathrm{sm}}(S)$ and then (ii) yields (i) trivially. It remains only to show (ii).

Claim 5.6.22. In order to prove (ii) we can assume that $\mathfrak{m}_0$ is a principal ideal.

Proof of the claim. We argue by induction on the number of generators of $\mathfrak{m}_0$. Thus, let $\varepsilon_1, ..., \varepsilon_k$ be a finite system of generators for $\mathfrak{m}_0$, with $k > 0$, and suppose that the assertion is known for all ideals generated by less than k elements. Let $\mathfrak{m}_0^{(n)}$ be the ideal generated by $\varepsilon_1^n, ..., \varepsilon_k^n$, and set $S_{(n)} := \mathrm{Spec}\,\mathscr{O}_S/\mathfrak{m}_0^{(n)} I$; for every $n \in \mathbb{N}$ there exists $N \in \mathbb{N}$ such that $\mathfrak{m}_0^N \subset \mathfrak{m}_0^{(n)} \subset \mathfrak{m}_0^n$, whence an isomorphism of prodiscrete spaces: $X^\wedge_{\mathrm{sm}}(S) \simeq \varprojlim_{n \in \mathbb{N}} X_{\mathrm{sm}}(S_{(n)})$. Thus, we can suppose that we are given a compatible system of sections $\sigma_n \in X_{\mathrm{sm}}(S_{(n)})$, and wish to show that, for every $n \in \mathbb{N}$, there exists $\sigma \in X_{\mathrm{sm}}(S)$ whose restriction to $S_{(n)}$ agrees with σ_n. Fix $N > 1$, set $T := \mathrm{Spec}\,\mathscr{O}_S/\varepsilon_1^N I$ and let $\mathfrak{m}_1 \subset \mathfrak{m}$ be the ideal generated by $\varepsilon_2, ..., \varepsilon_k$. Let also $T_{(n)} := T \times_S S_{(n)}$ for every $n \in \mathbb{N}$, and denote by $\sigma_{n|T} \in X_{\mathrm{sm}}(T_{(n)})$ the restriction of σ_n. Clearly $T_{(n)} \simeq \mathrm{Spec}\,\mathscr{O}_T/\mathfrak{m}_1^{(n)} \mathscr{O}_T$ for every $n \geq N$, and the pair $(\mathscr{O}_T, I\mathscr{O}_T)$ is henselian (cp. remark 5.1.10(v)). Hence, by inductive assumption, for every $n \geq N$ we can find $\sigma_T \in X(T)$ whose restriction to $T_{(n)}$ agrees with $\sigma_{n|T}$. We can then apply claim 5.6.21, in order to find a compatible system of sections $(\sigma_n' \in X(\mathrm{Spec}\,\mathscr{O}_S/\varepsilon_1^n I) \mid n \geq N)$, whose restriction to $\mathrm{Spec}\,\mathscr{O}_S/\varepsilon_1^{N-1} I$ agrees with the restriction of σ_T. Finally, if we assume that (ii) is known whenever $\mathfrak{m}_0$ is principal, we can find a section $\sigma \in X_{\text{sm}}(S)$ whose restriction to Spec $\mathscr{O}_S / \varepsilon_1^{N-1} \mathscr{O}_S$ agrees with σ'_N, whence the claim. $\Diamond$

For a given $\varepsilon \in \mathfrak{m}$, set $K(\varepsilon) := \bigcup_{n \in \mathbb{N}} \text{Ann}_{\mathscr{O}_S}(\varepsilon^n)$.

Claim 5.6.23. $K(\varepsilon)_* = \bigcup_{n \in \mathbb{N}} \text{Ann}_{\mathscr{O}_{S*}}(\varepsilon^n)$.

Proof of the claim. Clearly we have only to show the inclusion

$$K(\varepsilon)_* \subset \bigcup_{n \in \mathbb{N}} \text{Ann}_{\mathscr{O}_{S*}}(\varepsilon^n).$$

We apply the left exact functor $M \mapsto M_*$ to the left exact sequence $0 \to \text{Ann}_{\mathscr{O}_S}(\varepsilon^n) \to \mathscr{O}_S \xrightarrow{\varepsilon^n} \mathscr{O}_S$ to deduce that $\text{Ann}_{\mathscr{O}_S}(\varepsilon^n)_* = \text{Ann}_{\mathscr{O}_{S*}}(\varepsilon^n)$. However, $\varepsilon \cdot K(\varepsilon)_* \subset \bigcup_{n \in \mathbb{N}} \text{Ann}_{\mathscr{O}_S}(\varepsilon^n)_*$, so the claim follows easily. $\Diamond$

Claim 5.6.24. $(K(\varepsilon) \cap \varepsilon \mathscr{O}_S)_* \subset \text{nil}(\mathscr{O}_{S*})$ for every $n > 0$.

Proof of the claim. This follows from the obvious identity: $\text{Ann}_{\mathscr{O}_S}(\varepsilon^n) \cdot (\varepsilon \mathscr{O}_S)^n = 0$ by passing to almost elements and applying claim 5.6.23. $\Diamond$

Claim 5.6.25. If $n > 0$, every almost finitely generated subideal of $K(\varepsilon) \cap \varepsilon^n I$ is nilpotent.

Proof of the claim. Let $\mathscr{I}$ be such an ideal; we can find a finitely generated ideal $\mathscr{I}_0 \subset K(\varepsilon) \cap \varepsilon^n I$ such that $\varepsilon \mathscr{I} \subset \mathscr{I}_0 \subset \mathscr{I}$, so $\mathscr{I}^2 \subset \mathscr{I}_0$. From claim 5.6.24 we deduce that $\mathscr{I}_0$ is a nilpotent ideal, so the same holds for $\mathscr{I}$. $\Diamond$

Claim 5.6.26. In order to prove (ii) we can assume that $\mathfrak{m}_0 = \varepsilon V$, where $\varepsilon \in \mathfrak{m}$ is an $\mathscr{O}_S$-regular element.

Proof of the claim. Let us write $X = \text{Spec} \, S_P / J$, where $S_P := \text{Sym}_{\mathscr{O}_S}^\bullet P$ for some almost finitely projective $\mathscr{O}_S$-module P, and $J \subset S_P$ is an almost finitely generated ideal. By claim 5.6.22 we can assume that $\mathfrak{m}_0 = \varepsilon V$ for some $\varepsilon \in \mathfrak{m}$. Set $\overline{S} := \text{Spec} \, \mathscr{O}_S / K(\varepsilon)$ and $\overline{S}_n := \overline{S} \times_S S_n$ for every $n \in \mathbb{N}$. Let $\sigma^\wedge \in X_{\text{sm}}^\wedge(S)$; by definition $\sigma^\wedge$ is a compatible family of sections $\sigma_n : S_n \to X$ lying in the smooth locus of X over S. In turns, σ_n can be viewed as the datum of an $\mathscr{O}_S$-linear morphism $\tau_n : P \to \mathscr{O}_S / \varepsilon^n I$, such that the induced morphism of $\mathscr{O}_S$-algebras $\text{Sym}_{\mathscr{O}_S}^\bullet \tau : \text{Sym}_{\mathscr{O}_S}^\bullet P \to \mathscr{O}_S / \varepsilon^n I$ satisfies the condition: $\text{Sym}_{\mathscr{O}_S}^\bullet \tau(J) = 0$. For every $n \in \mathbb{N}$ we have a cartesian diagram of $\mathscr{O}_S$-algebras:

$$\begin{array}{ccc} \mathscr{O}_S / (K(\varepsilon) \cap \varepsilon^n I) & \longrightarrow & \mathscr{O}_{\overline{S}} \\ \downarrow & & \downarrow {\scriptstyle \pi_n} \\ \mathscr{O}_{S_n} & \xrightarrow{\ p_n\ } & \mathscr{O}_{\overline{S}_n}. \end{array}$$

Notice that ε is regular in $\mathscr{O}_{\overline{S}}$ and suppose that assertion (ii) holds for all almost finitely presented $\overline{S}$-schemes, especially for $\overline{X} := X \times_S \overline{S}$. Let $\overline{\sigma}_n : \overline{S}_n \to \overline{X}$ be the restriction of σ_n, for every $n \in \mathbb{N}$. By proposition 5.6.7(ii), the section $\overline{\sigma}_n$ lies in the smooth locus of $\overline{X}$ over $\overline{S}$; it then follows that, for every $n \in \mathbb{N}$, $\overline{\sigma}_n$ extends to a section $\overline{\sigma} : \overline{S} \to \overline{X}$ in the smooth locus of $\overline{X}$ over $\overline{S}$. By proposition 5.6.14, we then deduce that $\overline{\sigma}$ is even in the smooth locus of X over S, provided that $n > 0$. The datum of $\overline{\sigma}$ is equivalent to the datum of an $\mathscr{O}_{\overline{S}}$-linear morphism $\overline{\tau} : P \to \mathscr{O}_{\overline{S}}$ such that $\mathrm{Sym}^{\bullet}_{\mathscr{O}_S}\overline{\tau}(J) = 0$ and such that $\pi_n \circ \overline{\tau} = p_n \circ \tau_n$. The pair $(\tau_n, \overline{\tau})$ determines a morphism $\omega_n : P \to \mathscr{O}_S/(K(\varepsilon) \cap \varepsilon^n I)$, and by construction we have: $\mathrm{Sym}^{\bullet}_{\mathscr{O}_S}\omega_n(J) = 0$, *i.e.* ω_n induces a section $\sigma' : \mathrm{Spec}\,\mathscr{O}_S/(K(\varepsilon) \cap \varepsilon^n I) \to X$, and then σ' must lie in the smooth locus of X over S, in view of lemma 5.6.15. The obstruction to the existence of a lifting $\omega : P \to \mathscr{O}_S$ of ω_n, is a class $\alpha_n \in \mathrm{Ext}^1_{\mathscr{O}_S}(P, K(\varepsilon) \cap \varepsilon^n I)$. A simple verification shows that

$$(5.6.27) \qquad K(\varepsilon) \cap \varepsilon^n I = \varepsilon \cdot (K(\varepsilon) \cap \varepsilon^{n-1} I) \qquad \text{for all } n > 0.$$

From (5.6.27), an argument as in the proof of claim 5.1.20 allows to conclude that the image of α_n in $\mathrm{Ext}^1_{\mathscr{O}_S}(P, K(\varepsilon) \cap \varepsilon^{n-1} I)$ vanishes; however, this image is none other than α_{n-1}, so actually $\alpha_n = 0$, and the sought lifting can be found for every $n > 0$. By construction we have $\overline{J} := \mathrm{Sym}^{\bullet}_{\mathscr{O}_S}\omega(J) \subset K(\varepsilon) \cap \varepsilon^n I$. Now, ω induces a section $\sigma'' : \mathrm{Spec}\,\mathscr{O}_S/\overline{J} \to X$, which by construction extends σ'; moreover, the pair $(\mathscr{O}_S/\overline{J}, I/\overline{J})$ is henselian (cp. remark 5.1.10(v)), hence $\varepsilon^n I/\overline{J}$ is a tight radical ideal. Thus, using proposition 5.6.14 we derive that σ'' lies as well in the smooth locus of X over S. Since J is almost finitely generated, $\overline{J}$ is nilpotent in view of claim 5.6.25. Moreover, by (5.6.27) we can write $\overline{J} = \varepsilon \mathscr{J}$ for some ideal $\mathscr{J} \subset K(\varepsilon) \cap \varepsilon^{n-1} I$; $\mathscr{J}$ is nilpotent if $n \geq 2$, since in that case $\mathscr{J}^2 \subset \overline{J}$. Hence we can apply claim 5.6.21, to deduce that the restriction of σ'' to $\mathrm{Spec}\,\mathscr{O}_S/\mathscr{J}$ extends to a section $\sigma \in X_{\mathrm{sm}}(S)$; by construction σ extends σ_{n-1}. Since n can be taken to be arbitrarily large, the claim follows. $\Diamond$

So finally we suppose that $\mathfrak{m}_0 = tV$, with t an $\mathscr{O}_S$-regular element. Set $\mathscr{O}_S^\wedge := \lim_{n \in \mathbb{N}} \mathscr{O}_{S_n}$, $I^\wedge := \mathrm{Ker}(\mathscr{O}_S^\wedge \to \mathscr{O}_{S_0})$, and $S^\wedge := \mathrm{Spec}\,\mathscr{O}_S^\wedge$. We have a natural bijection

$$(5.6.28) \qquad X(S^\wedge) \simeq X^\wedge(S) := \lim_{n \in \mathbb{N}} X(S_n).$$

It follows from lemmata 5.6.11 and 5.6.12(ii) that (5.6.28) identifies the pro-discrete topology of $X^\wedge(S)$ with the $(t, I^\wedge)$-adic topology of $X(S^\wedge)$.

Claim 5.6.29. The homeomorphism (5.6.28) induces a bijection:

$$X_{\mathrm{sm}}(S^\wedge) \simeq X^\wedge_{\mathrm{sm}}(S).$$

Proof of the claim. Let $(\sigma_n \mid n \in \mathbb{N})$ be an element of $X^\wedge_{\mathrm{sm}}(S)$, and $\sigma^\wedge \in X(S^\wedge)$ the corresponding section. We have to show that $\sigma^\wedge$ lies in the smooth locus of X. Thus, we have a natural isomorphism: $\mathscr{O}_S^\wedge/tI^\wedge \simeq \mathscr{O}_S/tI$, and by assumption, the restriction of $\sigma^\wedge$ to $\mathrm{Spec}\,\mathscr{O}_S^\wedge/tI^\wedge$ lies in the smooth locus of X. We remark that the pair $(\mathscr{O}_S^\wedge, tI^\wedge)$ is tight henselian (the proof is as in remark 5.3.7; we leave the details to the reader). The assertion then follows from proposition 5.6.14. $\Diamond$

Under the standing assumptions, we have a cartesian diagram of V^a-algebras:

$$\begin{array}{ccc}
\mathscr{O}_S & \longrightarrow & \mathscr{O}_S[t^{-1}] \\
\downarrow & & \downarrow \\
\mathscr{O}_S^\wedge & \longrightarrow & \mathscr{O}_S^\wedge[t^{-1}]
\end{array}$$

whence a cartesian diagram of sets:

$$
\begin{array}{ccc}
X(S) & \longrightarrow & X_t(S_t) \\
\downarrow & & \downarrow{\scriptstyle \alpha} \\
X(S^\wedge) & \xrightarrow{\ \beta\ } & X_t(S_t^\wedge).
\end{array}
$$

Now, let $(\sigma_n \mid n \in \mathbb{N}) \in X_{\mathrm{sm}}^\wedge(S)$; by claim 5.6.29, the corresponding section $\sigma^\wedge$: $S^\wedge \to X$ lies in the smooth locus of X. Let $\sigma_t^\wedge \in X_{\mathrm{sm},t}(S_t^\wedge)$ be the restriction of $\sigma^\wedge$. By corollary 5.1.17(ii), the pair $(\mathscr{O}_{S*}, tI_*)$ is henselian; then by (5.6.8), lemma 5.6.11 and proposition 5.4.21, the restriction α_{sm} : $X_{\mathrm{sm},t}(S_t) \to X_{\mathrm{sm},t}(S_t^\wedge)$ has dense image for the $(t, I_*^\wedge)$-adic topology. Hence, we can approximate $\sigma_t^\wedge$ arbitrarily $(t, I_*^\wedge)$-adically close by a section of the form $\tau_t^\wedge := \alpha(\tau_t)$, where $\tau_t \in X_{\mathrm{sm},t}(S_t)$. Furthermore, β is an open imbedding, by lemma 5.6.12(i). Hence, if $\tau_t^\wedge$ is close enough to $\sigma^\wedge$, we can find $\tau^\wedge \in X(S^\wedge)$ such that $\beta(\tau^\wedge) = \tau_t^\wedge$. The pair $(\tau_t, \tau^\wedge)$ determines a unique section $\tau \in X(S)$, and by construction τ can be obtained as $S_\bullet$-adically close to $(\sigma_n \mid n \in \mathbb{N})$ as desired (terminology of (5.6.9)). Especially, in view of proposition 5.6.14, we can achieve that τ lies in the smooth locus of X, which concludes the proof of the theorem. $\qquad\Box$

5.7. **Quasi-projective almost schemes.** Let R be a V-algebra; we denote by $(R\text{-}\mathbf{Alg})^o_{\mathrm{fpqc}}$ the large fpqc site of *affine* R-schemes, and similarly for the site $(R^a\text{-}\mathbf{Alg})^o_{\mathrm{fpqc}}$. The localization functor $R\text{-}\mathbf{Alg} \to R^a\text{-}\mathbf{Alg}$ defines a morphism of sites:

$$
\mathrm{j} : (R^a\text{-}\mathbf{Alg})^o_{\mathrm{fpqc}} \to (R\text{-}\mathbf{Alg})^o_{\mathrm{fpqc}}.
$$

If $\mathscr{F}$ is any sheaf on $(R\text{-}\mathbf{Alg})^o_{\mathrm{fpqc}}$, then $\mathscr{F}^a := \mathrm{j}^*\mathscr{F}$ can be described as the sheaf associated to the presheaf $B \mapsto \mathscr{F}(B_*)$. Especially, we can regard any R-scheme X as a sheaf on $(R\text{-}\mathbf{Alg})^o_{\mathrm{fpqc}}$, and hence we obtain the *almost scheme X^a associated to X*. If X is affine, this notation agrees with that of (3.3.3), that is:

(5.7.1) $(\mathrm{Spec}\, S)^a = \mathrm{Spec}\, S^a$ for all R-algebras S.

For a general X, pick a Zariski hypercovering $Z_\bullet \to X$, where each Z_i is a disjoint union of affine R-schemes; then $X \simeq \underset{\Delta^o}{\mathrm{colim}}\, Z_\bullet$ as fpqc-sheaves, and thus $X^a \simeq \underset{\Delta^o}{\mathrm{colim}}\, Z_\bullet^a$ in the topos of sheaves on $(R^a\text{-}\mathbf{Alg})^o_{\mathrm{fpqc}}$. Furthermore, we have

$$
\mathrm{j}_*\mathrm{Spec}\, B = \mathrm{Spec}\, B_{!!} \otimes_{R_{!!}^a} R \qquad \text{for every } R^a\text{-algebra } B.
$$

If G is an R-group scheme, then G^a is clearly an R^a-group scheme, and if $Y \to X$ is any G-torsor over an R-scheme X (for the fpqc topology of X), then $Y^a \to X^a$ is a G^a-torsor.

Example 5.7.2. Let E be an R^a-module.

(i) We define the presheaf $\mathbb{P}_{R^a}(E)$ on $(R^a\text{-}\mathbf{Alg})^o_{\mathrm{fpqc}}$ as follows. For every affine R^a-scheme T, the T-sections of $\mathbb{P}(E)$ are the strictly invertible quotients of $E \otimes_{R^a} \mathscr{O}_T$, *i.e.* the equivalence classes of epimorphisms $E \otimes_{R^a} \mathscr{O}_T \to L$ where L is a strictly invertible $\mathscr{O}_T$-module (see definition (4.4.32)).

(ii) More generally, for every $r \in \mathbb{N}$ we have the presheaf $\mathrm{Grass}^r_{R^a}(E)$; its T-sections are the rank r almost $\mathscr{O}_T$-projective quotients $E \otimes_{R^a} \mathscr{O}_T \to P$.

By faithfully flat descent it is easy to verify that both these presheaves are in fact fpqc-sheaves.

5.7.3.   Let $\phi : \mathscr{G} \to \mathscr{F}$ be two objects in a topos T; recall that the *difference* $\mathscr{F} \setminus \mathscr{G}$ (relative to ϕ) is the largest subobject of $\mathscr{F}$ that has empty intersection with $\mathrm{Im}\,\phi$. Let now A be an R^a-algebra, $\mathscr{J} \subset A$ an almost finitely generated ideal. We define the *quasi-affine R^a-scheme* $\mathrm{Spec}\,A \setminus V(\mathscr{J})$ as the fpqc sheaf $\mathrm{Spec}\,A \setminus \mathrm{Spec}\,A/\mathscr{J}$.

Remark 5.7.4. In usual algebraic geometry, a quasi-affine scheme X can be realized in different ways as an open subscheme of an affine scheme, but the category of such open imbeddings has an initial object

$$\iota_X : X \to \mathrm{Spec}\,\Gamma(X, \mathscr{O}_X)$$

([27, Prop.5.1.2]). One can ask whether this holds in the almost setting. To make the question precise, let us remark that on the site $R^a\text{-}\mathbf{Alg}^\circ_{\mathrm{fpqc}}$ there is a tautological structure sheaf $\mathscr{O}_*$ of R-algebras; a quasi-affine scheme X is an object of the topos $\mathsf{T} := (R^a\text{-}\mathbf{Alg}^\circ)^\sim_{\mathrm{fpqc}}$, hence we can consider the R^a-algebra $\mathscr{O}(X) := \mathrm{Hom}_\mathsf{T}(X, \mathscr{O}_*)^a$. (Alternatively, one can work with the sheaf $\mathscr{O} := \mathscr{O}^a_*$ with values in R^a-algebras, and take its global sections (*i.e.* the limit) on $R^a\text{-}\mathbf{Alg}^\circ/X$.) Hence one would want to choose $\mathrm{Spec}\,\mathscr{O}(X)$ as a candidate initial object. However, to be rigorous, one must also deal with some set-theoretic nuisances. Namely, to define the fpqc site, one should consider only R^a-algebras in some universe U; then this site will be only a U'-site for a larger universe U'. Thus, $\mathscr{O}(X)$ is a priori only in U', but with some care one sees that it is actually in U, and independent of the universe, so that its spectrum is an object of the fpqc site, as required. Similarly, the set of morphisms between quasi-affine or quasi-projective almost schemes (to be defined below) is small and independent of the universe.

Now, there is a natural morphism $\iota_X : X \to \mathrm{Spec}\,\mathscr{O}(X)$; namely, to every affine R^a-scheme T and every morphism $f : T \to X$ one associates the induced map of R-algebras $f^\sharp : \mathscr{O}_*(X) \to \mathscr{O}_*(T)$, and then $\iota_X(T)(f) := \mathrm{Spec}\,(f^\sharp)^a : T \to \mathscr{O}(X)$ is a T-section of $\mathrm{Spec}\,\mathscr{O}(X)$. It is also easy to see that ι_X is an initial object in the category of morphisms from to X to affine R^a-schemes; indeed, such a morphism $g : X \to Y := \mathrm{Spec}\,B$ factors through $\mathrm{Spec}\,(g^\sharp)^a : \mathrm{Spec}\,\mathscr{O}(X) \to Y$. Thus the question arises, whether ι_X is always an "open imbedding" (with obvious meaning). A partial positive answer is given in remark 5.7.9(ii), with the aid of the following lemmata 5.7.6 and 5.7.8.

5.7.5.   Let A be an R^a-algebra, $J \subset A$ an almost finitely generated ideal. For an A-module M let $\tau_J M := \bigcup_{n \in \mathbb{N}} \mathrm{Ann}_M(J^n)$. We say that M is a J-power torsion module if $M = \tau_J M$.

Lemma 5.7.6. *The class of J-power torsion modules is a Serre subcategory $\tau_J\text{-}\mathbf{Mod}$ of $A\text{-}\mathbf{Mod}$, stable under colimits and $\mathrm{Tor}^A_i(M, -)$ functors, for all A-modules M.*

Proof. We check that τ_J-**Mod** is stable under extensions of A-modules; the other assertions are easier. Indeed, since the ideals J^n are also almost finitely generated, the functor $M \mapsto \mathrm{Ann}_M(J^n) = \mathrm{Hom}_A(A/J^n, M)$ commutes with filtered colimits (proposition 2.3.16(ii)), hence the same holds for the functor τ_J. If $M' \subset M$ and M' is annihilated by a power of J, it is easy to see that $\tau_J(M/M') = (\tau_J M)/M'$; therefore, if more generally, $M' = \tau_J M'$ we compute:

$$\tau_J(M/M') = \mathrm{colim}_{n\in\mathbb{N}} \tau_J(M/\mathrm{Ann}_{M'}(J^n)) = \mathrm{colim}_{n\in\mathbb{N}} (\tau_J M)/\mathrm{Ann}_{M'}(J^n)$$
$$= (\tau_J M)/M'$$

which implies the claim. $\square$

5.7.7.   Now, by [40, Ch.III, Prop.8], the localization functor

$$t : A\text{-}\mathbf{Mod} \to A\text{-}\mathbf{Mod}/\tau_J\text{-}\mathbf{Mod}$$

has a right adjoint, given by

$$s : [M] \mapsto \mathrm{colim}_{n\in\mathbb{N}} \mathrm{Hom}_A(J^n, M/\tau_J M).$$

Notice that s and t send algebras to algebras.

Lemma 5.7.8. *Let $\phi : A \to B$ be a morphism of R^a-algebras, $J \subset A$ an almost finitely generated ideal of A, and suppose that the induced morphism $t(\phi) : t(A) \to t(B)$ is an isomorphism. Then ϕ induces an isomorphism:*

$$\mathrm{Spec}\, B \setminus V(JB) \to \mathrm{Spec}\, A \setminus V(J).$$

Proof. One should check that if $\psi : A \to C$ is a morphism of R^a-algebras such that $\psi(J)C = C$, then ψ factors uniquely through ϕ. Indeed, using the Tor long exact sequences one sees that the morphism $\phi_C := 1_C \otimes_A \phi : C \to B \otimes_A C$ is an isomorphism modulo J-power torsion (*i.e.* it becomes an isomorphism after applying t). Then the condition $\psi(J)C = C$ implies that the kernel and cokernel of ϕ_C vanish. So ϕ_C is an isomorphism, and the assertion follows. $\square$

Remark 5.7.9. (i) Notice that, even when A is a usual ring, if $J \subset A$ is not finitely generated, the category of J-power torsion modules is not necessarily stable under extensions. In general, the construction of the smallest Serre subcategory of A-**Mod** containing the J-torsion modules and closed under colimits, is accomplished by a transfinite process.

(ii) In the situation of lemma 5.7.8, suppose that J is generated by finitely many almost elements, and set $X := \mathrm{Spec}\, A \setminus V(J)$. Then one can show that $\mathscr{O}(X) = s \circ t(A)$; applying lemma 5.7.8 with $B := \mathscr{O}(X)$ one deduces easily that ι_X is an open imbedding, which gives the announced partial answer to the question of remark 5.7.4. More generally, one obtains a positive answer if one assumes only that the stated property of J holds locally in the fpqc topology.

Lemma 5.7.10. *Let $f : T' \to T$ be a morphism of topoi, $\mathscr{G} \to \mathscr{F}$ a morphism of objects of T. Then:*

(i) $f^*(\mathscr{F} \setminus \mathscr{G}) \subset f^*\mathscr{F} \setminus f^*\mathscr{G}.$

    (ii) *If moreover, $\mathsf{T}' = \mathsf{T}/X$ for an object X of T, and $\mathfrak{f}$ is the localization morphism, then the foregoing inclusion is an equality.*

    (iii) *Let $\mathscr{F}' \to \mathscr{F}$ be another morphism of T, and set $\mathscr{G}' := \mathscr{F}' \times_{\mathscr{F}} \mathscr{G}$. Then*

$$\mathscr{F}' \setminus \mathscr{G}' = \mathscr{F}' \times_{\mathscr{F}} (\mathscr{F} \setminus \mathscr{G}).$$

Proof. (i) follows easily from the facts that the pull-back functor $\mathfrak{f}^*$ is left exact and sends empty objects to empty objects. (ii) and (iii) shall be left to the reader. □

Lemma 5.7.11. *Let B be a V^a-algebra and $g_1, ..., g_k \in B_*$. Then $\sum_{i=1}^{k} g_i B = B$ if and only if the natural morphism $B \to \prod_{i=1}^{k} B[g_i^{-1}]$ is faithfully flat.*

Proof. Supppose that $\sum_{i=1}^{k} g_i B = B$; we have to show that $\prod_{i=1}^{k} C[g_i^{-1}] \neq 0$ for every non-zero quotient C of B. Replacing B by C, we are reduced to showing that, if $B[g_i^{-1}] = 0$ for every $i \leq k$, then $B = 0$. However, the condition $B[g_i^{-1}] = 0$ implies that for every $\varepsilon \in \mathfrak{m}$ there exists $n_i \in \mathbb{N}$ such that $\varepsilon \cdot g_i^{n_i} = 0$; set $n := \max(n_i \mid i \leq k)$. We have $\sum_{i=1}^{k} g_i^n B = B$, hence $\varepsilon \cdot B = 0$. Since ε is arbitrary, the claim follows. Conversely, let $B' := \prod_{i=1}^{k} B[g_i^{-1}]$ and $I := \sum_{i=1}^{k} g_i B$; clearly $IB' = B'$, therefore $I = B$ provided B' is faithfully flat over B. □

Proposition 5.7.12. *If $Y \subset X$ is a finitely presented closed immersion of R-schemes, then:*

$$(X \setminus Y)^a = X^a \setminus Y^a.$$

Proof. Pick an affine Zariski covering $X = \bigcup_{i \in I} U_i$; then $\mathfrak{j}^* X = \bigcup_{i \in I} \mathfrak{j}^* U_i$. Using lemma 5.7.10(iii), we see that it suffices to show the assertion for the induced closed imbeddings $U_i \cap Y \to U$, hence we can assume that $X = \operatorname{Spec} S$ for some R-algebra S, and then $Y = V(J)$ for a finitely generated ideal $J \subset S$. Let $f_1, ..., f_k$ be a set of generators for J; then $X \setminus Y = \bigcup_{i=1}^{k} \operatorname{Spec} S[f_i^{-1}]$. Hence, in view of (5.7.1) and lemma 5.7.10(i) we reduce to showing:

Claim 5.7.13. $X^a \setminus Y^a \subset Z := \bigcup_{i=1}^{k} \operatorname{Spec} S[f_i^{-1}]^a$.

Proof of the claim. Indeed, for any affine R^a-scheme T, an element $\sigma \in X^a(T)$ is the same as a morphism of R^a-algebras $\sigma^{\sharp} : S^a \to \mathcal{O}_T$; the condition that $T \times_{X^a} Y^a = \varnothing$ (the empty object in the topos $(R^a\text{-}\mathbf{Alg}^\circ)^{\sim}_{\text{fpqc}}$) is equivalent to $\mathcal{O}_T \otimes_{S^a} S^a/J^a = 0$, i.e. $\sigma^{\sharp}(J^a)\mathcal{O}_T = \mathcal{O}_T$. By lemma 5.7.11, this is in turns equivalent to saying that the natural morphism $\phi : \mathcal{O}_T \to B := \prod_{i=1}^{k} \mathcal{O}_T[f_i^{-1}]$ is faithfully flat. Clearly the restriction $\sigma \circ \operatorname{Spec} \phi \in X^a(\operatorname{Spec} B)$ is an element of $Z(\operatorname{Spec} B)$. Since the assertion is local in the fpqc topology, the claim follows. □

5.7.14. Let E be an R-module, B an R^a-algebra; by definition, a section σ in $\operatorname{Grass}_R^r(E)(\operatorname{Spec} B_*)$ is a quotient map $\sigma : E \otimes_R B_* \to P$, onto a projective B_*-module P of rank r. To σ we associate the quotient $\sigma^a : E^a \otimes_R B \to P^a$; after passing to the associated sheaves we obtain a morphism:

(5.7.15) $\mathfrak{j}^* \operatorname{Grass}_R^r(E) \to \operatorname{Grass}_{R^a}^r(E^a)$ $\sigma \mapsto \sigma^a$.

Since $P \simeq P_*^a$, it is clear that the rule $\sigma \mapsto \sigma^a$ defines an injective map of presheaves, hence (5.7.15) is a monomorphism.

Lemma 5.7.16. *Suppose that E is a finitely generated R-module. Then* (5.7.15) *is an isomorphism.*

Proof. We have to show that (5.7.15) is an epimorphism of fpqc-sheaves; hence, let T be any affine R^a-scheme, $\sigma : E^a \otimes_{R^a} \mathscr{O}_T \to P$ a quotient morphism, with P almost projective over $\mathscr{O}_T$ of rank r. The contention is that there exists a faithfully flat morphism $f : U \to T$ such that $f^*\sigma$ is in the image of $\mathrm{Grass}^r_R(E)(U_*)$ (where $U_* := \mathrm{Spec}\,\mathscr{O}_{U*}$). By theorem 4.4.24 we can then reduce to the case when $P = \mathscr{O}^r_T$. We deduce an epimorphism $\Lambda^r_{\mathscr{O}_T}\sigma : \Lambda^r_{\mathscr{O}_T}(E^a \otimes_{R^a} \mathscr{O}_T) \to \Lambda^r_{\mathscr{O}_T}(\mathscr{O}^r_T) \simeq \mathscr{O}_T$. Let $e_1, ..., e_k$ be a finite set of generators of E, and for every subset $J := \{j_1, ..., j_r\} \subset \{1, ..., k\}$ of cardinality r, let $e_J := e_{j_1} \wedge ... \wedge e_{j_r} \in \Lambda^r_{\mathscr{O}_T}(E^a \otimes_{R^a} \mathscr{O}_T)_*$; it follows that the set $(t_J := \Lambda^r_{\mathscr{O}_T}\sigma(e_J) \mid J \subset \{1, ..., k\})$ generates $\mathscr{O}_T$. Set $U_J := \mathrm{Spec}\,\mathscr{O}_T[t_J^{-1}]$ for every $J \subset \{1, ..., k\}$ as above; by construction σ induces surjective maps $E \otimes_R \mathscr{O}_{T*}[t_J^{-1}] \to \mathscr{O}^r_{T*}[t_J^{-1}]$; after tensoring by $\mathscr{O}_T[t_J^{-1}]_*$ we obtain surjections $E \otimes_R \mathscr{O}_T[t_J^{-1}]_* \to \mathscr{O}_T[t_J^{-1}]^r_*$. In other words, the restriction $\sigma_{|U_J}$ lies in the image of $\mathrm{Grass}^r_R(E)(U_{J*})$ for every such J; on the other hand, lemma 5.7.11 says that the natural morphism $\amalg_J U_J \to T$ is a fpqc covering, whence the claim. $\qquad\square$

Lemma 5.7.16 says in particular that $\mathrm{Grass}^r_{R^a}(E^a)$ is an almost scheme associated to a scheme whenever E is finitely generated. Next we wish to define the projective R^a-scheme associated to a graded R^a-algebra.

Definition 5.7.17. (i) Let $A_\bullet := \oplus_{i \in \mathbb{N}} A_i$ be a graded R^a-algebra which is generated by A_1 over R^a, in the sense that the natural morphism of graded R^a-algebras $\mathrm{Sym}^\bullet_{R^a} A_1 \to A_\bullet$ is an epimorphism. We let $\mathrm{Proj}\,A_\bullet$ be the sheaf on $(R^a\text{-}\mathbf{Alg})^o_{\mathrm{fpqc}}$ defined as follows. For every affine R^a-scheme T, the T-points of $\mathrm{Proj}\,A_\bullet$ are the strictly invertible quotients $A_1 \otimes_{R^a} \mathscr{O}_T \to L$ of $A_1 \otimes_{R^a} \mathscr{O}_T$, inducing morphisms of graded R^a-algebras $A_\bullet \to \mathrm{Sym}^\bullet_{\mathscr{O}_T} L$.

(ii) A *quasi-projective almost scheme* is a sheaf on $(R^a\text{-}\mathbf{Alg})^o_{\mathrm{fpqc}}$ of the form $\mathrm{Proj}\,A_\bullet \setminus \mathrm{Proj}\,A_\bullet/\mathscr{J}_\bullet$, where $A_\bullet$ is as in (i), A_1 is an almost finitely generated R^a-module and $\mathscr{J}_\bullet \subset A_\bullet$ is an almost finitely generated graded ideal.

5.7.18.   Recall ([27, Ch.II, Prop.2.4.7(ii)]) that for an $\mathbb{N}$-graded ring $S_\bullet$, the scheme $\mathrm{Proj}\,S_\bullet$ depends only on the non-unital graded ring S_+ obtained by replacing S_0 by 0. This has an analogue in the present setting. First of all, if $A_\bullet$ is a graded R^a-algebra as in definition 5.7.17, and $A'_\bullet$ is obtained by replacing A_0 with R^a, then $\mathrm{Proj}\,A_\bullet = \mathrm{Proj}\,A'_\bullet$. Indeed every graded morphism $A'_\bullet \to \mathrm{Sym}^\bullet_{\mathscr{O}_T} L$ factors through $A_\bullet$, as $\mathrm{Ker}(R^a \to A_0)$ annihilates A_1, hence annihilates L and then annihilates also $\mathscr{O}_T$, if L is an invertible $\mathscr{O}_T$-module.

More generally, let us consider a commutative diagram of V^a-algebras where $A_\bullet$ (resp. $B_\bullet$) is an $\mathbb{N}$-graded R^a-algebra generated over R^a by A_1 (resp. over S^a by B_1), and $h_\bullet$ is a graded morphism. By [4, Exp.III, Prop.5.4] there is a natural equivalence of topoi

$$(S^a\text{-}\mathbf{Alg})^{\circ\sim}_{\mathrm{fpqc}} \xrightarrow{\sim} (R^a\text{-}\mathbf{Alg})^{\circ\sim}_{\mathrm{fpqc}}/\operatorname{Spec} S^a \quad : \quad F \mapsto f_! F$$

which assigns to a sheaf F on $(S^a\text{-}\mathbf{Alg})^{\circ}_{\mathrm{fpqc}}$ the natural map $f_! F \to \operatorname{Spec} S^a$ of sheaves on $(R^a\text{-}\mathbf{Alg})^{\circ}_{\mathrm{fpqc}}$, where the T-sections of $f_! F$ are given by the rule:

$$f_! F(T) := \coprod_{\phi \in \operatorname{Hom}_{R^a\text{-}\mathbf{Alg}}(S^a, T)} F(T)$$

for every R^a-algebra T. With this notation we have:

Proposition 5.7.19. *Suppose that $h_\bullet$ is an isomorphism in all degrees $> k$ (for some integer k). Then $h_\bullet$ induces an isomorphism*

$$\operatorname{Proj} h_\bullet : f_!(\operatorname{Proj} B_\bullet) \xrightarrow{\sim} \operatorname{Proj} A_\bullet.$$

Proof. For any morphism $\phi : T \to \operatorname{Spec} S^a$ of R^a-schemes, denote by $h_{1,\phi} : \mathscr{O}_T \otimes_{R^a} A_1 \to \mathscr{O}_T \otimes_{S^a} B_1$ the morphism deduced from h_1 and ϕ. Let $\gamma : \mathscr{O}_T \otimes_{S^a} B_1 \to L$ be a strictly invertible quotient defining a T-section of $\operatorname{Proj} B_\bullet$. The map $\operatorname{Proj} h_\bullet$ should assign to γ the composition $\gamma \circ h_{1,\phi}$. In order to see that this rule yields a well defined morphism $\operatorname{Proj} h_\bullet$, we need to show that, if $g_\bullet : B_\bullet \to \operatorname{Sym}^\bullet_{\mathscr{O}_T} L$ is a graded morphism such that B_1 generates the $\mathscr{O}_T$-module L, then the image of A_1 under $g_\bullet \circ h_\bullet$ also generates L. This is clear if $k = 0$. In general, B_n generates $L^{\otimes n}$, hence so does A_n for $n \in \mathbb{N}$ large enough. But by assumption, A_n is a quotient of $A_1^{\otimes n}$. Now, by lemma 4.1.23(i), the $\mathscr{O}_T$-submodule generated by the image of A_1 can be written in the form IL for an ideal $I \subset \mathscr{O}_T$; hence the image of A_n generates $I^n L^{\otimes n}$. If $I^n = \mathscr{O}_T$, then $I = \mathscr{O}_T$, whence the claim. To prove that $\operatorname{Proj} h_\bullet$ is an isomorphism, it suffices to show:

Claim 5.7.20. Let T be an R^a-scheme, L a strictly invertible $\mathscr{O}_T$-module, $g_\bullet : A_\bullet \to S_L := \operatorname{Sym}^\bullet_{\mathscr{O}_T} L$ a graded morphism such that $g_1(A_1)$ generates L. Then g factors uniquely through $h_\bullet$.

Proof of the claim. We can extend S_L to a $\mathbb{Z}$-graded algebra S'_L with $S'_{L,n} := \operatorname{Sym}^{-n}_{\mathscr{O}_T} L^*$ for $n < 0$. This is the splitting algebra $\operatorname{Split}(\mathscr{O}_T, L)$. Clearly it suffices to show the claim with S_L replaced by S'_L. By assumption the kernel and cokernel of $h_\bullet$ are annihilated by some power of A_+; arguing with the Tor long exact sequences, we deduce that the same holds for the kernel and cokernel of $\mathbf{1}_{S'_L} \otimes_{A_\bullet} h_\bullet : S'_L \to B_\bullet \otimes_{A_\bullet} S'_L$. But by assumption A_+ generates the unit ideal in S'_L, so the latter morphism is an isomorphism, which implies the claim. $\qquad\square$

Definition 5.7.21. Let $\phi : \mathscr{F} \to \mathscr{G}$ be a morphisms of sheaves on $(R^a\text{-}\mathbf{Alg})^{\circ}_{\mathrm{fpqc}}$. We say that ϕ is a *closed imbedding* if, for every affine R^a-scheme T and every section $T \to \mathscr{G}$, the induced morphism $T \times_\mathscr{G} \mathscr{F} \to T$ is a closed imbedding of affine R^a-schemes (see definition 4.5.5(i)).

Lemma 5.7.22. *With the notation of* (5.7.17), *the natural map of sheaves*

$$\mathrm{Proj}\, A_\bullet \to \mathbb{P}_{R^a}(A_1)$$

is a closed imbedding.

Proof. Let T be any affine R^a-scheme and $\sigma : T \to \mathbb{P}_{R^a}(A_1)$ a morphism of fpqc-sheaves. We have to show that $T' := T \times_{\mathbb{P}_{R^a}(A_1)} \mathrm{Proj}\, A_\bullet$ is a closed subscheme of T. By definition, $\sigma : \mathcal{O}_T \otimes_{R^a} A_1 \to L$ is a strictly invertible quotient; hence $\mathrm{Sym}_{\mathcal{O}_T}^\bullet \sigma$ induces a morphism $s : \mathrm{Sym}_{R^a}^\bullet A_1 \to \mathrm{Sym}_{\mathcal{O}_T}^\bullet L$ of graded R^a-algebras. Let U be an affine R^a-scheme and $U \to T'$ a morphism of fpqc-sheaves; set $\mathcal{K}_n := \mathrm{Ker}\,(\mathrm{Sym}_{R^a}^n A_1 \to A_\bullet)$ for every $n \in \mathbb{N}$ and denote by $\mathcal{K}_{T,n}$ the $\mathcal{O}_T$-submodule generated by $s(\mathcal{K}_n)$. By definition:

$$\mathcal{K}_{T,n} \subset \mathrm{Ker}\,(\mathrm{Sym}_{\mathcal{O}_T}^n L \to \mathcal{O}_U \otimes_{\mathcal{O}_T} \mathrm{Sym}_{\mathcal{O}_T}^n L). \qquad \text{for every } n \in \mathbb{N}$$

Since L is flat, this is the same as:

(5.7.23) $\mathcal{K}_{T,n} \subset \mathrm{Ker}\,(\mathcal{O}_T \to \mathcal{O}_U) \cdot \mathrm{Sym}_{\mathcal{O}_T}^n L.$ for every $n \in \mathbb{N}$

For every $n \in \mathbb{N}$ denote by $\mathrm{ev}_n : \mathrm{Sym}_{\mathcal{O}_T}^n L \otimes_{\mathcal{O}_T} \mathrm{Sym}_{\mathcal{O}_T}^n L^* \to \mathcal{O}_T$ the evaluation morphism of the $\mathcal{O}_T$-module $\mathrm{Sym}_{\mathcal{O}_T}^n L$. Then (5.7.23) is equivalent to:

$$\sum_{n \in \mathbb{N}} \mathrm{ev}_n(\mathcal{K}_{T,n} \otimes_{\mathcal{O}_T} \mathrm{Sym}_{\mathcal{O}_T}^n L^*) \subset \mathrm{Ker}\,(\mathcal{O}_T \to \mathcal{O}_U).$$

Conversely, let

(5.7.24) $$I := \sum_{n \in \mathbb{N}} \mathrm{ev}_n(\mathcal{K}_{T,n} \otimes_{\mathcal{O}_T} \mathrm{Sym}_{\mathcal{O}_T}^n L^*).$$

Then the restriction of σ to $\mathrm{Spec}\, \mathcal{O}_T/I$ is a section of $\mathrm{Proj}\, A_\bullet$. All in all, this shows that T' represents the closed subscheme $\mathrm{Spec}\, \mathcal{O}_T/I$ of T, whence the claim. $\square$

Lemma 5.7.25. *Let* $S_\bullet := \oplus_{i \in \mathbb{N}} S_i$ *be a graded R-algebra generated by S_1.*

  (i) *The natural map*

$$(\mathrm{Proj}\, S_\bullet)^a \to \mathrm{Proj}\, S_\bullet^a \qquad \sigma \mapsto \sigma^a$$

  is a monomorphism.

  (ii) *If S_1 is a finitely generated R-module, then the foregoing map is an isomorphism.*

Proof. (i) is proved by the same argument as in (5.7.14). To show (ii), let B be an R^a-algebra, and $\sigma : B \otimes_{R^a} S_1^a \to L$ a strictly invertible B-module quotient. By lemma 5.7.16 we know that there exists a faithfully flat B-algebra C such that the induced map $\sigma' : C_* \otimes_R S_1 \to (C \otimes_B L)_*$ is a rank one projective quotient. Suppose now that σ is a section of $\mathrm{Proj}\, S_\bullet^a$ and let $J := \mathrm{Ker}(\mathrm{Sym}_R^\bullet S_1 \to S_\bullet)$; by definition we have

$$\mathrm{Im}(B \otimes_{R^a} J^a \to B \otimes_{R^a} \mathrm{Sym}_{R^a}^\bullet S_1^a) \subset \mathrm{Ker}\, \mathrm{Sym}_B^\bullet \sigma$$

and since $\mathrm{Sym}_{C_*}^\bullet (C \otimes_B L)_*$ does not contain $\mathfrak{m}$-torsion, it follows that

$$\mathrm{Im}(C_* \otimes_R J \to C_* \otimes_R \mathrm{Sym}_R^\bullet S_1) \subset \mathrm{Ker}\, \mathrm{Sym}_{C_*}^\bullet \sigma'$$

which implies the claim. $\square$

Lemma 5.7.26. *Let $S_\bullet$ be a graded R-algebra generated by S_1 and $J_\bullet \subset S_\bullet$ a finitely generated graded ideal. Suppose moreover that S_1 is a finitely generated R-module. Then*

$$(\operatorname{Proj} S_\bullet \setminus \operatorname{Proj} S_\bullet / J_\bullet)^a \simeq \operatorname{Proj} S_\bullet^a \setminus \operatorname{Proj} S_\bullet^a / J_\bullet^a.$$

Proof. Follows from proposition 5.7.12 and lemma 5.7.25(ii). $\qquad\qquad\square$

5.7.27. Let $A_\bullet$ be as in definition 5.7.17(i) and $\mathscr{J}_\bullet := \oplus_{i \in \mathbb{N}} \mathscr{J}_i \subset A$ a graded ideal. We set $X := \operatorname{Proj} A_\bullet \setminus \operatorname{Proj} A_\bullet / \mathscr{J}_\bullet$ and $Y := \operatorname{Spec} A_\bullet \setminus \operatorname{Spec} A_\bullet / \mathscr{J}_\bullet A_+$, where $A_+ := \oplus_{i>0} A_i$; then Y represents the functor that assigns to every R^a-scheme T the set of all maps of R^a-algebras $\phi : A_\bullet \to \mathscr{O}_T$ such that

$$(5.7.28) \qquad\qquad \phi(\mathscr{J}_\bullet A_+) \cdot \mathscr{O}_T = \mathscr{O}_T.$$

Moreover, we have a map of R^a-schemes $\pi : Y \to X$, representing the natural transformation of functors that assigns to any T-section $\phi : A_\bullet \to \mathscr{O}_T$ of Y the induced map of $\mathscr{O}_T$-modules $\phi_1 : A_1 \otimes_R \mathscr{O}_T \to \mathscr{O}_T$. Condition (5.7.28) ensures that ϕ_1 is surjective and $\phi_1 \notin \operatorname{Proj}(A_\bullet / \mathscr{J}_\bullet)(T)$ whenever T is non-empty, hence this rule yields a well defined T-section of X. It is a standard fact that π is a $\mathbb{G}_m$-torsor. The action of $\mathbb{G}_m(T)$ on $Y(T)$ can be described explicitly as follows. To a given pair $(\phi, u) \in Y(T) \times \mathscr{O}_{T*}^\times$ one assigns the unique R^a-algebra map $\phi_u : A_\bullet \to \mathscr{O}_T$ such that $\phi_u(x) = u \cdot \phi(x)$ for every $x \in A_{1*}$.

Lemma 5.7.29. *Keep the notation of (5.7.27). The subsheaf*

$$Y_{\mathrm{sm}} := (\operatorname{Spec} A_\bullet)_{\mathrm{sm}} \setminus \operatorname{Spec}(A_\bullet / \mathscr{J}_\bullet A_+) \subset Y$$

is a subtorsor of Y for the natural $\mathbb{G}_m$-action on Y.

Proof. Clearly $\mathbb{G}_m = (\mathbb{G}_m)_{\mathrm{sm}}$; hence, by lemma 5.6.16, we have a natural identification:

$$(\mathbb{G}_m \times_{R^a} Y)_{\mathrm{sm}} \simeq \mathbb{G}_m \times Y_{\mathrm{sm}}.$$

However, the natural transformation $\mathbb{G}_m \times_{R^a} Y \to Y \times_X Y$ defined by the rule:

$$(u, \phi) \mapsto (\phi, \phi_u) \quad \text{for every local section } u \text{ of } \mathbb{G}_m \text{ and } \phi \text{ of } Y$$

is an isomorphism, therefore $(\mathbb{G}_m \times_{R^a} Y)_{\mathrm{sm}} \simeq (Y \times_{X^a} Y)_{\mathrm{sm}}$. Let T be an R^a-scheme, $\phi \in Y(T)$ and $u \in \mathbb{G}_m(T)$; it follows that $(\phi, \phi_u) \in (Y \times_X Y)_{\mathrm{sm}}(T)$ if and only if $\phi \in Y_{\mathrm{sm}}(T)$. Furthermore, it is clear that (ϕ, ϕ_u) is in the smooth locus of $Y \times_X Y$ if and only if (ϕ_u, ϕ) is (use the X-automorphism of $Y \times_X Y$ that swaps the two factors). However, the pair (ϕ_u, ϕ) can be written in the form (ψ, ψ_v) for $\psi = \phi_u$ and $v = u^{-1}$, so that ψ is in the smooth locus if and only if (ψ, ψ_v) is. *Ergo*, ϕ is in the smooth locus if and only if ϕ_u is, as required. $\qquad\square$

Definition 5.7.30. Let X and Y be as in (5.7.27). We define the *smooth locus of X* as the quotient (in the category of sheaves on $(R^a\text{-}\mathbf{Alg})^o_{\mathrm{fpqc}}$):

$$X_{\mathrm{sm}} := Y_{\mathrm{sm}} / \mathbb{G}_m.$$

The notation X_{sm} remains somewhat ambiguous, until one shows that the smooth locus of X does not depend on the choice of presentation of X as quotient of a quasi-affine almost scheme Y. While we do not completely elucidate this issue, we want at least to make the following remark:

Lemma 5.7.31. *In the situation of* (5.7.27), *suppose that $\mathscr{L}$ is a strictly invertible R^a-module, and set $B_\bullet := \oplus_{r\in\mathbb{N}} A_r \otimes_{R^a} \mathscr{L}^{\otimes r}$, $\mathscr{J}_{B_\bullet} := \oplus_{r\in\mathbb{N}} \mathscr{J}_r \otimes_{R^a} \mathscr{L}^{\otimes r}$, so $B_\bullet$ is a graded R^a-algebra and $\mathscr{J}_{B_\bullet} \subset B_\bullet$ a graded ideal. Then there are natural isomorphisms*

$$X \simeq X_B := \mathrm{Proj}(B_\bullet) \setminus \mathrm{Proj}(B_\bullet/\mathscr{J}_{B_\bullet}) \qquad X_{\mathrm{sm}} \simeq X_{B,\mathrm{sm}}.$$

Proof. We define as follows a natural isomorphism of functors $X \xrightarrow{\sim} X_B$. To every R^a-scheme T and every T-section $A_1 \otimes_{R^a} \mathscr{O}_T \to L$, we assign the strictly invertible quotient $A_1 \otimes_{R^a} \mathscr{L} \otimes_{R^a} \mathscr{O}_T \to \mathscr{L} \otimes_{R^a} L$. Set $Y_B := \mathrm{Spec}\, B_\bullet \setminus \mathrm{Spec}\,(B_\bullet/\mathscr{J}_{B_\bullet} B_+)$; it follows that $Y_B/\mathbb{G}_m \simeq X_B$. Hence, the identity $X_{\mathrm{sm}} \simeq X_{B,\mathrm{sm}}$ can be checked locally on the fpqc topology. However, $\mathscr{L}$ is locally free of rank one on $(R^a\text{-}\mathbf{Alg})^o_{\mathrm{fpqc}}$, and the claim is obvious in case $\mathscr{L}$ is free. $\qquad\square$

Theorem 5.7.32. *Suppose $\widetilde{\mathfrak{m}}$ has homological dimension ≤ 1, and that (R^a, I) is a henselian pair. Let P be an almost finitely generated projective R^a-module and $\mathscr{I}_\bullet, \mathscr{J}_\bullet \subset S_P := \mathrm{Sym}^\bullet_{R^a} P$ two graded ideals, with $\mathscr{J}_\bullet$ almost finitely generated. Set $X := \mathrm{Proj}(S_P/\mathscr{J}_\bullet) \setminus V(\mathscr{I}_\bullet)$. Then, for every finitely generated subideal $\mathfrak{m}_0 \subset \mathfrak{m}$, the image of $X_{\mathrm{sm}}(\mathrm{Spec}\, R^a/\mathfrak{m}_0 I)$ in $X_{\mathrm{sm}}(\mathrm{Spec}\, R^a/I)$ coincides with the image of $X_{\mathrm{sm}}(\mathrm{Spec}\, R^a)$.*

Proof. First of all, arguing as in the proof of claim 5.6.18 we reduce to the case where I is a tight ideal. Next, let $\overline{\sigma} : (P/\mathscr{J}_1) \otimes_{R^a} R^a/\mathfrak{m}_0 I \to \overline{L}$ be an $R^a/\mathfrak{m}_0 I$-section of X. By theorem 5.5.7 we can lift $\overline{L}$ to an almost finitely generated projective R^a-module L.

Claim 5.7.33. L is strictly invertible.

Proof of the claim. We have $\Lambda^2_{R^a} L \otimes_{R^a} R^a/\mathfrak{m}_0 I \simeq \Lambda^2_{R^a} \overline{L} = 0$ whence $\Lambda^2_{R^a} L = 0$ by Nakayama's lemma 5.1.7. Let $\mathscr{E}$ be the evaluation ideal of L; by proposition 2.4.28(i,iv) we have $\mathscr{E} \otimes_{R^a} R^a/\mathfrak{m}_0 I = R^a/\mathfrak{m}_0 I$, i.e $\mathscr{E} + \mathfrak{m}_0 I = R^a$. Set $M := R^a/\mathscr{E}$. It follows that $\mathfrak{m}_0 I M = M$, so that $M = 0$ by lemma 5.1.7, thus $\mathscr{E} = R^a$ and the claim follows from proposition 2.4.28(iv) and lemma 4.3.12. $\qquad\Diamond$

We set $S'_\bullet := \mathrm{Sym}^\bullet_{R^a}(P \otimes_{R^a} L^*)$, $\mathscr{I}'_\bullet := \oplus_{r\in\mathbb{N}} \mathscr{I}_r \otimes_{R^a} L^{*\otimes r}$, $\mathscr{J}'_\bullet := \oplus_{r\in\mathbb{N}} \mathscr{J}_r \otimes_{R^a} L^{*\otimes r}$, $B_\bullet := S'_\bullet/\mathscr{J}'_\bullet$, $Y' := \mathrm{Spec}\, B_\bullet \setminus \mathrm{Spec}\,(B_\bullet/\mathscr{I}'_\bullet B_+)$ and $X' := \mathrm{Proj}(B_\bullet) \setminus \mathrm{Proj}(B_\bullet/\mathscr{I}'_\bullet)$. By lemma 5.7.31 we have an isomorphism $\beta : X' \simeq X$ preserving the smooth loci; by inspecting the definition we find that $\beta^{-1}(\overline{\sigma})$ lies in the image of the projection $Y'_{\mathrm{sm}}(\mathrm{Spec}\, R^a/\mathfrak{m}_0 I) \to X'_{\mathrm{sm}}(\mathrm{Spec}\, R^a/\mathfrak{m}_0 I)$. Hence we can replace P by $P \otimes_{R^a} L^*$ and assume from start that $\overline{\sigma}$ lies in the image of the map $Y_{\mathrm{sm}}(\mathrm{Spec}\, R^a/\mathfrak{m}_0 I) \to X_{\mathrm{sm}}(\mathrm{Spec}\, R^a/\mathfrak{m}_0 I)$. Take $\overline{\tau} \in Y_{\mathrm{sm}}(\mathrm{Spec}\, R^a/\mathfrak{m}_0 I)$ that maps to $\overline{\sigma}$. By theorem 5.6.17(i), the image τ_0 of $\overline{\tau}$ in $(\mathrm{Spec}\, S_P/\mathscr{J}_\bullet)_{\mathrm{sm}}(\mathrm{Spec}\, R^a/I)$ lifts to a section $\tau \in (\mathrm{Spec}\, S_P/\mathscr{J}_\bullet)_{\mathrm{sm}}(\mathrm{Spec}\, R^a)$. To conclude, it suffices to show the following:

Claim 5.7.34. Suppose that I is tight. Then $\tau \in Y(\operatorname{Spec} R^a)$.

Proof of the claim. Let $A_\bullet := S_P/\mathscr{J}_\bullet$, so that $\tau : A \to R^a$ is a morphism of R^a-algebras, and set $J := \tau(A_+\mathscr{J})$; by assumption $\tau_0 \in Y(\operatorname{Spec} R^a/I)$, which means that $J \cdot (R^a/I) = (R^a/I)$, *i.e.* $J + I = R^a$. Set $M := R^a/J$; it then follows that $M = IM$, hence $M = 0$ by lemma 5.1.7; hence $JR^a = R^a$, which is the claim.   □

5.8. Lifting and descent of torsors. We begin this section with some generalities about group schemes and their linear representations, that are preliminary to our later results on liftings of torsors in an almost setting (theorems 5.8.19 and 5.8.21).

Furthermore, we apply the results of section 5.4 to derive a descent theorem for G-torsors, where G is a group scheme satisfying some fairly general conditions.

5.8.1. Let R be a ring, G an affine group scheme defined over R. We set $\mathscr{O}(G) := \Gamma(G, \mathscr{O}_G)$; so $\mathscr{O}(G)$ is an R-algebra and the multiplication map of G determines a structure of co-algebra on $\mathscr{O}(G)$. In (3.3.7) we have defined the notion of a (left or right) G-action on a quasi-coherent $\mathscr{O}_X$-module M, where X is a scheme acted on by G. We specialize now to the case where $X = \operatorname{Spec} R$, and G acts trivially on X. In such situation, a G-action on M is the same as a map of fpqc-sheaves, from the sheaf represented by G to the sheaf of automorphisms of M; *i.e.* the datum, for every R-scheme T, of a functorial group homomorphism:

(5.8.2) $$G(T) \to \operatorname{Aut}_{\mathscr{O}_T}(M \otimes_R \mathscr{O}_T).$$

Explicitly, given a map as in (5.8.2), take $T := G$; then the identity map $G \to G$ determines an $\mathscr{O}(G)$-linear automorphism β of $M \otimes_R \mathscr{O}(G)$, and one verifies easily that β fulfills the conditions of (3.3.7). Conversely, an automorphism β as in (3.3.7) extends uniquely to a well defined map of sheaves (5.8.2). Furthermore, the G-action on M can be prescribed by choosing, instead of a β as above, an R-linear map:

$$\gamma : M \to M \otimes_R \mathscr{O}(G)$$

satisfying certain identities analogous to (3.3.9) and (3.3.10); of course one can then recover the corresponding β by extension of scalars. One says that the pair (M, γ) is an $\mathscr{O}(G)$-*comodule*; more precisely, one defines right and left $\mathscr{O}(G)$-comodules, and the bijection just sketched sets up an equivalence from the category of R-modules with left G-actions to the category of right $\mathscr{O}(G)$-comodules. One says that an $\mathscr{O}(G)$-comodule is *finitely generated* (resp. *projective*) if its underlying R-module has the same property.

Lemma 5.8.3. *In the situation of* (5.8.1), *suppose that R is a Dedekind domain and that $\mathscr{O}(G)$ is flat over R. Then:*
  (i) *Every $\mathscr{O}(G)$-comodule is the filtered union of its finitely generated sub-$\mathscr{O}(G)$-comodules.*
  (ii) *Every finitely generated $\mathscr{O}(G)$-comodule is quotient of a finitely generated projective $\mathscr{O}(G)$-comodule.*

Proof. (i) is [72, §1.5, Cor.] and (ii) is [72, §2.2, Prop.3].   □

5.8.4. Let R be a ring, G be a flat group scheme of finite presentation over R, and suppose that G admits a closed imbedding as a subgroup scheme $G \subset \mathrm{GL}_n$. We suppose moreover that the quotient $X_G := \mathrm{GL}_n/G$ (for the right action of G on GL_n) is representable by a quasi-projective R-scheme, and that X_G admits an ample GL_n-equivariant line bundle. Then X_G is necessarily of finite presentation by [30, lemme 17.7.5]. Furthermore, X_G is smooth; indeed X_G is flat over $\mathrm{Spec}\,R$, hence smoothness can be checked on the geometric fibres over the points of $\mathrm{Spec}\,R$, which reduces to the case where R is an algebraically closed field. In such case, smoothness over R is the same as regularity and the latter follows since GL_n is regular and the quotient map $\mathrm{GL}_n \to X_G$ is faithfully flat. The following lemma 5.8.5 provides plenty of examples of the situation envisaged in this paragraph.

Lemma 5.8.5. *Suppose that R is a Dedekind domain and G is a flat affine group scheme of finite type over R. Then G fulfills the conditions of (5.8.4).*

Proof. The proof is obtained by assembling several references to the existing literature. To start with, let P be a finitely generated subrepresentation of the left (or right) regular representation of G on $\mathcal{O}(G)$ which generates $\mathcal{O}(G)$ as an R-algebra. By [19, 1.4.5] P is a finitely generated projective R-module and the action of G on P is faithful, in the sense that it gives a closed immersion: $G \to \mathrm{Aut}_R(P)$. Taking the direct sum with a trivial representation one gets a closed immersion of R-group schemes: $G \to \mathrm{GL}_{n,R}$. By a theorem of M.Artin (see [1, 3.1.1]) the quotient fppf sheaf GL_n/G is represented by an algebraic space X of finite presentation over $\mathrm{Spec}\,R$; X is separated, since it comes from a closed equivalence relation. Then by [1, Th.4.C], X is a scheme. Let K be the fraction field of R; by classical results of Chevalley and Chow, $X_K := GL_{n,K}/G_K$ is quasi-projective over K and by [65, VIII.2] it follows that GL_n/G is quasi-projective over $\mathrm{Spec}\,R$. The ample line bundle $\mathscr{L}$ on GL_n/G that one gets always has a GL_n-linearization. This follows from the proof of [73, lemma 1.2]. For our purpose it suffices to know that the weaker assertion that some tensor multiple of $\mathscr{L}$ is GL_n-linearizable. We may assume that $\mathscr{L}$ is trivial on the identity section $S := \mathrm{Spec}\,R \to \mathrm{GL}_{n,R}/G$.

Claim 5.8.6. $\mathrm{Ker}(\mathrm{Pic}\,\mathrm{GL}_{n,R} \xrightarrow{e^*} \mathrm{Pic}\,S) = 0$ (where $e : S \to \mathrm{GL}_{n,R}$ is the identity section).

Proof of the claim. $\mathrm{GL}_{n,K}$ is open in an affine space, hence it has trivial Picard group, hence every divisor on $\mathrm{GL}_{n,R}$ is linearly equivalent to a divisor whose irreducible components do not dominate S, *i.e.* a pull-back of a divisor on S, which gives the claim.   ◇

The assertion now follows from [65, VII, Prop.1.5].   □

Lemma 5.8.7. *Under the assumptions of (5.8.4), there is a GL_n-equivariant isomorphism of R-schemes:*

$$(5.8.8) \qquad X_G \xrightarrow{\sim} \mathrm{Proj}((\mathrm{Sym}_R^\bullet \mathscr{P})/\mathscr{I}_\bullet) \setminus V(\mathscr{J}_\bullet)$$

where $\mathscr{P}$ is a finitely generated projective GL_n-comodule, and $\mathscr{I}_\bullet, \mathscr{J}_\bullet \subset \mathrm{Sym}_R^\bullet \mathscr{P}$ are two finitely generated graded ideals, that are sub-$\mathcal{O}(G)$-comodules of $\mathrm{Sym}_R^\bullet \mathscr{P}$.

Proof. Let $\mathscr{L}$ be an ample GL_n-equivariant line bundle on X_G. Since X_G is of finite type, we can find $n \in \mathbb{N}$ large enough, so that $\mathscr{L}^{\otimes n}$ is very ample ([27, Ch.II, Cor.4.5.11]). We can then replace $\mathscr{L}$ by $\mathscr{L}^{\otimes n}$ and therefore assume that X_G admits a locally closed imbedding in $\mathbb{P}_R(\Gamma(X_G, \mathscr{L}))$. By [27, Ch.II, Prop.4.4.1] we can find a finitely generated R-submodule $W \subset \Gamma(X_G, \mathscr{L})$ such that X_G already imbeds into $\mathbb{P}_R(W)$. Let $f_1, ..., f_k$ be a finite set of generators of the R-module W. By lemma 5.8.3(i) we can find a finitely generated $\mathbb{Z}$-module $W_0 \subset \Gamma(X_G, \mathscr{L})$ which is a $\mathscr{O}(\mathrm{GL}_{n,\mathbb{Z}})$-comodule containing the $(f_i \mid i \leq k)$. Up to replacing W by the R-module generated by W_0, we can further assume that W is an $\mathscr{O}(\mathrm{GL}_n)$-comodule.

Claim 5.8.9. W is the quotient of a projective $\mathscr{O}(\mathrm{GL}_n)$-comodule of finite type L.

Proof of the claim. By the foregoing we can assume that W is generated by a finitely generated $\mathbb{Z}$-submodule $W_0 \subset W$ which is an $\mathscr{O}(\mathrm{GL}_{n,\mathbb{Z}})$-comodule. By lemma 5.8.3(ii) we can write W_0 as a quotient of a projective $\mathscr{O}(\mathrm{GL}_{n,\mathbb{Z}})$-comodule L_0; hence W is a quotient of $L := L_0 \otimes_\mathbb{Z} R$. $\diamondsuit$

It follows that X_G is a locally closed subscheme of $\mathbb{P}(L)$. Let Y be the schematic closure of the image of X_G; we can write $Y \simeq \mathrm{Proj}((\mathrm{Sym}_R^\bullet L)/I_\bullet)$, where $I_\bullet$ is the graded ideal of sections vanishing on Y, so $I_\bullet$ is an $\mathscr{O}(G)$-comodule. Furthermore, we have $Y \setminus X_G = V(J_\bullet)$, where $J_\bullet \subset \mathrm{Sym}_R^\bullet L$ is the graded ideal defined by the reduced scheme structure on $Y \setminus X_G$, also an $\mathscr{O}(G)$-comodule. Arguing as in the foregoing we can write $J_\bullet = \bigcup_\alpha J_\alpha$, where J_α runs over the filtered family of the finitely generated $\mathscr{O}(G)$-sub-comodules of J. By quasi-compactness one shows easily that we have equality of underlying sets: $Y \setminus X_G = V(J_\alpha) \cap Y$ for J_α large enough; we set $\mathscr{J}_\bullet := J_\alpha$. Likewise, as X_G is of finite presentation, we can find a finitely generated $\mathscr{O}(G)$-subcomodule $\mathscr{I}_\bullet \subset I_\bullet$ such that $(\mathbb{P}(L) \setminus V(\mathscr{J}_\bullet)) \cap V(I_\bullet) = (\mathbb{P}(L) \setminus V(\mathscr{J}_\bullet)) \cap V(\mathscr{I}_\bullet)$. □

5.8.10. Under the assumptions of (5.8.4), let T be a GL_n-torsor. (All torsors are right torsors, unless otherwise stated.) Then G acts on T via the imbedding $G \subset \mathrm{GL}_n$, and the functor T/G is representable by a smooth R-scheme.

Lemma 5.8.11. *The functor T/G : R-**Scheme** $\rightarrow$ **Set** is naturally isomorphic to the functor that assigns to every R-scheme S the set of all pairs (H, ω), where H is a G-torsor on S_{fpqc} and $\omega : H \times^G \mathrm{GL}_n \xrightarrow{\sim} T$ is an isomorphism of GL_n-torsors.*

Proof. Indeed, let $\sigma : S \rightarrow T/G$ be any morphism of R-schemes; set $H_\sigma := S \times_{T/G} T$, which is a G-torsor on S_{fpqc}. The natural morphism $H_\sigma \rightarrow T$ induces a well defined isomorphism of GL_n-torsors $\omega_\sigma : H_\sigma \times^G \mathrm{GL}_n \xrightarrow{\sim} T$. Conversely, given (H, ω) over S, we deduce a G-equivariant morphism $H \rightarrow T$; after taking quotients by the right action of G, we deduce a morphism $S \simeq H/G \rightarrow T/G$. It is easy to verify that these two rules define mutually inverse natural transformations of functors, whence the claim. □

5.8.12. The right GL_n-torsor T is endowed by a natural left action by the group $\mathrm{Aut}_{\mathrm{GL}_n}(T)$. The latter is a group scheme locally isomorphic to GL_n in the Zariski topology; especially it is smooth over R. After taking the quotient of T by the right action of G we deduce a left action of $\mathrm{Aut}_{\mathrm{GL}_n}(T)$ on T/G :

$$(5.8.13) \qquad\qquad \mathrm{Aut}_{\mathrm{GL}_n}(T) \times_R T/G \to T/G.$$

Theorem 5.8.14. *Resume the assumptions of proposition 5.4.21. Let G be a smooth group scheme over $\mathrm{Spec}\, R[t^{-1}]$ which satisfies (relative to $R[t^{-1}]$) the conditions of (5.8.4). Then the natural map*

$$H^1(\mathrm{Spec}\, R[t^{-1}]_{\mathrm{fpqc}}, G) \to H^1(\mathrm{Spec}\, R^\wedge[t^{-1}]_{\mathrm{fpqc}}, G)$$

is a bijection.

Proof. We begin with the following special case:

Claim 5.8.15. The theorem holds when $G = \mathrm{GL}_n$.

Proof of the claim. A GL_n torsor over a scheme X is the same as a locally free $\mathcal{O}_X$-module of rank n. Hence the assertion is just a restatement of corollary 5.4.41. ◊

For a given GL_n-torsor T, let $T^\wedge := T \times_{R[t^{-1}]} R^\wedge[t^{-1}]$ and denote by

$$H_T^1 \subset H^1(\mathrm{Spec}\, R[t^{-1}]_{\mathrm{fpqc}}, G)$$

the subset consisting of all classes of G-torsors H such that $H \times^G \mathrm{GL}_n \simeq T$; define likewise the subset $H_{T^\wedge}^1 \subset H^1(\mathrm{Spec}\, R^\wedge[t^{-1}]_{\mathrm{fpqc}}, G)$. According to claim 5.8.15 it suffices to show:

Claim 5.8.16. The restriction $H_T^1 \to H_{T^\wedge}^1$ is a bijection.

Proof of the claim. The morphism (5.8.13) defines (as in [51, Ch.VI,§2.5]) a groupoid

$$[T/G] := [T/G, \mathrm{Aut}_{\mathrm{GL}_n}(T) \times_R T/G, s, t, c, \iota]$$

of quasi-projective $R[t^{-1}]$-schemes; $[T/G]$ is locally isomorphic to the analogous groupoid $[X_G]$ in the Zariski topology, hence it satisfies the smoothness condition of theorem 5.4.37 (see 5.8.4) and we can therefore apply theorem 5.4.37. The assertion follows directly after one has remarked that, for every $R[t^{-1}]$-scheme S, the set $\pi_0([T/G](S))$ is in natural bijection with the set of isomorphism classes of G-torsors H on S_{fpqc} such that $H \times^G \mathrm{GL}_n \simeq T \times_{\mathrm{Spec}\, R[t^{-1}]} S$. □

5.8.17. Finally we come to the lifting problems for G^a-torsors. Let A be a V^a-algebra, set $R := A_*$ and let G a smooth affine group scheme of finite type over $\mathrm{Spec}\, R$. We suppose that G is a closed subgroup scheme of $\mathrm{GL}_{n,R}$, for some $n \in \mathbb{N}$. Clearly G^a is a group scheme over $S := \mathrm{Spec}\, A$.

Lemma 5.8.18. *Keep the notation of (5.8.17). The category of GL_n-torsors over S_{fpqc} is naturally equivalent to the category whose objects are the almost projective $\mathcal{O}_S$-modules of constant rank n, and whose morphisms are the linear isomorphisms.*

Proof. Let $\mathcal{O}_*$ be the structure sheaf of S_{fpqc}, i.e. the sheaf of rings defined by the rule: $T \mapsto \mathcal{O}_{T*}$. To a given GL_n-torsor P we assign the $\mathcal{O}_*$-module $F_P := P \times^{\mathrm{GL}_n} \mathcal{O}_*^n$ (for the natural left action of GL_n on $\mathcal{O}_*^n$). Locally in S_{fpqc} the sheaf F_P is a free $\mathcal{O}_*$-module of rank n, and the assignment $P \mapsto F_P$ is an equivalence from GL_n-torsors to the category of all such locally free $\mathscr{O}_*$-modules F of rank n, whose quasi-inverse is the rule: $F \mapsto \underline{\mathrm{Iso}}_{\mathscr{O}_*}(\mathscr{O}_*^n, F)$. On the other hand, theorem 4.4.24 says that any almost projective $\mathscr{O}_S$-module M of constant rank n defines a locally free $\mathscr{O}_*$-module on S_{fpqc}, by the rule $T \mapsto (\mathscr{O}_T \otimes_{\mathscr{O}_S} M)_*$, and by faithfully flat descent it is also clear that the resulting functor is an equivalence as well. □

Theorem 5.8.19. *Assume that* $\mathrm{hom.dim}_V \widetilde{\mathfrak{m}} \leq 1$. *Let S and G be as in* (5.8.17), *and $I \subset \mathscr{O}_S$ an ideal such that the pair $(\mathscr{O}_S, I)$ is henselian. Let also $\mathfrak{m}_0 \subset \mathfrak{m}$ be a finitely generated subideal, and set $S_n := \operatorname{Spec} \mathscr{O}_S / \mathfrak{m}_0^n I$ for $n = 0, 1$. Let P and Q be two G^a-torsors over S, and suppose that $\overline{\beta} : P \times_S S_1 \xrightarrow{\sim} Q \times_S S_1$ is an isomorphism of $G^a \times_S S_1$-torsors. Then $\beta \times_S 1_{S_0}$ lifts to an isomorphism $\beta : P \xrightarrow{\sim} Q$ of G^a-torsors.*

Proof. Let $\underline{\mathrm{Iso}}(P, Q)$ denote the functor that assigns to every affine S-scheme T the set of isomorphisms of $G^a \times_S T$-torsors $P \times_S T \xrightarrow{\sim} Q \times_S T$. It is easy to see that $\underline{\mathrm{Iso}}(P, Q)$ is a sheaf for the fpqc topology of S. Since, locally in S_{fpqc}, this sheaf is represented by the almost scheme G^a, it follows by faithfully flat descent that $\underline{\mathrm{Iso}}(P, Q)$ is represented by an affine S-scheme, which we denote by the same name. To the (right) G^a-torsor P we associate a left G^a-torsor P', whose underlying S-scheme is the same as P, and whose G^a-action is defined by the rule: $g \cdot x := xg^{-1}$ for every S-scheme T, every $g \in G^a(T)$ and every $x \in P(T)$. Furthermore, we let H denote the left and right G^a-torsor whose underlying S-scheme is G^a, and whose left and right G^a-actions are induced by the multiplication map $G^a \times_S G^a \to G^a$.

Claim 5.8.20. There is a natural isomorphism of sheaves on S_{fpqc}:

$$\omega : \underline{\mathrm{Iso}}(P, Q) \simeq Q \times^{G^a} H \times^{G^a} P'.$$

Proof of the claim. Recall that the meaning of the right-hand side is as follows. For every S-scheme T, one considers the presheaf whose T-sections are the equivalence classes of triples $(x, h, y) \in Q(T) \times H(T) \times P'(T)$, modulo the equivalence relation $\sim$ such that $(x, g_1 \cdot h \cdot g_2, y) \sim (x \cdot g_1, h, g_2 \cdot y)$ for every such (x, h, y) and every $g_1, g_2 \in G^a(T)$. Then $Q \times^{G^a} H \times^{G^a} P'$ is the sheaf associated to this presheaf. The sought isomorphism is defined as follows. Let T be an S-scheme, and β a T-section of $\underline{\mathrm{Iso}}(P, Q)$. We pick a covering morphism $U \to T$ in S_{fpqc} such that $P(U) \neq \varnothing$; let $y \in P(U)$; we set $\omega(\beta, y) := (\beta(y), 1, y)$. If now $z \in P(U)$ is any other section, we have $z = yg$ for some $g \in G^a(U)$, therefore: $\omega(\beta, y) = (\beta(y), g \cdot g^{-1}, y) \sim (\beta(y) \cdot g, 1, g^{-1}y) = (\beta(yg), 1, yg) = \omega(\beta, z)$, *i.e.* the equivalence class of $\omega(\beta, y)$ does not depend on the choice of y, hence we have a well defined U-section $\omega(\beta)$ of $Q \times^{G^a} H \times^{G^a} P'$. Furthermore, let $p_i : U \times_T U \to U$, $i = 1, 2$ be the two projections; the sections $p_1^* \omega(\beta, y)$ and $p_2^* \omega(\beta, y)$ differ by an element of $G^a(U \times_T U)$ so, by the same argument, they lie in the same equivalence class, which means that actually $\omega(\beta)$ comes from T, as required. We leave to the reader the verification that ω thus defined is an isomorphism. ◇

Composing the closed imbedding $G^a \subset \mathrm{GL}_{n,S}$ with the standard group homomorphism $\mathrm{GL}_{n,S} \subset SL_{n+1,S}$, we can view G^a as a closed subgroup scheme of $SL_{n,S}$, whence a closed G^a-equivariant imbedding of S-schemes $G^a \subset M_{n,S} :=$

Spec $\mathscr{O}_S[x_{ij} \mid i,j \leq n]$ (defined by a finitely generated ideal). From claim 5.8.20 we derive a closed imbedding of S-schemes:

$$\underline{\mathrm{Iso}}(P,Q) \subset X := Q \times^{G^a} M_{n,S} \times^{G^a} P'$$

(where G^a acts on the left and right on $M_{n,S}$ in the obvious way, and the imbedding is defined by an almost finitely generated ideal). X is the spectrum of a graded algebra, fpqc-locally isomorphic to a polynomial algebra, hence it is of the form Spec $(\mathrm{Sym}^{\bullet}_{\mathscr{O}_S} M)$, for an almost finitely generated projective $\mathscr{O}_S$-module M; especially, X is almost finitely presented over S, whence the same holds for $\underline{\mathrm{Iso}}(P,Q)$. Now, the given $\overline{\beta}$ gives a section of $\underline{\mathrm{Iso}}(P,Q)$ over S_1, so the theorem is an immediate consequence of theorem 5.6.17(i).   □

Theorem 5.8.21. *Keep the notation and assumptions of theorem 5.8.19, and suppose furthermore that I is tight, and that G fulfills the conditions of (5.8.4). Let P_0 be any G^a-torsor over S_0. Then there exists a G^a-torsor P over S with an isomorphism of G^a-torsors: $P_0 \simeq P \times_S S_0$.*

Proof. By assumption there exists a finitely generated subideal $\mathfrak{m}_0 \subset \mathfrak{m}$ and an integer $n > 0$ such that $I^n \subset \mathfrak{m}_0 \mathscr{O}_S$; then by theorem 3.3.38(ii) we can lift P_0 to a G-torsor over $S_1 := \mathscr{O}_S/\mathfrak{m}_0 I$. We form the GL_n-torsor $\overline{Q} := \overline{P} \times^{G^a} \mathrm{GL}_n$. By lemma 5.8.18, $\overline{Q}$ is the same as an almost projective $\mathscr{O}_{S_1}$-module of constant rank n; by theorem 5.5.7(i) we can then find an almost projective $\mathscr{O}_S$-module Q that lifts $\overline{Q}$, and then a standard application of Nakayama's lemma 5.1.7 shows that Q has constant rank equal to n (cp. the proof of theorem 5.7.32). Thus, Q is a GL_n-torsor with an isomorphism of GL_n-torsors $\omega : \overline{P} \times^{G^a} \mathrm{GL}_n \xrightarrow{\sim} Q \times_S S_1$. By (the almost version of) lemma 5.8.11, the datum of $(\overline{P}, \omega)$ is the same as the datum of a morphism $\overline{\sigma} : S_1 \to Q/G^a$ of sheaves on S_{fpqc}. The contention is that $\overline{\sigma} \times_S S_0$ lifts to a morphism $\sigma : S \to Q/G^a$. However, we have a natural isomorphism

$$Q/G^a \simeq Q \times^{\mathrm{GL}_n} (\mathrm{GL}_n/G^a) \simeq Q \times^{\mathrm{GL}_n} (\mathrm{GL}_n/G)^a.$$

By lemmata 5.7.26 and 5.8.7 we deduce an isomorphism:

$$Q/G^a \simeq \mathrm{Proj}((\mathrm{Sym}^{\bullet}_{\mathscr{O}_S} \mathscr{P}')/\mathscr{I}'_{\bullet}) \setminus V(\mathscr{J}'_{\bullet})$$

with $\mathscr{P}' := Q \times^{\mathrm{GL}_n} \mathscr{P}$, $\mathscr{I}'_{\bullet} := Q \times^{\mathrm{GL}_n} \mathscr{I}_{\bullet}$ and $\mathscr{J}'_{\bullet} := Q \times^{\mathrm{GL}_n} \mathscr{J}_{\bullet}$, where $\mathscr{P}$, $\mathscr{I}_{\bullet}$ and $\mathscr{J}_{\bullet}$ are as in (5.8.8). By lemma 3.2.25(ii),(iii) and remark 3.2.26(i),(ii) we see that $\mathscr{P}'$ is almost finitely generated projective and $\mathscr{I}'_{\bullet}$, $\mathscr{J}'_{\bullet}$ are almost finitely generated. Finally, Q/G^a is smooth since X_G is (see 5.8.4), so the existence of the section σ follows from theorem 5.7.32.   □

6. VALUATION THEORY

This chapter is an extended detour into valuation theory. The first two sections contain nothing new, and are only meant to gather in a single place some useful material that is known to experts, but for which satisfactory references are hard to find. The main theme of sections 6.3 through 6.5 is the study of the cotangent complex of an extension of valuation rings. One of the main results states that for any such extension $V \subset W$ the homology of $\mathbb{L}_{W/V}$ vanishes in degrees > 1 and is a torsion-free module in degree one (theorem 6.5.12). This assertion can be appreciated if one recalls that, for a finite type noetherian ring homomorphism $\phi : R \to S$, the same Tor-dimension condition holds for $\mathbb{L}_{S/R}$ if and only if ϕ is a local complete intersection. Hence, our theorem means that an extension of valuations behaves as an inductive limit of maps of local complete intersection. Another statement of the same nature is the following: suppose that k is a perfect field, and let W be a valuation ring containing k; then we show that $\Omega_{W/k}$ is a torsion-free W-module (corollary 6.5.21). Notice that both assertions would be an easy consequence of the existence of local uniformization (see the introduction of chapter 7); local uniformization was known classically for valuations centered at excellent local domains of equal characteristic zero and in some low dimensional cases, and is the subject of current research. Regardless of the current state of that research, our methods enable us to prove unconditionally at least these weaker results, as well as several other statements and variants for logarithmic differentials.

Furthermore, consider a finite separable extension $K \subset L$ of henselian valued fields of rank one; it is not difficult to see that the corresponding extension of valuation rings $K^+ \subset L^+$ is almost finite, hence one can define the different ideal $\mathscr{D}_{L^+/K^+}$ as in chapter 4. In case the valuation of K is discrete, it is well known that the length of the module of relative differentials Ω_{L^+/K^+} equals the length of $L^+/\mathscr{D}_{L^+/K^+}$; theorem 6.3.23 generalizes this identity to the case of arbitrary rank one valuations; notice that in this case the usual notion of length won't do, since the modules under consideration are only almost finitely generated.

Section 6.6 ties up with earlier work of Coates and Greenberg [21], in which the notion of *deeply ramified extension* of a local field was introduced, and applied to the study of p-divisible groups attached to abelian varieties defined over such p-adic fields. Essentially, section 2 of [21] rediscovers the results of Fresnel and Matignon [38], although via a different route, closer to the original treatment of Tate in [74]. In particular, an algebraic extension E of K is deeply ramified if and only if $\mathscr{D}_{E^+/K^+} = (0)$ according to the terminology of [38]. We adopt Coates and Greenberg's terminology for our section 6.6, and we give some complements which were not observed in [21]; notably, proposition 6.6.2, which we regard as the ultimate generalization of the one-dimensional case of the almost purity theorem. Our proof generalizes Faltings' method, which relied on the above mentioned relationship between differentials and the different ideal; of course, in the present setting we need to appeal to our theorem 6.3.23, rather than estimating lengths the way Faltings did. Finally, we extend the definition of deeply ramified extension to include valued fields of arbitrary rank.

6.1. **Ordered groups and valuations.** In this section we gather some generalities on valuations and related ordered groups, which will be used in later sections.

6.1.1.   As usual, a *valued field* $(K, | \cdot |_K)$ consists of a field K endowed with a surjective group homomorphism $| \cdot |_K : K^\times \to \Gamma_K$ onto a totally ordered abelian group $(\Gamma_K, \leq)$, such that

$$(6.1.2) \qquad |x + y|_K \leq \max(|x|_K, |y|_K)$$

whenever $x + y \neq 0$. We denote by 1 the neutral element of Γ_K, and the composition law of Γ_K will be denoted by: $(x, y) \mapsto x \cdot y$. It is customary to extend the map $| \cdot |_K$ to the whole of K, by adding a new element 0 to the set Γ_K, and setting $|0| := 0$.

One can then extend the ordering of Γ_K to $\Gamma_K \cup \{0\}$ by declaring that 0 is the smallest element of the resulting ordered set. In this way, (6.1.2) holds for every $x, y \in K$. The map $| \cdot |_K$ is called the *valuation* of K and Γ_K is its *value group*.

6.1.3.   An *extension of valued fields* $(K, | \cdot |_K) \subset (E, | \cdot |_E)$ consists of a field extension $K \subset E$, and a valuation $| \cdot |_E : E \to \Gamma_E \cup \{0\}$ together with an imbedding $j : \Gamma \subset \Gamma_E$, such that the restriction to K of $| \cdot |_E$ equals $j \circ | \cdot |_K$.

Example 6.1.4. Let $| \cdot | : K \to \Gamma \cup \{0\}$ be a valuation on the field K.

(i) Given a field extension $K \subset E$, it is known that there always exist valuations on E which extend $| \cdot |$ (cp. [16, Ch.VI, §1, n.3, Cor.3]).

(ii) If the field extension $K \subset E$ is algebraic and purely inseparable, then the extension of $| \cdot |$ is unique. (cp. [16, Ch.VI, §8, n.7, Cor.2]).

(iii) We can construct extensions of $| \cdot |$ on the polynomial ring $K[X]$, in the following way. Let Γ' be an ordered group with an imbedding of ordered groups $\Gamma \subset \Gamma'$. For every $x_0 \in K$, and every $\rho \in \Gamma'$, we define the *Gauss valuation*

$$| \cdot |_{(x_0, \rho)} : K[X] \to \Gamma' \cup \{0\}$$

centered at x_0 and with radius ρ (cp. [16, Ch.VI, §10, n.1, Lemma 1]) by the rule:

$$a_0 + a_1(X - x_0) + \dots + a_n(X - x_0)^n \mapsto \max\{|a_i| \cdot \rho^i \mid i = 0, 1, \dots, n\}.$$

(iv) The construction of (iii) can be iterated : for instance, suppose that we are given a sequence of elements $\rho := (\rho_1, \rho_2, \dots, \rho_k)$ of the ordered group Γ'. Then we can define a Gauss valuation $| \cdot |_{(0, \rho)}$ on the fraction field of $K[X_1, X_2, \dots, X_k]$, with values in Γ', by the rule: $\sum_{\alpha \in \mathbb{N}^k} a_\alpha X^\alpha \mapsto \max\{|a_\alpha| \cdot \rho^\alpha \mid \alpha \in \mathbb{N}^k\}$.

(v) Suppose again it is given an ordered group Γ' with an imbedding of ordered groups $\Gamma \subset \Gamma'$. Let $T \subset \Gamma'/\Gamma$ be a finite torsion subgroup, say $T \simeq \mathbb{Z}/n_1\mathbb{Z} \oplus \dots \oplus \mathbb{Z}/n_k\mathbb{Z}$. For every $i \leq k$, pick an element $\gamma_i \in \Gamma'$ whose class in Γ'/Γ generates the direct summand $\mathbb{Z}/n_i\mathbb{Z}$ of T. Let $x_i \in K$ such that $|x_i| = \gamma_i^{n_i}$. For every $i = 1, \dots, k$ pick an element y_i in a fixed algebraic closure K^a of K, such that $y_i^{n_i} = x_i$; then the field $E := K(y_1, \dots, y_k)$ has degree over K equal to the order of T, and it admits a unique valuation $| \cdot |_E$ extending $| \cdot |$. Of course, $|y_i|_E = \gamma_i$ for every $i \leq k$.

6.1.5.   We want to explain a construction that is a simultaneous generalization of examples 6.1.4(iv),(v). Suppose it is given a datum $\mathfrak{G} := (G, j, N, \leq)$ consisting of:

(a) an abelian group G with an imbedding $j : K^\times \hookrightarrow G$ such that $G/j(V^\times)$ is torsion-free;

(b) a subgroup N of $\overline{G} := G/j(V^\times)$ such that the natural map:

$$\Gamma \xrightarrow{\sim} K^\times/V^\times \to \Gamma_{\mathfrak{G}} := \overline{G}/N$$

is injective;

(c) a total ordering $\leq$ on $\Gamma_{\mathfrak{G}}$ such that the injective map: $\Gamma \to \Gamma_{\mathfrak{G}}$ is order-preserving.

Let us denote by $K[G]$ (resp. $K[K^\times]$) the group algebra over K of the abelian group G (resp. $K^\times$). Any element of $K[G]$ can be written uniquely as a formal linear combination $\sum_{g \in G} a_g \cdot [g]$, where $a_g \in K$ for every $g \in G$, and $a_g = 0$ for all but a finite number of $g \in G$. We augment $K[K^\times]$ over K via the K-algebra homomorphism

(6.1.6) $K[K^\times] \to K$: $[a] \mapsto a$ for every $a \in K^\times$.

Then we let $K[\mathfrak{G}] := K[G] \otimes_{K[K^\times]} K$, where the $K[K^\times]$-algebra structure on K is defined by the augmentation (6.1.6). It is easy to verify that $K[\mathfrak{G}]$ is the maximal quotient algebra of $K[G]$ that identifies the classes of $[g \cdot j(a)]$ and $a \cdot [g]$, for every $g \in G$ and $a \in K^\times$. Pick, for every class $\gamma \in G/K^\times$, a representative $g_\gamma \in G$. It follows that every element of $K[\mathfrak{G}]$ can be written uniquely as a formal K-linear combination $\sum_{\gamma \in G/K^\times} a_\gamma \cdot [g_\gamma]$. We define a map $|\cdot|_{\mathfrak{G}} : K[\mathfrak{G}] \to \Gamma_{\mathfrak{G}} \cup \{0\}$ by the rule:

(6.1.7) $$\sum_{\gamma \subset G/K^\times} a_\gamma \cdot [g_\gamma] \mapsto \max_{\gamma \in G/K^\times} |a_\gamma| \cdot |g_\gamma|$$

where $|g_\gamma| \in \Gamma_{\mathfrak{G}}$ denotes the class of g_γ. One verifies easily that $|\cdot|_{\mathfrak{G}}$ does not depend on the choice of representatives g_γ. Indeed, if $(h_\gamma \mid \gamma \in G/K^\times)$ is another choice then, for every $\gamma \in G/K^\times$ we have $g_\gamma = j(x_\gamma) \cdot h_\gamma$ for some $x_\gamma \in K^\times$; therefore $[g_\gamma] = x_\gamma \cdot [h_\gamma]$ and $|g_\gamma| = |x_\gamma| \cdot |h_\gamma|$.

Lemma 6.1.8. $K[\mathfrak{G}]$ *is an integral domain, and* $|\cdot|_{\mathfrak{G}}$ *extends to a valuation:*

$$|\cdot|_{\mathfrak{G}} : K(\mathfrak{G}) := \mathrm{Frac}(K[\mathfrak{G}]) \to \Gamma_{\mathfrak{G}} \cup \{0\}.$$

Proof. Let $(G_\alpha \mid \alpha \in I)$ be the filtered system of the subgroups G_α of G such that $K^\times \subset G_\alpha$ and $G_\alpha/K^\times$ is finitely generated. Each G_α defines a datum $\mathfrak{G}_\alpha := (G_\alpha, j, N \cap (G_\alpha/j(V^\times)), \leq)$, and clearly $K[\mathfrak{G}] = \operatorname*{colim}_{\alpha \in I} K[\mathfrak{G}_\alpha]$. We can therefore reduce to the case where $G/K^\times$ is finitely generated. Write $G/K^\times = T \oplus F$, where T is a torsion group and F is torsion-free. There exist unique subgroups $\widetilde{T}, \widetilde{F} \supset K^\times$ in G with $\widetilde{T}/K^\times = T$ and $\widetilde{F}/K^\times = F$. Let

$$\mathfrak{G}_T := (\widetilde{T}, j, \{0\}, \leq) \quad \text{and} \quad \mathfrak{G}_F := (\widetilde{F}, j, N \cap \widetilde{F}/j(V^\times), \leq)$$

be the corresponding data. The functor $H \mapsto K[H]$ from abelian monoids to K-algebras preserves colimits, since it is left adjoint to the forgetful functor from K-algebras to abelian monoids; it follows easily that $K[\mathfrak{G}] \simeq K[\mathfrak{G}_T] \otimes_K K[\mathfrak{G}_F]$. By inspecting (6.1.7), one can easily show that $K[\mathfrak{G}_T]$ is of the type of example 6.1.4(v) and $K[\mathfrak{G}_F]$ is of the type of example 6.1.4(iv). Especially, $K[\mathfrak{G}]$ is a domain, and $|\cdot|_\mathfrak{G}$ is a Gauss valuation of a Laurent polynomial algebra over the finite field extension $K[\mathfrak{G}_T]$ of K.   □

The next result shows that an arbitrary valuation is always "close" to some Gauss valuation.

Lemma 6.1.9. *Let* $(E, |\cdot|_E)$ *be a valued field extension of* $(K, |\cdot|)$. *Let* $x \in E \setminus K$, *and let* $(a_i \mid i \in I)$ *be a net of elements of* K *(indexed by the directed set* $(I, \leq)$*) with the following property. For every* $b \in K$ *there exists* $i_0 \in I$ *such that* $|x - a_i|_E \leq |x - b|_E$ *for every* $i \geq i_0$. *Let* $f(X) \in K[X]$ *be a polynomial that splits in* $K[X]$ *as a product of linear polynomials. Then there exists* $i_0 \in I$ *such that* $|f(x)|_E = |f(X)|_{(a_i, |x-a_i|_E)}$ *for every* $i \geq i_0$.

Proof. To prove the claim, it suffices to consider the case when $f(X) = X - b$ for some $b \in K$. However, by the ultrametric inequality we have

$$\max(|x - a_i|_E, |x - b|_E) = \max(|x - a_i|_E, |a_i - b|) \quad \text{for every } i \in I.$$

Then the assumption says that

$$|x - b|_E = \max(|x - a_i|_E, |b - a_i|) = |X - b|_{(a_i, |x-a_i|_E)}$$

for every sufficiently large $i \in I$.   □

6.1.10.   To see how to apply lemma 6.1.9, let us consider the case where K is algebraically closed and $E = K(X)$, the field of fractions of the free K-algebra on one generator, which we suppose endowed with some valuation $|\cdot|_E$ with values in Γ_E. We apply lemma 6.1.9 to the element $x := X \in E$. Suppose first that there exists an element $a \in K$ that minimizes the function $K \to \Gamma_E : b \mapsto |X - b|_E$. In this case the trivial net $\{a\}$ fulfills the condition of the lemma. Since every polynomial of $K[X]$ splits over K, we see that $|\cdot|_E$ *is the Gauss valuation centered at* a *and with radius* $|X - a|_E$. Suppose, on the other hand, that the function $b \mapsto |X - b|_E$ does not admit a minimum (then $|X - b|_E \in \Gamma$; cp. the proof of proposition 6.5.9). It will still be possible to choose a net of elements $\{a_i \mid i \in I\}$ fulfilling the conditions of lemma 6.1.9 (indexed, for instance, by a subset of the totally ordered set Γ). Then $|\cdot|_E$ is determined by the identity:

$$|f(X)|_E = \lim_{i \in I} |f(X)|_{(a_i, |X-a_i|_E)} \quad \text{for every } f(X) \in E.$$

6.1.11.   Given a valuation $|\cdot|$ on a field K, the subset $K^+ := \{x \in K \mid |x| \leq 1\}$ is a *valuation ring* of K, i.e., a subring of K such that, for every $x \in K \setminus \{0\}$, either $x \in K^+$ or $x^{-1} \in K^+$. The subset $(K^+)^\times$ of units of K^+ consists precisely of the elements $x \in K$ such that $|x| = 1$. Conversely, let V be a valuation ring of K with maximal ideal $\mathfrak{m}$; V induces a valuation $|\cdot|$ on K whose value group is

$\Gamma := K^\times / V^\times$ (then $|\cdot|$ is just the natural projection). The ordering on Γ is defined as follows. For given classes $\bar{x}, \bar{y} \in \Gamma$, we declare that $\bar{x} < \bar{y}$ if and only if $x/y \in \mathfrak{m}$.

Remark 6.1.12. (i) It follows easily from (6.1.11) that every finitely generated ideal of a valuation ring is principal. Indeed, if $a_1, ..., a_n$ is a set of generators for an ideal I, pick $i_0 \leq n$ such that $|a_{i_0}| = \max\{|a_i| \mid i \leq n\}$; then $I = (a_{i_0})$.

(ii) It is also easy to show that any finitely generated torsion-free K^+-module is free and any torsion-free K^+-module is flat (cp. [16, Ch.VI, §3, n.6, Lemma 1]). Hence every K^+-module is of Tor-dimension ≤ 1.

(iii) Let E be a field extension of the valued field $(K, |\cdot|_K)$. Then the integral closure W of K^+ in E is the intersection of all the valuation rings of E containing K^+ (cp. [16, Ch.VI, §1, n.3, Cor.3]). In particular, K^+ is integrally closed.

(iv) Furthermore, if E is an algebraic extension of K, then W is a *Prüfer domain*, that is, for every prime ideal $\mathfrak{p} \subset W$, the localization $W_\mathfrak{p}$ is a valuation ring. Moreover, the assignment $\mathfrak{m} \mapsto W_\mathfrak{m}$ establishes a bijection between the set of maximal ideals of W and the set of valuation rings V of E whose associated valuation $|\cdot|_V$ extends $|\cdot|_K$ (cp. [16, Ch.VI, §8, n.6, Prop.6]).

(v) Let R and S be local rings contained in a field K, $\mathfrak{m}_R$ and $\mathfrak{m}_S$ their respective maximal ideals. One says that R *dominates* S if $S \subset R$ and $\mathfrak{m}_S = S \cap \mathfrak{m}_R$. It is clear that the relation of dominance establishes a partial order structure on the set of local subrings of K. Then a local subring of K is a valuation ring of K if and only if it is maximal for the dominance relation (cp. [16, Ch.VI, §1, n.2, Th.1]).

(vi) Let K^+ be a valuation ring of K with maximal ideal $\mathfrak{m}$, and K^{+h} a henselization of K^+. One knows that K^{+h} is an ind-étale local K^+-algebra (cp. [66, Ch.VIII, Th.1]), hence it is integral and integrally closed (cp. [66, Ch.VII, §2, Prop.2]). Denote by K^h the field of fractions of K^{+h} and W the integral closure of K^+ in K^h. It follows that $W \subset K^{+h}$. Let $\mathfrak{m}^h$ be the maximal ideal of K^{+h}; since $\mathfrak{m}^h \cap K^+ = \mathfrak{m}$, we deduce that $\mathfrak{q} := \mathfrak{m}^h \cap W$ is a maximal ideal of W; then by (iv), $W_\mathfrak{q}$ is a valuation ring of K^h dominated by K^{+h}; by (v) it follows that $K^{+h} = W_\mathfrak{q}$, in particular this shows that the henselization of a valuation ring is again a valuation ring. The same argument works also for strict henselizations.

The following lemma provides a simple method to construct extensions of valuation rings, which is sometimes useful.

Lemma 6.1.13. *Let* $(K, |\cdot|)$ *be a valued field,* κ *the residue field of* K^+, R *a* K^+*-algebra which is finitely generated free as a* K^+*-module, and suppose that* $R \otimes_{K^+} \kappa$ *is a field. Then* R *is a valuation ring, and the morphism* $K^+ \to R$ *induces an isomorphism of value groups* $\Gamma_K \xrightarrow{\sim} \Gamma_R$.

Proof. Let $e_1, ..., e_n$ be a K^+-basis of R. Let us define a map $|\cdot|_R : R \to \Gamma_R \cup \{0\}$ in the following way. Given $x \in R$, write $x = \sum_{i=1}^n x_i \cdot e_i$; then $|x|_R := \max\{|x_i| \mid i = 1, ..., n\}$. If $|x| = 1$, then the image $\bar{x}$ of x in $R \otimes_{K^+} \kappa$ is not zero, hence it is invertible by hypothesis. By Nakayama's lemma it follows easily that x itself is invertible in R. Hence, every element y of R can be written in the form $y = u \cdot b$, where $u \in R^\times$ and $b \in K^+$ is an element such that $|b| = |y|_R$. It follows easily that R is an integral domain. Moreover, it is also clear that, given any

$x \in \operatorname{Frac}(R) \setminus \{0\}$, either $x \in R$ or $x^{-1} \in R$, so R is indeed a valuation ring and $|\cdot|_R$ is its valuation. □

Lemma 6.1.14. *Every finitely presented torsion K^+-module M is isomorphic to a direct sum of the form*

$$(K^+/a_1 K^+) \oplus \dots \oplus (K^+/a_n K^+)$$

where $a_1, \dots, a_n \in K^+$. More precisely, if $F \xrightarrow{\phi} M$ is any surjection from a free K^+-module F of rank n, then there is a basis $e_1, \dots, e_n$ of F and elements $a_1, \dots, a_n \in K^+ \setminus \{0\}$ such that $\operatorname{Ker} \phi = (K^+ a_1 e_1) \oplus \dots \oplus (K^+ a_n e_n)$.

Proof. We proceed by induction on the rank n of F. For $n = 1$ the claim follows easily from remark 6.1.12(i). Suppose $n > 1$; first of all, $S := \operatorname{Ker}(\phi)$ is finitely generated by [16, Ch.I, §2, n.8, lemme 9]. Then S is a free K^+-module, in light of remark 6.1.12(ii); its rank is necessarily equal to n, since $S \otimes_{K^+} K = F \otimes_{K^+} K$.

The image of the evaluation map $S \otimes_{K^+} F^* \to K^+$ given by $f \otimes \alpha \mapsto \alpha(f)$ is a finitely generated ideal $I \neq 0$ of K^+, hence it is principal, by remark 6.1.12(i). Let $\sum_i f_i \otimes \alpha_i$ be an element whose image generates I; this means that $\sum_i \alpha_i(f_i)$ is a generator of I, hence one of the terms in the sum, say $\alpha_1(f_1)$, is already a generator. The map $\alpha_1 : S \to I$ is surjective onto a free rank one K^+-module, therefore it splits, which shows that $S = (f_1 K^+) \oplus (S \cap \operatorname{Ker} \alpha_1)$. In particular, $S' := S \cap \operatorname{Ker} \alpha_1$ is a finitely generated torsion-free, hence free K^+-module. Let $e_1, \dots, e_n$ be a basis of F; then $f_1 = \sum_{i=1}^n a_i \cdot e_i$ for some $a_i \in K^+$. Consider the projection $\pi_i : F \to K^+$ such that $\pi_i(e_j) = \delta_{ij}$ for $j = 1, \dots, n$; clearly $\pi_i(f_1) = a_i \in I$. This shows that $f_1 = \alpha_1(f_1) \cdot g$ for some $g \in F$. It follows that $\alpha_1(g) = 1$, whence $F = (gK^+) \oplus \operatorname{Ker} \alpha_1$. Set $F' := \operatorname{Ker} \alpha_1$; we have shown that $M \simeq (K^+/\alpha_1(f_1)K^+) \oplus (F'/S')$. But F' is a free K^+-module of rank $n - 1$, hence we conclude by induction. □

6.1.15.   In later sections we will be concerned with almost ring theory in the special case where the basic setup $(V, \mathfrak{m})$ (see 2.1.1) consists of a valuation ring V of rank one. In preparation for this, we fix the following terminology, which will stand throughout the rest of this work. If V is a rank one valuation ring, then the *standard setup* attached to V is the pair $(V, \mathfrak{m})$ where $\mathfrak{m} := V$ in case the value group of V is isomorphic to $\mathbb{Z}$, and otherwise $\mathfrak{m}$ is the maximal ideal of V.

6.1.16.   Let $K \to \Gamma_K \cup \{0\}$ be a valuation of rank one on the field K, and K^+ its valuation ring. We consider the category K^{+a}-**Mod** relative to the standard setup $(K^+, \mathfrak{m})$. The topological group $\operatorname{Div}(K^{+a})$ of non-zero *fractional ideals* of K^{+a} is the subspace of $\mathscr{I}_{K^{+a}}(K^a)$ which consists of all the non-zero submodules $I \neq K^a$ of the K^{+a}-module K^a. The group structure is induced by the multiplication of fractional ideals.

Remark 6.1.17. One verifies easily that $\operatorname{Div}(K^{+a})$ is isomorphic to the group $D(K^+)$ defined in [16, Ch.VII, §1]. Note that $D(K^+)$ is a group by Theorem 1 of *loc.cit.*

The structure of the ideals of K^+ can be largely read off from the value group Γ. In order to explain this, we are led to introduce some notions for general ordered abelian groups.

6.1.18. First of all, we endow an ordered group Γ with the uniform structure defined in the following way. For every $\gamma \in \Gamma$ such that $\gamma > 1$, the subset of $\Gamma \times \Gamma$ given by $E(\gamma) := \{(\alpha, \beta) \mid \gamma^{-1} < \alpha^{-1} \cdot \beta < \gamma\}$ is an entourage for the uniform structure, and when $\Gamma \neq \{1\}$ the subsets of this kind form a fundamental system of entourages. Let $\Gamma^\wedge$ be the completion of Γ for this uniform structure. We also endow $\{0\} \cup \Gamma^\wedge$ with the unique topology such that $\Gamma^\wedge$ is an open subspace and such that the subsets of the form $U_\gamma := \{0\} \cup \{\alpha \in \Gamma^\wedge \mid \alpha < \gamma\}$ form a fundamental system of open neighborhoods of 0 when γ ranges over the elements of $\Gamma^\wedge$.

Lemma 6.1.19. *With the notation of (6.1.16), there exists a natural isomorphism of topological groups:* $\mathrm{Div}(K^{+a}) \xrightarrow{\sim} \Gamma_K^\wedge$.

Proof. We only indicate how to construct the morphism, and leave the details to the reader. In light of remark 6.1.12(i) and the proof of claim 6.3.10 (which does not depend on lemma 6.1.19), for every non-zero ideal $I \subset K^{+a}$ we can find a net $\{J_i \mid i \in S\}$ of principal ideals converging to I (for some filtered ordered set $(S, \leq)$). Let $\gamma_i \in \Gamma_K$ be the value of a generator of J_i. One verifies that the net $\{\gamma_i \mid i \in S\}$ converges in $\Gamma_K^\wedge$ to some element $\widehat{\gamma}$. Then we assign: $I \mapsto \widehat{\gamma}$. One verifies that this rule is well-defined and that it extends uniquely to the whole of $\mathrm{Div}(K^{+a})$. $\square$

Definition 6.1.20. Let Γ be any ordered abelian group with neutral element 1.

(i) We denote by $\Gamma^! \subset \Gamma$ the subset of all the $\gamma \in \Gamma$ such that $\gamma \leq 1$.
(ii) A subgroup Δ of Γ is said to be *convex* if it satisfies the following property. If $x \in \Delta^+$ and $1 > y > x$, then $y \in \Delta$. The set $\mathrm{Spec}\,\Gamma$ of all the convex subgroups of Γ will be called the *spectrum* of Γ. We define the *convex rank* of Γ as the supremum $\mathrm{c.rk}(\Gamma)$ over the lengths r of the chains $0 \subsetneq \Delta_1 \subsetneq \ldots \subsetneq \Delta_r := \Gamma$, such that all the Δ_i are convex subgroups. In general $\mathrm{c.rk}(\Gamma) \in \mathbb{N} \cup \{\infty\}$, but we will mainly encounter situations for which the convex rank is a positive integer. It is easy to see that the convex rank is always less than or equal to the usual rank, defined as $\mathrm{rk}(\Gamma) := \dim_\mathbb{Q}(\Gamma \otimes_\mathbb{Z} \mathbb{Q})$. To keep the two apart, we call *rational rank* the latter.

Example 6.1.21. (i) If Γ is an ordered abelian group, there exists a unique ordered group structure on $\Gamma \otimes_\mathbb{Z} \mathbb{Q}$ such that the natural map $\Gamma \to \Gamma \otimes_\mathbb{Z} \mathbb{Q}$ is order-preserving. Indeed, if Γ is the value group of a valuation $|\cdot|$ on a field K, and $|\cdot|_{K^a}$ is any extension of $|\cdot|_K$ to the algebraic closure K^a of K, then it is easy to see (*e.g.* using example 6.1.4(v)) that $\Gamma_{K^a} \simeq \Gamma \otimes_\mathbb{Z} \mathbb{Q}$.

(ii) Furthermore, let $K^s \subset K^a$ be the separable closure of K; we claim that $|\cdot|_{K^a}$ maps K^s surjectively onto Γ_{K^a}. Indeed, if $a \in K^a$ is inseparable over K^s, then the minimal polynomial $m(X) \in K^s[X]$ of a is of the form $X^{p^m} - b$ for some $b \in K^s$. For $c \in K^\times$, let $m_c(X) \in K^s[X]$ be the polynomial $m(X) + cX$; if a' is a root of $m_c(X)$, then $a' \in K^s$; moreover, $|(a - a')^{p^m}|_{K^a} = |c \cdot a'|_{K^a}$, hence for $|c|_{K^a}$ sufficiently small we have $|a|_{K^a} = |a'|_{K^a}$.

(iii) For any valued field $(K, |\cdot|)$, and every $\gamma \in \Gamma_K$, let $U_\gamma := \{x \in K \mid |x| < \gamma\}$. One defines the *valuation topology* on K as the unique group topology such that the family $(U_\gamma \mid \gamma \in \Gamma)$ is a fundamental system of open neighborhoods of 0. The argument in (ii) shows more precisely that, if the valuation of K is non-trivial, K^s is dense in K^a for the valuation topology of $(K^a, |\cdot|_{K^a})$.

(iv) If $\Delta \subset \Gamma$ is any subgroup, then (cp. [16, Ch.VI, §10, n.2, Prop.3])

$$\mathrm{c.rk}(\Gamma) \leq \mathrm{c.rk}(\Delta) + \mathrm{rk}(\Gamma/\Delta).$$

(v) A subgroup $\Delta \subset \Gamma$ is convex if and only if there is an ordered group structure on Γ/Δ such that the natural map $\Gamma \to \Gamma/\Delta$ is order-preserving. Then the ordered group structure with this property is unique.

(vi) If $\mathrm{c.rk}(\Gamma) = 1$, we can find an order-preserving imbedding

$$\rho : (\Gamma, \cdot, \leq) \hookrightarrow (\mathbb{R}, +, \leq).$$

Indeed, pick an element $g \in \Gamma$ with $g > 1$. For every $h \in \Gamma$, and every positive integer n, there exists a largest integer $k(n)$ such that $g^{k(n)} < h^n$. Then $(k(n)/n \mid n \in \mathbb{N})$ is a Cauchy sequence and we let $\rho(h) := \lim_{n \to \infty} k(n)/n$. One verifies easily that ρ is an order-preserving group homomorphism, and since the convex rank of Γ equals one, it follows that ρ is injective.

6.1.22. There is an inclusion-reversing bijection from the set of convex subgroups of the value group Γ of a valuation $|\cdot|$ to the set of prime ideals of its valuation ring K^+. This bijection assigns to a convex subgroup $\Delta \subset \Gamma$, the prime ideal

$$\mathfrak{p}_\Delta := \{x \in K^+ \mid \gamma > |x| \text{ for every } \gamma \in \Delta\}.$$

Conversely, to a prime ideal $\mathfrak{p}$, there corresponds the convex subgroup

$$\Delta_\mathfrak{p} := \{\gamma \in \Gamma \mid \gamma > |x| \text{ for all } x \in \mathfrak{p}\}.$$

The value group of the valuation ring $K_\mathfrak{p}^+$ is (naturally isomorphic to) $\Gamma/\Delta_\mathfrak{p}$. Furthermore, $K^+/\mathfrak{p}$ is a valuation ring of its field of fractions, with $\Delta_\mathfrak{p}$ as value group.

6.1.23. The *rank of a valuation* is defined as the convex rank of its value group. It is clear from (6.1.22) that this is the same as the Krull dimension of the associated valuation ring.

For any field extension $F_1 \subset F_2$, denote by $\mathrm{tr.deg}(F_2/F_1)$ the transcendence degree of F_2 over F_1. Let E be a field extension of the valued field K, and $|\cdot|_E : E^\times \to \Gamma_E$ an extension of the valuation $|\cdot|_K : K^\times \to \Gamma_K$ of K to E. Let κ (resp. $\kappa(E)$) be the residue field of the valuation ring of $(K, |\cdot|)$ (resp. of $(E, |\cdot|)$). Then we have the inequality:

$$\mathrm{rk}(\Gamma_E/\Gamma_K) + \mathrm{tr.deg}(\kappa(E)/\kappa) \leq \mathrm{tr.deg}(E/K)$$

(cp. [16, Ch.VI, §10, n.3, Cor.1]).

6.1.24. The image of $K^+ \setminus \{0\}$ in Γ is the monoid Γ^+. Henceforth, all the monoids under consideration shall be *unitary*, unless explicitly stated otherwise. The submonoids of Γ^+ are in bijective correspondence with the multiplicative subsets of $K^+ \setminus \{0\}$ which contain $(K^+)^\times$. The bijection is exhibited by the following "short exact sequence" of monoids:

$$ 1 \to (K^+)^\times \to K^+ \setminus \{0\} \overset{\pi}{\to} \Gamma^+ \to 1. $$

Then, to a monoid $M \subset \Gamma^+$ one assigns the multiplicative subset $\pi^{-1}(M)$.

6.1.25. Let us say that a submonoid N of a monoid M is *convex* if the following holds. If $\gamma, \delta \in M$ and $\gamma \cdot \delta \in N$, then $\gamma \in N$ and $\delta \in N$. For every submonoid N there is a smallest convex submonoid N^{con} such that $N \subset N^{\mathrm{con}}$. One deduces a natural bijection between convex submonoids of Γ^+ and prime ideals of K^+, by assigning on one hand, to a convex monoid M, the ideal $\mathfrak{p}(M) := K^+ \cap \pi^{-1}(M)$, and on the other hand, to a prime ideal $\mathfrak{p}$, the convex monoid $M(\mathfrak{p}) := \pi(K^+ \setminus \mathfrak{p})$.

6.1.26. The subsets of the form $M \setminus N$, where N is a convex submonoid of the monoid M, are the first examples of ideals in a monoid. More generally, one says that a subset $I \subset M$ is an *ideal* of M, if $I \cdot M \subset I$. Then we say that I is a *prime ideal* if I is a proper ideal such that, for every $x, y \in M$ with $x \cdot y \in I$, we have either $x \in I$ or $y \in I$. Equivalently, an ideal I is a prime ideal if and only if $M \setminus I$ is a submonoid; in this case $M \setminus I$ is necessarily a convex submonoid. For a monoid M, let us denote by $\operatorname{Spec} M$ the set of all the prime ideals of M. Taking into account (6.1.22), we derive bijections

$$ \operatorname{Spec} \Gamma \overset{\sim}{\to} \operatorname{Spec} K^+ \overset{\sim}{\to} \operatorname{Spec} \Gamma^+ \quad : \quad \Delta \mapsto \mathfrak{p}_\Delta \mapsto \pi(\mathfrak{p}_\Delta) = \Gamma^+ \setminus \Delta^+. $$

Furthermore, the bijection $\operatorname{Spec} K^+ \overset{\sim}{\to} \operatorname{Spec} \Gamma^+$ extends to an inclusion-preserving bijection between the ideals of K^+ and the ideals of Γ^+.

In the sequel, it will be sometimes convenient to study a monoid via the system of its finitely generated submonoids. In preparation for this, we want to delve a little further into the theory of general commutative monoids.

Definition 6.1.27. Let M be a commutative monoid.
  (i) We say that M is *integral* if we have $a = b$, whenever $a, b, c \in M$ and $a \cdot c = b \cdot c$.
  (ii) We say that M is *free* if it isomorphic to $\mathbb{N}^{(I)}$ for some index set I. In this case, a minimal set of generators for M will be called a *basis*.

6.1.28. Let **Mnd** be the category of commutative monoids. The natural forgetful functor $\mathbb{Z}\text{-}\mathbf{Mod} \to \mathbf{Mnd}$ admits a left adjoint functor $M \mapsto M^{\mathrm{gp}}$. Given a monoid M, the abelian group M^{gp} can be realized as the set of equivalence classes of pairs $(a, b) \in M \times M$, where $(a, b) \sim (a', b')$ if there exists $c \in M$ such that $a \cdot b' \cdot c = a' \cdot b \cdot c$; the multiplication is defined termwise, and the unit of the adjunction is the map $\phi : M \to M^{\mathrm{gp}} : a \mapsto (a, 1)$ for every $a \in M$. It is easy to see that ϕ is injective if and only if M is integral.

6.1.29. The category **Mnd** admits arbitrary limits and colimits. In particular, it admits direct sums. The functor $M \mapsto M^{\mathrm{gp}}$ commutes with colimits, but does not commute with limits (*e.g.* consider the fibre product $\mathbb{N} \times_{\mathbb{Z}} (-\mathbb{N})$). The forgetful functor $\mathbb{Z}$-**Mod** $\rightarrow$ **Mnd** commutes with limits and colimits.

Theorem 6.1.30. *Let Δ be an ordered abelian group, $N \subset \Delta^+$ a finitely generated submonoid. Then there exists a free finitely generated submonoid $N' \subset \Delta^+$ such that $N \subset N'$.*

Proof. Since N is a submonoid of a group, it is integral, so $N \subset N^{\mathrm{gp}}$. The group homomorphism $N^{\mathrm{gp}} \subset \Delta$ induced by the imbedding $N \subset \Delta$ is injective as well. The verification is straightforward, using the description of N^{gp} in (6.1.28). Then N^{gp} inherits a structure of ordered group from Δ, and we can replace Δ by N^{gp}, thereby reducing to the case where Δ is finitely generated and N spans Δ. Thus, in our situation, the convex rank of Δ is finite; we will argue by induction on c.rk(Δ).

• Suppose then that c.rk$(\Delta) = 1$. In this case we will argue by induction on the rank n of Δ. If $n = 1$, then one has only to observe that $\mathbb{Z}^+$ is a free monoid. Suppose next that $n = 2$; in this case, let $g_1, g_2 \in \Delta$ be a basis. We can suppose that $g_1 < g_2 < 1$; indeed, if $g_1 > 1$, we can replace it by g_1^{-1}; then, since c.rk$(\Delta) = 1$, we can find an integer k such that $g_2' := g_2 \cdot g_1^k < 1$ and $g_2' > g_1$; clearly g_1, g_2' is still a basis of Δ. We define inductively a sequence of elements $g_i \in \Delta^+$, for every $i \geq 1$, such that every successive pair is a basis, in the following way. Suppose that $i > 2$ and that the elements $g_3 < g_4 < \ldots < g_{i-1}$ have already been assigned; let $k_i := \sup\{n \in \mathbb{N} \mid g_{i-2} \cdot g_{i-1}^{-n} \leq 1\}$; notice that, since the convex rank of Δ equals 1, we have $k_i < \infty$. We set $g_i := g_{i-2} \cdot g_{i-1}^{-k_i}$.

Claim 6.1.31. $g_i^{\mathbb{N}} \cdot g_{i+1}^{\mathbb{N}} \subset g_{i+1}^{\mathbb{N}} \cdot g_{i+2}^{\mathbb{N}}$ for every $i \geq 1$, and $\Delta^+ = \bigcup_{i \geq 1}(g_i^{\mathbb{N}} \cdot g_{i+1}^{\mathbb{N}})$.

Proof of the claim. The first assertion is obvious. We prove the second assertion. Let $g \in \Delta^+ \setminus \{1\}$; for every $i \geq 1$ we can write $g = g_i^{a_i} \cdot g_{i+1}^{b_i}$ for unique $a_i, b_i \in \mathbb{Z}$. Notice that either a_i or b_i is strictly positive. Suppose that either a_{i+1} or b_{i+1} is not in $\mathbb{N}$; we show that in this case

$$(6.1.32) \qquad |a_{i+1}| + |b_{i+1}| < |a_i| + |b_i|.$$

Indeed, we must have either $a_i < 0$ and $b_i > 0$, or $a_i > 0$ and $b_i < 0$. However, $a_{i+1} = a_i \cdot k_{i+2} + b_i$ and $b_{i+1} = a_i$; thus, if $a_i < 0$, then $b_{i+1} < 0$, and consequently $a_{i+1} > 0$, whence

$$(6.1.33) \qquad |a_{i+1}| < |b_i|$$

and if $a_i > 0$, then $a_{i+1} < 0$, so again (6.1.33) holds. From (6.1.32) it now follows that eventually a_i and b_i become both positive. $\Diamond$

Since N is finitely generated, claim 6.1.31 shows that $N \subset g_i^{\mathbb{N}} \cdot g_{i+1}^{\mathbb{N}}$ for $i > 0$ sufficiently large, so the claim follows in this case.

• Next, suppose that c.rk$(\Delta) = 1$ and $n := \mathrm{rk}(\Delta) > 2$. Write $\Delta = H \oplus G$ for two subgroups such that $\mathrm{rk}(H) = n - 1$ and $G = g^{\mathbb{Z}}$ for some $g \in \Delta$.

Claim 6.1.34. For every $\delta \in \Delta^+ \setminus \{1\}$ we can find $a, b \in H$ such that $\delta < a \cdot g^{-1} < 1$ and $\delta < b^{-1} \cdot g < 1$.

Proof of the claim. Let $\rho : \Delta \hookrightarrow \mathbb{R}$ be an order-preserving imbedding as in example 6.1.21(vi); since $\operatorname{rk}(H) > 1$, it is easy to see that $\rho(H)$ is dense in $\mathbb{R}$. The claim is an immediate consequence. ◇

Let $g_1, ..., g_k$ be a set of generators for N (and we assume that $g_i \neq 1$ for every $i \leq k$). For every $i \leq k$ we can write $g_i = h_i \cdot g^{n_i}$ for unique $h_i \in H$ and $n_i \in \mathbb{Z}$. It follows easily from claim 6.1.34 that there exists $a, b \in H$ such that $a < g < b$ and $g_i < (b^{-1} \cdot g)^{n_i} < 1$ (resp. $g_i < (a^{-1} \cdot g)^{n_i} < 1$) for every i such that $n_i > 0$ (resp. $n_i < 0$). Then for $n_i \geq 0$ set $h_i' := h_i \cdot b^{n_i}$, for $n_i < 0$ set $h_i' := h_i \cdot a^{n_i}$. Notice that $h_i' < 1$. Set $h_0 := a \cdot b^{-1}$ and let M be the submonoid of H^+ spanned by $h_0, h_1', ..., h_k'$.

Claim 6.1.35. $M^{\mathrm{gp}} = H$.

Proof of the claim. Notice that

$$(6.1.36) \quad \begin{aligned} g_i &= h_i' \cdot (g \cdot a^{-1})^{n_i} & \text{if} \quad n_i < 0 \\ g_i &= h_i' \cdot (g \cdot b^{-1})^{n_i} = h_i' \cdot h_0^{n_i} \cdot (g \cdot a^{-1})^{n_i} & \text{if} \quad n_i \geq 0. \end{aligned}$$

Since by assumption N spans Δ, we deduce that $\Delta = M^{\mathrm{gp}} \oplus (g \cdot a^{-1})^{\mathbb{Z}}$, whence the claim. ◇

By the inductive assumption, we can imbed M in a free submonoid $L \subset H^+$. By claim 6.1.35, any basis of L consists of precisely $n - 1$ elements; let $l_1, ..., l_{n-1}$ be such a basis. We can write $h_0 = \prod_{i=1}^t l_{k_i}$ for integers $k_1, ..., k_t \in \{1, ..., n-1\}$. Notice that $h_0 \leq b^{-1} \cdot g$ and let $s < t$ be the largest integer such that $\prod_{i=1}^s l_{k_i} > g \cdot b^{-1}$; set $l' := g \cdot b^{-1} \cdot \prod_{i=1}^s l_{k_i}^{-1}$ and $l'' := g^{-1} \cdot b \cdot \prod_{i=1}^{s+1} l_{k_i}$.

Claim 6.1.37. The submonoid L' generated by $\{l_1, ..., l_{n-1}, l', l''\} \setminus \{l_{k_{s+1}}\}$ contains N.

Proof of the claim. Indeed, since $l' \cdot l'' = l_{k_{s+1}}$, it follows that $L \subset L'$; moreover, $g \cdot b^{-1} = l' \cdot \prod_{i=1}^s l_{k_i}$ and $g^{-1} \cdot a = l'' \cdot \prod_{i=s+2}^t l_{k_i}$ so $g \cdot b^{-1}, g^{-1} \cdot a \in L'$. Now the claim follows from (6.1.36). ◇

Now, it is clear that $L' \subset \Delta^+$; since moreover L' spans Δ and is generated by n elements, it follows that L' is a free monoid, so the proof is concluded in case $\mathrm{c.rk}(\Delta) = 1$.

• Finally, suppose $\mathrm{c.rk}(\Delta) > 1$ and pick a convex subgroup $0 \neq \Delta_0 \subsetneq \Delta$; then the ordering on Δ induces a unique ordering on Δ/Δ_0 such that the projection map $\pi : \Delta \to \Delta/\Delta_0$ is order-preserving. Let $N_0 := \pi(N)$. By induction, N_0 can be imbedded into a finitely generated free submonoid F_0 of $(\Delta/\Delta_0)^+$. By lifting a minimal set of generators of F_0 to elements $f_1, ..., f_n \in \Delta^+$, we obtain a free finitely generated monoid $F \subset \Delta^+$ with $\pi(F) = F_0$. Now, choose a finite set S of generators for N; we can partition $S = S_1 \cup S_2$, where $S_1 = S \cap \Delta_0$ and $S_2 = S \setminus \Delta_0$. By construction, for every $x \in S_2$ there exist integers $k_{i,x} \geq 0$ ($i = 1, ..., n$) such that $y_x := x \cdot \prod_{i=1}^n f_i^{-k_{i,x}} \in \Delta_0$. Let $g := \max\{y_x \mid x \in S_2\}$; if $g < 1$, let $e_i := f_i$, otherwise let $e_i := f_i \cdot g$ for every $i \leq n$. Since Δ_0 is convex, we have in any case: $e_i < 1$ for $i \leq n$. Moreover, the elements $z_x := x \cdot \prod_{i=1}^n e_i^{-k_{i,x}}$ are contained in Δ_0^+. By induction, the submonoid of Δ_0^+ generated by $S_1 \cup \{z_x \mid x \in$

S_2} is contained in a free finitely generated monoid $F' \subset \Delta_0^+$. One verifies easily that $N' := F \cdot F'$ is a free monoid. Clearly $N \subset N'$, so the assertion follows.   □

Remark 6.1.38. Another proof of theorem 6.1.30 can be found in [32, Th.2.2]. Moreover, this theorem can also be deduced from the resolution of singularities of toric varieties ([53, Ch.I, Th.11]).

6.2. Basic ramification theory. This section is a review of some basic ramification theory in the setting of general valuation rings and their algebraic extensions.

6.2.1. Throughout this section we fix a valued field $(K, | \cdot |)$. Its value group and valuation ring will be denoted by Γ and K^+, the residue field of K^+ will be denoted by κ. If $(E, | \cdot |_E)$ is any valued field extension of K, we will denote by E^+ the valuation ring of E, by $\kappa(E)$ its residue field and by Γ_E its value group (so $\Gamma_K = \Gamma$). Furthermore, we let K^a be an algebraic closure of K, and K^s the separable closure of K contained in K^a.

6.2.2. Let $E \subset K^a$ be a finite extension of K. Let W be the integral closure of K^+ in E; by remark 6.1.12(iv), to every maximal ideal $\mathfrak{p}$ of W we can associate a valuation $| \cdot |_{\mathfrak{p}} : E^\times \to \Gamma_{\mathfrak{p}}$ extending $| \cdot |$, and (up to isomorphisms of value groups) every extension of $| \cdot |$ to E is obtained in this way. Set $\kappa(\mathfrak{p}) := W/\mathfrak{p}$; it is known that $\sum_{\mathfrak{p} \in \mathrm{Max}(W)}(\Gamma_{\mathfrak{p}} : \Gamma) \cdot [\kappa(\mathfrak{p}) : \kappa] \leq [E : K]$ (cp. [16, Ch.VI §8, n.3 Th.1]).

6.2.3. Suppose now that E is a Galois extension of K. Then $\mathrm{Gal}(E/K)$ acts transitively on $\mathrm{Max}(W)$. For a given $\mathfrak{p} \in \mathrm{Max}(W)$, the *decomposition subgroup* $D_{\mathfrak{p}} \subset \mathrm{Gal}(E/K)$ of $\mathfrak{p}$ is the stabilizer of $\mathfrak{p}$. Then $\kappa(\mathfrak{p})$ is a normal extension of κ and the natural morphism $D_{\mathfrak{p}} \to \mathrm{Aut}(\kappa(\mathfrak{p})/\kappa)$ is surjective; its kernel $I_{\mathfrak{p}}$ is the *inertia subgroup* at $\mathfrak{p}$ (cp. [16, Ch.V, §2, n.2, Th.2] for the case of a finite Galois extension; the general case is obtained by passage to the limit over the family of finite Galois extensions of K contained in E).

6.2.4. If now E is a finite Galois extension of K, then it follows easily from (6.2.2) and (6.2.3) that the integers $(\Gamma_{\mathfrak{p}} : \Gamma)$ and $[\kappa(\mathfrak{p}) : \kappa]$ are independent of $\mathfrak{p}$, and therefore, if W admits n maximal ideals, we have : $n \cdot (\Gamma_{\mathfrak{p}} : \Gamma) \cdot [\kappa(\mathfrak{p}) : \kappa] \leq [E : K]$.

Lemma 6.2.5. *Let $K^{+\mathrm{sh}}$ be a strict henselization of K^+; then $K^{+\mathrm{sh}}$ is a valuation ring and $\Gamma_{K^{+\mathrm{sh}}} = \Gamma$.*

Proof. It was shown in remark 6.1.12(vi) that $K^{+\mathrm{sh}}$ is a valuation ring. To show the second assertion, let R be more generally any integrally closed local domain; the (strict) henselization of R can be constructed as follows (cp. [66, Ch.X, §2, Th.2]). Let $F := \mathrm{Frac}(R)$, F^s a separable closure of F, $\mathfrak{p}$ any maximal ideal of the integral closure W of R in F^s, D and I respectively the decomposition and inertia subgroups of $\mathfrak{p}$; let W^D (resp. W^I) be the subring of elements of W fixed by D (resp. by I) and set $\mathfrak{p}^D := W^D \cap \mathfrak{p}$, $\mathfrak{p}^I := W^I \cap \mathfrak{p}$. Then the localization $R^{\mathrm{h}} := (W^D)_{\mathfrak{p}^D}$ (resp. $R^{\mathrm{sh}} := (W^I)_{\mathfrak{p}^I}$) is a henselization (resp. strict henselization) of R. Now, let us make $R := K^+$, so $F := K$ and $F^s := K^s$; let $E \subset K^s$ be any finite Galois extension of K; $W_E := W \cap E$ is the integral closure of K^+ in E; let D_E (resp.

I_E) be the image of D (resp. of I) in $\mathrm{Gal}(E/K)$, $E' := E^{D_E}$, $E'' := E^{I_E}$. Let
$\mathfrak{p}' := \mathfrak{p} \cap E'$; it then follows from [71, Ch.VI,§12, Th.23] that $(\Gamma_{\mathfrak{p}'} : \Gamma) = 1$.
Clearly the value group Γ_{K^h} of K^{+h} is the filtered union of all such $\Gamma_{\mathfrak{p}'}$, so we
deduce $\Gamma_{K^h} = \Gamma$. Therefore, in order to prove the lemma, we can assume that
$K = K^h$. In this case $\mathrm{Gal}(E/K)$ coincides with the decomposition subgroup of
$\mathfrak{p} \cap E$ and I_E is a normal subgroup of $\mathrm{Gal}(K/E)$ such that $(\mathrm{Gal}(K/E) : I_E)$
equals the cardinality n of $\mathrm{Aut}(\kappa(E)/\kappa)$. By the definition of I_E it follows that
the natural map : $\mathrm{Aut}(\kappa(E)/\kappa) \to \mathrm{Aut}(\kappa(E'')/\kappa)$ is an isomorphism. We derive :
$[E'' : K] = n \le [\kappa(E'') : \kappa]$; then from (6.2.4) we obtain $\Gamma_{E''} = \Gamma$ and the claim
follows. □

6.2.6. We suppose now that K^+ is a *henselian* valuation ring. Then, on any alge-
braic extension $E \subset K^a$ of K, there is a unique valuation $|\cdot|_E$ extending $|\cdot|$, and
thus a unique inertia subgroup, which we denote simply by I. By remark 6.1.12(iv),
E^+ is the integral closure of K^+ in E.

Remark 6.2.7. (i) In the situation of (6.2.6), the inequality of (6.2.2) simplifies to :

$$[\kappa(E) : \kappa] \cdot (\Gamma_E : \Gamma) \le [E : K].$$

(ii) Sometimes this inequality is actually an equality; this is for instance the case
when the valuation of K is discrete and the extension $K \subset E$ is finite and separable
(cp. [16, Ch.VI, §8, n.5, Cor.1]).

(iii) However, even when the valuation of K is discrete, it may happen that the
inequality (i) is strict, if E is inseparable over K. As an example, let κ be a perfect
field of characteristic $p > 0$, and choose a power series $f(T) \in \kappa[[T]]$ which is
transcendental over the subfield $\kappa(T)$. Endow $F := \kappa(T^{1/p}, f(T))$ with the T-adic
valuation, and let K be the henselization of F. Then the residue field of K is κ and
the valuation of K is discrete. Let $E := K[f(T)^{1/p}]$. Then $[E : K] = p$, $\Gamma_E = \Gamma$
and $\kappa(E) = \kappa$.

6.2.8. For a field F, we denote by $\mu(F)$ the torsion subgroup of $F^\times$. Let E be a
finite Galois extension of K (with K^+ still henselian). One defines a pairing

$$(6.2.9) \qquad I \times (\Gamma_E/\Gamma_K) \to \mu(\kappa(E))$$

in the following way. For $(\sigma, \gamma) \in I \times \Gamma_E$, let $x \in E^\times$ such that $|x| = \gamma$; then let
$(\sigma, \gamma) \mapsto \sigma(x)/x (\mod \mathfrak{m}_E)$. One verifies easily that this definition is independent
of the choice of x; moreover, if $x \in K^\times$, then σ acts trivially on x, so the definition
is seen to depend only on the class of γ in Γ_E/Γ_K.

6.2.10. Suppose furthermore that κ is separably closed. Then the inertia subgroup
coincides with the Galois group $\mathrm{Gal}(E/K)$ and moreover $\mu(\kappa(E)) = \mu(\kappa)$. The
pairing (6.2.9) induces a group homomorphism

$$(6.2.11) \qquad \mathrm{Gal}(E/K) \to \mathrm{Hom}_{\mathbb{Z}}(\Gamma_E/\Gamma_K, \mu(\kappa)).$$

Let p be the characteristic exponent of κ. For a group G, let us denote by $G^{(p)}$ the
maximal abelian quotient of G that does not contain p-torsion.

Proposition 6.2.12. *Under the assumptions of* (6.2.10), *the map* (6.2.11) *is surjective and its kernel is a p-group.*

Proof. Let $n := [E : K]$. Notice that $\mu(\kappa)$ does not contain p-torsion, hence every homomorphism $\Gamma_E/\Gamma_K \to \mu(\kappa)$ factors through $(\Gamma_E/\Gamma_K)^{(p)}$. Let m be the order of $(\Gamma_E/\Gamma_K)^{(p)}$. Let us recall the definition of the Kummer pairing: one takes the Galois cohomology of the exact sequence of $\mathrm{Gal}(K^{\mathrm{s}}/K)$-modules

$$1 \to \mu_m \to (K^{\mathrm{s}})^\times \xrightarrow{(-)^m} (K^{\mathrm{s}})^\times \to 1$$

and applies Hilbert 90, to derive an isomorphism

$$K^\times/(K^\times)^m \simeq H^1(\mathrm{Gal}(K^{\mathrm{s}}/K), \mu_m).$$

Since $(m, p) = 1$ and κ is separably closed, the equation $X^m = 1$ admits m distinct solutions in κ. Since K^+ is henselian, these solutions lift to roots of 1 in K, i.e., $\mu_m \subset K^\times$, whence $H^1(\mathrm{Gal}(K^{\mathrm{s}}/K), \mu_m) \simeq \mathrm{Hom}_{\mathrm{cont}}(\mathrm{Gal}(K^{\mathrm{s}}/K), \mu_m)$. By working out the identifications, one checks easily that the resulting group isomorphism

$$K^\times/(K^\times)^m \simeq \mathrm{Hom}_{\mathrm{cont}}(\mathrm{Gal}(K^{\mathrm{s}}/K), \mu_m)$$

can be described as follows. To a given $a \in K^\times$, we assign the homomorphism

$$\mathrm{Gal}(K^{\mathrm{s}}/K) \to \mu_m : \sigma \mapsto \sigma(a^{1/m})/a^{1/m}.$$

Notice as well that, since κ is separably closed, more generally every equation of the form $X^m = u$ admits m distinct solutions in κ, provided $u \neq 0$; again by the henselian property we deduce that every unit of K^+ is an m-th power in $K^\times$; therefore $K^\times/(K^\times)^m \simeq \Gamma_K/m\Gamma_K$.

Dualizing, we obtain an isomorphism

$$\mathrm{Hom}_{\mathbb{Z}}(\Gamma_K/m\Gamma_K, \mu_m) \simeq \mathrm{Hom}_{\mathbb{Z}}(\mathrm{Hom}_{\mathrm{cont}}(\mathrm{Gal}(K^{\mathrm{s}}/K), \mu_m), \mu_m).$$

However,

$$\mathrm{Hom}_{\mathrm{cont}}(\mathrm{Gal}(K^{\mathrm{s}}/K), \mu_m) = \operatorname*{colim}_{H \subset \mathrm{Gal}(K^{\mathrm{s}}/K)} \mathrm{Hom}_{\mathbb{Z}}(\mathrm{Gal}(K^{\mathrm{s}}/K)/H, \mu_m)$$

where H runs over the cofiltered system of open normal subgroups of $\mathrm{Gal}(K^{\mathrm{s}}/K)$ such that $\mathrm{Gal}(K^{\mathrm{s}}/K)/H$ is abelian with exponent dividing m. It follows that

$$\mathrm{Hom}_{\mathbb{Z}}(\mathrm{Hom}_{\mathrm{cont}}(\mathrm{Gal}(K^{\mathrm{s}}/K), \mu_m), \mu_m) \simeq \lim_{H \subset \mathrm{Gal}(K^{\mathrm{s}}/K)} \mathrm{Gal}(K^{\mathrm{s}}/K)/H$$

where the right-hand side is a quotient of $\mathrm{Gal}(K^{\mathrm{s}}/K)^{(p)}$. Hence, we have obtained a surjective group homomorphism

$$(6.2.13) \quad \mathrm{Gal}(K^{\mathrm{s}}/K) \to \mathrm{Hom}_{\mathbb{Z}}(\frac{1}{m}\Gamma_K/\Gamma_K, \mu_m) \xrightarrow{\sim} \mathrm{Hom}_{\mathbb{Z}}(\frac{1}{m}\Gamma_K/\Gamma_K, \mu(\kappa)).$$

(Since Γ_K is torsion-free, we can identify naturally $\Gamma_K/m\Gamma_K$ to the subgroup $m^{-1}\Gamma_K/\Gamma_K \subset (\Gamma_K \otimes_{\mathbb{Z}} \mathbb{Q})/\Gamma_K$.) Let $j : \Gamma_E/\Gamma_K \hookrightarrow m^{-1}\Gamma_K/\Gamma_K$ be the inclusion map. One verifies directly from the definitions, that the maps (6.2.11) and (6.2.13) fit into a commutative diagram

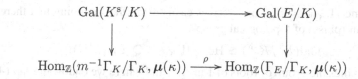

where the top map is the natural surjection, and $\rho := \mathrm{Hom}_{\mathbb{Z}}(j, \boldsymbol{\mu}(\kappa))$. Finally, an easy application of Zorn's lemma shows that ρ is surjective, and therefore, so is (6.2.11).

It remains to show that the kernel H of (6.2.11) is a p-group. Suppose that $\sigma \in H$ and nevertheless p does not divide the order l of σ; then we claim that the K-linear map $\phi : E \to E$ given by $x \mapsto \sum_{i=0}^{l-1} \sigma^i(x)$ is an isometry. Indeed, $\phi(x) = l \cdot x + \sum_{i=1}^{l-1}(\sigma^i(x) - x)$; it suffices then to remark that $|l \cdot x| = |x|$ and $|\sigma^i(x) - x| < |x|$ if $x \neq 0$, since $\sigma^i \in H$ for $i = 0, ..., l-1$. Next, for every $x \in E$ we can write $0 = \sigma^l(x) - x = \phi(x - \sigma(x))$; hence $\sigma(x) = x$, that is, σ is the neutral element of $\mathrm{Gal}(E/K)$, as asserted. $\qquad\qquad\qquad\square$

Corollary 6.2.14. *Keep the assumptions of* (6.2.10), *and suppose moreover that* $(p, [E : K]) = 1$. *Then* $\Gamma_E/\Gamma_K \simeq \mathrm{Hom}_{\mathbb{Z}}(\mathrm{Gal}(E/K), \boldsymbol{\mu}(K))$. *Moreover, if* $\Gamma_E/\Gamma_K \simeq \mathbb{Z}/q_1\mathbb{Z} \oplus ... \oplus \mathbb{Z}/q_k\mathbb{Z}$, *then there exist* $a_1, ..., a_k \in K$ *with* $E = K[a_1^{1/q_1}, ..., a_k^{1/q_k}]$ *an extension of the type of example* 6.1.4(v).

Proof. To start out, since $(p, [E : K]) = 1$, proposition 6.2.12 tells us that (6.2.11) is an isomorphism. In particular, $\mathrm{Gal}(E/K)$ is abelian, and $(\Gamma_E : \Gamma_K) \geq [E : K]$, whence $(\Gamma_E : \Gamma_K) = [E : K]$ by remark 6.2.7(ii). Therefore $\mathrm{Gal}(E/K) \simeq \mathbb{Z}/q_1\mathbb{Z} \oplus ... \oplus \mathbb{Z}/q_k\mathbb{Z}$ and E is a compositum of cyclic extensions $E_1, ..., E_k$ of degree $q_1, ..., q_k$. It follows as well that $\Gamma_E/\Gamma_K \simeq \mathrm{Hom}_{\mathbb{Z}}(\mathrm{Gal}(E/K), \boldsymbol{\mu}(\kappa))$, so the first assertion holds; furthermore the latter holds also for every extension of K contained in E. We deduce :

Claim 6.2.15. The Galois correspondence establishes an inclusion preserving bijection from the subgroups of Γ_E containing Γ_K, to the subfields of E containing K.

To prove the second assertion, we are thus reduced to the case where E is a cyclic extension, say $\mathrm{Gal}(E/K) \simeq \mathbb{Z}/q\mathbb{Z}$, with $(q, p) = 1$. Let $\gamma \in \Gamma_E$ be an element whose class in Γ_E/Γ_K is a generator; we can find $a \in K$ such that $|a| = \gamma^q$. Let $E' := K[a^{1/m}]$ and $F := E \cdot E'$. Since Γ_F is torsion-free, one sees easily that its subgroups Γ_E and $\Gamma_{E'}$ coincide. However, F satisfies again the assumptions of the corollary, therefore claim 6.2.15 applies to F, and yields $E = E'$. $\qquad\square$

Definition 6.2.16. Let $(K, |\cdot|)$ be a valued field. We denote by $K^{+\mathrm{sh}}$ the strict henselisation of K^+ and set $K^{\mathrm{sh}} := \mathrm{Frac}(K^{+\mathrm{sh}})$. The *maximal tame extension* K^{t} of K in its separable closure K^{s} is the union of all the finite Galois extensions E of K^{sh} inside K^{s}, such that $([E : K^{\mathrm{sh}}], p) = 1$. Notice that, by corollary 6.2.14, every such extension is abelian and the compositum of two such extensions is again of the same type, so the family of all such finite extension is filtered, and therefore their union is their colimit, so the definition makes sense.

6.2.17. Since $\Gamma_{K^{\mathrm{sh}}} = \Gamma_K$, one verifies easily from the foregoing that there is a natural isomorphism of topological groups:

$$\mathrm{Gal}(K^{\mathrm{t}}/K^{\mathrm{sh}}) \simeq \mathrm{Hom}_{\mathbb{Z}}(\Gamma_K \otimes_{\mathbb{Z}} \mathbb{Q}/\Gamma_K, \boldsymbol{\mu}^{(p)})$$

where μ denotes the group of roots of 1 in K^{sh} and where we endow the right-hand side with the pointwise convergence topology.

6.2.18. Let $E \subset K^{\mathrm{a}}$ be any algebraic extension of K. Then it is easy to check that $E^{\mathrm{t}} = E \cdot K^{\mathrm{t}}$. Indeed, the description in the proof of lemma 6.2.5 gives that $E^{\mathrm{sh}} = E \cdot K^{\mathrm{sh}}$, and $\mathrm{Gal}(K^{\mathrm{s}}/K^{\mathrm{t}})$ is the kernel of the homomorphism (defined as above) $\mathrm{Gal}(K^{\mathrm{s}}/K^{\mathrm{sh}}) \to \mathrm{Hom}_{\mathbb{Z}}(\Gamma_K \otimes_{\mathbb{Z}} \mathbb{Q}, \mu(\kappa))$, which restricts to the corresponding homomorphism defining E^{t}, so $\mathrm{Gal}(E^{\mathrm{s}}/E^{\mathrm{t}}) = \mathrm{Gal}(E^{\mathrm{s}}/E^{\mathrm{sh}}) \cap \mathrm{Gal}(K^{\mathrm{s}}/K^{\mathrm{t}})$, *i.e.* $K^{\mathrm{t}} \cdot E^{\mathrm{sh}} = E^{\mathrm{t}}$.

6.3. **Algebraic extensions.** In this section we return to almost rings: we suppose it is given a valued field $(K, | \cdot |)$ of rank one, and then we will study exclusively the almost ring theory relative to the standard setup attached to K^+ (see (6.1.15)). Especially, if the valuation of K is discrete, $\mathfrak{m} = K^+$; in this case, if we wish to refer to the maximal ideal of K^+, we will use the notation $\mathfrak{m}_K$. For an extension E of K, we will use the notation of (6.2.1). Furthermore, we will denote W_E the integral closure of K^+ in E.

Lemma 6.3.1. *Let R be a ring and $0 \to M_1 \to M_2 \to M_3 \to 0$ a short exact sequence of finitely generated R-modules, and suppose that the* Tor-*dimension of M_3 is ≤ 1. Then $F_0(M_2) = F_0(M_1) \cdot F_0(M_3)$.*

Proof. We can find epimorphisms $\phi_i : R^{n_i} \to M_i$ for $i \leq 3$, with $n_2 = n_1 + n_3$, fitting into a commutative diagram with exact rows:

$$
\begin{array}{ccccccccc}
0 & \longrightarrow & R^{n_1} & \longrightarrow & R^{n_2} & \longrightarrow & R^{n_3} & \longrightarrow & 0 \\
& & \downarrow{\scriptstyle \phi_1} & & \downarrow{\scriptstyle \phi_2} & & \downarrow{\scriptstyle \phi_3} & & \\
0 & \longrightarrow & M_1 & \longrightarrow & M_2 & \longrightarrow & M_3 & \longrightarrow & 0.
\end{array}
$$

Let $N_i := \mathrm{Ker}\,\phi_i$ ($i \leq 3$). By the snake lemma we have a short exact sequence: $0 \to N_1 \to N_2 \xrightarrow{\pi} N_3 \to 0$. Since the Tor-dimension of the M_3 is ≤ 1, it follows that N_3 is a flat R-module.

Claim 6.3.2. $\Lambda_R^{n_3+1} N_3 = 0$.

Proof of the claim. Since N_3 is flat, the antisymmetrization operator $\overline{a}_k : \Lambda_R^k N_3 \to N_3^{\otimes k}$ is injective for every $k \geq 0$ (cp. the proof of proposition 4.3.26). On the other hand, $\Lambda_R^{n_3+1} R^{n_3+1} = 0$, thus it suffices to show that the natural map $j^{\otimes k} : N_3^{\otimes k} \to (R^{n_3})^{\otimes k}$ is injective for every $k \geq 0$. This is clear for $k = 0$. Suppose that injectivity is known for $j^{\otimes k}$; we have $j^{\otimes k+1} = (1_{R^{\otimes k}} \otimes_R j) \circ (j^{\otimes k} \otimes_R 1_{N_3})$. Since N_3 is flat, we conclude by induction on k. $\Diamond$

Next recall that, for every $k \geq 0$ there are exact sequences

(6.3.3) $\qquad N_1 \otimes_R \Lambda_R^k N_2 \to \Lambda_R^{k+1} N_2 \xrightarrow{\pi^{\wedge k+1}} \Lambda_R^{k+1} N_3 \to 0.$

(To show that such sequences are exact, one uses the universality of $\Lambda_R^{k+1} N_3$ for $(k+1)$-linear alternating maps to R-modules.) From (6.3.3) and claim 6.3.2, a simple argument by induction on k shows that the natural maps $\psi : \Lambda_R^k N_1 \otimes_R \Lambda_R^{n_3} N_2 \to \Lambda_R^{k+n_3} N_2$ are surjective. Finally, by definition, we have

$$F_0(N_i) = \mathrm{Im}(\Lambda_R^{n_i} N_i \xrightarrow{j_i^{\wedge n_i}} \Lambda_R^{n_i} R^{n_i} \xrightarrow{\sim} R).$$

To conclude, it suffices therefore to remark that the diagram:

(6.3.4)
$$
\begin{array}{ccc}
\Lambda_R^{n_1} N_1 \otimes_R \Lambda_R^{n_3} N_2 & \xrightarrow{\psi} & \Lambda_R^{n_2} N_2 \\
{\scriptstyle 1_{\Lambda_R^{n_1} N_1} \otimes \pi^{\wedge k+1}}\downarrow & & \downarrow{\scriptstyle j_2^{\wedge n_2}} \\
\Lambda_R^{n_1} N_1 \otimes_R \Lambda_R^{n_3} N_3 & \longrightarrow & \Lambda_R^{n_2} R^{n_2}
\end{array}
$$

commutes. We leave to the reader the task of verifying that the commutativity of (6.3.4) boils down to a well-known identity for determinants of matrices. $\qquad\square$

Remark 6.3.5. (i) In view of remark 6.1.12(ii), lemma 6.3.1 applies especially to a short exact sequence of finitely generated K^+-modules.

(ii) By the usual density arguments (cp. the proof of proposition 2.3.25), it then follows that lemma 6.3.1 holds true *verbatim*, even when we replace R by K^+ and the R-modules M_1, M_2, M_3 by uniformly almost finitely generated K^{+a}-modules.

Proposition 6.3.6. *Let M be an almost finitely generated K^{+a}-module.*

(i) *M is almost finitely presented.*

(ii) *M is uniformly almost finitely generated if and only if $\Lambda^p M = 0$ for some $p \geq 0$.*

(iii) *If M is uniformly almost finitely generated, then*

$$F_i(M) = \prod_{j>i} \mathrm{Ann}_{K^{+a}}(\Lambda^j M)$$

for every $i \geq 0$ (notice that the product is finite in view of (ii)).

Proof. (i) is reduced to the case of a finitely generated module and then, using lemma 2.3.18, to the case of a module generated by one almost element, which is easy.

(ii): We show more precisely that the isomorphism class of M is in $\mathscr{U}_n(K^{+a})$ if and only if $\Lambda_{K^{+a}}^{n+1} M = 0$. The direction $\Rightarrow$ is clear. For $\Leftarrow$ we may assume that the valuation on K is not discrete; by (i), for every $\varepsilon \in \mathfrak{m}$ there is a morphism $f : N^a \to M$, with N a finitely presented K^+-module such that Ker f and Coker f are annihilated by ε. So there exists $g : M \to N^a$ such that $g \circ f$ and $f \circ g$ are multiplication by ε^2. If now $\Lambda_{K^{+a}}^{n+1} M = 0$, we deduce

(6.3.7)
$$\varepsilon^{2(n+1)}(\Lambda_{K^+}^{n+1} N)^a = 0.$$

If N_t is the torsion submodule of N, then N/N_t is free, so the exact sequence $0 \to N_t \to N \to N/N_t \to 0$ splits and N_t has the structure given by lemma

6.1.14, from which one sees that (6.3.7) is equivalent to the assertion that $\varepsilon^{2(n+1)} N$ is generated by n elements.

(iii) is reduced to the "classical" finitely presented case by the usual density argument. □

Proposition 6.3.8. *Suppose that K^+ is a valuation ring of rank one. Let E be a finite separable extension of K. Then W_E^a and $\Omega_{W_E^a/K^{+a}}$ are uniformly almost finitely generated K^{+a}-modules which admit the uniform bounds $[E : K]$ and respectively $[E : K]^2$. Moreover, W_E^a is an almost projective K^{+a}-module.*

Proof. In view of the presentation (2.5.28), the assertion for $\Omega_{W_E^a/K^{+a}}$ is an immediate consequence of the assertion for W_E^a. The trace pairing $t_{E/K} : E \times E \to K$ is perfect since E is separable over K. Let $e_1, ..., e_n$ be a basis of the K-vector space E and $e_1^*, ..., e_n^*$ the dual basis under the trace morphism, so that $t_{E/K}(e_i \otimes e_j^*) = \delta_{ij}$ for every $i, j \leq n$. We can assume that $e_i \in W_E$ and we can find $a \in K^+ \setminus \{0\}$ such that $a \cdot e_i^* \in W_E$ for every $i \leq n$. Let $w \in W_E$; we can write $w = \sum_{i=1}^n a_i \cdot e_i$ for some $a_i \in K$. We have $t_{E/K}(w \otimes a \cdot e_j^*) \in K^+$ for every $j \leq n$; on the other hand, $t_{E/K}(w \otimes a \cdot e_j^*) = a \cdot a_j$. Thus, if we let $\phi : K^n \to E$ be the isomorphism $(x_1, ..., x_n) \mapsto \sum_{i=1}^n x_i \cdot e_i$, we see that

$$(6.3.9) \qquad (K^+)^n \subset \phi^{-1}(W_E) \subset a^{-1} \cdot (K^+)^n.$$

We can write W_E as the colimit of the family $\mathscr{W}$ of all its finitely generated K^+-submodules containing $e_1, ..., e_n$; if $W_0 \in \mathscr{W}$, then W_0 is a free K^+-module by remark 6.1.12(ii); then it is clear from (6.3.9) that the rank of W_0 must be equal to n. In case the valuation of K is discrete, one of these W_0 equals W_E, and the proposition follows. In the non-discrete case, it suffices to show:

Claim 6.3.10. Let $\varepsilon \in \mathfrak{m}$; there exists $W_0 \in \mathscr{W}$ such that $\varepsilon \cdot W_E \subset W_0$.

Proof of the claim. Indeed, suppose that this is not the case. Then we can find an infinite sequence of finitely generated submodules $\oplus_{i=1}^n e_i \cdot K^+ \subset W_0 \subset W_1 \subset W_2 \subset ... \subset W_E$ such that $\varepsilon \cdot W_{i+1} \not\subset W_i$ for every $i \geq 0$. From (6.3.9) and lemma 6.3.1 it follows easily that $F_0((K^+)^n/a \cdot (K^+)^n) \subset F_0(W_{k+1}/W_0) = \prod_{i=0}^k F_0(W_{i+1}/W_i)$ for every $k \geq 0$. However,

$$F_0(W_{i+1}/W_i) \subset \operatorname{Ann}_{K^+}(W_{i+1}/W_i) \subset \varepsilon \cdot K^+$$

for every $i \geq 0$. We deduce that $|a|^n < |\varepsilon|^k$ for every $k \geq 0$, which is absurd, since the valuation of K has rank one. □

6.3.11.   Suppose that the valuation ring of K has rank one. Let $K \subset E \subset F$ be a tower of finite separable extensions. Let $\mathfrak{p} \subset W_E$ be any prime ideal; then $W_{E,\mathfrak{p}}$ is a valuation ring (see remark 6.1.12(iii)), and $W_{F,\mathfrak{p}}$ is the integral closure of $W_{E,\mathfrak{p}}$ in F. It then follows from proposition 6.3.8 and remark 6.1.12(ii) that $W_{F,\mathfrak{p}}^a$ is an almost finitely generated projective $W_{E,\mathfrak{p}}^a$-module; we deduce that W_F^a is an almost finitely generated projective W_E^a-module, therefore we can define the different ideal of W_F^a over W_E^a. To ease notation, we will denote it by $\mathscr{D}_{W_F/W_E}$. If $|\cdot|_F$ is a valuation of F extending $|\cdot|$, then $F^+ = W_{F,\mathfrak{p}}$ for some prime ideal

$\mathfrak{p} \subset W_F$; moreover, if $|\cdot|_E$ is the restriction of $|\cdot|_F$ to E, then $E^+ = W_{E,\mathfrak{q}}$, where $\mathfrak{q} = \mathfrak{p} \cap W_E$. For this reason, we are led to define $\mathscr{D}_{F^+/E^+} := (\mathscr{D}_{W_F/W_E})_{\mathfrak{p}}$.

Lemma 6.3.12. *Let $K \subset E \subset F$ be a tower of finite separable extensions. Then:*

(i) *The W_E^a-module $(W_E^a)^*$ is invertible.*

(ii) $\mathscr{D}_{W_E/K^+} \cdot \mathscr{D}_{W_F/W_E} = \mathscr{D}_{W_F/K^+}.$

Proof. In view of proposition 4.1.25, (ii) follows from (i). We show (i): from proposition 6.3.8 we can find, for every $\varepsilon \in \mathfrak{m}$, a finitely generated K^+-submodule $M \subset W_E$ such that $\varepsilon \cdot W_E \subset M$. By remark 6.1.12(ii) it follows that M is a free K^+-module, so the same holds for $M^* := \operatorname{Hom}_{K^+}(M, K^+)$. The scalar multiplication $M^* \to M^* : \phi \mapsto \varepsilon \cdot \phi$ factors through a map $M^* \to W_E^*$, and if we let N be the W_E-module generated by image of the latter map, then $\varepsilon \cdot W_E^* \subset N$. Furthermore, for every prime ideal $\mathfrak{p} \subset W_E$, the localization $N_{\mathfrak{p}}$ is a torsion-free $W_{E,\mathfrak{p}}$-module; since $W_{E,\mathfrak{p}}$ is a valuation ring, it follows that $N_{\mathfrak{p}}$ is free of finite rank, again by remark 6.1.12(ii). Hence, N is a projective W_E-module. In particular, this shows that $(W_E^a)^*$ is almost finitely generated projective as a W_E^a-module. To show that $(W_E^a)^*$ is also invertible, it will suffice to show that the rank of N equals one. However, the rank of N can be computed as $\dim_E N \otimes_{W_E} E$. We have $N \otimes_{W_E} E = W_E^* \otimes_{K^+} K = \operatorname{Hom}_K(E, K)$, so the assertion follows by comparing the dimensions of the two sides. $\qquad\qquad\square$

Proposition 6.3.13. *Suppose that $(K, |\cdot|)$ has rank one. Let $K \subset E$ be a finite field extension such that $l := [E : K]$ is a prime. Let p be the characteristic exponent of κ and suppose that either:*

(a) $l \neq p$ and $K = K^{\mathrm{sh}}$, or

(b) $l = p$ and $K = K^{\mathrm{t}}$, or

(c) *the valuation $|\cdot|$ is discrete and henselian, and E is separable over K, or*

(d) *the valuation $|\cdot|$ is discrete and henselian, and E is inseparable over K.*

Then :

(i) *In case (a), (b) or (d) holds, there exists $x \in E \setminus K$ such that E^+ is the filtered union of a family of finite K^+-subalgebras of the form $E_i^+ := K^+[a_i x + b_i]$, ($i \in \mathbb{N}$) where $a_i, b_i \in K$ are elements with $|a_i x + b_i| \leq 1$.*

(ii) *In case (c) holds, there exists an element $x \in E^+$ such that $E^+ = K^+[x]$.*

(iii) *Furthermore, if E is a separable extension of K, then $H_j(\mathbb{L}_{E^+/K^+}) = 0$ for every $j > 0$.*

(iv) *If E is an inseparable extension of K, then $H_j(\mathbb{L}_{E^+/K^+}) = 0$ for every $j > 1$, and moreover $H_1(\mathbb{L}_{E^+/K^+})$ is a torsion-free E^+-module.*

Proof. Let us first show how assertions (iii) and (iv) follow from (i) and (ii). Indeed, since the cotangent complex commutes with colimits of algebras, by (i) and (ii) we reduce to dealing with an algebra of the form $K^+[w]$ for $w \in E^+ \setminus K^+$. Such an algebra is a complete intersection K^+-algebra, quotient of the free algebra $K^+[X]$ by the ideal $I \subset K^+[X]$ generated by the minimal polynomial $m(X)$ of w. In view of [50, Ch.III, Cor.3.2.7], one has a natural isomorphism in $\mathsf{D}(K^+[w]\text{-}\mathbf{Mod})$

$$\mathbb{L}_{K^+[w]/K^+} \simeq (0 \to I/I^2 \xrightarrow{\delta} \Omega_{K^+[X]/K^+} \otimes_{K^+[X]} K^+[w] \to 0).$$

If we identify $\Omega_{K^+[X]/K^+} \otimes_{K^+[X]} K^+[w]$ to the rank one free $K^+[w]$-module generated by dX, then δ is given explicitly by the rule: $f(X) \mapsto f'(w)dX$, for every $f(X) \in (m(X))$. However, E is separable over K if and only if $m'(w) \neq 0$. Hence δ is injective if and only if E is separable over K, which proves (iii). If E is inseparable over K, then δ vanishes identically by the same token. This shows (iv).

We prove (ii). Since the valuation is discrete, we have either $e := (\Gamma_E : \Gamma_K) = l$ or $f := [\kappa(E) : \kappa] = l$ (see remark 6.2.7(ii)). If $e = l$, then pick any uniformizer $a \in E$; every element of E can be written as a sum $\sum_{i=0}^{l-1} x_i \cdot a^i$ with $x_i \in K$ for every $i < l$. Then it is easy to see that such a sum is in E^+ if and only $x_i \in K^+$ for every $i < l$. In other words, $E^+ = K^+[a]$. In case $f = l$, we can write $\kappa(E) = \kappa[\bar{u}]$ for some unit $u \in (E^+)^\times$; moreover, $\mathfrak{m}_E = \mathfrak{m}_K E^+$; then $K^+[u] + \mathfrak{m}_K E^+ = E^+$; since in this case E^+ is a finite K^+-module, we deduce $E^+ = K^+[u]$ by Nakayama's lemma.

We prove (i). Suppose that (a) holds; then by corollary 6.2.14 it follows that $\Gamma_E/\Gamma_K \simeq \mathbb{Z}/l\mathbb{Z}$ and $E = K[a^{1/l}]$ for some $a \in K$. Hence:

$$(6.3.14) \qquad |a^{i/l}| \notin \Gamma_K \text{ for every } i = 1, ..., l - 1.$$

We can suppose that the valuation of K is not discrete, otherwise we fall back on case (c); then, for every $\varepsilon \in \mathfrak{m}$, there exists $b_\varepsilon \in K$ such that $|\varepsilon| < |b_\varepsilon^l \cdot a| < 1$. Let $x_0, ..., x_{l-1} \in K$ and set $w := \sum_{i=0}^{l-1} x_i \cdot a^{i/l}$. Clearly every element of E can be written in this form. From (6.3.14) we derive that the values $|x_i \cdot a^{i/l}|$ such that $x_i \neq 0$ are all distinct. Hence, $|w| = \max_{0 \le i < l} |x_i \cdot a^{i/l}|$. Suppose now that $w \in E^+$; it follows that $|x_i \cdot a^{i/l}| \le 1$ for $i = 0, ..., l - 1$, and in fact $|x_i \cdot a^{i/l}| < 1$ for $i \neq 0$. Take $\varepsilon \in \mathfrak{m}$ such that $|\varepsilon| > |x_i \cdot a^{i/l}|$ for every $i \neq 0$. A simple calculation shows that $|x_i \cdot b_\varepsilon^{-i}| < 1$ for every $i \neq 0$, in other words, $w \in K^+[b_\varepsilon \cdot a^{1/l}]$, which proves the claim in this case.

In order to deal with (b) and (d) we need some preparation. Let us set:

$$\rho(x) := \inf_{a \in K} |x - a| \in \Gamma_E^\wedge \cup \{0\} \qquad \text{for any } x \in E \setminus K.$$

We consider (b). Notice that the hypothesis $K = K^t$ implies that the valuation of K is not discrete. Fix $x \in E \setminus K$; for any $y \in E \setminus K$ we can write $y = f(x)$ for some $f(X) := b_0 + b_1 X + ... + b_d X^d \in K[X]$ with $d := \deg f(X) < p$. The degree of the minimal Galois extension F of K containing all the roots of $f(X)$ divides $d!$, hence $F \subset K^t = K$. In other words, we can write $y = a_k \cdot \prod_{i=1}^d (x - \alpha_i)$ for some $\alpha_1, ..., \alpha_d \in K$.

• Case I : there exist $x \in E \setminus K$ and $a \in K$ with $|x - a| = \rho(x)$. In this case, replacing x by $x - a$ we may achieve that $|x| \le |x - a|$ for every $a \in K$. Then the constant sequence $(a_n := 0 \mid n \in \mathbb{N})$ fulfills the condition of lemma 6.1.9. Thus, if $y = f(x)$ as above is in E^+, we must have $|f(X)|_{(0,\rho(x))} \le 1$; in other terms:

$$(6.3.15) \qquad |b_i| \cdot \rho(x)^i \le 1 \quad \text{for every } i \le d.$$

Now, if $\rho(x) \in \Gamma_K$, we can find $c \in K$ such that $x_0 := x \cdot c$ still generates E and $|x_0| = 1$, whence $|b_i/c^i| \leq 1$ for every $i \leq 1$; however, $y = b_0 + (b_1/c) \cdot x_0 + (b_2/c^2)x_0^2 + \dots + (b_d/c^d)x_0^d$, thus $y \in K^+[x_0]$, so in this case, E^+ itself is one of the E_i^+.

In case $\rho(x) \notin \Gamma_K$, since anyway Γ_K is of rank one and not discrete, we can find a sequence of elements $c_1, c_2, \dots \in K$ such that, letting $x_i := x \cdot c_i$, we have

$$|x_j - a| \geq |x_j|; \quad |x_j| < |x_{j+1}| < 1 \quad \text{for every } a \in K, j \in \mathbb{N} \quad \text{and} \quad |x_j| \to 1.$$

Claim 6.3.16. If $|x| \notin \Gamma_K$, then $|x^l| \notin \Gamma_K$ for every $0 < l < p$.

Proof of the claim. Indeed, suppose that $|x^l| \in \Gamma_K$ for some $0 < l < p$; since Γ_K is l-divisible, we can multiply x by some $a \in K$ to have $|x^l| = 1$, therefore $|x| = 1$, a contradiction. $\diamondsuit$

From (6.3.15) and claim 6.3.16 we deduce that actually $|b_i| \cdot \rho(x)^i < 1$ whenever $i > 0$. It follows that, for j sufficiently large, we will have $1 > |x_j^i| > |b_i| \cdot \rho(x)^i$ for every $i > 0$. Writing $y = b_0 + (b_1/c_j)x_j + (b_2/c_j^2)x_j^2 + \dots + (b_d/c_j^d)x_j^d$ we deduce $y \in K^+[x_j]$, therefore the sequence of K^+-subalgebras $K^+[c_i \cdot x]$ will do in this case.

- **Case II** : the infimum $\rho(x)$ is not attained for any $x \in E \setminus K$. In this case, since the valuation is not discrete and of rank 1, we can find, for every $x \in E \setminus K$, a sequence of elements $a_0, a_1, a_2, \dots \in K$ such that

(6.3.17) $\qquad \gamma_j := |x - a_j| \to \rho(x) \quad \text{and} \quad \gamma_j > \gamma_{j+1} \quad \text{for every } j \in \mathbb{N}.$

(Here the convergence is relative to the topology of $\Gamma_E \cup \{0\}$ defined in (6.1.18).) In particular, for j sufficiently large we will have $|x| > |x - a_j|$, therefore $|x| = |a_j|$. This shows:

(6.3.18) $\qquad\qquad \Gamma_E = \Gamma_K.$

Now, pick $x \in E \setminus K$ and any sequence of elements $a_i \in K$ such that (6.3.17) holds; it is clear that $(a_i \mid i \in \mathbb{N})$ fulfills the condition of lemma 6.1.9. Consequently

(6.3.19) $\qquad\qquad |y| = |f(X)|_{(a_j, \gamma_j)} \quad \text{for every sufficiently large } j.$

Let $f(X) = b_{0,j} + b_{1,j}(X - a_j) + \dots + b_{d,j}(X - a_j)^d$. (6.3.19) says that $|b_{i,j}| \cdot \gamma_j^i \leq 1$ whenever j is sufficiently large. However, from (6.3.18) we know that $\gamma_j \in \Gamma_K$. Pick $c_j \in K$ such that $|c_j| = \gamma_j^{-1}$ and set $x_j := c_j(x - a_j)$. It follows that $|b_{i,j}/c_j^i| \leq 1$ and $y = b_{0,j} + (b_{1,j}/c_{1,j})x_j + \dots + (b_{d,j}/c_j^d)x_j^d$. Hence $y \in K^+[x_j]$. It is then easy to verify that the family of all such K^+-subalgebras is filtered by inclusion, and thus conclude the proof under assumption (b).

At last, we turn to case (d). Notice that in this case $l = p$ and $E \subset K^{1/p}$. Moreover, it is easy to see that, if Γ_E/Γ_K is non-trivial, then it is of order p, and if the residue field extension is non-trivial, then it is of degree p. These two subcases are treated as case (c), using [16, Ch.6, §8, Th.2.]. Hence we can assume that $\Gamma_E = \Gamma_K$ and $\kappa(E) = \kappa$. Take $x \in E \setminus K$ with $|x| = 1$; let $a \in \mathfrak{m}_K$ be a uniformizer.

Claim 6.3.20. There exists a sequence of elements $(b_n \mid n \in \mathbb{N})$ such that $b_n \in K$ and $|x - b_n| = a^n$ for every $n \in \mathbb{N}$.

Proof of the claim. The sequence is constructed inductively, starting with $b_0 := 0$. Next, for given $n \in \mathbb{N}$, suppose that b_n has been found with the sought property; since $\kappa(E) = \kappa$, we can find an element $c \in K^+$ such that $c \equiv (x - b_n)/a^n$ (mod $\mathfrak{m}_K$). Set $b_{n+1} := b_n + c \cdot a^n$; then $|x - b_{n+1}| \leq |a^{n+1}|$; if the inequality is strict, it suffices to replace b_{n+1} by $b_{n+1} + a^{n+1}$. $\diamond$

Claim 6.3.20 shows that $\rho(x) = 0$, and the sequence $(b_n \mid n \in \mathbb{N})$ converges to x in the valuation topology. Let $y \in E \setminus K$; we can write $y = f(x)$ for a polynomial $f(X) \in K[X]$ of degree $d < p$. Let F be the minimal field extension of K that contains all the roots of $f(X)$. Notice that $[F : K]$ divides $d!$, hence F is separable over K, and $[E \cdot F : F] = p$. Let $f(X) = c \cdot \prod_{i=0}^{d}(X - \alpha_i)$ be the factorization of $f(X)$ in $F[X]$. By lemma 6.1.9 we deduce that, for every sufficiently large $n \in \mathbb{N}$ we have: $|y| = |f(X)|_{(b_n,|x-b_n|)}$, where $|\cdot|_{(b_n,|x-b_n|)}$ is the Gauss valuation on $F(X)$. One then argues as in the proof of case (b), to show that $y \in E_n^+ := K^+[c_n(x - b_n)]$, with $c_n \in K$ such that $|c_n(x - b_n)| = 1$. Again, it is easy to verify that $E_i^+ \subset E_{i+1}^+$ for every $i \in \mathbb{N}$, so the proof is complete. $\square$

Corollary 6.3.21. *Let E be a finite field extension of K of prime degree l.*

(i) *If E satisfies condition* (a) *of proposition 6.3.13, and the valuation of K is not discrete (but still of rank one), then $\Omega_{E^+/K^+} = 0$, $\mathbb{L}_{E^+/K^+} \simeq 0$ and $\mathscr{D}_{E^+/K^+} = E^{+a}$.*

(ii) *If E satisfies condition* (c) *of proposition 6.3.13, then $H_i(\mathbb{L}_{E^+/K^+}) = 0$ for every $i > 0$ and $F_0(\Omega_{E^+/K^+}) = \mathscr{D}_{E^+/K^+}$.*

Proof. (i): Since condition (a) holds, proposition 6.3.13 and its proof show that there exists $a \in K$ such that E^+ is the increasing union of all K^+-subalgebras of the form $E_b^+ := K^+[b \cdot a^{1/l}]$, where $b \in K^+$ ranges over all elements such that $|b^l \cdot a| < 1$. Consequently, $\Omega_{E^+/K^+} = \operatorname*{colim}_b \Omega_{E_b^+/K^+}$, and $\mathbb{L}_{E^+/K^+} = \operatorname*{colim}_b \mathbb{L}_{E_b^+/K^+}$. Then, again from proposition 6.3.13 it follows that $H_j(\mathbb{L}_{E^+/K^+}) = 0$ for every $j > 0$. Hence, in order to show the first two assertions, it suffices to show that the filtered system of the $\Omega_{E_b^+/K^+}$ is essentially zero. However, the E_b^+-module $\Omega_{E_b^+/K^+}$ is generated by $\omega_b := d(b \cdot a^{1/l})$, and clearly $l \cdot (b^l \cdot a)^{(l-1)/l} \cdot \omega_b = 0$. Since $(l, p) = 1$, it follows that $(b^l \cdot a)^{(l-1)/l} \cdot \omega_b = 0$. On the other hand, for $|b| < |c|$ we can write: $\omega_b = b \cdot c^{-1} \cdot \omega_c$. Therefore, the image of ω_b in $\Omega_{E_c^+/K^+}$ vanishes whenever $|b \cdot c^{-1}| < |c^l \cdot a|^{(l-1)/l}$, *i.e.* whenever $|b \cdot a^{1/l}| < |c^l \cdot a| < 1$. Since the valuation of K is of rank one and not discrete, such a c can always be found. To show the last stated equality, let us recall the following general fact (for whose proof we refer to [66, Ch.VII, §1]).

Claim 6.3.22. Suppose that $E = K[w]$ for some $w \in K^{s+}$, and let $f(X) \in K[X]$ be its minimal polynomial; the elements $1, w, w^2, ..., w^{l-1}$ form a basis of the K-vector space E. Let $S := \{e_1^*, ..., e_n^*\}$ be the corresponding dual basis under the trace pairing; then the bases S and $S' := \{w^{l-1}/f'(w), w^{l-2}/f'(w), ..., 1/f'(w)\}$ span the same K^+-submodule of E.

Let us take $w = b \cdot a^{1/l}$ for some $b \in V$ such that $|b^l \cdot a| < 1$. It follows from claim 6.3.22 that $(\mathscr{D}_{E^+/K^+})^{-1} \subset f'(w)^{-1} \cdot E^{+a}$, whence $f'(w) \in \mathscr{D}_{E^+/K^{+*}}$.

However, $f'(w) = l \cdot w^{l-1}$, and from the definition of w we see that $|f'(w)|$ can be made arbitrarily close to 1, by choosing $|b|$ closer and closer to $|a|^{-1/l}$.

(ii): The claim about the cotangent complex is just a restatement of proposition 6.3.13(iii),(iv). By proposition 6.3.13(ii) we can write $E^+ = K^+[w]$ for some $w \in E^+$. Let $f(x) \in K^+[X]$ be the minimal polynomial of w. Claim 6.3.22 implies that $\mathscr{D}_{E^+/K^+} = (f'(w))$; a standard calculation yields $\Omega_{E^+/K^+} \simeq E^+/(f'(w))$, so the assertion holds. $\qquad\square$

Theorem 6.3.23. *Let* $(E, | \cdot |_E)$ *be a finite separable valued field extension of* $(K, | \cdot |)$ *and suppose that* K^+ *has rank one. Then* $F_0(\Omega_{E^{+a}/K^{+a}}) = \mathscr{D}_{E^+/K^+}$ *and* $H_i(\mathbb{L}_{E^+/K^+}) = 0$ *for* $i > 0$.

Proof. We begin with a few reductions:

Claim 6.3.24. We can assume that E is a Galois extension of K.

Proof of the claim. Indeed, let $(L, | \cdot |_L)$ be a Galois valued field extension of K extending $(E, |\cdot|_E)$. We obtain by transitivity ([50, II.2.1.2]) a distinguished triangle

$$(6.3.25) \qquad \mathbb{L}_{L^+/E^+}[-1] \to \mathbb{L}_{E^+/K^+} \otimes_{E^+} L^+ \to \mathbb{L}_{L^+/K^+} \to \mathbb{L}_{L^+/E^+}.$$

Suppose that the theorem is already known for the Galois extensions $K \subset L$ and $E \subset L$. Then (6.3.25) implies that $H_i(\mathbb{L}_{E^+/K^+}) = 0$ for $i > 0$ and moreover provides a short exact sequence

$$0 \to \Omega_{E^+/K^+} \otimes_{E^+} L^+ \to \Omega_{L^+/K^+} \to \Omega_{L^+/E^+} \to 0.$$

However, on one hand, by lemma 6.3.12(ii) the different is multiplicative in towers of extensions, and the other hand, the Fitting ideal F_0 is multiplicative for short exact sequences, by virtue of remark 6.3.5(ii), so the claim follows. $\qquad\diamond$

Claim 6.3.26. We can assume that K^+ is strictly henselian.

Proof of the claim. Let $K^{+\mathrm{sh}}$ be the strict henselisation of K^+ and $K^{\mathrm{sh}} := \mathrm{Frac}(K^{+\mathrm{sh}})$. By inspecting the definitions, we see that it suffices to show that

$$(6.3.27) \quad F_0(\Omega^a_{W_E/K^+}) = \mathscr{D}_{W_E/K^+} \qquad \text{and} \qquad H_i(\mathbb{L}_{W_E/K^+}) = 0 \text{ for } i > 0.$$

It is known that $K^{+\mathrm{sh}}$ is an ind-étale extension of K^+, therefore $W_E \otimes_{K^+} K^{+\mathrm{sh}}$ is a reduced normal semilocal integral and flat $K^{+\mathrm{sh}}$-algebra, whence a product of reduced normal local integral and flat $K^{+\mathrm{sh}}$-algebras $W_1,..., W_k$. Each such W_i is necessarily the integral closure of $K^{+\mathrm{sh}}$ in $E_i := \mathrm{Frac}(W_i)$. It follows that

$$\mathbb{L}_{W_E/K^+} \otimes_{K^+} K^{+\mathrm{sh}} \simeq \mathbb{L}_{W_E \otimes_{K^+} K^{+\mathrm{sh}}/K^{+\mathrm{sh}}} \simeq \oplus_{i=1}^k \mathbb{L}_{W_i/K^{+\mathrm{sh}}}.$$

Furthermore:

$$\mathscr{D}_{W_E/K^+} \otimes_{K^{+a}} (K^{+\mathrm{sh}})^a \simeq \oplus_{i=1}^k \mathscr{D}_{E_i^+/K^{+\mathrm{sh}}}$$

and similarly for the modules of differentials. We remark as well that the formation of Fitting ideals commutes with arbitrary base changes. In conclusion, it is clear that (6.3.27) holds if and only if the theorem holds for each extension $K^{\mathrm{sh}} \subset E_i$. $\qquad\diamond$

Let p be the characteristic exponent of the residue field κ of K.

Claim 6.3.28. Suppose $K = K^{\mathrm{sh}}$. We can assume that $\mathrm{Gal}(E/K)$ is a p-group.

Proof of the claim. Indeed, let P be the kernel of (6.2.11). By proposition 6.2.12, P is p-group; let L be the fixed field of P. Then $L \subset K^t$ and, by virtue of corollary 6.2.14, we see that L admits a chain of subextensions $K := L_0 \subset L_1 \subset ... \subset L_k := L$ such that each $L_i \subset L_{i+1}$ satisfies condition (a) of proposition 6.3.13. Then, by corollary 6.3.21 it follows that the assertions of the theorem are already known for the extensions $L_i \subset L_{i+1}$. From here, using transitivity of the cotangent complex and multiplicativity of the different in towers of extensions, and of the Fitting ideals for short exact sequences, one shows that the assertions hold also for the extension $K \subset L$ (cp. the proof of claim 6.3.24). Now, if the assertions are known to hold as well for the extension $L \subset E$, again the same argument proves them for $K \subset E$. $\Diamond$

Claim 6.3.29. *The theorem holds if the valuation of K is discrete.*

Proof of the claim. By claim 6.3.28, we can suppose that $\mathrm{Gal}(E/K)$ is a p-group. Hence, we can find a sequence of subextensions $E_0 := K \subset E_1 \subset E_2 \subset ... \subset E_n := E$ with $[E_{i+1} : E_i] = p$, for every $i = 0, ..., n - 1$. Arguing like in the proof of claim 6.3.28 we see that it suffices to prove the claim for each of the extensions $E_i \subset E_{i+1}$. In this case we are left to dealing with an extension $K \subset E$ of degree p, which is taken care of by corollary 6.3.21(ii). $\Diamond$

Claim 6.3.30. *Suppose $K = K^{\mathrm{sh}}$, that $\mathrm{Gal}(E/K)$ is a p-group and that the valuation of K is not discrete. Let L be a finite Galois extension of K such that $([L : K], p) = 1$. Then the natural map $E^+ \otimes_{K^+} L^+ \to (E \cdot L)^+$ is an isomorphism.*

Proof of the claim. By corollary 6.2.14 we know that L admits a tower of subextensions of the form $K := L_0 \subset L_1 \subset ... \subset L_k := L$, such that, for each $i \leq k$ we have $L_{i+1} = L_i[a^{1/l}]$ for some $a \in L_i$ and some prime $l \neq p$. By induction on i, we can then reduce to the case where $L = K[a^{1/l}]$ for some $a \in K$ and a prime $l \neq p$. Under the above assumptions, we must have $E \cap L = K$, hence $a^{1/l} \notin E$. Then $E \cdot L = E[a^{1/l}]$ and by proposition 6.3.13 and its proof, $(E \cdot L)^+$ is the filtered union of all its subalgebras of the form $E^+[b \cdot a^{1/l}]$, where $b \in E$ ranges over all the elements such that $|b^l \cdot a| < 1$. However, since the valuation of K is not discrete and has rank one, Γ_K is dense in Γ_E, and consequently the subfamily consisting of the $E^+[b \cdot a^{1/l}]$ with $b \in K$ is cofinal. Finally, for $b \in K$ we have $E^+[b \cdot a^{1/l}] \simeq E^+ \otimes_{K^+} K^+[b \cdot a^{1/l}]$. By taking colimits, it follows that $(E \cdot L)^+ \simeq E^+ \otimes_{K^+} L^+$. $\Diamond$

Claim 6.3.31. *We can assume that K is equal to K^t.*

Proof of the claim. By claim 6.3.29 we may assume that the valuation of K is not discrete, and by claim 6.3.26 we may also assume that $K = K^{\mathrm{sh}}$, in which case K^t is the filtered union of all the finite Galois extension L of K such that $([L : K], p) = 1$. Then $K^{t+} = \bigcup_L L^+$ and $(E \cdot K^t)^+ = \bigcup_L (E \cdot L)^+$, where L ranges over all such extensions. By claim 6.3.28 we can also assume that $\mathrm{Gal}(E/K)$ is a p-group, in which case, by claim 6.3.30, we have $E^+ \otimes_{K^+} L^+ \xrightarrow{\sim} (E \cdot L)^+$ for every L as above. Taking colimit, we get $E^+ \otimes_{K^+} K^{t+} \xrightarrow{\sim} (E \cdot K^t)^+$. Since K^{t+} is faithfully flat over K^+, this shows that, in order to prove the theorem, we can replace K by K^t; however, by (6.2.18) we have $(K^t)^t = K^t$, whence the claim. ◇

After this preparation, we are ready to finish the proof of the theorem. We are reduced to considering a Galois extension E of $K = K^t$ such that $\operatorname{Gal}(E/K)$ is a p-group. Then, arguing as in the proof of claim 6.3.29, we can further reduce to dealing with an extension $K \subset E$ of degree p; furthermore, the condition $K = K^t$ still holds, by virtue of (6.2.18). In this situation, condition (b) of proposition 6.3.13 is fulfilled, hence $H_j(\mathbb{L}_{E^+/K^+}) = 0$ for $j > 0$, by proposition 6.3.13(iii). It remains to show the identity $F_0(\Omega_{E^{+a}/K^{+a}}) = \mathscr{D}_{E^+/K^+}$. By proposition 6.3.13(i), there exists $x \in E$ such that E^+ is the filtered union of a family of finite K^+-subalgebras $E_i^+ := K^+[a_i x + b_i]$ ($i \in \mathbb{N}$) of E^+. Let $f(X) \in K^+[X]$ be the minimal polynomial of x. By construction of E_i^{+a}, it is clear that they form a Cauchy net in $\mathscr{I}_{K^{+a}}(E^{+a})$ converging to E^{+a}. Then lemma 2.5.27 says that the net $\{\Omega_{E_i^+/K^+}^a \otimes_{E_i^{+a}} E^{+a} \mid i \in \mathbb{N}\}$ converges to $\Omega_{E^{+a}/K^{+a}}$ in $\mathscr{M}(E^{+a})$. In particular,

$$F_0(\Omega_{E^+/K^+}) = \lim_{i \to \infty} F_0(\Omega_{E_i^+/K^+} \otimes_{E_i^+} E^+).$$

The minimal polynomial of $a_i x_i + b_i$ is $f_i(X) := f(a_i^{-1}(X - b_i))$, therefore: $\Omega_{E_i^+/K^+} = E_i^+/(f_i'(a_i x_i + b_i)) = E_i^+/(a_i^{-1} f'(x))$. Consequently, $F_0(\Omega_{E^+/K^+}) = \lim_{i \to \infty} (a_i^{-1} f'(x))$. On the other hand, claim 6.3.22 yields: $\mathscr{D}_{E_i^+/K^+} = (a_i^{-1} f'(x))$ for every $i \in \mathbb{N}$. Then the claim follows from lemma 4.1.28. □

The final theorem of this section completes and extends theorem 6.3.23 to include valuations of arbitrary rank.

Theorem 6.3.32. *Let* $(K, |\cdot|)$ *be any valued field and* $(E, |\cdot|_F)$ *any algebraic valued field extension of* $(K, |\cdot|)$. *We have :*

(i) $H_i(\mathbb{L}_{E^+/K^+}) = 0$ *for* $i > 1$ *and* $H_1(\mathbb{L}_{E^+/K^+})$ *is a torsion-free* E^+-*module.*

(ii) *If moreover,* E *is separable over* K, *then* $H_i(\mathbb{L}_{E^+/K^+}) = 0$ *for* $i > 0$.

Proof. Let us show first how to deduce (ii) from (i). Indeed, suppose that E is separable over K. Then $\mathbb{L}_{E/K} \simeq 0$. However, by (i), the natural map

$$H_1(\mathbb{L}_{E^+/K^+}) \to H_1(\mathbb{L}_{E^+/K^+}) \otimes_{K^+} K \simeq H_1(\mathbb{L}_{E/K})$$

is injective, so the assertion follows.

In order to prove (i), let us write K as the filtered union of its subfields L_α that are finitely generated over the prime field. For each such L_α, let $K_\alpha := (L_\alpha)^a \cap K$ and $E_\alpha := (L_\alpha)^a \cap E$. Then E_α is an algebraic extension of K_α and K (resp. E) is the filtered union of the K_α (resp. E_α). It follows easily that we can replace the extension $K \subset E$ by the extension $K_\alpha \subset E_\alpha$, thereby reducing to the case where the transcendence degree of K over its prime field is finite. Similarly, we reduce easily to the case where E is a finite extension of K. In this situation, the rank r of K is finite (cp. (6.1.23)). We argue by induction on r. Suppose first that $r = 1$. We can split into a tower of extensions $K \subset K^s \cap E \subset E$; then, by using transitivity (cp. the proof of claim 6.3.24), we reduce easily to prove the assertion for the extensions $K \subset K^s \cap E$ and $K^s \cap E \subset E$. However, the first case is already covered by theorem 6.3.23, so we can assume that E is purely inseparable over K. In this case, we can further split E into a tower of extensions of degree equal to p; thus we reduce to the case where $[E : K] = p$. We apply transitivity to the tower $K \subset K^t \subset E \cdot K^t = E^t$: by proposition 6.3.13(iv) we know that $H_i(\mathbb{L}_{E^{t+}/K^{t+}})$ vanishes for $i > 1$ and is torsion-free for $i = 1$; by theorem 6.3.23, we have $H_i(\mathbb{L}_{K^{t+}/K^+}) = 0$ for $i > 0$, therefore $H_i(\mathbb{L}_{E^{t+}/K^+})$ vanishes for $i > 1$ and is torsion-free for $i = 1$. Next we apply transitivity to the tower $K \subset E \subset E^t$: by theorem 6.3.23 we have $H_i(\mathbb{L}_{E^{t+}/E^+}) = 0$ for $i > 0$, and the claim follows easily.

Next suppose that $r > 1$, and that the theorem is already known for ranks $< r$. Arguing as in the proof of claim 6.3.26, we can even reduce to the case where K^+ is henselian, and then E^+ is the integral closure of K^+ in E. Let $\mathfrak{p}_r := (0) \subset \mathfrak{p}_{r-1} \subset \dots \subset \mathfrak{p}_0$ be the chain of prime ideals of K^+, and for every $i \leq r$ let $\mathfrak{q}_i$ be the unique prime ideal of E^+ lying over $\mathfrak{p}_i$. The valuation ring $E^+_{\mathfrak{q}_1}$ has rank $r - 1$, thus, by inductive assumption, the desired assertions are known for the extension $K^+_{\mathfrak{p}_1} \subset E^+_{\mathfrak{q}_1}$. It suffices therefore to show that $H_i(\mathbb{L}_{E^+/K^+}) \subset H_i(\mathbb{L}_{E^+_{\mathfrak{q}_1}/K^+_{\mathfrak{p}_1}})$ for every $i > 0$. Pick $a \in \mathfrak{p}_0 \setminus \mathfrak{p}_1$. Then $K^+_{\mathfrak{p}_1} = K^+[a^{-1}]$ and $E^+_{\mathfrak{q}_1} = E^+[a^{-1}]$ and $\mathbb{L}_{E^+_{\mathfrak{q}_1}/K^+_{\mathfrak{p}_1}} = \mathbb{L}_{E^+/K^+} \otimes_{K^+} K^+[a^{-1}]$. Hence, we are reduced to show that multiplication by a is injective on the homology of $\mathbb{L}_{E^+/K^+}$. Let

$$R := K^+/aK^+ \quad \text{and} \quad R_E := E^+ \otimes_{K^+} R.$$

We have a short exact sequence $0 \to K^+ \xrightarrow{a} K^+ \to R \to 0$, therefore, after tensoring by $\mathbb{L}_{E^+/K^+}$, a distinguished triangle:

$$\mathbb{L}_{E^+/K^+} \xrightarrow{a} \mathbb{L}_{E^+/K^+} \to \mathbb{L}_{E^+/K^+} \overset{\mathbb{L}}{\otimes}_{K^+} R \to \sigma \mathbb{L}_{E^+/K^+}.$$

On the other hand, according to remark 6.1.12(ii), E^+ is flat over K^+, therefore $\mathbb{L}_{E^+/K^+} \overset{\mathbb{L}}{\otimes}_{K^+} R \simeq \mathbb{L}_{R_E/R}$ (by [50, II.2.2.1]). Consequently, it suffices to show that $H_i(\mathbb{L}_{R_E/R}) = 0$ for $i \geq 2$. However, $R = (K^+/\mathfrak{p}_1) \otimes_{K^+} R$, and $R_E = (E^+/\mathfrak{q}_1) \otimes_{K^+} R$; moreover, $E^+/\mathfrak{q}_1$ is the integral closure of the valuation ring $K^+/\mathfrak{p}_1$ in the finite field extension $\mathrm{Frac}(E^+/\mathfrak{q}_1)$ of $\mathrm{Frac}(K^+/\mathfrak{p}_1)$. Therefore we can replace K^+ by $K^+/\mathfrak{p}_1$ and E^+ by $K^+/\mathfrak{q}_1$. This turns us back to the case where $r = 1$. Then the vanishing of $H_i(\mathbb{L}_{E^+/K^+})$ for $i \geq 2$ yields the vanishing of $H_i(\mathbb{L}_{R_E/R})$ for $i > 2$. Moreover, since $H_1(\mathbb{L}_{E^+/K^+})$ is torsion-free, multiplication by a on $H_1(\mathbb{L}_{E^+/K^+})$ is injective, therefore $H_2(\mathbb{L}_{R_E/R})$ vanishes as well. $\square$

6.4. **Logarithmic differentials.** In this section K^+ is a valuation ring of arbitrary rank. We keep the notation of (6.2.1). We start by reviewing some facts on logarithmic structures, for which the general reference is [52].

6.4.1.   Let $\mathbf{Mnd}_X$ (reps. $\mathbb{Z}\text{-}\mathbf{Mod}_X$) be the category of sheaves of commutative monoids (resp. of abelian groups) on a topological space X. The forgetful functor $\mathbb{Z}\text{-}\mathbf{Mod}_X \to \mathbf{Mnd}_X$ admits a left adjoint functor $\underline{M} \mapsto \underline{M}^{\mathrm{gp}}$. If $\underline{M}$ is a sheaf of monoids, $\underline{M}^{\mathrm{gp}}$ is the sheaf associated to the presheaf defined by : $U \mapsto \underline{M}(U)^{\mathrm{gp}}$ for every open subset $U \subset X$.

The functor $\Gamma : \mathbf{Mnd}_X \to \mathbf{Mnd}$ that associates to every sheaf of monoids its global sections, admits a left adjoint $\mathbf{Mnd} \to \mathbf{Mnd}_X : M \mapsto M_X$. For a monoid M, M_X is the sheaf associated to the constant presheaf with value M.

6.4.2. Recall that a *pre-log structure* on a scheme X is a morphism of sheaves of commutative monoids : $\alpha : \underline{M} \to \mathcal{O}_X$, where the monoid structure of $\mathcal{O}_X$ is induced by multiplication of local sections. We denote by $\mathbf{pre\text{-}log}_X$ the category of pre-log structures on X.

To a monoid M and a morphism of monoids $\phi : M \to \Gamma(X, \mathcal{O}_X)$, one can associate a pre-log structure $\phi_X : M_X \to \mathcal{O}_X$ by composing the induced morphism of constant sheaves $M_X \to \Gamma(X, \mathcal{O}_X)_X$ with the counit of the adjunction $\Gamma(X, \mathcal{O}_X)_X \to \mathcal{O}_X$.

6.4.3. To a morphism $\phi : Y \to X$ of schemes, one can associate a pair of adjoint functors:

$$\phi^* : \mathbf{pre\text{-}log}_X \to \mathbf{pre\text{-}log}_Y \qquad \phi_* : \mathbf{pre\text{-}log}_Y \to \mathbf{pre\text{-}log}_X.$$

Namely, let $(\underline{M}, \alpha : \underline{M} \to \mathcal{O}_X)$ (resp. $(\underline{N}, \beta : \underline{N} \to \mathcal{O}_Y)$) be a pre-log structure on X (resp. on Y) and

$$\phi^\flat : \mathcal{O}_X \to \phi_* \mathcal{O}_Y \qquad \phi^\sharp : \phi^{-1} \mathcal{O}_X \to \mathcal{O}_Y$$

the natural morphisms (corresponding to each other under the adjunction (ϕ^{-1}, ϕ_*) on sheaves of $\mathbb{Z}$-modules); then

$$\phi^{-1} \underline{M} \xrightarrow{\phi^{-1}\alpha} \phi^{-1} \mathcal{O}_X \xrightarrow{\phi^\sharp} \mathcal{O}_Y$$

defines $\phi^*(\underline{M}, \alpha : \underline{M} \to \mathcal{O}_X)$ and $\phi_*(\underline{N}, \beta : \underline{N} \to \mathcal{O}_Y)$ is the projection on the second factor $\gamma : \phi_* \underline{N} \times_{\phi_* \mathcal{O}_Y} \mathcal{O}_X \to \mathcal{O}_X$, which makes commute the cartesian diagram

$$\begin{array}{ccc}
\phi_* \underline{N} \times_{\phi_* \mathcal{O}_Y} \mathcal{O}_X & \xrightarrow{\gamma} & \mathcal{O}_X \\
\downarrow & & \downarrow{\phi^\flat} \\
\phi_* \underline{N} & \xrightarrow{\phi_* \beta} & \phi_* \mathcal{O}_Y.
\end{array}$$

6.4.4. A pre-log structure α is said to be a *log structure* if $\alpha^{-1}(\mathcal{O}_X^\times) \xrightarrow{\sim} \mathcal{O}_X^\times$. We denote by $\mathbf{log}_X$ the category of log structures on X. The forgetful functor

$$\mathbf{log}_X \to \mathbf{pre\text{-}log}_X : \underline{M} \mapsto \underline{M}^{\mathrm{pre\text{-}log}}$$

admits a left adjoint

(6.4.5) $$\mathbf{pre\text{-}log}_X \to \mathbf{log}_X : \underline{M} \mapsto \underline{M}^{\log}$$

and the resulting diagram:

$$(6.4.6) \quad \begin{array}{ccc} \alpha^{-1}(\mathscr{O}_X^\times) & \longrightarrow & \underline{M} \\ \downarrow & & \downarrow \\ \mathscr{O}_X^\times & \longrightarrow & \underline{M}^{\log} \end{array}$$

is cocartesian in the category of pre-log structures. From this, one can easily verify that the counit of the adjunction : $(\underline{M}^{\text{pre-log}})^{\log} \to \underline{M}$ is an isomorphism for every log structure $\underline{M}$.

6.4.7.   The category $\log_X$ admits arbitrary colimits; indeed, since the counit of the adjunction (6.4.5) is an isomorphism, it suffices to construct such colimits in the category of pre-log structures, and then apply the functor $(-) \mapsto (-)^{\log}$ which preserves colimits, since it is a left adjoint. In particular, $\log_X$ admits arbitrary direct sums, and for any family $(\underline{M}_i \mid i \in I)$ of pre-log structures we have $(\oplus_{i \in I} \underline{M}_i)^{\log} \simeq \oplus_{i \in I} \underline{M}_i^{\log}$.

6.4.8.   For any morphism of schemes $Y \to X$ we remark that, if $(\underline{M}, \alpha)$ is a log structure on Y, then the pre-log structure $\phi_*(\underline{M}, \alpha)$ is actually a log structure (this can be checked on the stalks). We deduce a pair of adjoint functors (ϕ^*, ϕ_*) for log structures, as in (6.4.3). These are formed by composing the corresponding functors for pre-log structures with the functor (6.4.5).

6.4.9.   We say that a log structure $\underline{M}$ is *regular* if $\underline{M} = (M_X, \phi_X)^{\log}$ for some free monoid M, and the associated morphism of monoids $\phi : M \to \Gamma(X, \mathscr{O}_X)$ maps M into the set of elements of $\Gamma(X, \mathscr{O}_X)$ that are non-zero-divisors on $\mathscr{O}_{X,x}$ for every $x \in X$.

6.4.10.   For an $\mathscr{O}_X$-module $\mathscr{F}$, denote by $\mathbf{Hom}_{\mathscr{O}_X}(\mathscr{F}, *)$ the category of all homomorphisms of $\mathscr{O}_X$-modules $\mathscr{F} \to \mathscr{A}$ (for any $\mathscr{O}_X$-module $\mathscr{A}$). A morphism from $\mathscr{F} \to \mathscr{A}$ to $\mathscr{F} \to \mathscr{B}$ is a morphism $\mathscr{A} \to \mathscr{B}$ of $\mathscr{O}_X$-modules which induces the identity on $\mathscr{F}$. This category admits arbitrary colimits.

6.4.11.   Given a pre-log structure $\alpha : \underline{M} \to \mathscr{O}_X$, one defines the sheaf of *logarithmic differentials* $\Omega_{X/\mathbb{Z}}(\log \underline{M})$ as the quotient of the $\mathscr{O}_X$-module $\Omega_{X/\mathbb{Z}} \oplus (\mathscr{O}_X \otimes_{\mathbb{Z}_X} \underline{M}^{\text{gp}})$ by the $\mathscr{O}_X$-submodule generated by the local sections of the form $(d\alpha(m), -\alpha(m) \otimes m)$, for every local section m of $\underline{M}$. (The meaning of this is that one adds to $\Omega_{X/\mathbb{Z}}$ the logarithmic differentials $\alpha(m)^{-1} d\alpha(m)$.) For every local section m of $\underline{M}$, we denote by $d\log(m)$ the image of $1 \otimes m$ in $\Omega_{X/\mathbb{Z}}(\log \underline{M})$. The assignment $\underline{M} \mapsto (\Omega_{X/\mathbb{Z}} \to \Omega_{X/\mathbb{Z}}(\log \underline{M}))$ defines a (covariant) functor :

$$\Omega : \mathbf{pre\text{-}log}_X \to \mathbf{Hom}_{\mathscr{O}_X}(\Omega_{X/\mathbb{Z}}, *).$$

Lemma 6.4.12. *Let X be a scheme.*

  (i) *The functor Ω commutes with all colimits.*
  (ii) *The functor Ω factors through the functor* (6.4.5).

(iii) *Let $j : U \to X$ be a formally étale morphism of schemes and $\underline{M}$ a log structure on X. Then the natural morphism:*

$$j^*\Omega_{X/\mathbb{Z}}(\log \underline{M}) \to \Omega_{U/\mathbb{Z}}(\log j^*\underline{M})$$

is an isomorphism.

(iv) *If $\underline{M}$ is a regular log structure, then $\Omega(\underline{M})$ is a monomorphism of $\mathcal{O}_X$-modules.*

(v) *If $\underline{M}$ is the pre-log structure associated to a morphism of monoids $\phi : M \to \Gamma(X, \mathcal{O}_X)$, the sheaf $\Omega_{X/\mathbb{Z}}(\log \underline{M})$ is quasi-coherent, and if $X = \mathrm{Spec}(A)$ is affine, its module of global sections is the quotient of $\Omega_{A/\mathbb{Z}} \oplus (A \otimes_{\mathbb{Z}} M^{\mathrm{gp}})$ by the A-submodule generated by the elements of the form $(d\phi(m), -\phi(m) \otimes m)$, for all $m \in M$.*

Proof. (i): It is clear that Ω commutes with filtered colimits. Thus, to show that it commutes with all colimits, it suffices to show that it commutes with finite direct sums and with coequalizers. We consider first direct sums. We have to show that, for any two pre-log structures $\underline{M}_1$ and $\underline{M}_2$, the natural morphism

$$\Omega_{X/\mathbb{Z}}(\log \underline{M}_1) \underset{\Omega_{X/\mathbb{Z}}}{\mathrm{II}} \Omega_{X/\mathbb{Z}}(\log \underline{M}_2) \to \Omega_{X/\mathbb{Z}}(\log \underline{M}_1 \oplus \underline{M}_2)$$

is an isomorphism. Notice that the functor $(-) \mapsto (-)^{\mathrm{gp}}$ of (6.4.1) commutes with colimits, since it is a left adjoint. It follows that the diagram

$$
\begin{array}{ccc}
\Omega_{X/\mathbb{Z}} & \longrightarrow & \Omega_{X/\mathbb{Z}} \oplus (\mathcal{O}_X \otimes_{\mathbb{Z}x} \underline{M}_1^{\mathrm{gp}}) \\
\downarrow & & \downarrow \\
\Omega_{X/\mathbb{Z}} \oplus (\mathcal{O}_X \otimes_{\mathbb{Z}x} \underline{M}_2^{\mathrm{gp}}) & \longrightarrow & \Omega_{X/\mathbb{Z}} \oplus (\mathcal{O}_X \otimes_{\mathbb{Z}x} (\underline{M}_1 \oplus \underline{M}_2)^{\mathrm{gp}})
\end{array}
$$

is cocartesian. Thus, we are reduced to show that the kernel of the map

$$\Omega_{X/\mathbb{Z}} \oplus (\mathcal{O}_X \otimes_{\mathbb{Z}x} \underline{M}^{\mathrm{gp}}) \to \Omega_{X/\mathbb{Z}}(\log \underline{M})$$

is generated by the images of the kernels of the corresponding maps relative to $\underline{M}_1$ and $\underline{M}_2$. However, any section of $\underline{M}_1 \oplus \underline{M}_2$ can be written locally in the form $x \cdot y$ for two local sections x of $\underline{M}_1$ and y of $\underline{M}_2$. Then we have :

$$(d\alpha(xy), -\alpha(xy) \otimes xy) = (\alpha(x) \cdot d\alpha(y) + \alpha(y) \cdot d\alpha(x), -\alpha(x) \cdot \alpha(y) \otimes xy)$$
$$= \alpha(x) \cdot (d\alpha(y), -\alpha(y) \otimes y) + \alpha(y) \cdot (d\alpha(x), -\alpha(x) \otimes x)$$

so the claim is clear. Next, suppose that $\phi, \psi : \underline{M} \to \underline{N}$ are two morphisms of pre-log structures. Let : $\alpha : Q \to \mathcal{O}_X$ be the coequalizer of ϕ and ψ. Clearly, Q is the coequalizer of ϕ and ψ in the category of sheaves of monoids. The functor $\underline{M} \mapsto \underline{M}^{\mathrm{gp}}$ preserves colimits, so Q^{gp} is the cokernel of $\beta := \phi^{\mathrm{gp}} - \psi^{\mathrm{gp}}$. Moreover, clearly we have $\mathrm{Coker}(\beta) \otimes_{\mathbb{Z}x} \mathcal{O}_X \simeq \mathrm{Coker}(\beta \otimes_{\mathbb{Z}x} \mathcal{O}_X)$; the claim follows as above.

(ii): Let us apply the functor Ω to the cocartesian diagram (6.4.6). In view of (i), the resulting diagram of $\mathcal{O}_X$-modules is cocartesian. However, it is easy to check that $\Omega_{X/\mathbb{Z}}(\log \alpha^{-1}(\mathcal{O}_X^\times)) \simeq \Omega_{X/\mathbb{Z}}(\log \mathcal{O}_X^\times) \simeq \Omega_{X/\mathbb{Z}}$. The assertion follows directly.

(iii): One uses [30, Ch.IV, Cor. 17.2.4]; the details will be left to the reader.

(iv): By (ii), the functor Ω descends to a functor

(6.4.13) $$\log_X \to \mathbf{Hom}_{\mathscr{O}_X}(\Omega_{X/\mathbb{Z}}, *).$$

Since the unit of the adjunction (6.4.5) is an isomorphism, it follows easily that (6.4.13) commutes with all colimits of log structures. Hence, to verify that $\Omega(\underline{M})$ is a monomorphism when $\underline{M}$ is regular, we are immediately reduced to the case when $\underline{M}$ is the regular log structure associated to a morphism of monoids $\phi : \mathbb{N} \to \Gamma(X, \mathscr{O}_X)$. Let $f := \phi(1)$. It is easy to check that in this case, the diagram

$$
\begin{array}{ccc}
\mathscr{O}_X & \xrightarrow{\ f\ } & \mathscr{O}_X \\
{\scriptstyle df}\downarrow & & \downarrow{\scriptstyle d\log f} \\
\Omega_{X/\mathbb{Z}} & \xrightarrow{\ \beta\ } & \Omega_{X/\mathbb{Z}}(\log \underline{M})
\end{array}
$$

is cocartesian. By assumption, multiplication by f is a monomorphism of $\mathscr{O}_X$-modules, so the assertion follows.

(v) is easy and shall be left to the reader. $\qquad\square$

6.4.14.   This general formalism will be applied here to the following situation. We consider the submonoid $M := K^+ \setminus \{0\}$ of K^+. The imbedding $M \subset K^+$ induces a log structure on $\operatorname{Spec} K^+$, which we call the *total log structure* on K^+. More generally, we consider the natural projection $\pi : M \to \Gamma^+$ (see (6.1.24)); then for every submonoid $N \subset \Gamma^+$, we have a log structure $\underline{N}$ corresponding to the imbedding $\pi^{-1}(N) \subset K^+$. To ease notation, we will denote by $\Omega_{K^+/\mathbb{Z}}(\log N)$ the corresponding K^+-module of logarithmic differentials.

Proposition 6.4.15. *In the situation of (6.4.14), let $\Delta \subset \Gamma$ be any subgroup, N a prime ideal of Δ^+ (terminology of (6.1.26)) and suppose that the convex rank of $\Sigma := \Delta/(\Delta^+ \setminus N)^{\mathrm{gp}}$ equals one. Then we have a short exact sequence*

$$0 \to \Omega_{K^+/\mathbb{Z}}(\log \Delta^+ \setminus N) \xrightarrow{\ j\ } \Omega_{K^+/\mathbb{Z}}(\log \Delta^+) \xrightarrow{\ \rho\ } \Sigma \otimes_{\mathbb{Z}} (K^+/\pi^{-1}(N) \cdot K^+) \to 0.$$

Proof. Let us first remark that the assumptions and the notation make sense : indeed, since N is a prime ideal of Δ^+, it follows that $M := \Delta^+ \setminus N$ is a convex submonoid of Δ^+, hence $M = (M^{\mathrm{gp}})^+$ and M^{gp} is a convex subgroup of Δ (cp. (6.1.25)), therefore Σ is an ordered group (cp. example (6.1.21)(v)), and hence it makes sense to say that its convex rank equals one.

Let us show that j is injective. We can write Δ^+ as the colimit of the filtered family of its finitely generated submonoids F_α. For each such F_α, theorem 6.1.30 gives us a free finitely generated submonoid $L_\alpha \subset \Delta^+$ such that $F_\alpha \subset L_\alpha$. Clearly Δ^+ is the colimit of the L_α (ordered by inclusion), and M is the colimit of the $M_\alpha := M \cap L_\alpha$. Thus $\underline{\Delta}^+$ is the colimit of the $\underline{L}_\alpha$ and $\underline{M}$ is the colimit of the $\underline{M}_\alpha$. Let S_α be a basis of L_α. Since M is convex in Δ, we see that M_α is free with basis $S_\alpha \cap M$ and $L_\alpha = M_\alpha \oplus N_\alpha$, where N_α is the free submonoid spanned by $S_\alpha \setminus M$. For each $e \in S_\alpha \setminus M$, pick arbitrarily an element $x_e \in K^+$ such that $|x_e| = e$. The map $e \mapsto x_e$ can be extended to a map of monoids $N_\alpha \to K^+$, and then to a pre-log structure $\nu_\alpha : (N_\alpha)_{\mathrm{Spec}V} \to \mathcal{O}_{\mathrm{Spec}K^+}$. Clearly we have an isomorphism of pre-log structures: $\underline{L_\alpha} = \underline{M_\alpha} \oplus \nu_\alpha$. Since the formation of logarithmic differentials commutes with colimits of monoids, we are reduced to showing that the analogous map

$$j_\alpha : \Omega_{K^+/\mathbb{Z}}(\log M_\alpha) \to \Omega_{K^+/\mathbb{Z}}(\log L_\alpha)$$

is injective. By lemma 6.4.12(i), we have an epimorphism

$$\mathrm{Ker}(\Omega(\nu_\alpha) : \Omega_{K^+/\mathbb{Z}} \to \Omega_{K^+/\mathbb{Z}}(\log \nu_\alpha)) \to \mathrm{Ker}\, j_\alpha.$$

By lemma 6.4.12(iv), the map $\Omega(\nu_\alpha)$ is injective, whence the assertion.

Next we proceed to show how to construct ρ. Define a map

$$\widetilde{\rho} : X := \Omega_{K^+/\mathbb{Z}} \oplus ((\pi^{-1}\Delta)^{\mathrm{gp}} \otimes_{\mathbb{Z}} K^+) \to \Sigma \otimes_{\mathbb{Z}} (K^+/\pi^{-1}(N) \cdot K^+)$$

by the rule : $(\omega, a \otimes b) \mapsto \overline{\pi(a)} \otimes \overline{b}$, for any $\omega \in \Omega_{K^+/\mathbb{Z}}, a \in \pi^{-1}\Delta, b \in K^+$.

Claim 6.4.16. $\mathrm{Ker}\,\widetilde{\rho}$ contains the kernel of the surjection $X \to \Omega_{K^+/\mathbb{Z}}(\log \Delta^+)$.

Proof of the claim. It suffices to show that $(0, a \otimes a) \in \mathrm{Ker}\,\widetilde{\rho}$ whenever $\pi(a) \notin (\Delta^+ \setminus N)^{\mathrm{gp}}$. However, $\pi(a) \notin (\Delta^+ \setminus N)^{\mathrm{gp}} \Leftrightarrow \pi(a) \notin \Delta^+ \setminus N \Leftrightarrow \pi(a) \in N \Leftrightarrow a \in \pi^{-1}(N)$, so the claim follows. $\Diamond$

By claim 6.4.16 we deduce that $\widetilde{\rho}$ descends to the map ρ as desired. It is now obvious that ρ is surjective and that its kernel contains the image of j. To conclude the proof, it suffices to show that the cokernel of j is annihilated by $\pi^{-1}N$. A slightly stronger result holds, namely:

Claim 6.4.17. $\pi^{-1}N$ annihilates $\mathrm{Coker}(i : \Omega_{K^+/\mathbb{Z}} \to \Omega_{K^+/\mathbb{Z}}(\log \Delta^+))$.

Proof of the claim. It suffices to show that $\pi^{-1}(N)$ annihilates the classes in $\mathrm{Coker}\, i$ of the elements $d\log(e)$, for every $e \in \pi^{-1}\Delta^+$. Let $a \in \pi^{-1}(N)$. Since the convex rank of Σ equals one, and $\pi(e) \leq 1$, there exists $k \geq 0$ and $b \subset K^+$ such that $e = a^k \cdot b$ and $1 \geq |b| > |a|$. Hence $d\log(e) = d\log(a^k \cdot b) = k \cdot d\log(a) + d\log(b)$, and it is clear that a annihilates the images in $\mathrm{Coker}\, i$ of each of the terms of this expression. $\Box$

Corollary 6.4.18. *In the situation of (6.4.14), we have :*

(i) *The natural map $\beta_{K^+} : \Omega_{K^+/\mathbb{Z}} \to \Omega_{K^+/\mathbb{Z}}(\log \Gamma^+)$ is injective.*

(ii) *Suppose moreover that K^+ has finite rank. Denote by*

$$\Delta_r := \Gamma \supset \Delta_{r-1} \supset ... \supset \Delta_0 := 0$$

the ascending chain of all convex subgroups of Γ. Then $\mathrm{Coker}\, \beta_{K^+}$ admits a finite ascending filtration indexed by the totally ordered set $\mathrm{Spec}\,\Gamma$, with:

$$\mathrm{Fil}_{\Delta_0}(\mathrm{Coker}\, \beta_{K^+}) = 0 \qquad \mathrm{Fil}_{\Delta_r}(\mathrm{Coker}\, \beta_{K^+}) = \mathrm{Coker}\, \beta_{K^+}$$

and:

$$\mathrm{gr}_{\Delta_i}(\mathrm{Coker}\, \beta_{K^+}) \simeq (\Delta_i/\Delta_{i-1}) \otimes_{\mathbb{Z}} (K^+/\mathfrak{p}_{\Delta_{i-1}})$$

for every $\Delta_i \in \mathrm{Spec}\,\Gamma \setminus \{\Delta_0\}$ (notation of (6.1.22)).

(iii) *Suppose furthermore that $(K, |\cdot|) \subset (E, |\cdot|_E)$ is an extension of valued fields of finite rank, and that the value group Γ of $|\cdot|$ is divisible. Then $\operatorname{Coker}\beta_{K^+}$ is a $\mathbb{Q}$-vector space and the induced map*

$$\phi : E^+ \otimes_{K^+} \operatorname{Coker}\beta_{K^+} \to \operatorname{Coker}\beta_{E^+}$$

is injective.

Proof. (i): Since the formation of differentials and logarithmic differentials commutes with filtered colimits of $\mathbb{Z}$-algebras and log structures, we can reduce to the case where K is a field of finite type over its prime field. In this case the convex rank of Γ is finite, so the assertion follows from proposition 6.4.15 and an easy induction.

(ii) is a straightforward consequence of proposition 6.4.15.

(iii): Under our current assumptions, K^+ and E^+ are valuation rings of finite Krull dimension, by (6.1.23). Let $\Delta_s := \Gamma_E \supset \Delta_{s-1} \supset \ldots \supset \Delta_0 := 0$ be the ascending chain of convex subgroups of Γ_E. Let $\operatorname{Fil}_\bullet(\operatorname{Coker}\beta_{E^+})$ (resp. $\operatorname{Fil}_\bullet(\operatorname{Coker}\beta_{K^+})$) be the finite filtration indexed by the totally ordered set $\operatorname{Spec}\Gamma_E$ (resp. $\operatorname{Spec}\Gamma$), provided by (ii). Since it is preferable to work with a single indexing set, we use the surjections

$$\operatorname{Spec}\Gamma_E \xrightarrow{j_\Gamma^*} \operatorname{Spec}\Gamma \ : \ \Delta_i \mapsto \Delta_i \cap \Gamma \qquad \operatorname{Spec}E^+ \xrightarrow{j^*} \operatorname{Spec}K^+ \ : \ \mathfrak{p}_i \mapsto \mathfrak{p}_i \cap K^+$$

to replace by $\operatorname{Spec}\Gamma_E$ the indexing of the filtration on $\operatorname{Coker}\beta_{K^+}$; of course in this way some of the graded subquotients become trivial, but we do not mind. We remark also that j^* and j_Γ^* admit canonical sections; on prime spectra this is the right inverse mapping

(6.4.19) $$\operatorname{Spec}K^+ \to \operatorname{Spec}E^+ \qquad \mathfrak{q} \mapsto \mathfrak{q} \cdot E^+.$$

Indeed, since the value group of K^+ is divisible, one sees easily that, for any prime ideal $\mathfrak{q}$ of K^+, the extension $\mathfrak{q} \cdot E^+$ is a radical, hence prime ideal. With this notation we can write down the identities:

$$\operatorname{gr}_{\Delta_i}(\operatorname{Coker}\beta_{K^+}) \simeq (j_\Gamma^* \Delta_i / j_\Gamma^* \Delta_{i-1}) \otimes_{\mathbb{Z}} (K^+/j^* \mathfrak{p}_{\Delta_{i-1}})$$

for every $\Delta_i \in \operatorname{Spec}\Gamma_E \setminus \{\Delta_0\}$. The first assertion of (iii) follows already. Furthermore, the map ϕ respects these filtrations. If now $\mathfrak{p}_{\Delta_{i-1}} \in \operatorname{Spec}E^+$ is in the image of (6.4.19), then clearly the map

$$\operatorname{gr}_{\Delta_i}\phi : E^+ \otimes_{K^+} \operatorname{gr}_{\Delta_i}(\operatorname{Coker}\beta_{K^+}) \to \operatorname{gr}_{\Delta_i}(\operatorname{Coker}\beta_{E^+})$$

is induced by the natural maps

$$\psi_i : j_\Gamma^* \Delta_i / j_\Gamma^* \Delta_{i-1} \subset \Delta_i / \Delta_{i-1} \quad \text{and} \quad E^+ \otimes_{K^+} (K^+/j^* \mathfrak{p}_{\Delta_{i-1}}) \xrightarrow{\sim} E^+/\mathfrak{p}_{\Delta_{i-1}}.$$

Since $j_\Gamma^* \Delta_i / j_\Gamma^* \Delta_{i-1}$ is uniquely divisible, ψ_i is universally injective, therefore $\operatorname{gr}_{\Delta_i}\phi$ is injective in this case. On the other hand, if $\mathfrak{p}_{\Delta_{i-1}}$ is not in the image of (6.4.19), we have $\operatorname{gr}_{\Delta_i}(\operatorname{Coker}\beta_{K^+}) = 0$, so $\operatorname{gr}_{\Delta_i}\phi$ is trivially injective in this case as well. Since the map $\operatorname{gr}_\bullet\phi$ is injective, the same holds for ϕ, which is the contention. $\qquad\qquad\square$

6.5. Transcendental extensions.
In this section we extend the results of section 6.3 to the case of arbitrary extensions of valued fields.

6.5.1. We fix the following notation throughout this section. For a valued field extension $(K, |\cdot|) \subset (E, |\cdot|_E)$, we let

$$\rho_{E^+/K^+} : E^+ \otimes_{K^+} \Omega_{K^+/\mathbb{Z}}(\log \Gamma^+) \to \Omega_{E^+/\mathbb{Z}}(\log \Gamma_E^+)$$

be the natural morphism. One of the main results of this section states that ρ_{E^+/K^+} is injective with torsion-free cokernel when K is algebraically closed (theorem 6.5.20(ii)) or when K has characteristic zero (lemma 6.5.16). Lurking behind the results of this sections there should be some notion of "logarithmic cotangent complex", which however is not currently available.

6.5.2. Let $\mathfrak{G} := (G, j, N, \leq)$ be a datum as in (6.1.5). We wish to study the total log structure of the valued field $(K(\mathfrak{G}), |\cdot|_\mathfrak{G})$. We consider the morphism of monoids

(6.5.3) $$G \to K[\mathfrak{G}] \quad : \quad g \mapsto [g].$$

Let $K[\mathfrak{G}]^+$ be the subring of the elements $x \in K[\mathfrak{G}]$ such that $|x|_\mathfrak{G} \leq 1$. Let $\pi : G \to \Gamma_\mathfrak{G}$ be the projection; for every submonoid $M \subset \Gamma_\mathfrak{G}^+$, the preimage $\pi^{-1}M$ is a submonoid of G, and the restriction of (6.5.3) induces a morphism of monoids $\pi^{-1}(M) \to K[\mathfrak{G}]^+$, whence a pre-log structure $\pi^{-1}M_X$ on $X := \operatorname{Spec} K[\mathfrak{G}]^+$ (see (6.4.2)). To ease notation, we set $\underline{M} := (\pi^{-1}M_X)^{\log}$ and we will write $\Omega_{K[\mathfrak{G}]^+/\mathbb{Z}}(\log M)$ for the module of global sections of the associated sheaf of log differentials.

Lemma 6.5.4. *Resume the notation of (6.1.5). Then the natural diagram*

$$\begin{array}{ccc}
K^\times \otimes_\mathbb{Z} K[\mathfrak{G}]^+ & \xrightarrow{\ \alpha\ } & G \otimes_\mathbb{Z} K[\mathfrak{G}]^+ \\
{\scriptstyle \beta}\downarrow & & \downarrow{\scriptstyle \eta} \\
\Omega_{K^+/\mathbb{Z}}(\log \Gamma^+) \otimes_{K^+} K[\mathfrak{G}]^+ & \xrightarrow{\ \lambda\ } & \Omega_{K[\mathfrak{G}]^+/\mathbb{Z}}(\log \Gamma_\mathfrak{G}^+)
\end{array}$$

is cocartesian.

Proof. Here $\alpha := j \otimes_\mathbb{Z} 1_{K[\mathfrak{G}]^+}$ (notation of (6.1.5)); β (resp. η) is the map:

$$x \otimes y \mapsto d\log(x) \otimes y \quad (\text{resp. } g \otimes y \mapsto y \cdot d\log[g])$$

for every $x \in K^\times$ (resp. $g \in G$) and $y \in K[\mathfrak{G}]^+$. Notice that the definition of η makes implicit use of the natural identification:

(6.5.5) $$(\pi^{-1}\Gamma_\mathfrak{G}^+)^{\mathrm{gp}} \xrightarrow{\sim} G.$$

Let P be the push out of α and β; P is a $K[\mathfrak{G}]^+$-module and the maps λ and η induce a $K[\mathfrak{G}]^+$-linear homomorphism $\phi : P \to \Omega_{K[\mathfrak{G}]^+/\mathbb{Z}}(\log \Gamma_\mathfrak{G}^+)$.

Claim 6.5.6. ϕ is surjective.

Proof of the claim. Indeed, by lemma 6.4.12(v), every element of the target of ϕ is sum of terms of the form $\omega := y \cdot d\log(a[g])$, where $y \in K[\mathfrak{G}]^+$, $g \in G$ and a is an element of $K^\times$ such that $|a[g]|_\mathfrak{G} \leq 1$. However, $\omega = y \cdot d\log a + y \cdot d\log[g] = \lambda(d\log(a) \otimes y) + \eta(g \otimes y)$. $\diamond$

In view of claim 6.5.6 we need only find a left inverse ψ for ϕ. Such ψ is equivalent to the datum of a compatible pair (δ, ϑ) consisting of a derivation

$\delta : K[\mathfrak{G}]^+ \to P$ and a $\mathbb{Z}$-linear map $\vartheta : (\pi^{-1}\Gamma_{\mathfrak{G}}^+)^{\mathrm{gp}} \to P$. In light of (6.5.5), the compatibility between δ and ϑ amounts to the identity: $[g] \cdot \vartheta(g) = \delta([g])$ for every $g \in \pi^{-1}\Gamma_{\mathfrak{G}}^+$. Every element of P can be represented by a pair (x, y) where $x \in \Omega_{K^+/\mathbb{Z}}(\log \Gamma^+) \otimes_{K^+} K[\mathfrak{G}]^+$ and $y \in G \otimes_{\mathbb{Z}} K[\mathfrak{G}]^+$; with this notation the left inverse property of ψ forces the identities:

$$(6.5.7) \qquad \vartheta(g) := (0, g \otimes 1) \qquad \delta([h]) := (0, h \otimes [h])$$

for every $g \in G$ and $h \in \pi^{-1}\Gamma_{\mathfrak{G}}^+$. Conversely, one checks easily that (6.5.7) does indeed define a $K[\mathfrak{G}]^+$-linear homomorphism ψ with the sought properties. $\qquad \square$

6.5.8.   Let $K(\mathfrak{G})^+$ be the valuation ring of the valuation $|\cdot|_{\mathfrak{G}}$. It is easy to see that $K(\mathfrak{G})^+ = K[\mathfrak{G}]_{\mathfrak{p}}^+$, where $\mathfrak{p}$ is the ideal of elements $x \in K[\mathfrak{G}]$ such that $|x|_{\mathfrak{G}} < 1$. It then follows from lemma 6.4.12(v) that the diagram of lemma 6.5.4 remains cocartesian when we replace everywhere $K[\mathfrak{G}]^+$ by $K(\mathfrak{G})^+$.

Proposition 6.5.9. *Suppose that K is algebraically closed, let $(E, |\cdot|_E)$ be a purely transcendental valued field extension of $(K, |\cdot|)$ with $\mathrm{tr.deg}(E/K) = 1$. Then:*

   (i) *Ω_{E^+/K^+} is a torsion-free E^+-module and $H_i(\mathbb{L}_{E^+/K^+}) = 0$ for $i > 0$.*

   (ii) *ρ_{E^+/K^+} (notation of (6.5.1)) is injective with torsion-free cokernel.*

   (iii) *Suppose that $\Gamma_E = \Gamma$. Then the natural diagram*

$$
\begin{array}{ccc}
\Omega_{K^+/\mathbb{Z}} \otimes_{K^+} E^+ & \longrightarrow & \Omega_{E^+/\mathbb{Z}} \\
\downarrow & & \downarrow \\
\Omega_{K^+/\mathbb{Z}}(\log \Gamma^+) \otimes_{K^+} E^+ & \xrightarrow{\;\rho_{E^+/K^+}\;} & \Omega_{E^+/\mathbb{Z}}(\log \Gamma_E^+)
\end{array}
$$

is cocartesian.

Proof. Let $\mathfrak{m}_E$ be the maximal ideal of E^+ and $X \in E$ such that $E = K(X)$. Following (6.1.10), we distinguish two cases, according to whether or not there exists an element $a \in K$ which minimizes the function $K \to \Gamma_E : b \mapsto |X - b|_E$.

   • Suppose first that such an element does not exist. We pick a net $(a_i \mid i \in I)$ satisfying the conditions of lemma 6.1.9, relative to the element $x := X$. We can even assume that the mapping $(I, \leq) \to (\Gamma_E, \leq)$ given by $i \mapsto |X - a_i|$ is decreasing. For a given $b \in K$, choose $i \in I$ such that $|X - a_i|_E < |X - b|_E$; then we have $|a_i - b| = |X - b|_E$ and it follows easily that $\Gamma_E = \Gamma_K$ in this case. Then, for every $i \in I$ we can find $b_i \in K$ such that $|X - a_i|_E = |b_i|$. Let $f(X)/g(X) \in E^+$ be the quotient of two elements $f(X), g(X) \in K[X]$. By lemma 6.1.9, we have $|f(X)/g(X)|_{(a_i, |b_i|)} \leq 1$ and $\gamma := |g(X)|_E = |g(X)|_{(a_i, |b_i|)}$ for every sufficiently large $i \in I$. Pick $a \in K$ such that $|a| = \gamma$. Arguing as in the proof of case (b) of proposition 6.3.13(i), we deduce that

$$a^{-1} \cdot g(X), a^{-1} \cdot f(X) \in A_i := K^+[(X - a_i)/b_i]$$

and if we let $\mathfrak{p}_i := A_i \cap \mathfrak{m}_{E^+}$, then $f(X)/g(X) \in E_i^+ := A_{i,\mathfrak{p}_i}$. It is also easy to see that the family of the K^+-algebras E_i^+ is filtered by inclusion. Clearly $\Omega_{E_i^+/K^+}$ is a free E_i^+-module of rank one, and $H_n(\mathbb{L}_{E_i^+/K^+}) = 0$ for every $n > 0$, so (i) follows easily in this case. Notice that, since $\Gamma = \Gamma_E$, the log structure $\underline{\Gamma}_E^+$ on $\operatorname{Spec} E^+$ (notation of (6.4.14)) is the log structure associated to the morphism of monoids $K^+ \setminus \{0\} \to E^+$. It follows easily that, for every $i \in I$, we have a cocartesian diagram

(6.5.10)
$$
\begin{array}{ccc}
\Omega_{K^+/\mathbb{Z}} \otimes_{K^+} A_i & \xrightarrow{\ \alpha_i\ } & \Omega_{A_i/\mathbb{Z}} \\
\downarrow & & \downarrow \\
\Omega_{K^+/\mathbb{Z}}(\log \Gamma^+) \otimes_{K^+} A_i & \longrightarrow & \Omega_{A_i/\mathbb{Z}}(\log \Gamma^+)
\end{array}
$$

where moreover, α_i is split injective; the diagram of (iii) is obtained from (6.5.10), by localizing at $\mathfrak{p}_i$ and taking colimits over I; since both operations preserve colimits, we get (ii) and (iii) in this case.

• Finally, suppose that there exists an element $a \in K$ such that $|X - a|$ is minimal; we can replace X by $X - a$, and thus assume that $a = 0$. By (6.1.10) it follows that $|\cdot|_E$ is a Gauss valuation; then this case can be realized as the valuation $|\cdot|_\mathfrak{G}$ associated to the datum $\mathfrak{G} := (K^\times \oplus \mathbb{Z}, j, N, \leq)$, where j is the obvious imbedding, and N is either $\mathbb{Z}$ or $\{0\}$, depending on whether $|X|_E \in \Gamma$ or otherwise. In either case, (6.5.8) tells us that the map of (ii) is split injective, with cokernel isomorphic to E^+, so (ii) holds. Suppose first that $|X|_F \in \Gamma$. Then we can find $b \in K$ such that $|X/b|_E = 1$, and one verifies easily that E^+ is the localization of $A := K^+[X/b]$ at the prime ideal $\mathfrak{m}_K \cdot A$. Clearly (6.5.10) remains cocartesian when we replace A_i by A and α_i by the corresponding map α, so (iii) follows easily. (i) is likewise obvious in this case. In case $|X|_E \notin \Gamma$, we distinguish three cases. First, suppose that $|X|_E < |b|$ for every $b \in K^\times$. Then $K^+[X/b] \subset E^+$ for every $b \in K^\times$, and indeed it is easy to check that E^+ is the filtered union of its K^+-subalgebras of the form $K^+[X/b]_{\mathfrak{p}_b}$, where $\mathfrak{p}_b$ is the prime ideal generated by $\mathfrak{m}_K$ and X/b. Again (i) follows. The second case, when $|X|_E > |b|$ for every $b \in K^\times$, is reduced to the former, by replacing X with X^{-1}. It remains only to consider the case where there exist $a_0, b_0 \in K$ such that $|a_0| < |X|_E < |b_0|$; then we can find a net $(a_i, b_i \mid i \in I)$ consisting of pairs of elements of $K^\times$, such that $|a_i| < |X|_E < |b_i|$ for every $i \in I$, and moreover, for every $a, b \in K^\times$ such that $|a| < |X| < |b|$, there exists $i_0 \in I$ with $|a| < |a_i|$ and $|b_i| < |b|$ whenever $i \geq i_0$. In such a situation, one verifies easily that E^+ is the filtered union of its K^+-subalgebras $E_i^+ := K^+[a_i/X, X/b_i]_{\mathfrak{p}_i}$, where $\mathfrak{p}_i$ is the prime ideal generated by $\mathfrak{m}_K$ and the elements $a_i/X, X/b_i$. Each E_i^+ is a complete intersection K^+-algebra, isomorphic to a localization of $R_i^+ := K^+[X, Y]/(XY - a_i/b_i)$. It follows already that $\mathbb{L}_{E_i^+/K^+}$ is acyclic in degrees > 0. To conclude it suffices to show:

Claim 6.5.11. $\Omega_{E_i^+/K^+}$ is a torsion-free E_i^+-module.

Proof of the claim. Set $R_i := R_i^+ \otimes_{K^+} K$; clearly

$$
\Omega_{R_i^+/K^+} \simeq R_i^+ dX \oplus R_i^+ dY/(X dY + Y dX)
$$

is a free K^+-module with basis $\{X dX\} \cup \{X^n dX, Y^n dY \mid n \in \mathbb{N}\}$, hence the natural map $\Omega_{R_i^+/K^+} \to \Omega_{R_i/K}$ is injective. However, $\Omega_{R_i/K}$ is a free R_i-module, so $\Omega_{R_i^+/K^+}$ is a torsion-free R_i^+-module. Since $\Omega_{E_i^+/K^+}$ is a localization of the latter, the claim follows. $\square$

Theorem 6.5.12. *Let* $(K, |\cdot|) \subset (E, |\cdot|_E)$ *be any extension of valued fields. Then*

   (i) $H_i(\mathbb{L}_{E^+/K^+}) = 0$ *for* $i > 1$ *and* $H_1(\mathbb{L}_{E^+/K^+})$ *is a torsion-free* E^+-*module.*

   (ii) *If* K *is perfect, then* $H_i(\mathbb{L}_{E^+/K^+}) = 0$ *for every* $i > 0$.

Proof. Let us show first how to deduce (ii) from (i). Indeed, we reduce easily to the case where E is finitely generated over K. Then, if K is perfect, we can find a subextension $F \subset E$ which is purely transcendental over K, and such that E is separable algebraic over F; by transitivity, we deduce that $\mathbb{L}_{E/K} \simeq E \otimes_F \mathbb{L}_{F/K}$; moreover $H_i(\mathbb{L}_{F/K}) = 0$ for $i > 0$; by (i) we know that $H_1(\mathbb{L}_{E^+/K^+})$ imbeds into $H_1(\mathbb{L}_{E^+/K^+}) \otimes_{E^+} E \simeq H_1(\mathbb{L}_{E/K})$, so the assertion follows.

   To show (i), let $|\cdot|_{E^a}$ be a valuation on the algebraic closure E^a of E, which extends $|\cdot|_E$; recall that E^{a+} is a faithfully flat K^+-module by remark 6.1.12(ii). We apply transitivity to the tower $K^+ \subset E^+ \subset E^{a+}$ to see that the theorem holds for the extension $(K, |\cdot|) \subset (E, |\cdot|_E)$ if it holds for $(K, |\cdot|) \subset (E^a, |\cdot|_{E^a})$ and for $(E, |\cdot|_E) \subset (E^a, |\cdot|_{E^a})$. For the latter extension the assertion is already known by theorem 6.3.32(i), so we are reduced to prove the theorem for the case $(K, |\cdot|) \subset (E^a, |\cdot|_{E^a})$. Similarly, we apply transitivity to the tower $K^+ \subset K^{a+} \subset E^{a+}$ to reduce to the case where both K and E are algebraically closed. Then we can write E as the filtered union of the algebraic closures E_i^a of its subfields E_i finitely generated over K, thereby reducing to prove the theorem for the extensions $K \subset E_i^a$; hence we can assume that tr.deg(E/K) is finite. Again, by transitivity, we further reduce to the case where the transcendence degree of E over K equals one. In this case, we can pick an element $X \in E$ transcendental over K, and write $E = K(X)^a$. Using once more transitivity, we reduce to show the assertion for the purely transcendental extension $K \subset K(X)$, in which case proposition 6.5.9(i) applies, and concludes the proof. $\square$

Lemma 6.5.13. *Let* $R \to S$ *be a ring homomorphism.*

   (i) *Suppose that* $\mathbb{F}_p \subset R$, *denote by* $\Phi_R : R \to R$ *the Frobenius endomorphism of* R, *and define similarly* Φ_S. *Let* $R_{(\Phi)} := \Phi_R^* R$ *and* $S_{(\Phi)} := \Phi_S^* S$ *(cp. (3.5.7)). Suppose moreover that the natural morphism :*

$$R_{(\Phi)} \overset{\mathbf{L}}{\otimes}_R S \to S_{(\Phi)}$$

is an isomorphism in $\mathsf{D}(R\text{-}\mathbf{Mod})$. *Then* $\mathbb{L}_{S/R} \simeq 0$ *in* $\mathsf{D}(s.S\text{-}\mathbf{Mod})$.

   (ii) *Suppose that* S *is a flat* R-*algebra and let* p *be a prime integer,* $b \in R$ *a non-zero-divisor such that* $p \cdot R \subset b^p \cdot R$. *Suppose moreover that the Frobenius endomorphisms of* $R' := R/b^p \cdot R$ *and* $S' := S/b^p \cdot S$ *are surjective. Then the natural morphism :*

$$\mathbb{L}_{S/R} \to \mathbb{L}_{S[b^{-1}]/R[b^{-1}]}$$

is an isomorphism in $\mathsf{D}(s.S\text{-}\mathbf{Mod})$.

Proof. (i): Let $P^\bullet := P_R^\bullet(S)$ be the standard simplicial resolution of S by free R-algebras. Then $\mathbb{L}_{S/R} \simeq \Omega_{P\bullet/R} \otimes_{P\bullet} S$. Let $\Phi_{P\bullet} : P^\bullet \to P_{(\Phi)}^\bullet$ be the termwise Frobenius endomorphism of the simplicial algebra $P^\bullet$. As usual, we can write $\Phi_{P\bullet} = \Phi_{P\bullet/R} \circ (\Phi_R \otimes_R 1_{P\bullet})$, where the relative Frobenius

$$\Phi_{P\bullet/R} : R_{(\Phi)} \otimes_R P^\bullet \to P_{(\Phi)}^\bullet$$

is a morphism of simplicial $R_{(\Phi)}$-algebras. Concretely, if $P^k = R[X_i \mid i \in I]$ is a free algebra on generators $(X_i \mid i \in I)$, then

(6.5.14) $\qquad \Phi_{P^k/R}(X_i) = X_i^p \quad$ for every $i \in I$.

Under the assumption of the lemma, $\Phi_{P\bullet/R}$ is a quasi-isomorphism of simplicial $R_{(\Phi)}$-algebras. It then follows from [50, Ch.II, Prop.1.2.5.3,1.2.6.2] that $\Phi_{P\bullet/R}$ induces an isomorphism

(6.5.15) $\qquad R_{(\Phi)} \otimes_R \mathbb{L}_{S/R} \xrightarrow{\sim} \mathbb{L}_{S_{(\Phi)}/R_{(\Phi)}} \simeq \mathbb{L}_{S/R}.$

However, (6.5.14) shows that (6.5.15) is represented by a map of simplicial complexes which is termwise the zero map, so the claim follows.

(ii): Under the stated assumptions, the Frobenius map induces an isomorphism of R-algebras: $R/b \cdot R \xrightarrow{\sim} R'_{(\Phi)}$ (resp. of S-algebras: $S/b \cdot S \xrightarrow{\sim} S'_{(\Phi)}$). Thus the map

$$S' \otimes_{R'} (R')_{(\Phi)} \to S'_{(\Phi)} \quad : \quad x \otimes y \mapsto \Phi_{S'}(x) \cdot y$$

is an isomorphism. Since moreover S' is a flat R'-algebra, we see that the assumption of (i) is satisfied, whence $\mathbb{L}_{S'/R'} \simeq 0$. If we now tensor the short exact sequence $0 \to R \to R \to R' \to 0$ by $\mathbb{L}_{S/R}$, we obtain a distinguished triangle

$$\mathbb{L}_{S/R} \xrightarrow{b^p} \mathbb{L}_{S/R} \to \mathbb{L}_{S/R} \overset{L}{\otimes}_R R' \to \sigma \mathbb{L}_{S/R}.$$

However, $\mathbb{L}_{S/R} \overset{L}{\otimes}_R R' \simeq \mathbb{L}_{S'/R'}$ by [50, II.2.2.1], so we have shown that b^p acts as an isomorphism on $\mathbb{L}_{S/R}$. In other words, $\mathbb{L}_{S/R} \simeq \mathbb{L}_{S/R} \otimes_S S[b^{-1}] \simeq \mathbb{L}_{S[b^{-1}]/R[b^{-1}]}$, as claimed. $\qquad\square$

Lemma 6.5.16. *Let* $(K, |\cdot|) \subset (E, |\cdot|_E)$ *be an extension of valued fields and suppose that* $\mathbb{Q} \subset \kappa(K)$. *Then the map* ρ_{E^+/K^+} *of (6.5.1) is injective with torsion-free cokernel.*

Proof. To begin with, let $(F, |\cdot|_F)$ be any valued field extension of $(E, |\cdot|_E)$. We remark that:

(6.5.17) $\qquad \rho_{F^+/E^+} \circ (\rho_{E^+/K^+} \otimes_{E^+} 1_{F^+}) = \rho_{F^+/K^+}.$

Claim 6.5.18. Suppose moreover that E is an algebraic extension of K. Then ρ_{E^+/K^+} is an isomorphism.

Proof of the claim. Applying (6.5.17), with $F := E^a$, we reduce easily to prove the claim in case E is algebraically closed. Let K^{sh} be the field of fractions of the strict henselization of K^+ (which we see as imbedded in E^+). Let $j : \text{Spec } K^{sh+} \to \text{Spec } K^+$ be the morphism induced by the imbedding $K \subset K^{sh}$; In view of lemma 6.2.5, the log structure $\underline{\Gamma}_{K^{sh}}^+$ on $\text{Spec } K^{sh+}$ (notation of (6.4.14)) equals $j^*\underline{\Gamma}^+$.

Since moreover $K^{\mathrm{sh}+}$ is local ind-étale over K^+, we deduce from 6.4.12(iii) that $\rho_{K^{\mathrm{sh}+}/K^+}$ is an isomorphism. Then arguing as in the foregoing, we see that it suffices to prove the claim for the case when $K = K^{\mathrm{sh}}$. Since everything in sight commutes with filtered unions of field extensions, we can even reduce to the case where E is a finite (Galois) extension of K. Then, by corollary 6.2.14, this case can be realized as the extension associated to some datum $\mathfrak{G} := (G, j, N, \leq)$, where moreover $G/K^\times$ is a finite abelian group. Since by assumption $\mathbb{Q} \subset K[\mathfrak{G}]^+$, the claim follows by lemma 6.5.4 and (6.5.8). $\diamond$

Now, if $K \subset E$ is an arbitrary extension, we can apply (6.5.17) with $F := E^{\mathrm{a}}$ and claim 6.5.18 to the extension $E \subset E^{\mathrm{a}}$ to reduce to the case where E is algebraically closed. Then we can apply again claim 6.5.18 to the extension $K \subset K^{\mathrm{a}}$ and (6.5.17) with $E := K^{\mathrm{a}}$ and $F := E$, to reduce to the case where also K is algebraically closed. Then, by the usual argument we reduce to the case of an extension of finite transcendence degree, and even to the case of transcendence degree equal to one. We factor the latter as a tower of extensions $K \subset K(X) \subset E$, where X is transcendental over K, hence E algebraic over $K(X)$. So finally we are reduced to the case $E = K(X)$, in which case we conclude by proposition 6.5.9(ii). $\square$

Theorem 6.5.19. *Let* $|\cdot|_{K^{\mathrm{s}}}$ *be a valuation on the separable closure K^{s} of K, extending the valuation of K. Then the map $\rho := \rho_{K^{\mathrm{s}+}/K^+}$ is injective.*

Proof. Suppose first that Γ is divisible. In this case $\Gamma = \Gamma_{K^{\mathrm{s}}}$, and therefore the log structure $\underline{\Gamma}^+_{K^{\mathrm{s}}}$ on $\operatorname{Spec} K^{\mathrm{s}}$ is the log structure associated to the morphism of monoids $K^+ \setminus \{0\} \to K^{\mathrm{s}+}$. It follows easily that the commutative diagram:

$$
\begin{array}{ccc}
\Omega_{K^+/\mathbb{Z}} \otimes_{K^+} K^{\mathrm{s}+} & \xrightarrow{\quad \alpha \quad} & \Omega_{K^{\mathrm{s}+}/\mathbb{Z}} \\
{\scriptstyle \beta_{K^+} \otimes 1_{K^{\mathrm{s}+}}} \big\downarrow & & \big\downarrow {\scriptstyle \beta_{K^{\mathrm{s}+}}} \\
\Omega_{K^+/\mathbb{Z}}(\log \Gamma^+) \otimes_{K^+} K^{\mathrm{s}+} & \xrightarrow{\quad \rho \quad} & \Omega_{K^{\mathrm{s}+}/\mathbb{Z}}(\log \Gamma^+_{K^{\mathrm{s}}})
\end{array}
$$

is cocartesian (where β_{K^+} and $\beta_{K^{\mathrm{s}+}}$ are the maps of corollary 6.4.18(i)). By theorem 6.3.32(ii), α is injective, hence the same holds for ρ.

In case Γ is not necessarily divisible, let us choose a datum

$$
\mathfrak{G} := (G, j : (K^{\mathrm{s}})^\times \to G, N, \leq)
$$

as in (6.1.5), such that $G := (K^{\mathrm{s}})^\times \oplus F$, where F is a torsion-free abelian group (whose composition law we write in multiplicative notation) and N is the graph of a surjective group homomorphism $\phi : F \to (K^{\mathrm{s}})^\times$. Notice that in this case $\Gamma_{\mathfrak{G}} \simeq \Gamma_{K^{\mathrm{s}}} \simeq \Gamma \otimes_{\mathbb{Z}} \mathbb{Q}$, and the restriction to F of the projection $G \to \Gamma_{\mathfrak{G}}$ is the map $x \mapsto |\phi(x^{-1})|_{K^{\mathrm{s}}}$. Let now $H := K^\times \oplus F$ and define a new datum

$$
\mathfrak{H} := (H, j_{|K^\times} : K^\times \to H, H \cap N, \leq)
$$

Since ϕ is surjective, clearly we still have $\Gamma_{\mathfrak{H}} \simeq \Gamma \otimes_{\mathbb{Z}} \mathbb{Q}$. Notice as well that $K^{\mathrm{s}}(\mathfrak{G})$ is separable algebraic over $K(\mathfrak{H})$. Set:

$$
\rho_{\mathfrak{H}} := \rho_{K(\mathfrak{H})^+/K^+} \qquad \rho_{\mathfrak{G}} := \rho_{K^{\mathrm{s}}(\mathfrak{G})^+/K^{\mathrm{s}+}} \qquad \rho_{\mathfrak{G}/\mathfrak{H}} := \rho_{K^{\mathrm{s}}(\mathfrak{G})^+/K(\mathfrak{H})^+}.
$$

We consider the diagram :

$$\begin{array}{ccc}
\Omega_{K^+/\mathbb{Z}}(\log\Gamma^+) \otimes_{K^+} K^s(\mathfrak{G})^+ & \xrightarrow{\rho\otimes 1_{K^s(\mathfrak{G})^+}} & \Omega_{K^{s+}/\mathbb{Z}}(\log\Gamma^+_{K^s}) \otimes_{K^{s+}} K^s(\mathfrak{G})^+ \\
\downarrow{\rho_{\mathfrak{H}}\otimes 1_{K^s(\mathfrak{G})^+}} & & \downarrow{\rho_{\mathfrak{G}}} \\
\Omega_{K(\mathfrak{H})^+/\mathbb{Z}}(\log\Gamma^+_{\mathfrak{H}}) \otimes_{K(\mathfrak{H})^+} K^s(\mathfrak{G})^+ & \xrightarrow{\rho_{\mathfrak{G}/\mathfrak{H}}} & \Omega_{K^s(\mathfrak{G})^+/\mathbb{Z}}(\log\Gamma^+_{\mathfrak{G}}).
\end{array}$$

Since F is torsion-free, it follows easily from (6.5.8) and lemma 6.5.4 that $\rho_{\mathfrak{H}}$ and $\rho_{\mathfrak{G}}$ are injective with torsion-free cokernels. Hence, in order to prove that ρ is injective, it suffices to show that $\rho_{K^s(\mathfrak{G})^+/K(\mathfrak{H})^+}$ is. Finally, let E be a separable closure of $K(\mathfrak{H})$ and choose a valuation on E which extends the valuation of $K^s(\mathfrak{G})$; we notice that $\operatorname{Ker}\rho_{K^s(\mathfrak{G})^+/K(\mathfrak{H})^+} \subset \operatorname{Ker}\rho_{E^+/K(\mathfrak{H})^+}$. Therefore, we can replace K by $K(\mathfrak{H})$ and reduce to the case where Γ is divisible, which has already been dealt with. □

Theorem 6.5.20. *Let* $(K,|\cdot|) \subset (E,|\cdot|_E)$ *be an extension of valued fields, with* K *algebraically closed. Then:*

   (i) Ω_{E^+/K^+} *is a torsion-free* E^+*-module.*
   (ii) ρ_{E^+/K^+} *is injective with torsion-free cokernel.*

Proof. By the usual argument we can reduce to the case where $\operatorname{tr.deg}(E/K)$ is finite and K is the algebraic closure of an extension of finite type of its prime field.

   (i): Pick a valuation $|\cdot|_{E^a}$ of the field E^a extending $|\cdot|_E$. By theorem 6.5.12(ii) we have an exact sequence:

$$0 \to H_1(\mathbb{L}_{E^a/E}) \to E^{a+} \otimes_{E^+} \Omega_{E^+/K^+} \to \Omega_{E^{a+}/K^+}$$

where the leftmost term is torsion-free by theorem 6.3.32, so it suffices to show that Ω_{E^{a+}/K^+} is torsion-free, and we can therefore assume that E is algebraically closed. In this case, if now $\operatorname{char}(K) > 0$, it follows that the Frobenius endomorphism of E^+ is surjective; then, for any $a \in E^+$ we can write $da = d(a^{1/p})^p = p \cdot a^{(p-1)/p} \cdot da^{1/p} = 0$, so actually $\Omega_{E^+/K^+} = 0$. In case $\operatorname{char}(K) = 0$ and $\operatorname{char}(\kappa(K)) = p > 0$, let us pick an element $b \in K^+$ such that $|b^p| = |p|$. Since K and E are algebraically closed, the Frobenius endomorphisms on K^+/b^pK^+ and E^+/b^pE^+ are surjective, so $\mathbb{L}_{E^+/K^+} \simeq \mathbb{L}_{E^+[b^{-1}]/K^+[b^{-1}]}$ by lemma 6.5.13(ii). Now, $K^+[b^{-1}]$ is the valuation ring of a valuation $|\cdot|'$ on K, which extends to a valuation $|\cdot|'_E$ on E^+ whose valuation ring is $E^+[b^{-1}]$. Furthermore, the residue fields of these valuations are fields of characteristic zero. Hence, we have reduced the proof of the theorem to the case where $\kappa(K) \supset \mathbb{Q}$. By lemma 6.5.16 we have a commutative diagram with exact rows:

$$\begin{array}{ccccccc}
E^+ \otimes_{K^+} \Omega_{K^+/\mathbb{Z}} & \longrightarrow & \Omega_{E^+/\mathbb{Z}} & \longrightarrow & \Omega_{E^+/K^+} & \longrightarrow & 0 \\
\downarrow{1_{E^+}\otimes\beta_{K^+}} & & \downarrow{\beta_{E^+}} & & \downarrow{\gamma} & & \\
0 \longrightarrow E^+ \otimes_{K^+} \Omega_{K^+/\mathbb{Z}}(\log\Gamma^+) & \longrightarrow & \Omega_{E^+/\mathbb{Z}}(\log\Gamma^+_E) & \longrightarrow & \operatorname{Coker}(\rho_{E^+/K^+}) & \longrightarrow & 0
\end{array}$$

where β_{K^+} and β_{E^+} are the maps of corollary 6.4.18(i). By virtue of lemma 6.5.16, it suffices to show that γ is injective. Since β_{E^+} is injective by corollary 6.4.18(i), the theorem follows from the snake lemma and corollary 6.4.18(iii).

(ii): Let $|\cdot|_{E^s}$ be an extension of the valuation $|\cdot|_E$ to a separable closure E^s of E. We have $\operatorname{Ker}\rho_{E^+/K^+} \subset \operatorname{Ker}\rho_{E^{s+}/K^+}$, and even $\operatorname{Coker}\rho_{E^+/K^+} \subset \operatorname{Coker}\rho_{E^{s+}/K^+}$, by theorem 6.5.19. Thus we can replace E by E^s and suppose that E is separably closed, hence Γ_E is divisible, by example 6.1.21(ii). By corollary 6.4.18(i) we have a commutative diagram with exact rows :

$$0 \to E^+ \otimes_{K^+} \Omega_{K^+/\mathbb{Z}} \to E^+ \otimes_{K^+} \Omega_{K^+/\mathbb{Z}}(\log\Gamma^+) \to E^+ \otimes_{K^+} \operatorname{Coker}\beta_{K^+} \to 0$$

$$\downarrow{\alpha} \qquad\qquad \downarrow{\rho_{E^+/K^+}} \qquad\qquad \downarrow{\gamma}$$

$$0 \longrightarrow \Omega_{E^+/\mathbb{Z}} \longrightarrow \Omega_{E^+/\mathbb{Z}}(\log\Gamma_E^+) \longrightarrow \operatorname{Coker}\beta_{E^+} \longrightarrow 0.$$

By theorem 6.5.12(ii), the map α is injective. The same holds for γ, in view of corollary 6.4.18(iii). It follows already that ρ_{E^+/K^+} is injective.

Since both Γ and Γ_E are divisible, corollary 6.4.18(iii) says that $\operatorname{Coker}\beta_{K^+}$ and $\operatorname{Coker}\beta_{E^+}$ are $\mathbb{Q}$-vector spaces, hence the same holds for $\operatorname{Coker}\gamma$. Now, if the characteristic of K is positive, it follows that $\operatorname{Coker}\beta_{K^+} = \operatorname{Coker}\beta_{E^+} = 0$, in which case $\operatorname{Coker}\rho_{E^+/K^+} \simeq \Omega_{E^+/K^+}$ and the sought assertion follows from (i).

Suppose then that K has characteristic zero. By (i), $\operatorname{Coker}\alpha$ is a torsion-free E^+-module, hence in this case the foregoing implies that $\operatorname{Coker}\rho_{E^+/K^+}$ is a torsion-free $\mathbb{Z}$-module, and thus we are reduced to showing that $\mathbb{Q}\otimes_{\mathbb{Z}}\operatorname{Coker}\rho_{E^+/K^+}$ is a torsion-free E^+-module. However, $\mathbb{Q}\otimes_{\mathbb{Z}}\operatorname{Coker}\rho_{E^+/K^+} \simeq \operatorname{Coker}\rho_{E_{\mathbb{Q}}^+/K_{\mathbb{Q}}^+}$, where $E_{\mathbb{Q}}^+ := E^+ \otimes_{\mathbb{Z}}\mathbb{Q}$ and $K_{\mathbb{Q}}^+ := K^+ \otimes_{\mathbb{Z}}\mathbb{Q}$ are valuation rings with residue fields of characteristic zero. But the assertion to prove is already known in this case, by lemma 6.5.16. $\qquad\square$

Corollary 6.5.21. *Let $(K,|\cdot|)$ be a valued field, and k a perfect field such that $k \subset K^+$. Then $\Omega_{K^+/k}$ and $\operatorname{Coker}\rho_{K^+/k}$ are torsion-free K^+-modules.*

Proof. We have $k^a \subset K^{sh+}$; let $E := k^a \cdot K \subset K^{sh}$ and denote by $j : \operatorname{Spec} E^+ \to \operatorname{Spec} K^+$ the morphism induced by the imbedding $K \subset E$. By lemma 6.2.5 the natural map : $j^*\underline{\Gamma}^+ \to \underline{\Gamma}_E^+$ is an isomorphism of log structures; moreover $\Omega_{k^a/k} = 0$, since k is perfect. Hence $\Omega_{K^+/k} \subset \Omega_{E^+/k^a}$; furthermore, lemma 6.4.12(iii) says that $\operatorname{Coker}\rho_{K^+/k} \subset \operatorname{Coker}\rho_{E^+/k^a}$. Then the assertion follows from theorem 6.5.20. $\qquad\square$

Remark 6.5.22. Notice that corollary 6.5.21 is a straightforward consequence of a standard (as yet unproven) conjecture on the existence of resolution of singularities over perfect fields.

6.6. Deeply ramified extensions. We keep the notation of section 6.3. We borrow the notion of deeply ramified extension of valuation rings from the paper [21], even though our definition applies more generally to valuations of arbitrary rank.

Definition 6.6.1. Let $(K, |\cdot|)$ be a valued field, $|\cdot|_{K^s}$ a valuation on K^s which extends $|\cdot|$. We say that $(K, |\cdot|)$ is *deeply ramified* if $\Omega_{K^{s+}/K^+} = 0$. Notice that the definition does not depend on the choice of the extension $|\cdot|_{K^s}$.

Proposition 6.6.2. *Let $(K, |\cdot|)$ be a valued field whose valuation has rank one, $|\cdot|_{K^s}$ an extension of $|\cdot|$ to K^s. Then the following conditions are equivalent:*

    (i) *$(K, |\cdot|)$ is deeply ramified;*
    (ii) *The morphism of almost algebras $(K^+)^a \to (K^{s+})^a$ is weakly étale;*
    (iii) *$(\Omega_{K^{s+}/K^+})^a = 0$.*

Moreover, the above conditions imply that the valuation of K is not discrete.

Proof. We leave to the reader the verification that (iii) (and, *a fortiori*, (i)) can hold only in case the valuation of K is not discrete.

Let $K^{+\mathrm{sh}}$ be a strict henselization of K^+ contained in K^{s+}, and K^{sh} its fraction field. It is easy to check that $(K, |\cdot|)$ is deeply ramified if and only if $(K^{\mathrm{sh}}, |\cdot|_{K^{\mathrm{sh}}})$ is. Moreover, in view of lemma 3.1.2(iv), condition (ii) holds for $(K, |\cdot|)$ if and only if it holds for $(K^{\mathrm{sh}}, |\cdot|_{K^{\mathrm{sh}}})$, and similarly for condition (iii). Hence we can assume that K^+ is strictly henselian. It is also clear that (i)$\Rightarrow$(iii)$\Leftarrow$(ii). To show that (iii)$\Rightarrow$(ii), let $E \subset K^s$ be a finite separable extension of K and set $E^+ := K^{s+} \cap E$; it follows from theorem 6.3.23 that the natural map $\Omega_{E^+/K^+} \otimes_{E^+} K^{s+} \to \Omega_{K^{s+}/K^+}$ is injective; thus, if (iii) holds, we deduce that $(\Omega_{E^+/K^+})^a = 0$ for every finite separable extension E of K. Again by theorem 6.3.23 we derive that $\mathscr{D}_{E^+/K^+} = E^{+a}$ for every such E. Finally, lemmata 6.3.12(i) and 4.1.27 show that E^{+a} is étale over K^{+a}, whence (ii). Next, let κ be the residue field of K^+; we point out:

Claim 6.6.3. Suppose that $\mathrm{char}(\kappa) = 0$ and that $|\cdot|$ is not discrete (and still of rank one). Then $(K, |\cdot|)$ is deeply ramified.

Proof of the claim. Arguing as in the foregoing we reduce to the case where $K = K^{\mathrm{sh}}$. Then, under the stated assumptions, every finite extension E of K factors as a tower of Kummer extensions of prime degree, therefore $\Omega_{E^+/K^+} = 0$ by corollary 6.3.21(i), whence the claim. $\Diamond$

In view of claim 6.6.3, in order to show that (iii)$\Rightarrow$(i) we can assume that $\mathrm{char}(\kappa) = p > 0$. Hence, let us choose $a \in \mathfrak{m} \setminus \{0\}$ such that $|a| \geq |p|$. It follows easily from example 6.1.21(iii) that every element $x \in K^{s+}$ can be written in the form $x = y^p + a \cdot z$ for some $y, z \in K^{s+}$. Hence $dx = p \cdot dy^{p-1} + a \cdot dz$, which means that $\Omega_{K^{s+}/K^+} = a \cdot \Omega_{K^{s+}/K^+}$, and the contention follows. $\square$

Lemma 6.6.4. *Let $(K, |\cdot|)$ be a valued field such that $\mathbb{Q} \subset K$, and $p := \mathrm{char}(\kappa) > 0$. Let $(K^s, |\cdot|_{K^s})$ be an extension of the valuation $|\cdot|$ to a separable closure of K. Denote by T the K^{s+}-torsion submodule of $\Omega_{K^{s+}/\mathbb{Z}}$. Then $T \simeq K^s/K^{s+}$.*

Proof. Let $(\mathbb{Q}^a, |\cdot|_{\mathbb{Q}^a})$ be the restriction of $|\cdot|_{K^s}$ to the algebraic closure of $\mathbb{Q}$ in K^s. From theorems 6.5.12(ii) and 6.5.20(i) it follows easily that $T \simeq \Omega_{\mathbb{Q}^{a+}/\mathbb{Z}} \otimes_{\mathbb{Q}^{a+}} K^{s+}$, hence we can suppose that $K = \mathbb{Q}$. From theorem 6.3.23 we deduce that the natural map $\mathbb{Q}^{a+} \otimes_{E^+} \Omega_{E^+/\mathbb{Z}} \to \Omega_{\mathbb{Q}^{a+}/\mathbb{Z}}$ is injective for every subextension $E \subset \mathbb{Q}^a$. For every $n \in \mathbb{N}$ and every subextension $E \subset \mathbb{Q}^a$, let $E_n := E(\zeta_{p^n})$, where ζ_{p^n} is any primitive p^n-th root of 1 and set $E_\infty := \bigcup_{n>0} E_n$.

Claim 6.6.5. For every finite subextension $E \subset \mathbb{Q}^a$, there exists $n \in \mathbb{N}$ such that the image of $E_n^+ \otimes_{E^+} \Omega_{E^+/\mathbb{Z}}$ in $\Omega_{E_n^+/\mathbb{Z}}$ is included in the image of $E_n^+ \otimes_{\mathbb{Q}_n^+} \Omega_{\mathbb{Q}_n^+/\mathbb{Z}}$.

Proof of the claim. For every $n \in \mathbb{N}$, E_n^+ is a discrete valuation ring and $\kappa(E_n)$ is a finite separable extension of $\kappa(\mathbb{Q}) = \mathbb{F}_p$; from the exact sequence

$$\mathfrak{m}_{E_n}/\mathfrak{m}_{E_n}^2 \to \Omega_{E_n^+/\mathbb{Z}} \otimes_{E_n^+} \kappa(E_n) \to \Omega_{\kappa(E_n)/\mathbb{F}_p} = 0$$

we deduce that $\Omega_{E_n^+/\mathbb{Z}}$ is a (torsion) cyclic E_n^+-module. By comparing the annihilators of the modules under consideration, one obtains easily the claim.   ◇

A standard calculation shows that $\Omega_{\mathbb{Q}_\infty^\pm/\mathbb{Z}} \simeq \mathbb{Q}_\infty/\mathbb{Q}_\infty^+$. This, together with claim 6.6.5 implies the lemma.   □

Proposition 6.6.6. *Keep the notation and assumptions of proposition 6.6.2 and suppose moreover that the characteristic p of the residue field κ of K^+ is positive and that the valuation on K is not discrete. Let $(K^\wedge, |\cdot|^\wedge)$ be the completion of $(K, |\cdot|)$ for the valuation topology. Then the following conditions are equivalent:*

(i) *$(K, |\cdot|)$ is deeply ramified.*

(ii) *The Frobenius endomorphism of $K^{\wedge+}/pK^{\wedge+}$ is surjective.*

(iii) *For some $b \in K^+ \setminus \{0\}$ such that $1 > |b| \geq |p|$, the Frobenius endomorphism on $(K^+/bK^+)^a$ is an epimorphism.*

(iv) *$\Omega_{K^+/\mathbb{Z}}(\log \Gamma^+)$ is a K^+-divisible K^+-module.*

(v) *$\Omega_{K^+/\mathbb{Z}}$ is a K^+-divisible K^+-module.*

(vi) *$\Omega_{K^+/\mathbb{Z}}^a$ is a K^{+a}-divisible K^{+a}-module, (i.e., for every $x \in K^+ \setminus \{0\}$ we have $\Omega_{K^+/\mathbb{Z}}^a = x \cdot \Omega_{K^+/\mathbb{Z}}^a$).*

(vii) *Coker $\rho_{K^{s+}/K^+} = 0$ (notation of (6.5.1)).*

(viii) *Coker$(\rho_{K^{s+}/K^+})^a = 0$.*

Proof. To start out we remark, slightly more generally:

Claim 6.6.7. Let $(F, |\cdot|_F)$ be a rank one valued field with non-discrete valuation. Then the map

$$\beta_{F^+}^a : \Omega_{F^+/\mathbb{Z}}^a \to \Omega_{F^+/\mathbb{Z}}(\log \Gamma^+)^a$$

is an isomorphism (for the standard setup associated to $(F, |\cdot|_F)$).

Proof of the claim. It is an immediate consequence of corollary 6.4.18(ii).   ◇

Claim 6.6.7 – applied to $(K, |\cdot|)$ and to any extension of $|\cdot|$ to K^s – implies directly that (i)⇒(viii); combining with proposition 6.6.2 we deduce also the converse: (viii)⇒(i). Similarly, claim 6.6.7 yields (iv)⇒(vi), and clearly (v)⇒(vi) as well.

Suppose that (i) holds; then by proposition 6.6.2 it follows that the morphism $(K^+)^a \to (K^{s+})^a$ is weakly étale, so the same holds for the morphism $(K^+/bK^+)^a \to (K^{s+}/bK^{s+})^a$, for every $b \in K^+$ with $|b| \geq |p|$. In view of example 6.1.21(iii), one sees that the Frobenius endomorphism on K^{s+}/bK^{s+} is an epimorphism. Using theorem 3.5.13(ii) we deduce that the Frobenius endomorphism on $(K^+/bK^+)^a$ is an epimorphism as well. This shows that (i)⇒(iii).

To show that (iii)⇒(ii), let us choose $\varepsilon \in \mathfrak{m} \setminus \{0\}$ such that $|\varepsilon^p| > |b|$; by hypothesis, for every $x \in K^+$ there exists $y \in K^+$ such that $\varepsilon^p \cdot x - y^p \in bK^+$. It follows easily that the Frobenius endomorphism is surjective on $K^+/(b \cdot \varepsilon^{-p})K^+$. Replacing b by $b \cdot \varepsilon^{-p}$ we can assume that the Frobenius endomorphism is surjective on K^+/bK^+. Let $b_1 \in K^+$ such that $1 > |b_1^p| \geq |b|$; we let $\mathrm{Fil}_1^\bullet(K^+/pK^+)$ (resp. $\mathrm{Fil}_2^\bullet(K^+/pK^+)$) be the b_1-adic (resp. b_1^p-adic) filtration on K^+/pK^+. The group topology on K^+/pK^+ defined by the filtrations $\mathrm{Fil}_i^\bullet(K^+/pK^+)$ ($i = 1, 2$) is the same as the one induced by the valuation topology of K^+; moreover, one verifies easily that the Frobenius endomorphism defines a morphism of filtered abelian groups $\mathrm{Fil}_1^\bullet(K^+/pK^+) \to \mathrm{Fil}_2^\bullet(K^+/pK^+)$ and that the associated morphism of graded abelian groups is surjective. It then follows from [16, Ch.III, §2, n.8, Cor.2] that (ii) holds.

Next suppose that (ii) holds; choose $b \in K^+$ such that $1 > |b| > |b^p| \geq |p|$; by hypothesis, the Frobenius endomorphism on K^+/b^pK^+ is surjective; the same holds for the Frobenius map on K^{s+}/b^pK^{s+}, in view of example 6.1.21(iii). Hence, the assumptions of lemma 6.5.13(ii) are fulfilled, and we deduce that $\mathbb{L}_{K^{s+}/K^+} \simeq \mathbb{L}_{K^{s+}[b^{-1}]/K^+[b^{-1}]}$. Since the valuation $|\cdot|$ has rank one, we have $K^+[b^{-1}] = K$, and consequently $\Omega_{K^{s+}/K^+} \simeq \Omega_{K^s/K} = 0$, which is (i).

Furthermore, (ii) implies that $\Omega_{(K^+/b^pK^+)/\mathbb{Z}} = 0$ (since $dx^p = px^{p-1}dx = 0$). Let $I := b^pK^+$; it follows that the natural map

$$I/I^2 \to (K^+/b^pK^+) \otimes_\mathbb{Z} \Omega_{K^+/\mathbb{Z}}$$

is surjective, i.e. $\Omega_{K^+/\mathbb{Z}} = b^p \cdot \Omega_{K^+/\mathbb{Z}} + K^+ db^p \subset b^p \cdot \Omega_{K^+/\mathbb{Z}}$, which implies (v).

Next, by corollary 6.4.18(ii), we have $\Omega_{K^+/\mathbb{Z}}(\log \Gamma^+)/\Omega_{K^+/\mathbb{Z}} \simeq \kappa \otimes_\mathbb{Z} \Gamma$, and this last term vanishes since (ii) implies that $\Gamma = \Gamma^p$. This shows that (ii)⇒(iv) as well. To proceed further we will need the following :

Claim 6.6.8. Let $C := \mathrm{Coker}(\rho_{K^{s+}/K^+})$ (notation of (6.5.1)). Then:

   (i) C is a K^+-torsion module.
   (ii) $C^a \simeq (\Omega_{K^{s+}/K^+})^a$.
   (iii) $\Omega_{K^{s+}/\mathbb{Z}}$ is a K^{s+}-divisible module.

Proof of the claim. In view of example 6.1.21(iii), $(K^s, |\cdot|_{K^s})$ satisfies condition (iii), hence the third assertion follows from the implications (iii)⇔(ii)⇒(v), which have already been shown. Furthermore, it is clear that Ω_{K^{s+}/K^+} is a torsion K^+-module and therefore the first assertion follows easily from corollary 6.4.18. The second assertion follows from claim 6.6.7. ◇

Suppose now that (vi) holds. We consider first the case where $\mathbb{F}_p \subset K^+$; in this case $\Omega_{K^{s+}/\mathbb{Z}}$ is a torsion-free K^{s+}-module according to corollary 6.5.21. Let $b \in K^+ \setminus \{0\}$ be any element; by theorem 6.5.19, claim 6.6.8(iii) and the snake lemma we deduce that the b-torsion submodule $C[b]^a := \mathrm{Ker}(C^a \to C^a : x \mapsto b \cdot x)$ is isomorphic to the almost module defined by the cokernel of the scalar multiplication by b on the module $K^{s+} \otimes_{K^+} \Omega_{K^+/\mathbb{Z}}(\log \Gamma^+)$; the latter vanishes by claim 6.6.7, under assumption (vi), and by claim 6.6.8(i) we have $C^a = \bigcup_{b \in K^+ \setminus \{0\}} C[b]^a$, whence $C^a = 0$, which is equivalent to (i) by claim 6.6.8(ii) and proposition 6.6.2.

Finally, in case K^+ is of mixed characteristic, denote by T (resp. T') the K^+-torsion submodule of $\Omega_{K^{s+}/\mathbb{Z}}$ (resp. of $K^{s+} \otimes_{K^+} \Omega_{K^+/\mathbb{Z}}$) and define $T[b]$ (resp. $T'[b]$) as its b-torsion submodule, for any $b \in K^+$. Claim 6.6.7 shows that T^a is isomorphic to the K^{+a}-torsion submodule of $(\Omega_{K^{s+}/\mathbb{Z}}(\log \Gamma^+_{K^s}))^a$, and similarly for $(T')^a$; moreover, from (vi) and the snake lemma we obtain a short exact sequence $0 \to T'[b]^a \to T[b]^a \to C[b]^a \to 0$ for every $b \in K^+ \setminus \{0\}$, whence a short exact sequence $0 \to (T')^a \to T^a \to C^a \to 0$. Under (vi), $(T')^a$ is a divisible module; however, it is clear from lemma 6.6.4 that the only divisible $(K^{s+})^a$-submodules of T^a are 0 and T^a. Consequently, in light of claim 6.6.8(ii) and proposition 6.6.2, in order to prove that (vi)$\Rightarrow$(i), it suffices to show that $(T')^a \neq 0$. In turn, this is implied by the following :

Claim 6.6.9. The image in $\Omega_{K^+/\mathbb{Z}}(\log \Gamma)$ of $d\log(p) \in \Omega_{\mathbb{Q}^+/\mathbb{Z}}(\log \Gamma^+_{\mathbb{Q}})$, has annihilator pK^+.

Proof of the claim. By theorem 6.5.19 it suffices to consider the image of $d\log(p)$ in $\Omega_{K^{s+}/\mathbb{Z}}(\log \Gamma^+_{K^s})$. Then, by theorem 6.5.20(ii), we reduce to the case $K^s = \overline{\mathbb{Q}}$. Then, once more by theorem 6.5.19, it suffices to look at the annihilator of $d\log(p)$ in $\Omega_{\mathbb{Q}^+/\mathbb{Z}}(\log \Gamma^+_{\mathbb{Q}})$ itself, where $\mathbb{Q}^+ = \mathbb{Z}_{(p)}$ and the claim follows. $\diamondsuit$

Thus far we have shown that the conditions (i)-(vi) and (viii) are all equivalent. Furthermore, (i) implies (vii) because $\beta_{K^{s+}}$ is an isomorphism (by corollary 6.4.18(iii)); since obviously (vii)$\Rightarrow$(viii), we are done. $\square$

Remark 6.6.10. By inspecting the proof of proposition 6.6.6, we see that the argument for (ii)$\Rightarrow$(i) still goes through for valued fields $(K, |\cdot|)$ of arbitrary rank and characteristic $p > 0$.

Lemma 6.6.11. *Let $(K, |\cdot|)$ be a valued field and $b \in K$ an element with $0 < |b| < 1$. Denote by $\mathfrak{q}(b)$ the radical of the ideal bK^+ and set $\mathfrak{p}(b) := \bigcap_{r>0} b^r \cdot K^+$. Then $\mathfrak{p}(b)$ and $\mathfrak{q}(b)$ are consecutive prime ideals, i.e. $\mathfrak{p}(b)$ is strictly contained in $\mathfrak{q}(b)$ and there are no prime ideals strictly between $\mathfrak{p}(b)$ and $\mathfrak{q}(b)$. Equivalently, the ring $W(b) := (K^+/\mathfrak{p}(b))_{\mathfrak{q}(b)}$ is a valuation ring of rank one and the image of b is topologically nilpotent in the valuation topology of $W(b)$.*

Proof. It is easy to verify that $\mathfrak{p}(b)$ and $\mathfrak{q}(b)$ are prime ideals, and using (6.1.22) one deduces that $W(b)$ is a valuation ring of rank one, which means that $\mathfrak{p}(b)$ and $\mathfrak{q}(b)$ are consecutive, $\square$

Theorem 6.6.12. *Let $(K, |\cdot|)$ be a valued field, $(K^\wedge, |\cdot|^\wedge)$ its completion. The following conditions are equivalent :*

   (i) *$(K, |\cdot|)$ is deeply ramified.*
   (ii) *For every valued extension $(E, |\cdot|_E)$ of $(K, |\cdot|)$, for every $b \in K^+ \setminus \{0\}$ and for every $i > 0$ we have $H_i(\mathbb{L}_{(E^+/bE^+)/(K^+/bK^+)}) = 0$.*
   (iii) *For every pair of consecutive prime ideals $\mathfrak{p} \subset \mathfrak{q} \subset K^+$, the valuation ring $(K^+/\mathfrak{p})_{\mathfrak{q}}$ is deeply ramified.*
   (iv) *For every pair of convex subgroups $H_1 \subset H_2 \subset \Gamma$, the quotient H_2/H_1 is not isomorphic to $\mathbb{Z}$, and moreover, if $p := \mathrm{char}(\kappa) > 0$ and the valuation is non-trivial, the Frobenius endomorphism on $K^{\wedge+}/pK^{\wedge+}$ is surjective.*

Proof. To show that (ii)⇒(i), we take $E := K^s$ and we choose a valuation on K^s extending $|\cdot|$. Then, by arguing as in the proof of lemma 6.5.13(ii), we deduce from (ii) that the scalar multiplication by b on Ω_{K^{s+}/K^+} is injective. Since the latter is a torsion K^{s+}-module, we deduce (i). To show (i)⇒(ii), we reduce first to the case where $E = E^a$; indeed, let $|\cdot|_{E^a}$ be a valuation on E^a extending $|\cdot|_E$ and suppose that the sought vanishing is known for the extension $(K, |\cdot|) \subset (E^a, |\cdot|_{E^a})$; by transitivity, it then suffices to show :

Claim 6.6.13. $H_i(\mathbb{L}_{(E^{a+}/bE^{a+})/(E^+/bE^+)}) = 0$ for every $i > 1$.

Proof of the claim. By [50, II.2.2.1] we have

$$\mathbb{L}_{(E^{a+}/bE^{a+})/(E^+/bE^+)} \simeq \mathbb{L}_{E^{a+}/E^+} \overset{\mathbf{L}}{\otimes}_{E^{a+}} E^{a+}/bE^{a+}$$

whence a spectral sequence

$$E_{pq}^2 := \mathrm{Tor}_p^{E^{a+}}(H_q(\mathbb{L}_{E^{a+}/E^+}), E^{a+}/bE^{a+}) \Rightarrow H_{p+q}(\mathbb{L}_{(E^{a+}/bE^{a+})/(E^+/bE^+)}).$$

Since E^{a+}/bE^{a+} is an E^{a+}-module of Tor-dimension ≤ 1, we see that $E_{pq}^2 = 0$ for every $p > 1$; furthermore, by theorem 6.3.32(i), it follows that $E_{pq}^2 = 0$ whenever $p, q > 0$ or $q > 1$, so the claim follows. $\diamond$

Thus, we can suppose that E is algebraically closed. A spectral sequence analogous to the foregoing computes $H_i(\mathbb{L}_{(E^+/bE^+)/(K^{a+}/bK^{a+})})$, and using theorems 6.5.12(ii) and 6.5.20, we find that the latter vanishes for $i > 0$. Consequently, by applying transitivity to the tower of extensions $K^+/bK^+ \subset K^{a+}/bK^{a+} \subset E^+/bE^+$, we reduce to show the assertion for the case $E = K^a$. However, by example 6.1.21(iii) we have $K^{a+}/bK^{a+} \simeq K^{s+}/bK^{s+}$ for every $b \in K^+ \setminus \{0\}$, so we can further reduce to the case $E = K^s$. In this case, one concludes the proof by another spectral sequence argument, this time using assumption (i) and theorem 6.3.32(ii) to show that the relevant terms E_{pq}^2 vanish.

To show that (iii)⇒(iv), we consider two subgroups $H_1 \subset H_2$ as in (iv); if c.rk$(H_2/H_1) > 1$, then clearly H_2/H_1 cannot be isomorphic to $\mathbb{Z}$, so we can assume that H_1 and H_2 are consecutive, so that the corresponding prime ideals are too (see (6.1.22)). In this case, (iii) and proposition 6.6.2 show that H_2/H_1 is not isomorphic to $\mathbb{Z}$, which is the first assertion of (iv). To prove the second assertion, it will suffice to show the following :

Claim 6.6.14. Suppose that $p := \mathrm{char}(\kappa) > 0$ and that (iii) holds. Then, for every $b \in K^+ \setminus \{0\}$ with $1 > |b| \geq |p|$, the Frobenius endomorphism on K^+/bK^+ is surjective.

Proof of the claim. For such a b as above, define $\mathfrak{p}(b)$, $\mathfrak{q}(b)$ and $W(b)$ as in lemma 6.6.11; then $W(b)$ is a valuation ring of rank one, so it is deeply ramified by assumption (iii). Then, by proposition 6.6.6 it follows that the Frobenius endomorphism is surjective on $W(b)/b \cdot W(b) \simeq K_{\mathfrak{q}(b)}^+/bK_{\mathfrak{q}(b)}^+$. We remark that $bK_{\mathfrak{q}(b)}^+ \subset K^+$; there follows a natural imbedding: $K^+/bK_{\mathfrak{q}(b)}^+ \subset W(b)/b \cdot W(b)$, commuting with the Frobenius maps. It is then easy to deduce that the Frobenius endomorphism is surjective on $K^+/bK_{\mathfrak{q}(b)}^+$. Moreover, by proposition 6.6.2, the valuation of $W(b)$

is not discrete, hence its value group is isomorphic to a dense subgroup of $(\mathbb{R}, \geq)$ (see example 6.1.21(vi)); therefore, by (6.1.22) and example 6.1.21(v), we deduce that there exists an element $c \in K^+$ such that $|b| > |c^{3p}|$ and $|b| < |c^{2p}|$. These inequalities have been chosen so that $c^p K^+_{\mathfrak{q}(b)} \subset K^+$ and $bK^+_{\mathfrak{q}(b)} \subset c^{2p} K^+_{\mathfrak{q}(b)}$, whence $bK^+_{\mathfrak{q}(b)} \subset c^p K^+$, and finally we conclude that the Frobenius endomorphism induces a surjection: $K^+/cK^+ \to K^+/c^p K^+$. We let $\mathrm{Fil}^\bullet_1(K^+/bK^+)$ (resp. $\mathrm{Fil}^\bullet_2(K^+/bK^+)$) be the c-adic (resp. c^p-adic) filtration on K^+/bK^+. The foregoing implies that the Frobenius endomorphism induces a morphism of filtered modules $\mathrm{Fil}^\bullet_1(K^+/bK^+) \to \mathrm{Fil}^\bullet_2(K^+/bK^+)$ which is surjective on the associated graded modules; by [16, Ch.III, §2, n.8, Cor.2] the claim follows. $\Diamond$

Next, assume (iv) and let $W := (K^+/\mathfrak{p})_\mathfrak{q}$, for two consecutive prime ideals $\mathfrak{p} \subset \mathfrak{q} \subset K^+$. Suppose first that the residue characteristic of W is $p > 0$. Then, by assumption the Frobenius map is surjective on K^+/bK^+, whenever $b \in K^+ \setminus \{0\}$ and $|b| \geq |p|$; we deduce easily that the Frobenius endomorphism is surjective on W/bW, which implies (iii), in view of proposition 6.6.6. In case that the residue characteristic of W is zero, (iii) follows as well, in view of claim 6.6.3.

(i)$\Rightarrow$(iii): Indeed, let $\mathfrak{p} \subset \mathfrak{q}$ be as in (iii); we need to show that $(K^+/\mathfrak{p})_\mathfrak{q}$ is deeply ramified. After replacing K^+ by $K^+_\mathfrak{q}$ we can assume that $\mathfrak{q}$ is the maximal ideal of K^+. The ring $k^+ := K^+/\mathfrak{p}$ is a valuation ring; let $k := \mathrm{Frac}(k^+)$, $|\cdot|_k$ the valuation on k corresponding to k^+, and $(k', |\cdot|_{k'})$ a finite separable valued extension of $(k, |\cdot|_k)$. It suffices to show that $\Omega_{k'^+/k^+} = 0$. We have $k' \simeq k[X]/(f(X))$ for some irreducible monic polynomial $f(X) \in k[X]$; let $\widetilde{f}(X) \in K^+_\mathfrak{p}[X]$ be a lifting of $f(X)$ to a monic polynomial. Then $E^+ := K^+_\mathfrak{p}[X]/(\widetilde{f}(X))$ is the integral closure of K^+ in the finite separable extension $E := \mathrm{Frac}(E^+)$ of K, and $E^+/\mathfrak{p}E^+ \simeq k'$, so that E^+ is a valuation ring, by lemma 6.1.13. Furthermore, the preimage of k'^+ in E^+ is a valuation ring R of E with $R \cap K = K^+$ and $R/\mathfrak{p}R \simeq k'^+$. From (i) and theorem 6.3.32(ii) we deduce that $\Omega_{R/K^+} = 0$, whence $\Omega_{k'^+/k^+} = 0$ as required.

Finally we show that (iv) implies (i). We distinguish several cases. The case when $p := \mathrm{char}(K) > 0$ has already been dealt with, in view of remark 6.6.10. Next suppose that $\mathrm{char}(\kappa) = 0$; we will adapt the argument given for the rank one case to prove corollary 6.3.21(i). As usual, we reduce to the case where K is strictly henselian; it suffices to show that $\Omega_{E^+/K^+} = 0$ for every finite extension $(E, |\cdot|_E)$ of K. Then E factors as a tower of subextensions $E_0 := K \subset E_1 \subset E_2 \subset ... \subset E_n := E$ such that $E_{i+1} = E_i[b_i^{1/l_i}]$ for every $i = 0, ..., n-1$, where $l_i := [E_{i+1} : E_i]$ is a prime number and $b_i \in E_i^\times$ such that $|b_i| \notin l_i \cdot \Gamma_{E_i}$. It is easy to see that assumption (iv) is inherited by every finite algebraic extension of K, hence we can reduce to the case $E = K[b^{1/l}]$, with $b \in K^\times$, $|b| \notin l \cdot \Gamma_K$. One verifies as in the proof of proposition 6.3.13(i) that E^+ consists of the elements of the form $\sum_{i=0}^{l-1} x_i \cdot b^{i/l}$ such that $x_i \in K$ and $|x_i \cdot b^{i/l}|_E \leq 1$ for every $i = 0, ..., l-1$ and we have to show that $d(x_i \cdot b^{i/l}) = 0$ for every $i \leq l-1$. We may assume that $i > 0$, and up to replacing b by $b^i \cdot x_i^l$, we can obtain that $b \in K^+$; we have then to verify that $db^{1/l} = 0$. Define $\mathfrak{p}(b), \mathfrak{q}(b)$ as in lemma 6.6.11, so that $\mathfrak{p}(b)$ and $\mathfrak{q}(b)$ are consecutive prime ideals of K^+, therefore by assumption $(K^+/\mathfrak{p}(b))_{\mathfrak{q}(b)}$ is a non discrete rank one valuation ring. Then, using (6.1.22) and example 6.1.21(v), we deduce that there exists an element $c \in K^+$ such that $|b| > |c^{l+1}|$ and $|b| < |c^l|$. We can write $b = x \cdot c^l$ for some $x \in K^+$, whence $db^{1/l} = c \cdot dx^{1/l}$. However, $|c| \le |b|^{(l-1)/l(l+1)} \le |b^{1/l} \cdot c^{-1}|^{l-1} = |x|^{(l-1)/l}$. Since $x^{(l-1)/l} \cdot dx^{1/l} = 0$, the claim follows. Finally, suppose that $p := \operatorname{char}(\kappa) > 0$ and $\operatorname{char}(K) = 0$. Arguing as in the previous case, we produce an element $b \in K^+$ such that $|b^p| > |p|$ and $|b^{p+1}| < |p|$. The Frobenius map is surjective on $K^+/b^p K^+$ by assumption, and on $K^{\mathrm{s}+}/b^p K^{\mathrm{s}+}$ by example 6.1.21(iii), hence $\Omega_{K^{\mathrm{s}+}/K^+} \simeq \Omega_{K^{\mathrm{s}+}[1/p]/K^+[1/p]}$, by lemma 6.5.13(ii). Now it suffices to remark that $K^+[1/p]$ is a valuation ring with residue field of characteristic zero, so we are reduced to the previous case, and the proof is concluded.   □

Remark 6.6.15. By inspection of the proof, it is easy to check that condition (ii) of theorem 6.6.12 is equivalent to the following. There exists a subset $S \subset K^+ \setminus \{0\}$ such that the convex subgroup generated by $|S| := \{|s| \mid s \in S\}$ equals Γ_K and $H_i(\mathbb{L}_{E^+/K^+} \otimes_{K^+} K^+/s \cdot K^+) = 0$ for every valued field extension $(E, |\cdot|_E)$ of $(K, |\cdot|)$, every $s \in S$ and every $i > 0$.

Corollary 6.6.16. *Let* $(K, |\cdot|) \subset (E, |\cdot|_E)$ *be an algebraic extension of valued fields.*

   (i) *If K is deeply ramified then E is deeply ramified.*
   (ii) *If E is finite over K and E is deeply ramified, then K is deeply ramified.*

Proof. (i): This holds by definition for separable extensions. It remains to consider the case that E is purely inseparable over K and K is of characteristic $p > 0$. We may also assume that the valuation of K is non-trivial (every field with a trivial valuation is deeply ramified). By theorem 6.6.12 the condition that K is deeply ramified is equivalent to the condition that the completion for the valuation topology $K^\wedge$ is perfect. As $K^\wedge$ is perfect, E embeds in $K^\wedge$, so $K^\wedge$ serves as a completion of E and $E^\wedge$ is perfect, whence the claim, again by theorem 6.6.12.

   (ii): By the equivalence (i)⇔(iii) of theorem 6.6.12, one reduces the assertion to the case of rank 1 valuations. In the equal characteristic 0 case, one has to show that the valuation of K is not discrete, which holds because its value group is of finite index in the value group of E. In the equal characteristic $p > 0$ case : by [16, Ch. VI, §8, Prop.2], $E^\wedge$ is a finite extension of $K^\wedge$ and in this case it is easily seen that $K^\wedge$ is perfect if and only if $E^\wedge$ is perfect. Finally, in the unequal characteristic $(0, p)$ case : let E^{s} be a separable closure of E with an extension of the valuation. It also serves as a separable closure of K. By the definition and an exact sequence of differential modules, $\Omega_{E^{\mathrm{s}+}/K^+}$ is generated by the image of Ω_{E^+/K^+}. The latter is almost finitely generated (proposition 6.3.8) and torsion, hence annihilated by some power of p. Hence $\Omega_{E^{\mathrm{s}+}/K^+}$ is both p-divisible and annihilated by a power of p, so it is 0.   □

In this final chapter we bring into the picture p-adic analytic geometry and formal schemes. The first three sections develop a theory of the *analytic cotangent complex* : we show how to attach a complex $\mathbb{L}^{an}_{X/Y}$ to any morphism of locally finite type $\phi : X \to Y$ of formal schemes or of R.Huber's adic spaces. This complex is obtained via *derived completion* from the usual cotangent complex of the morphism of ringed spaces underlying ϕ. We prove that $\mathbb{L}^{an}_{X/Y}$ controls the *analytic* deformation theory of the morphism ϕ, in the same way as the usual cotangent complex computes the deformations of the map of ringed spaces underlying ϕ. We hope the reader will agree with us that these sections – though largely independent from the rest of the monograph – are not misplaced, in view of the prominence of the cotangent complex construction throughout our work. Some of what we do here had been already anticipated in an appendix of André's treatise [2, Supplément].

The main result of the remaining sections 7.4 and 7.5 is a kind of weak purity statement valid for affinoid varieties over a deeply ramified valued field of rank one (theorem 7.4.17). The occurence of analytic geometry in purity issues (and in p-adic Hodge theory at large) is rather natural; indeed, the literature on the subject is littered with indications of the relevance of analytic varieties, and already in [74], Tate explicitly asked for a p-adic Hodge theory in the framework of rigid varieties. We elect instead to use the language of adic spaces, introduced by R.Huber in [48]. Adic spaces are generalizations of Zariski-Riemann spaces, that had already made a few cameo appearences in earlier works on rigid analytic geometry. Recall that Zariski-Riemann spaces were introduced originally by Zariski in his quest for the resolution of singularities of algebraic varieties. Zariski's idea was to attach to any reduced and irreducible variety X defined over a field k, the ringed space $(\widetilde{X}, \mathscr{O}_{\widetilde{X}})$ defined as the projective limit of the cofiltered system of all blow-up maps $X_\alpha \to X$; so a point of $\widetilde{X}$ is a compatible system $\widetilde{x} := (x_\alpha \in X_\alpha \mid X_\alpha \to X)$. It is easy to verify that, for any such $\widetilde{x}$, the stalk $\mathscr{O}_{\widetilde{X}, \widetilde{x}}$ is a valuation ring dominating the image of $\widetilde{x}$ in X. The strategy to construct a regular model for X was broken up in two stages : first one sought to show that for any point $\widetilde{x} \in \widetilde{X}$ one can find some x_α under $\widetilde{x}$ which is non-singular in the blow-up X_α. This first stage goes under the name of local uniformization; translated in algebraic terms, this implies that for every valuation ring v containing k of the field $k(X)$ of rational functions on X, there is a complete model of $k(X)$ on which the center of v is a non-singular point (a model is a reduced irreducible k-scheme Y of finite type such that $k(Y) = k(X)$). Local uniformization and the quasi-compactness of the Zariski-Riemann space imply that there is a "finite resolving system", *i.e.* a finite number of models such that every valuation of $k(X)$ has a non-singular center on one of them (notice that non-singularitiy of a point x_α extends to an open neighborhood U_α of x_α in X_α, so we achieve uniformization not just for $\widetilde{x}$, but for all the valuations contained in the preimage $\widetilde{U}$ of U_α, which is open in $\widetilde{X}$).

The second step is to try to reduce the number of models in a finite resolving system; restricting to open subsets of the Zariski-Riemann space, one reduces this to the question of going from a resolving system of cardinality two to a non-singular model. This is what Zariski called the "fundamental theorem" (see [78, §2.5]). Zariski showed local uniformization for all valuation rings containing the field $\mathbb{Q}$, but for the fundamental theorem he could only find proofs in dimensions ≤ 3.

Our theorem 7.4.17 is directly inspired by Zariski's strategy : rather than looking at the singularities of a variety, we try to resolve the singularities of an étale covering $Y \to X$ of smooth affinoid adic spaces over a deeply ramified non-archimedean field K (so the map is singular only on the special fibre of a given integral model defined over the valuation ring of K). We assume that X admits generically étale coordinates $t_1, ..., t_d \in \mathcal{O}_X(X)$, and the role of Zariski's $\tilde{X}$ is played by the projective system of all finite coverings of the form $X_n := X[t_1^{1/n}, ..., t_d^{1/n}]$, where n ranges over the positive integers. We have to show that the étale covering $Y_n := X_n \times_X Y \to X_n$ becomes "less and less" singular (that is, on its special fibre) as n grows; so we need a numerical invariant that quantifies the singularity of the covering : this is the *discriminant* $\mathfrak{d}_{Y/X}$. The analogue of local uniformization is our proposition 7.4.13, whose proof uses the results of section 6.6. Next, in order to exploit the quasi-compactness of the affinoid adic spaces Y_n, we have to show that the estimates furnished by proposition 7.4.13 "spread around" in an open subset; to this aim we prove that the discriminant function is semicontinuous : this is the purpose of section 7.5. Notice that in our case we do not need an analogue of Zariski's difficult fundamental theorem; this is because the discriminant function always decreases when n grows : what makes the problem of resolution of singularities much harder is that a non-singular point x_α on a blow up model X_α may be dominated by singular points x_β on some further blow up $X_\beta \to X_\alpha$ (so, for instance, one cannot trivially reduce the cardinality of a resolving system by forming joins of the various models of the system). Besides, our spaces are not varieties, but adic spaces, *i.e.* morally we work only "at the level of the Zariski-Riemann space" and we do not need – as for Zariski's problem – to descend to (integral) models of our spaces; so we are dealing exclusively with valuation rings (or mild extensions thereof), rather than more complicated local rings.

In essence, this is the complete outline of the method; however, our theorem is – alas – much weaker than Faltings' and does not yield by itself the kind of Galois cohomology vanishings that are required to deduce comparison theorems for the cohomology of algebraic varieties; we explain more precisely the current status of the question in (7.5.29).

Throughout this chapter we fix a valued field $(K, | \cdot |)$ with valuation of rank one, complete for its valuation topology. We also let a be a topologically nilpotent element in $K^\times$.

7.1. Derived completion functor. Let A be a complete K^+-algebra of topologically finite presentation. For any A-module M, we denote by $M^\wedge$ the (separated) a-adic completion of M.

Proposition 7.1.1. *Let A be as in* (7.1).

(i) *Every finitely generated A-module which is torsion-free as a K^+-module, is finitely presented.*

(ii) *A is a coherent ring.*

(iii) *Let N be a finitely generated A-module, $N' \subset N$ a submodule. Then there exists an integer $c \geq 0$ such that*

$$(7.1.2) \qquad\qquad a^k N \cap N' \subset a^{k-c} N'$$

for every $k \geq c$. In particular, the topology on N' induced by the a-adic topology on N, agrees with the a-adic topology of N'.

(iv) *Every finitely generated A-module is a-adically complete and separated.*

(v) *Every submodule of a free A-module F of finite type is closed for the a-adic topology of F.*

(vi) *Every A-algebra of topologically finite type is separated.*

Proof. (i) is an easy consequence of [13, Lemma 1.2]. To show (ii), one chooses a presentation $A := K^+\langle T_1, ..., T_n\rangle/I$ for some finitely generated ideal I, and then reduces to prove the statement for $K^+\langle T_1, ..., T_n\rangle$, in which case it follows from (i). Next, let N, N' be as in (iii) and define T to be the K^+-torsion submodule of $N'' := N/N'$; clearly T is an A-submodule, and the A-module N''/T is K^+-torsion-free, therefore is finitely presented by (i). Since N is finitely generated, this implies that $M := \mathrm{Ker}(N \to N''/T)$ is finitely generated. Hence, there exists an integer $c \geq 0$ such that $a^c M \subset N'$. If now $k \geq c$, we have $a^k N \cap N' \subset a^k N \cap M = a^k M \subset a^{k-c} N'$, which shows (iii). Next let us show:

Claim 7.1.3. Assertion (iv) holds for every finitely presented A-module.

Proof of the claim. Let N be a finitely presented A-module and choose a presentation $0 \to K \to A^n \to N \to 0$. By (ii), K is again finitely presented, and by (iii), the topology on K induced by the a-adic topology on A^n coincides with the a-adic topology of K. Hence, after taking a-adic completion, we obtain a short exact sequence : $0 \to K^\wedge \to A^n \to N^\wedge \to 0$ (see [58, Th.8.1]). It follows that the natural map $K \to K^\wedge$ is injective, which shows that the map $N \to N^\wedge$ is surjective for every finitely presented A-module N. In particular, this holds for K, whence $K \simeq K^\wedge$, and $N \simeq N^\wedge$, as claimed.   ◇

Finally, let M be a submodule of A^n. By (iii), the topology on M induced by A^n coincides with the a-adic topology. Consequently, if M is finitely presented, then M is complete for the a-adic topology by claim 7.1.3, hence complete as a subspace of A^n, hence closed in A^n. For an arbitrary M, define $\overline{M} := \bigcup_{n>0}(M : a^n)$; then $\overline{M}$ is a submodule of A^n and $A/\overline{M}$ is torsion-free as a K^+-module, so it is finitely presented by (i), therefore $\overline{M}$ is finitely presented by (ii). It follows that $a^c \overline{M} \subset M$ for some $c \geq 0$, whence $a^k A^n \cap \overline{M} = a^k A^n \cap M$ for every $k \geq c$. By the foregoing, $\overline{M}$ is complete, so $a^k A^n \cap \overline{M}$ is also complete, and finally M is complete, hence closed. This settles (v) and (iv) follows as well. (vi) is an immediate consequence of (v).   □

Lemma 7.1.4. *Let $A \to B$ be a map of K^+-algebras of topologically finite presentation. Then B is of topologically finite presentation as an A-algebra. More precisely, if $\phi : A\langle T_1, ..., T_n \rangle \to B$ is any surjective map, $\mathrm{Ker}\, \phi$ is finitely generated.*

Proof. By proposition 7.1.1(vi), B is complete and separated, hence we can find a surjective map $\phi : A\langle T_1, ..., T_n \rangle \to B$. It remains to show that $\mathrm{Ker}\, \phi$ is finitely generated for any such ϕ. We can write $A := K^+\langle T_{n+1}, ..., T_m \rangle / I$ for some finitely generated ideal I, and thus reduce to the case where $A = K^+$ and $\phi : K^+\langle T_1, ..., T_n \rangle \to B$. We will need the following :

Claim 7.1.5. Let $\alpha : K^+\langle Y_1, ..., Y_{r+s} \rangle \to B$ be a surjective map and denote by $\beta : K^+\langle Y_1, ..., Y_r \rangle \to K^+\langle Y_1, ..., Y_{r+s} \rangle$ the natural imbedding. Suppose that $\gamma := \alpha \circ \beta$ is surjective as well. Then $\mathrm{Ker}\, \alpha$ is finitely generated if and only if $\mathrm{Ker}\, \gamma$ is finitely generated.

Proof of the claim. For $i = r+1, ..., r+s$, choose $f_i \in K^+\langle Y_1, ..., Y_r \rangle$ such that $\gamma(f_i) = \alpha(Y_i)$. We define a surjective map $\delta : K^+\langle Y_1, ..., Y_{r+s} \rangle \to K^+\langle Y_1, ..., Y_r \rangle$ by setting $\delta(Y_i) := Y_i$ for $i \leq r$ and $\delta(Y_i) := f_i$ for $i > r$. Clearly $\gamma \circ \delta = \alpha$. There follows a short exact sequence $0 \to \mathrm{Ker}\, \delta \to \mathrm{Ker}\, \alpha \to \mathrm{Ker}\, \gamma \to 0$. However, $\mathrm{Ker}\, \delta$ is the closure of the ideal I generated by $Y_i - f_i$ for $i = r+1, ..., r+s$. By proposition 7.1.1(v), we deduce that $\mathrm{Ker}\, \delta = I$, and the claim follows easily. $\diamond$

By hypothesis there is at least one surjection $\psi : K^+\langle Y_1, ..., Y_r \rangle \to B$ with finitely generated kernel. Let $\mu : B \widehat{\otimes}_{K^+} B \to B$ be the multiplication map and set $\theta := \mu \circ (\phi \widehat{\otimes}_{K^+} \psi) : K^+\langle T_1, ..., T_n, Y_1, ..., Y_r \rangle \to B$. Applying twice claim 7.1.5 we deduce first that $\mathrm{Ker}\, \theta$ is finitely generated, and then that $\mathrm{Ker}\, \phi$ is too, as required. $\square$

Lemma 7.1.6. *Let F be a flat A-module. Then:*

 (i) *$F^\wedge$ is a flat A-module.*

 (ii) *For every finitely presented A-module M, the natural map*

$$(7.1.7) \qquad M \otimes_A F^\wedge \to (M \otimes_A F)^\wedge$$

 is an isomorphism.

Proof. To begin with, we claim that the functor $N \mapsto (N \otimes_A F)^\wedge$ is exact on the abelian category of finitely presented A-modules. Indeed, let $\underline{E} := (0 \to N' \to N \to N'' \to 0)$ be an exact sequence of finitely presented A-modules; we have to show that $(\underline{E} \otimes_A F)^\wedge$ is still exact. Obviously $\underline{E} \otimes_A F$ is exact, so the assertion will follow by [58, Th.8.1(ii)], once we know:

Claim 7.1.8. The topology on $N' \otimes_A F$ induced by the imbedding into $N \otimes_A F$ agrees with the a-adic topology.

Proof of the claim. By proposition 7.1.1(iii), we can find $c \geq 0$ such that (7.1.2) holds. Since F is flat, we derive

$$a^k(N \otimes_A F) \cap (N' \otimes_A F) \subset a^{k-c}(N' \otimes_A F)$$

which implies the claim. $\diamond$

(ii): Clearly (7.1.7) is an isomorphism in case M is a free module of finite type. For a general M, one chooses a resolution $\underline{R} := (A^n \to A^m \to M \to 0)$; by the foregoing, the sequence $(\underline{R} \otimes_A F)^{\wedge}$ is still exact, so one concludes by applying the 5-lemma to the map of complexes $\underline{R} \otimes_A F^{\wedge} \to (\underline{R} \otimes_A F)^{\wedge}$.

(i): We have to show that, for every injective map of A-modules $f : N' \to N$, $f \otimes_A 1_{F^{\wedge}}$ is still injective. By the usual reductions, we can assume that both N and N' are finitely presented. In view of (ii), this is equivalent to showing that the induced map $(N' \otimes_A F)^{\wedge} \to (N \otimes_A F)^{\wedge}$ is injective, which is already known. $\square$

7.1.9.   We will need to consider the left derived functor of the a-adic completion functor, which we denote:

(7.1.10) $\mathsf{D}^-(A\text{-}\mathbf{Mod}) \to \mathsf{D}^-(A\text{-}\mathbf{Mod})$: $(K^{\bullet}) \mapsto (K^{\bullet})^{\wedge}$.

As usual, it can be defined by completing termwise bounded above complexes of projective A-modules. However, the following lemma shows that it can also be computed by arbitrary flat resolutions.

Lemma 7.1.11. *Let* $\phi : K_1^{\bullet} \to K_2^{\bullet}$ *be a quasi-isomorphism of bounded above complexes of flat A-modules and denote by $(K_i^{\bullet})^{\wedge}$ the termwise a-adic completion of $K_i^{\bullet}$ ($i = 1, 2$). Then the induced morphism*

(7.1.12) $(K_1^{\bullet})^{\wedge} \to (K_2^{\bullet})^{\wedge}$

is a quasi-isomorphism.

Proof. Since $K_1^{\bullet}$ and $K_2^{\bullet}$ are termwise flat, we deduce quasi-isomorphisms

$$\phi_n : K_{1,n}^{\bullet} := K_1^{\bullet} \otimes_A A/a^n A \to K_{2,n}^{\bullet} := K_2^{\bullet} \otimes_A A/a^n A$$

for every $n \in \mathbb{N}$. The map of inverse system of complexes $(K_{1,n}^{\bullet})_{n \in \mathbb{N}} \to (K_{2,n}^{\bullet})_{n \in \mathbb{N}}$ can be viewed as a morphism of complexes of objects of the abelian category $(A\text{-}\mathbf{Mod})^{\mathbb{N}}$ of inverse systems of A-modules. As such, it induces a morphism $(\phi_n)_{n \in \mathbb{N}}$ in the derived category $\mathsf{D}((A\text{-}\mathbf{Mod})^{\mathbb{N}})$, and it is clear that $(\phi_n)_{n \in \mathbb{N}}$ is a quasi-isomorphism. Let

$$R\lim : \mathsf{D}((A\text{-}\mathbf{Mod})^{\mathbb{N}}) \to \mathsf{D}(A\text{-}\mathbf{Mod})$$

be the right derived functor of the inverse limit functor $\lim$: $(A\text{-}\mathbf{Mod})^{\mathbb{N}} \to A\text{-}\mathbf{Mod}$. We remark that, for every $j \in \mathbb{Z}$, the inverse systems $(K_{i,n}^j)_{n \in \mathbb{N}}$ ($i = 1, 2$) are acyclic for the functor $\lim$, since their transition maps are surjective. We derive that $R\lim(K_{i,n}^{\bullet})_{n \in \mathbb{N}} \simeq (K_i^{\bullet})^{\wedge}$, and, under this identification, the morphism (7.1.12) is the same as $R\lim(\phi_n)_{n \in \mathbb{N}}$. Since the latter preserves quasi-isomorphisms, the claim follows. $\square$

7.1.13.   We denote by $\mathsf{D}^-(A\text{-}\mathbf{Mod})^{\wedge}$ the essential image of the functor (7.1.10).

Corollary 7.1.14. (i) *For any object $K^{\bullet}$ of $\mathsf{D}^-(A\text{-}\mathbf{Mod})$, the natural morphism*

$$(K^{\bullet})^{\wedge} \to ((K^{\bullet})^{\wedge})^{\wedge}$$

is a quasi-isomorphism.

(ii) $\mathsf{D}^-(A\text{-}\mathbf{Mod})^{\wedge}$ *is a full triangulated subcategory of* $\mathsf{D}^-(A\text{-}\mathbf{Mod})$.

Proof. Notice first that there are two natural morphisms as in (i), which coincide : namely, for any complex E in $\mathsf{D}^-(A\text{-}\mathbf{Mod})$ one has a natural morphism $u_E : E \to E^\wedge$; then one can take either $(u_K)^\wedge$ or $u_{K^\wedge}$. Now, (i) is an immediate consequence of lemmata 7.1.11 and 7.1.6. It follows from (i) that $\mathsf{D}^-(A\text{-}\mathbf{Mod})^\wedge$ is the full subcategory of $\mathsf{D}^-(A\text{-}\mathbf{Mod})$ consisting of the objects K for which u_K is an isomorphism,which implies (ii). $\qquad\square$

We will need some generalities on pseudo-coherent complexes of R-modules (for an arbitrary ring R), which we borrow from [11, Exp.I]. In our situation, the definitions can be simplified somewhat, since we are only concerned with complexes of sheaves over the one-point site that are pseudo-coherent relative to the subcategory of free R-modules of finite type.

7.1.15.   For given $n \in \mathbb{Z}$, one says that a complex $K^\bullet$ of R-modules is *n-pseudo-coherent* if there exists a quasi-isomorphism $E^\bullet \to K^\bullet$ where $E^\bullet$ is a complex bounded above such that E^i is a free R-module of finite type for every $i \geq n$. One says that $K^\bullet$ is *pseudo-coherent* if it is n-pseudo-coherent for every $n \in \mathbb{Z}$.

7.1.16.   Let $K^\bullet$ be a n-pseudo-coherent (resp. pseudo-coherent) complex of R-modules, and $F^\bullet \to K^\bullet$ a quasi-isomorphism. Then $F^\bullet$ is n-pseudo-coherent (resp. pseudo-coherent) ([11, Exp.I, Prop.2.2(b)]). It follows that that the pseudo-coherent complexes form a (full) subcategory $\mathsf{D}(R\text{-}\mathbf{Mod})_{\mathrm{coh}}$ of $\mathsf{D}(R\text{-}\mathbf{Mod})$.

7.1.17.   Furthermore, let $X \to Y \to Z \to X[1]$ be a distinguished triangle in $\mathsf{D}^-(R\text{-}\mathbf{Mod})$. If X and Z are n-pseudo-coherent (resp. pseudo-coherent), then the same holds for Y ([11, Exp.I, Prop.2.5(b)]).

Lemma 7.1.18. *Let* $n, p \in \mathbb{N}$, $K^\bullet$ *a n-pseudo-coherent complex in* $\mathsf{D}^{\leq 0}(R\text{-}\mathbf{Mod})$, *and* $\mathscr{F}_p$ *one of the functors* $\otimes_R^p$, Sym_R^p, Λ_R^p, Γ_R^p *defined in* [50, I.4.2.2.6]. *Then* $L\mathscr{F}_p(K^\bullet)$ *is an n-pseudo-coherent complex.*

Proof. It is well known that $\mathscr{F}_p$ sends free R-modules of finite type to free R-modules of finite type. $K^\bullet$ can be represented by a complex of free R-modules, concentrated in non positive degrees, with finitely generated terms in degrees $\geq n$. The assertion of the lemma can then be checked by inspecting the definitions. We omit the details. $\qquad\square$

7.1.19.   Let $K^\bullet$ be a pseudo-coherent complex. By ([11, Exp.I, Prop.2.7]) there exists a quasi-isomorphism $E^\bullet \to K^\bullet$ where $E^\bullet$ is a bounded above complex of free R-modules of finite type.

7.1.20.   Suppose now that R is coherent; then we deduce easily that a complex $K^\bullet$ of R-modules is pseudo-coherent if and only if $H^i(K^\bullet)$ is a coherent R-module for every $i \in \mathbb{Z}$ and $H^i(K^\bullet) = 0$ for every sufficiently large $i \in \mathbb{Z}$ ([11, Exp.I, Cor.3.5]). Especially, for any R-module M of finite presentation, the complex $M[0]$ consisting of the module M placed in degree zero is pseudo-coherent.

7.1.21. For instance, take $R = A$ a K^+-algebra A as in (7.1); then from (7.1.19) and proposition 7.1.1(ii) it follows also that $\mathsf{D}^-(A\text{-}\mathbf{Mod})_{\mathrm{coh}} \subset \mathsf{D}^-(A\text{-}\mathbf{Mod})^\wedge$; especially, the natural morphism $M[0] \to M[0]^\wedge$ is a quasi-isomorphism for every finitely presented A-module M. More generally, we have:

Lemma 7.1.22. *Let A be as in (7.1) and M an A-module. The natural map*

$$(7.1.23) \qquad H_0(M[0]^\wedge) \to M^\wedge$$

is an isomorphism whenever one of the following conditions holds:

(a) *M is finitely generated.*
(b) *M is annihilated by a power of a.*
(c) *M is a-divisible.*

Proof. Quite generally, (7.1.23) is a surjection that induces an isomorphism modulo $a^n A$ for every $n \in \mathbb{N}$, and therefore an isomorphism on the a-adic completions. The kernel of (7.1.23) can be identified with:

$$(7.1.24) \qquad \lim_{n \in \mathbb{N}}{}^1 \mathrm{Tor}_1^A(M, A/a^n A)$$

(see [77, Th.3.5.8]). For every $n \in \mathbb{N}$, the A-module $\mathrm{Tor}_1^A(M, A/a^n A)$ is a quotient of $M_n := \mathrm{Ker}(a^n : M \to M)$, and the transition maps in the inverse system defining (7.1.24) are induced by the multiplication by $a : M_{n+1} \to M_n$. It follows that these transition maps are surjective in case M is a-divisible, whence the assertion in case (c) holds. Condition (a) and proposition 7.1.1(iii) imply that there exists $n \in \mathbb{N}$ such that $M_k = M_n$ for all $k \geq n$; the same holds trivially under condition (b), and then we deduce easily that the above inverse system is essentially zero when either (a) or (b) holds; the claim now follows from [77, Prop.3.5.7].   □

Lemma 7.1.25. *Let $A \to B$ be a map of complete K^+-algebras of topologically finite presentation. For every object $K^\bullet$ of $\mathsf{D}^-(A\text{-}\mathbf{Mod})$, the natural morphism*

$$(K^\bullet \overset{\mathbf{L}}{\otimes}_A B)^\wedge \to (K^{\bullet\wedge} \overset{\mathbf{L}}{\otimes}_A B)^\wedge$$

is a quasi-isomorphism.

Proof. We can suppose that $K^\bullet$ is a complex of free A-modules. Then we are reduced to showing that, for every free A-module F, the natural map $(F \otimes_A B)^\wedge \to (F^\wedge \otimes_A B)^\wedge$ is an isomorphism. We leave this task to the reader.   □

Definition 7.1.26. Let $\phi : A \to B$ be a map of complete K^+-algebras of topologically finite presentation. The B-module of *analytic differentials relative to ϕ* is defined as $\Omega^{\mathrm{an}}_{B/A} := \Omega^\wedge_{B/A}$. The *analytic cotangent complex* of ϕ is the complex

$$\mathbb{L}^{\mathrm{an}}_{B/A} := (\mathbb{L}_{B/A})^\wedge.$$

Notice that $\mathbb{L}^{\mathrm{an}}_{B/A}$ is defined here via the standard resolution $P_A(B) \to B$ of [50, Ch.I, §1.5], and it is therefore well defined as a *complex* of B-modules, not just as an object in the derived category $\mathsf{D}^-(B\text{-}\mathbf{Mod})^\wedge$. This will be essential in order to globalize the construction to formal schemes, in (7.2.1), and to adic spaces, in definition 7.2.32.

Lemma 7.1.27. *Let* $A \to B$ *be a continuous map of* K^+-*algebras of topologically finite presentation. The following holds:*

   (i) $\Omega^{\mathrm{an}}_{B/A}$ *is a finitely presented* B-*module.*

   (ii) *The natural map* $\phi_{B/A} : \Omega_{B/A} \to \Omega^{\mathrm{an}}_{B/A}$ *is surjective with* K^+-*divisible kernel.*

   (iii) *The natural map* $H_0(\mathbb{L}^{\mathrm{an}}_{B/A}) \to \Omega^{\mathrm{an}}_{B/A}$ *is an isomorphism.*

Proof. One writes $B = B_0/I$ with $B_0 := A\langle T_1, ..., T_n\rangle$ and $I \subset B_0$ finitely generated. Then $\Omega^{\mathrm{an}}_{B_0/A} \simeq \oplus_{i=1}^n B_0 dT_i$, from which one sees easily that $\phi_{B_0/A}$ is onto. Next, proposition 7.1.1(iv) implies that

$$\mathrm{Coker}(d : I/I^2 \to B \otimes_{B_0} \Omega^{\mathrm{an}}_{B_0/A}) = \Omega^{\mathrm{an}}_{B/A}$$

which shows that (i) holds and that $\phi_{B/A}$ is onto as well, and yields a surjective map $B \otimes_{B_0} \mathrm{Ker}\, \phi_{B_0/A} \to \mathrm{Ker}\, \phi_{B/A}$. This allows to reduce (ii) to the case where $B = B_0$. In this case, $\mathrm{Ker}\, \phi_{B/A}$ is generated by the terms of the form $\delta(f) := df - \sum_{i=1}^n (\partial f/\partial T_i) \cdot dT_i$, where f ranges over all the elements of B_0. For given $f \in B_0$, we can write $f = f_0 + af_1$, with $f_0 \in A[T_1, ..., T_n]$, $f_1 \in B_0$. It follows easily that $\delta(f) = \delta(af_1) = a \cdot \delta(f_1)$, whence the claim.

   To show (iii), first we remark that there is a natural identification

$$H_0(\mathbb{L}^{\mathrm{an}}_{B/A}) = H_0(\Omega_{B/A}[0]^\wedge)$$

whose proof we shall leave to the reader. We are thus reduced to consider the natural map $H_0(\Omega_{B/A}[0]^\wedge) \to \Omega^{\mathrm{an}}_{B/A}$. Let $M := \mathrm{Ker}\, \phi_{B/A}$; according to (ii) and lemma 7.1.22 we have $H_0(M[0]^\wedge) = M^\wedge = 0$. Since the derived completion functor is triangulated and preserves $\mathsf{D}^{\leq 0}(A\text{-}\mathbf{Mod})$, we conclude that the natural map $H_0(\Omega_{B/A}[0]^\wedge) \to H_0(\Omega^{\mathrm{an}}_{B/A}[0]^\wedge)$ is an isomorphism. But in view of (i) and lemma 7.1.22, the target of this map is naturally identified with $\Omega^{\mathrm{an}}_{B/A}$, whence the claim. $\qquad\square$

Remark 7.1.28. Lemma 7.1.27(ii) is generalized by the following observation. Let $K_{B/A} := \mathrm{Cone}(\psi_{B/A} : \mathbb{L}_{B/A} \to \mathbb{L}^{\mathrm{an}}_{B/A})[1]$; one has: $K_{B/A} \overset{\mathbf{L}}{\otimes}_B B/aB \simeq 0$. Indeed, directly on the definition of $\mathbb{L}^{\mathrm{an}}_{B/A}$ one sees that $\psi_{B/A} \overset{\mathbf{L}}{\otimes}_B 1_{B/aB}$ is an isomorphism.

Proposition 7.1.29. *Let* $\phi : A \to B$ *be a map of complete* K^+-*algebras of topologically finite presentation, and suppose that* ϕ *is formally smooth for the* a-*adic topology. Then there is a natural quasi-isomorphism*

$$\mathbb{L}^{\mathrm{an}}_{B/A} \simeq \Omega^{\mathrm{an}}_{B/A}[0].$$

Proof. For every $n \in \mathbb{N}$, set $A_n := A/a^n \cdot A$ and $B_n := B/a^n \cdot B$. The hypothesis on ϕ implies that $\phi_n := \phi \otimes_A 1_{A_n}$ is of finite presentation and formally smooth for the discrete topology, therefore

(7.1.30) $\qquad\qquad \mathbb{L}_{B_n/A_n} \simeq \Omega_{B_n/A_n}[0] \simeq \Omega_{B/A} \otimes_A A_n[0]$

for every $n \in \mathbb{N}$. Moreover, ϕ is flat by [13, Lemma 1.6], hence $\mathbb{L}_{B_n/A_n} \simeq \mathbb{L}_{B/A} \otimes_A A_n$. On the other hand, for every $i \in \mathbb{Z}$ there is a short exact sequence (cp. [77, Th.3.5.8])

$$0 \to \lim_{n \in \mathbb{N}} {}^1 H^{i-1}(\mathbb{L}_{B/A} \otimes_A A_n) \to H^i(\mathbb{L}_{B/A}^{an}) \to \lim_{n \in \mathbb{N}} H^i(\mathbb{L}_{B/A} \otimes_A A_n) \to 0.$$

In view of (7.1.30), the inverse system $(H^{i-1}(\mathbb{L}_{B/A} \otimes_A A_n))_{n \in \mathbb{N}}$ vanishes for $i \neq 1$ and has surjective transition maps for $i = 1$, hence its $\lim^1$ vanishes for every $i \in \mathbb{Z}$, and the claim follows easily. $\qquad\square$

Proposition 7.1.31. *Let $\phi : A \to B$ be a surjective map of complete K^+-algebras of topologically finite presentation. Then $\mathbb{L}_{B/A}$ is a pseudo-coherent complex, in particular it lies in $\mathsf{D}^-(B\text{-}\mathbf{Mod})^\wedge$, and $\mathbb{L}_{B/A} \simeq \mathbb{L}_{B/A}^{an}$.*

Proof. First of all, notice that by lemma 7.1.4, B is of finite presentation, hence it is coherent as an A-module. Let $P := P_A(B)$ be the standard simplicial resolution of B by free A-algebras ([50, Ch.I, §1.5]). We obtain a morphism of simplicial B-algebras $\phi : B \otimes_A P \to B$ by tensoring with B the augmentation $P \to B$ (here B is regarded as a constant simplicial algebra). By the foregoing, P is pseudo-coherent, hence $P \otimes_A B$ lies in $\mathsf{D}(B\text{-}\mathbf{Mod})_{coh}$. Let $J := \operatorname{Ker} \phi$. The short exact sequence of complexes $0 \to J \to P \otimes_A B \to B \to 0$ is split, therefore J is also pseudo-coherent. Recall that we have natural isomorphisms: $J^i/J^{i+1} \xrightarrow{\sim} \operatorname{Sym}_B^i(\mathbb{L}_{B/A})$ for every $i \in \mathbb{N}$ (where $J^0 := B \otimes_A P$ and $\operatorname{Sym}_B^0(\mathbb{L}_{B/A}) := B$) ([50, Ch.III,§3.3]). Furthermore, we have (see *loc.cit.*) :

$$(7.1.32) \qquad H_n(J^i) = 0 \quad \text{for every } n, i \in \mathbb{N} \text{ such that } i > n.$$

We prove by induction on n that $\mathbb{L}_{B/A}$ is n-pseudo-coherent for every $n \leq 1$. If $n = 1$ there is nothing to prove. Suppose that the claim is known for the integer n. It then follows by lemma 7.1.18 that J^i/J^{i+1} is n-pseudo-coherent for every $i > 0$. However, it follows from (7.1.32) that J^i is n-pseudo-coherent as soon as $i > -n$. Hence, by (7.1.17) (and an easy induction), we deduce that J^i is n-pseudo-coherent for every $i \in \mathbb{N}$. Hence $J^2[1]$ is $(n-1)$-pseudo-coherent; if we now apply (7.1.17) to the distinguished triangle $J \to \mathbb{L}_{B/A} \to J^2[1] \to J[1]$, we deduce that $\mathbb{L}_{B/A}$ is $(n-1)$-pseudo-coherent. $\qquad\square$

Theorem 7.1.33. *Let $A \to B \to C$ be maps of complete K^+-algebras of topologically finite presentation. Then:*

(i) *$\mathbb{L}_{B/A}^{an}$ lies in $\mathsf{D}^-(B\text{-}\mathbf{Mod})_{coh}$.*

(ii) *There is a natural distinguished triangle in $\mathsf{D}^-(C\text{-}\mathbf{Mod})$:*

$$(7.1.34) \qquad C \otimes_B \mathbb{L}_{B/A}^{an} \to \mathbb{L}_{C/A}^{an} \to \mathbb{L}_{C/B}^{an} \to C \otimes_B \mathbb{L}_{B/A}^{an}[1].$$

Proof. (i): By lemma 7.1.4 we can find a surjection $B_0 := A\langle T_1, ..., T_n \rangle \to B$ from a topologically free A-algebra onto B. If we apply transitivity to the sequence of maps $A \to B_0 \to B$ and take the (derived) completion of the resulting distinguished triangle, we end up with the triangle:

$$(B \otimes_{B_0} \mathbb{L}_{B_0/A})^\wedge \to \mathbb{L}_{B/A}^{an} \to \mathbb{L}_{B/B_0}^{an} \to (B \otimes_{B_0} \mathbb{L}_{B_0/A})^\wedge[1].$$

We know already from proposition 7.1.31 that $\mathbb{L}_{B/B_0}$ is pseudo-coherent, hence it coincides with $\mathbb{L}_{B/B_0}^{an}$. Lemma 7.1.25 yields a quasi-isomorphism:

$$(B \otimes_{B_0} \mathbb{L}_{B_0/A})^\wedge \xrightarrow{\sim} (B \otimes_{B_0} \mathbb{L}_{B_0/A}^{an})^\wedge.$$

In view of proposition 7.1.29, $\mathbb{L}_{B_0/A}^{an}$ is a free B_0-module of finite rank in degree zero, in particular it is pseudo-coherent, so the same holds for $(B \otimes_{B_0} \mathbb{L}_{B_0/A})^\wedge$, and taking into account (7.1.17), the claim follows.

(ii): If we apply transitivity to the sequence of maps $A \to B \to C$, and then we complete the distinguished triangle thus obtained, we obtain (7.1.34), except that the first term is replaced by $(C \otimes_B \mathbb{L}_{B/A})^\wedge$, which we can also write as $(C \otimes_B \mathbb{L}_{B/A}^{an})^\wedge$, in view of lemma 7.1.25. However, by (i), $\mathbb{L}_{B/A}^{an}$ is pseudo-coherent, so it remains such after tensoring by C; in particular $C \otimes_B \mathbb{L}_{B/A}^{an}$ is already complete, and the claim follows. $\qquad\square$

7.2. Cotangent complex for formal schemes and adic spaces.

In this section we show how to globalize the definition of the analytic cotangent complex introduced in section 7.1. We consider two kinds of globalization : first we define the cotangent complex of a morphism $\mathfrak{f} : \mathfrak{X} \to \mathfrak{Y}$ of formal schemes locally of finite presentation over $\operatorname{Spf} K^+$; then we will define the cotangent complex for a morphism of adic spaces locally of finite type over $\operatorname{Spa}(K, K^+)$.

7.2.1.   Let $\mathfrak{f} : \mathfrak{X} \to \mathfrak{Y}$ be a morphism of formal schemes locally of finite presentation over $\operatorname{Spf} K^+$, and suppose that $\mathfrak{Y}$ is separated. For every affine open subset $U \subset \mathfrak{X}$, the small category F_U of all affine open subsets $V \subset \mathfrak{Y}$ with $\mathfrak{f}(U) \subset V$, is cofiltered under inclusion (or else it is empty). For every $V \in F_U$, $\mathscr{O}_{\mathfrak{Y}}(V)$ is a K^+-algebra of topologically finite presentation, hence the induced morphism $\mathscr{O}_{\mathfrak{Y}}(V) \to \mathscr{O}_{\mathfrak{X}}(U)$ is of the kind considered in definition 7.1.26. We set

$$L(U/\mathfrak{Y}) := \operatorname*{colim}_{V \in F_U^o} \mathbb{L}_{\mathscr{O}_{\mathfrak{X}}(U)/\mathscr{O}_{\mathfrak{Y}}(V)}^{an}.$$

Definition 7.2.2. The mapping $U \mapsto L(U/\mathfrak{Y})$ defines a complex of presheaves on a cofinal family of affine open subsets of $\mathfrak{X}$. By applying degreewise the construction of [26, Ch.0, §3.2.1], we can extend the latter to a complex of presheaves of $\mathscr{O}_{\mathfrak{X}}$-modules on $\mathfrak{X}$. We define the *analytic cotangent complex* $\mathbb{L}_{\mathfrak{X}/\mathfrak{Y}}^{an}$ of the morphism $\mathfrak{f} : \mathfrak{X} \to \mathfrak{Y}$ as the complex of sheaves associated to this complex of presheaves (this means that we form degreewise the associated sheaf, and we consider the resulting complex).

7.2.3.   More generally, if $\mathfrak{Y}$ is not necessarily separated, we can choose an affine covering $\mathfrak{Y} = \bigcup_{i \in I} \mathfrak{U}_i$ and the construction above applies to the restrictions $\mathfrak{V}_i := \mathfrak{f}^{-1}(\mathfrak{U}_i) \to \mathfrak{U}_i$; since the definition of $\mathbb{L}_{\mathfrak{V}_i/\mathfrak{U}_i}^{an}$ is local on $\mathfrak{V}_i$, one can then glue them into a single cotangent complex $\mathbb{L}_{\mathfrak{X}/\mathfrak{Y}}^{an}$.

Lemma 7.2.4. *Let $\mathbb{L}_{\mathfrak{X}/\mathfrak{Y}}$ denote the (usual) cotangent complex of the morphism*

$$(f, f^\sharp) : (\mathfrak{X}, \mathscr{O}_{\mathfrak{X}}) \to (\mathfrak{Y}, \mathscr{O}_{\mathfrak{Y}})$$

of ringed spaces; there exists a natural map of complexes

(7.2.5) $$\mathbb{L}_{\mathfrak{X}/\mathfrak{Y}} \to \mathbb{L}_{\mathfrak{X}/\mathfrak{Y}}^{an}$$

inducing an isomorphism

(7.2.6) $$\mathbb{L}_{\mathfrak{X}/\mathfrak{Y}} \otimes_{\mathcal{O}_{\mathfrak{X}}} \mathcal{O}_{\mathfrak{X}}/a^n\mathcal{O}_{\mathfrak{X}} \xrightarrow{\sim} \mathbb{L}_{\mathfrak{X}/\mathfrak{Y}}^{an} \otimes_{\mathcal{O}_{\mathfrak{X}}} \mathcal{O}_{\mathfrak{X}}/a^n\mathcal{O}_{\mathfrak{X}}$$

for every $n \in \mathbb{N}$.

Proof. It suffices to construct (7.2.5) in case $\mathfrak{Y}$ is affine. According to [50, Ch.II, (1.2.3.6)] and [50, Ch.II, (1.2.3.4)], the complex $\mathbb{L}_{\mathfrak{X}/\mathfrak{Y}}$ is naturally isomorphic to the sheafification of the complex of presheaves defined by the rule:

$$U \mapsto \operatorname*{colim}_{V \in F_U^\circ} \mathbb{L}_{\mathcal{O}_{\mathfrak{X}}(U)/\mathcal{O}_{\mathfrak{Y}}(V)}$$

and then it is clear how to define 7.2.5. From the construction it is obvious that (7.2.6) is an isomorphism. $\qquad\square$

It is occasionally important to know that both $\mathbb{L}_{\mathfrak{X}/\mathfrak{Y}}^{an}$ and the morphism (7.2.5) are well defined in the category of complexes of $\mathcal{O}_{\mathfrak{X}}$-modules (not just in its derived category).

7.2.7.   Furthermore, denote by $\Omega_{\mathfrak{X}/\mathfrak{Y}}^{an}$ the sheaf of *analytic relative differentials* for the morphism $\mathfrak{f}$, which is defined as $(\mathcal{I}/\mathcal{I}^2)_{|\mathfrak{X}}$, where $\mathcal{I}$ is the ideal defining the diagonal imbedding $\mathfrak{X} \to \mathfrak{X} \times_{\mathfrak{Y}} \mathfrak{X}$. We see easily that there is a natural isomorphism

(7.2.8) $$H_0(\mathbb{L}_{\mathfrak{X}/\mathfrak{Y}}^{an}) \simeq \Omega_{\mathfrak{X}/\mathfrak{Y}}^{an}.$$

Lemma 7.2.9. *Let $\mathfrak{X} := \operatorname{Spf} A$ be an affine formal scheme finitely presented over $\operatorname{Spf} K^+$. For every $f \in A$, let $\mathfrak{D}(f) := \{x \in \mathfrak{X} \mid f \notin \mathfrak{m}_x\}$ (where $\mathfrak{m}_x$ is the maximal ideal of $\mathcal{O}_{\mathfrak{X},x}$). The natural map $A \to \Gamma(\mathfrak{D}(f), \mathcal{O}_{\mathfrak{X}})$ is flat.*

Proof. Since $\Gamma(\mathfrak{D}(f), \mathcal{O}_{\mathfrak{X}})$ is the a-adic completion of A_f, the lemma follows from lemma 7.1.6(i). $\qquad\square$

Proposition 7.2.10. *Let $\mathfrak{f} : \mathfrak{X} \to \mathfrak{Y}$ be a morphism of formal schemes locally of finite type over $\operatorname{Spf}(K^+)$. Then:*

 (i) *$\mathbb{L}_{\mathfrak{X}/\mathfrak{Y}}^{an}$ is a pseudo-coherent complex of $\mathcal{O}_{\mathfrak{X}}$-modules.*

 (ii) *If $\mathfrak{f}$ is a closed imbedding, then (7.2.5) is a quasi-isomorphism.*

 (iii) *If $\mathfrak{f}$ is formally smooth, then (7.2.8) induces a quasi-isomorphism:*

$$\mathbb{L}_{\mathfrak{X}/\mathfrak{Y}}^{an} \xrightarrow{\sim} \Omega_{\mathfrak{X}/\mathfrak{Y}}^{an}[0].$$

Proof. (i): According to [11, Exp.I, Prop.2.11(b)], it suffices to show that $H_i(\mathbb{L}_{\mathfrak{X}/\mathfrak{Y}}^{an})$ is a coherent $\mathcal{O}_{\mathfrak{X}}$-module for every $i \in \mathbb{N}$. To this aim, let $U \subset \mathfrak{X}$ be an affine open subset such that the family F_U (notation of (7.2.1)) is not empty; pick any $V \in F_U$. After replacing $\mathfrak{X}$ by U, we can suppose that $U = \mathfrak{X}$ and $\mathfrak{Y}$ is separated. Set $A := \Gamma(\mathfrak{X}, \mathcal{O}_{\mathfrak{X}})$ and let $L_i^\triangle$ be the coherent $\mathcal{O}_{\mathfrak{X}}$-module associated to the coherent A-module $L_i := H_i(\mathbb{L}_{\mathcal{O}_{\mathfrak{X}}(\mathfrak{X})/\mathcal{O}_{\mathfrak{Y}}(V)}^{an})$ (cp. [26, Ch.I, §10.10.1], where this concept is discussed in the case of locally noetherian formal schemes; we note that [26, Ch.I, 10.8.8, 10.10.2, 10.10.3] extend to our case). By the definition of $\mathbb{L}_{\mathfrak{X}/\mathfrak{Y}}^{an}$ we deduce natural morphisms of $\mathcal{O}_{\mathfrak{X}}$-modules:

$$\alpha_i : L_i^\triangle \to H_i(\mathbb{L}^{an}_{\mathfrak{X}/\mathfrak{Y}}) \qquad \text{for every } i \in \mathbb{N}.$$

Claim 7.2.11. α_i an isomorphism for every $i \in \mathbb{N}$.

Proof of the claim. It suffices to show that α_i induces an isomorphism on the stalks. To this aim, we remark first that the natural map $L_i \to H_i(L(\mathfrak{X}/\mathfrak{Y}))$ is an isomorphism. Indeed, it suffices to consider another open subset $V' \subset F_{\mathfrak{X}}$ with $V' \subset V$; we have $\mathbb{L}^{an}_{\mathscr{O}_{\mathfrak{Y}}(V')/\mathscr{O}_{\mathfrak{Y}}(V)} \simeq 0$ by proposition 7.1.29, and then it follows by transitivity (theorem 7.1.33(ii)) that the map $L_i \to H_i(\mathbb{L}^{an}_{\mathscr{O}_{\mathfrak{X}}(\mathfrak{X})/\mathscr{O}_{\mathfrak{Y}}(V')})$ is an isomorphism. More generally, this argument shows that, for every affine open subset $U' \subset \mathfrak{X}$, the natural map $H_i(\mathbb{L}^{an}_{\mathscr{O}_{\mathfrak{X}}(U')/\mathscr{O}_{\mathfrak{Y}}(V)}) \to H_i(L(U'/\mathfrak{Y}))$ is an isomorphism. However, on one hand we have $(L_i^\triangle)_x \simeq L_i \otimes_A \mathscr{O}_{\mathfrak{X},x}$. On the other hand, we have ([26, Ch.0, §3.2.4]) :

$$(7.2.12) \qquad H_i(\mathbb{L}^{an}_{\mathfrak{X}/\mathfrak{Y}})_x \simeq \operatorname*{colim}_{x \in U'} H_i(L(U'/\mathfrak{Y}))$$

where the colimit ranges over the set S of all affine open neighborhoods of x in $\mathfrak{X}$. We can replace S by the cofinal subset of all open neighborhoods of the form $\mathfrak{D}(f)$ (for $f \in A$ such that $f \notin \mathfrak{m}_x$). Then, lemma 7.2.9, together with another easy application of transitivity allows to identify the right-hand side of (7.2.12) with $H_i(L(\mathfrak{X}/\mathfrak{Y})) \otimes_A \mathscr{O}_{\mathfrak{X},x}$, whence the claim. $\diamond$

Now (i) follows directly from claim 7.2.11 and theorem 7.1.33(i). (ii) and (iii) are immediate consequences of proposition 7.1.31, respectively proposition 7.1.29. $\square$

Proposition 7.2.13. *Let* $\mathfrak{X} \xrightarrow{f} \mathfrak{Y} \xrightarrow{g} 3$ *be two morphisms of formal schemes locally of finite presentation over* $\operatorname{Spf} K^+$. *There is a natural distinguished triangle in* $D^-(\mathscr{O}_{\mathfrak{X}}\text{-}\mathbf{Mod})$

$$(7.2.14) \qquad Lf^*\mathbb{L}^{an}_{\mathfrak{Y}/3} \to \mathbb{L}^{an}_{\mathfrak{X}/3} \to \mathbb{L}^{an}_{\mathfrak{X}/\mathfrak{Y}} \to Lf^*\mathbb{L}^{an}_{\mathfrak{Y}/3}[1].$$

Proof. For simplicity, we explain the construction of (7.2.14) under the assumption that the formal schemes are separated. As explained in [50, Ch.II, §2.1], for every sequence of ring homomorphisms $A \to B \to C$, the transitivity triangle is induced by a functorial exact sequence of complexes $\mathbb{L}_{C/B/A}$ of flat C-modules (a "true triangle" in *loc. cit.*). Suppose now that A, B, C are complete K^+-algebras of topologically finite type; upon a-adic completion, one deduces a true triangle $\mathbb{L}^\wedge_{C/B/A}$. Then, to every sequence of affine open subsets

$$(7.2.15) \qquad U \subset \mathfrak{X} \qquad V \subset \mathfrak{Y} \qquad W \subset 3$$

such that $f(U) \subset V$ and $g(V) \subset W$, one can associate the true triangle

$$\mathbb{L}^\wedge_{\mathscr{O}_{\mathfrak{X}}(U)/\mathscr{O}_{\mathfrak{Y}}(V)/\mathscr{O}_3(W)}.$$

Since the construction is functorial in all arguments, one derives a presheaf of true triangles on a cofinal family of open subsets of $\mathfrak{X}$, which we can then sheafify in the usual manner. The resulting distinguished triangle in $D^-(\mathscr{O}_{\mathfrak{X}}\text{-}\mathbf{Mod})$ gives rise to (7.2.14). In case the formal schemes are not necessarily separated, one can still perform the construction on all triples of open separated formal subschemes $U \to V \to W$ as in (7.2.15) and then glue over $\mathfrak{X}$. $\square$

In the following we wish to define the cotangent complex of a morphism of adic spaces (studied in [48] and [49]). For simplicity, we will restrict to adic spaces of topologically finite type over $\mathrm{Spa}(K, K^+)$, which suffice for our applications. For the convenience of the reader we recall a few basic definitions from [49].

7.2.16. An *f-adic ring* is a topological ring A that admits an open subring A_0 such that the induced topology on A_0 is pre-adic and defined by a finitely generated ideal $I \subset A_0$. As an example, every K-algebra of topologically finite type is an f-adic ring. A subring A_0 with the above properties is called a *ring of definition* for A, and I is an *ideal of definition*. One denotes by A° the open subring of power-bounded elements of A.

7.2.17. Let A and B be complete f-adic rings and $\phi : A \to B$ a ring homomorphism. One says that ϕ is *of topologically finite type* if there exist rings of definition $A_0 \subset A$ and $B_0 \subset B$ such that $\phi(A_0) \subset B_0$, the restriction $A_0 \to B_0$ factors through a quotient map (*i.e.* open and surjective) $A_0\langle T_1, ..., T_n \rangle \to B_0$ and B is finitely generated over $A \cdot B_0$.

7.2.18. An *affinoid ring* is a pair $A = (A^\triangleright, A^+)$ consisting of an f-adic ring $A^\triangleright$ and a subring $A^+ \subset A^\triangleright$ which is open, integrally closed in $A^\triangleright$ and contained in the subring A°. A^+ is called the *subring of integral elements* of A.

7.2.19. The *completion* of an affinoid ring $A = (A^\triangleright, A^+)$ is the pair

$$A^\wedge := ((A^\triangleright)^\wedge, (A^+)^\wedge)$$

(it turns out that $(A^+)^\wedge$ is integrally closed in $(A^\triangleright)^\wedge$).

A homomorphism $\phi : (A^\triangleright, A^+) \to (B^\triangleright, B^+)$ of affinoid rings is a ring homomorphism $\phi^\triangleright : A^\triangleright \to B^\triangleright$ such that $\phi(A^+) \subset B^+$. One says that ϕ is *of topologically finite type* if $\phi^\triangleright$ is of topologically finite type and there exists an open subring $C \subset B^+$ such that B^+ is the integral closure of C, $\phi(A^+) \subset C$ and the induced map $A^+ \to C$ is of topologically finite type (cp. (7.2.17)).

7.2.20. For instance, (K, K^+) is an affinoid ring, complete for its valuation topology; notice that in this case we have $K^+ = K^\circ$. A complete f-adic ring of topologically finite type over K is the same as a K-algebra of topologically finite type. Furthermore, suppose that A and B are complete f-adic rings of topologically finite type over K, and let $f : A \to B$ be a continuous ring homomorphism. Then there is a unique subring of integral elements B^+ such that $f : (A, A^\circ) \to (B, B^+)$ is a morphism of affinoid rings of topologically finite type; namely one must take $B^+ := B^\circ$. Especially, $A^+ := A^\circ$ is the only ring of integral elements of A such that the affinoid ring (A, A^+) is of topologically finite type over (K, K^+) ([48, Prop.2.4.15]).

7.2.21.   Given an arbitrary ring A, a *valuation* on A is a map

$$|\cdot| : A \to \Gamma \cup \{0\}$$

where Γ is an ordered abelian group whose composition law we denote multiplicatively, and the ordering is extended to $\Gamma \cup \{0\}$ as usual. Then $|\cdot|$ is required to satisfy the usual conditions, namely: $|x \cdot y| = |x| \cdot |y|$ and $|x + y| \leq \max(|x|, |y|)$ for every $x, y \in A$, and $|0| = 0$, $|1| = 1$. As in the case of valued fields, one also requires that Γ is generated by $|A| \setminus \{0\}$.

7.2.22.   Now, let A be an f-adic ring, and $|\cdot| : A \to \Gamma \cup \{0\}$ a valuation on A. For every $\gamma \in \Gamma$, let $U_\gamma := \{\alpha \in \Gamma \mid \alpha < \gamma\} \cup \{0\}$. We endow $\Gamma \cup \{0\}$ with the topology which restricts to the discrete topology on Γ, and which admits $(U_\gamma \mid \gamma \in \Gamma)$ as a fundamental system of open neighborhoods of 0.

   We say that $|\cdot|$ is *continuous* if it is continuous with respect to the above topology on $\Gamma \cup \{0\}$. One denotes by $\mathrm{Cont}(A)$ the set of all (equivalence classes of) continuous valuations on A. Given $a, b \in A$, let $U(a/b) \subset \mathrm{Cont}(A)$ be the subset of all valuations $|\cdot|$ such that $|a| \leq |b| \neq 0$. $\mathrm{Cont}(A)$ is endowed with the topology which admits the collection $(U(a/b) \mid a, b \in A)$ as a sub-basis. With this topology, $\mathrm{Cont}(A)$ is a spectral topological space (see [49, 1.1.13] for the definition of spectral space). In particular, this implies that $\mathrm{Cont}(A)$ admits a basis of quasi-compact open subsets. Such a basis is provided by the *rational subsets*, defined as follows. A subset $U \subset \mathrm{Cont}(A)$ is called rational if there exist $f_1, ..., f_n, g \in A$ such that the ideal $J := f_1 A + ... + f_n A$ is open in A and U consists of all $|\cdot| \in \mathrm{Cont}(A)$ such that $|f_i| \leq |g| \neq 0$ for every $i = 1, ..., n$. (Notice that, if we restrict to f-adic rings containing K, asking for J to be an open ideal is the same as requiring that $J = A$.) Given $f_1, ..., f_n, g \in A$ with the above property, we denote by $R(f_1/g, ..., f_n/g)$ the corresponding rational subset.

7.2.23.   If $A := (A^\triangleright, A^+)$, then one defines the subset

$$\mathrm{Spa}\, A := \{|\cdot| \in \mathrm{Cont}(A^\triangleright) \mid |a| \leq 1 \text{ for every } a \in A^+\} \subset \mathrm{Cont}(A^\triangleright).$$

$\mathrm{Spa}\, A$, endowed with the subspace topology, is called the *adic spectrum* of the affinoid ring A. $\mathrm{Spa}\, A$ is a pro-constructible subset of $\mathrm{Cont}(A)$, hence it is a spectral space too. Any continuous map $A \to B$ of affinoid rings induces in the obvious way a continuous map on adic spectra: $\mathrm{Spa}\, B \to \mathrm{Spa}\, A$.

7.2.24.   For any affinoid ring A, one can endow $X := \mathrm{Spa}\, A$ with a presheaf $\mathscr{O}_X$ of topological rings, as follows. First of all, for any $f_1, ..., f_n, g \in A^\triangleright$ as in (7.2.22), one defines an affinoid ring $A(f_1/g, ..., f_n/g)$, such that

$$A(f_1/g, ..., f_n/g)^\triangleright := (A^\triangleright)_g$$

and $A(f_1/g, ..., f_n/g)^+$ is the integral closure of the subring $A^+[f_1/g, ..., f_n/g]$ in $A(f_1/g, ..., f_n/g)^\triangleright$. If $B \subset A^\triangleright$ is a ring of definition and $I \subset B$ an ideal of definition, let $B(f_1/g, ..., f_n/g)$ be the subring of $(A^\triangleright)_g$ generated by B and $f_1/g, ..., f_n/g$; we endow $B(f_1/g, ..., f_n/g)$ with the pre-adic topology defined by the ideal $I \cdot B(f_1/g, ..., f_n/g)$; then the f-adic topology on $A(f_1/g, ..., f_n/g)^\triangleright$ is defined to be the unique ring topology for which $B(f_1/g, ..., f_n/g)$ is a ring of definition. Next, let $A\langle f_1/g, ..., f_n/g\rangle := A(f_1/g, ..., f_n/g)^\wedge$ (cp. (7.2.19)). After these preliminaries, one sets:

$$\mathscr{O}_X(R(f_1/g, ..., f_n/g)) := A\langle f_1/g, ..., f_n/g\rangle^\triangleright.$$

In this way, $\mathscr{O}_X$ is well defined on every rational subset. One can then extend the definition to an arbitrary open subset of $\mathrm{Spa}\, A$, following [26, Ch.0, §3.2.1]. It is not difficult to check that, for every open subset $U \subset \mathrm{Spa}\, A$, and every $x \in U$, any valuation $|\cdot|_x$ in the equivalence class x extends to the whole of $\mathscr{O}_X(U)$, hence to the stalk $\mathscr{O}_{X,x}$. One denotes by $\mathscr{O}_X^+$ the sub-presheaf defined by the rule:

$$\mathscr{O}_X^+(U) := \{f \in \mathscr{O}_X(U) \mid |f|_x \leq 1 \text{ for every } x \in U\}.$$

In the cases of interest, the presheaf $\mathscr{O}_X$ is a sheaf (and $\mathscr{O}_X^+$ is therefore a subsheaf). In such cases, one can show that, for every rational subset $R(f_1/g, ..., f_n/g)$, the natural map

$$A\langle f_1/g, ..., f_n/g\rangle^+ \to \mathscr{O}_X^+(R(f_1/g, ..., f_n/g))$$

is an isomorphism of topological rings.

This holds notably when $A^\triangleright$ is a K-algebra of topologically finite type. One calls the datum $(\mathrm{Spa}\, A, \mathscr{O}_{\mathrm{Spa}\, A}, \mathscr{O}_{\mathrm{Spa}\, A}^+)$ an *affinoid adic space*. General adic spaces are obtained as usual, by gluing affinoids. Adic spaces form a category, whose morphisms $f : X \to Y$ are the morphisms of topologically locally ringed spaces $(X, \mathscr{O}_X) \to (Y, \mathscr{O}_Y)$ which induce morphisms of locally ringed spaces $(X, \mathscr{O}_X^+) \to (Y, \mathscr{O}_Y^+)$.

7.2.25. Let $f : X \to Y$ be a morphism of adic spaces. One says that f is *locally of finite type* if for every $x \in X$ there exist open affinoid subspaces $U \subset X$, $V \subset Y$ such that $x \in U$, $f(U) \subset V$ and the induced morphism of affinoid rings $(\mathscr{O}_Y(V), \mathscr{O}_Y^+(V)) \to (\mathscr{O}_X(U), \mathscr{O}_X^+(U))$ is of topologically finite type.

7.2.26. A morphism $f : X \to Y$ between adic spaces (defined over $\mathrm{Spa}(K, K^+)$) is called *smooth* (resp. *unramified*, resp. *étale*) if f is locally of finite type and if, for any affinoid ring A, any ideal I of $A^\triangleright$ with $I^2 = \{0\}$ and any morphism $\mathrm{Spa}\, A \to Y$, the mapping

$$\mathrm{Hom}_Y(\mathrm{Spa}\, A, X) \to \mathrm{Hom}_Y(\mathrm{Spa}\, A/I, X)$$

is surjective (resp. injective, resp. bijective).

7.2.27. In [49, §1.9] it is shown how to associate functorially to every formal scheme $\mathfrak{X}$ (say locally of finite presentation over $\mathrm{Spf}\, K^+$) an adic space $d(\mathfrak{X})$, together with a morphism of topologically ringed spaces $\lambda : d(\mathfrak{X}) \to \mathfrak{X}$, characterized by a certain universal property which we won't spell out here, but that includes the condition that $\mathrm{Im}(\mathscr{O}_\mathfrak{X} \to \lambda_* \mathscr{O}_{d(\mathfrak{X})}) \subset \mathscr{O}_{d(\mathfrak{X})}^+$. If $\mathfrak{X} = \mathrm{Spf}\, A_0$ for a K^+-algebra A_0 of topologically finite type, then $d(\mathfrak{X}) = \mathrm{Spa}\, A$, where A is the affinoid ring $(A_0 \otimes_{K^+} K, A^+)$, with A^+ defined as the integral closure of the image of A_0 in $A_0 \otimes_{K^+} K$. Moreover, $\mathfrak{X}$ is quasi-compact if and only if $d(\mathfrak{X})$ is.

7.2.28. Let $\mathfrak{X}$ be a formal scheme of finite presentation over Spf K^+. The collection $\mathscr{C}_{\mathfrak{X}}$ of all morphisms $\mathfrak{f} : \mathfrak{X}' \to \mathfrak{X}$ of formal schemes of finite presentation over Spf K^+ such that $d(\mathfrak{f})$ is an isomorphism, forms an essentially small cofiltered category (with morphisms given as usual by the commutative diagrams). It is shown in [48, §3.9] that there is a natural isomorphism of topologically ringed spaces

$$(7.2.29) \qquad (d(\mathfrak{X}), \mathscr{O}^+_{d(\mathfrak{X})}) \xrightarrow{\sim} \lim_{(\mathfrak{X}' \to \mathfrak{X}) \in \mathscr{C}} (\mathfrak{X}', \mathscr{O}_{\mathfrak{X}'}).$$

(Actually, the argument in *loc.cit.* is worked out only in the case of noetherian formal schemes, but it is not difficult to adapt it to the present situation.)

7.2.30. Let $f : A \to B$ be a morphism of affinoid rings of topologically finite type over (K, K^+) (especially $A^+ = A^\circ$ and $B^+ = B^\circ$, see (7.2.20)). We let $\mathscr{C}_f$ be the filtered family consisting of all the pairs (A_0, B_0) of K^+-algebras of topologically finite presentation, such that A_0 (resp. B_0) is an open subalgebra of A° (resp. of B°) and $f(A_0) \subset B_0$. The *analytic cotangent complex* of the morphism f is the complex of B-modules

$$\mathbb{L}^{\mathrm{an}}_{B/A} := K \otimes_{K^+} \mathbb{L}^+_{B/A} \qquad \text{where} \qquad \mathbb{L}^+_{B/A} := \operatorname*{colim}_{(A_0, B_0) \in \mathscr{C}_f} \mathbb{L}^{\mathrm{an}}_{B_0/A_0}.$$

7.2.31. Let $f : X \to Y$ be a morphism of adic spaces locally of finite type over $\mathrm{Spa}(K, K^+)$, and suppose that Y is separated. For every affinoid open subset $U \subset X$, the small category F_U of all affinoid open subsets $V \subset Y$ with $f(U) \subset V$, is cofiltered under inclusion (or else it is empty). For every $V \in F_U$, $(\mathscr{O}_Y(V), \mathscr{O}_Y(V)^+)$ is an affinoid (K, K^+)-algebra of topologically finite type, hence the induced morphism $\mathscr{O}_Y(V) \to \mathscr{O}_X(U)$ is of the kind considered in (7.2.30). We set

$$L(U/Y) := \operatorname*{colim}_{V \in F^\circ_U} \mathbb{L}^+_{\mathscr{O}_X(U)/\mathscr{O}_Y(V)}.$$

Definition 7.2.32. The mapping $U \mapsto L(U/Y)$ defines a complex of presheaves on a cofinal family of affinoid open subsets of X. By applying degreewise the construction of [26, Ch.0, §3.2.1], we can extend the latter to a complex of presheaves of $\mathscr{O}^+_X$-modules on X. We let $\mathbb{L}^+_{X/Y}$ be the complex of sheaves associated to this complex of presheaves (cp. definition (7.2.2)) and we define the *analytic cotangent complex* of the morphism $f : X \to Y$ as the complex $\mathbb{L}^{\mathrm{an}}_{X/Y} := K \otimes_{K^+} \mathbb{L}^+_{X/Y}$. The definitions can be extended to the case of a morphism $f : X \to Y$ where Y is not necessarily separated: one argues as in (7.2.3).

7.2.33. Let X^+ denote the ringed space $(X, \mathscr{O}^+_X)$ and define likewise Y^+. Just as in the proof of lemma 7.2.4, by inspecting the construction we obtain a natural map of complexes (resp. a morphism in $\mathrm{D}^-(\mathscr{O}_X\text{-}\mathbf{Mod})$)

$$(7.2.34) \qquad \mathbb{L}_{X^+/Y^+} \to \mathbb{L}^+_{X/Y} \qquad (\text{resp. } \mathbb{L}_{X/Y} \to \mathbb{L}^{\mathrm{an}}_{X/Y}).$$

(The second morphism in (7.2.34) is defined by the quasi-isomorphism of complexes $K \otimes_{K^+} \mathbb{L}_{X^+/Y^+} \xrightarrow{\sim} \mathbb{L}_{X/Y}$ and the natural map $K \otimes_{K^+} \mathbb{L}_{X^+/Y^+} \to \mathbb{L}^{\mathrm{an}}_{X/Y}$.)

7.2.35.   Let A, B be complete f-adic rings and $A \to B$ a ring homomorphism of topologically finite type. We refer to [49, §1.6] for the construction of a *universal A-derivation of B*, which is a continuous A-derivation $d : B \to \Omega^{an}_{B/A}$ from B to a complete topological B-module $\Omega^{an}_{B/A}$, universal for A-derivations $B \to M$ to complete topological B-modules M. The construction of $\Omega^{an}_{B/A}$ can be globalized to a sheaf of relative differentials $\Omega^{an}_{X/Y}$ for any morphism of adic spaces $X \to Y$ locally of finite type. In the setup of definition 7.2.32 one checks easily using lemma 7.1.27(iii) that:

$$(7.2.36) \qquad H_0(\mathbb{L}^{an}_{X/Y}) \simeq \Omega^{an}_{X/Y}.$$

We set $\Omega_{X/Y} := H_0(\mathbb{L}_{X/Y})$; moreover, we define

$$\Omega^{na}_{X/Y} := \mathrm{Ker}(\Omega_{X/Y} \to \Omega^{an}_{X/Y}) \qquad (\text{resp. } \Omega^{na}_{B/A} := \mathrm{Ker}(\Omega_{B/A} \to \Omega^{an}_{B/A}))$$

for a morphism of adic spaces $X \to Y$ (resp of complete f-adic rings $A \to B$) as above. ("na" stands for "not analytic".)

Lemma 7.2.37. *Let $A \to B$ be a map of K-algebras of topologically finite type, M a finitely generated B-module. Then:*

(i) *Every A-linear derivation $B \to M$ is continuous.*

(ii) $\Omega^{an}_{B/A}$ *is the largest finitely generated quotient B-module of $\Omega_{B/A}$.*

(iii) $\mathrm{Hom}_B(\Omega^{na}_{B/A}, M) = 0$.

Proof. A derivation $B \to M$ induces a ring homomorphism $B \to B \oplus M$ (with the usual ring structure on the target). Since $B \oplus M$ is again of topologically finite type over K, assertion (i) follows from the following well known:

Claim 7.2.38. Every map $A \to B$ from a noetherian K-Banach algebra A to K-algebra B of topologically finite type, is continuous.

Proof of the claim. This is [12, §6.1.3, Th.1]. $\Diamond$

(ii) is an immediate consequence of (i). In order to prove (iii), it suffices to show that $\Omega^{na}_{B/A}$ does not admit any non-zero finitely generated quotient. However, if $N \subset \Omega^{na}_{B/A}$ is a B-submodule such that $\Omega^{na}_{B/A}/N$ is a finitely generated B-module, then $\Omega_{B/A}/N$ is also finitely generated, hence $N = \Omega^{na}_{B/A}$. $\square$

Proposition 7.2.39. *Let $X \xrightarrow{f} Y \xrightarrow{g} Z$ be two morphisms of adic spaces locally of finite type over $\mathrm{Spa}(K, K^+)$. There is a natural distinguished triangle in $D^-(\mathcal{O}_X\text{-}\mathbf{Mod})$*

$$(7.2.40) \qquad Lf^*\mathbb{L}^{an}_{Y/Z} \to \mathbb{L}^{an}_{X/Z} \to \mathbb{L}^{an}_{X/Y} \to Lf^*\mathbb{L}^{an}_{Y/Z}[1].$$

Proof. Mutatis mutandis, this is the same as the proof of proposition 7.2.13, so we can leave the details to the reader. $\square$

Remark 7.2.41. In fact, the proof of proposition 7.2.39 shows that (7.2.40) is represented by a *functorial* true triangle (see [50, Ch.I, §3.2.4]). This will be important in the sequel, when we will need to compute the truncation $\tau_{[-1}\mathbb{L}^{an}_{X/Z}$ of the analytic cotangent complex in the situation contemplated in proposition 7.2.48.

Theorem 7.2.42. *Let* $\mathfrak{f} : \mathfrak{X} \to \mathfrak{Y}$ *(resp.* $f : X \to Y$, *resp.* $\phi : A_0 \to B_0$) *be a morphism of formal schemes (resp. of adic spaces, resp. of K^+-algebras) locally of finite presentation over* $\operatorname{Spf} K^+$ *(resp. locally of finite type over* $\operatorname{Spa}(K, K^+)$, *resp. of topologically finite presentation over K^+) such that the induced morphism* $d(\mathfrak{f}) : d(\mathfrak{X}) \to d(\mathfrak{Y})$ *is smooth (resp. such that f is smooth, resp. such that $d(\operatorname{Spf} \phi)$ is smooth). Then:*

(i) $\mathbb{L}^{\mathrm{an}}_{\mathfrak{X}/\mathfrak{Y}} \otimes_{K^+} K \simeq \Omega^{\mathrm{an}}_{\mathfrak{X}/\mathfrak{Y}}[0] \otimes_{K^+} K$ *in* $\mathsf{D}^-((\mathscr{O}_{\mathfrak{X}} \otimes_{K^+} K)\text{-}\mathbf{Mod})$.

(ii) $\mathbb{L}^{\mathrm{an}}_{X/Y} \simeq \Omega^{\mathrm{an}}_{X/Y}[0]$ *in* $\mathsf{D}^-(\mathscr{O}_X\text{-}\mathbf{Mod})$.

(iii) $\mathbb{L}^{\mathrm{an}}_{B_0/A_0} \otimes_{K^+} K \simeq \Omega^{\mathrm{an}}_{B_0/A_0} \otimes_{K^+} K[0]$ *in* $\mathsf{D}(B\text{-}\mathbf{Mod})$.

Proof. After the usual reductions, both assertions (i) and (ii) come down to (iii). To show (iii), we write $B_0 = C_0/I_0$, where $C_0 := A_0\langle T_1, ..., T_n\rangle$ and I_0 is some finitely generated ideal. Set $I := I_0 \otimes_{K^+} K$, $A := A_0 \otimes_{K^+} K$ and let $\mathfrak{n}$ be a maximal ideal of $C := C_0 \otimes_{K^+} K$ with $I \subset \mathfrak{n}$.

Claim 7.2.43. $I_\mathfrak{n}$ is generated by a regular sequence of elements of the ring $C_\mathfrak{n}$.

Proof of the claim. Let $\mathfrak{p} := \mathfrak{n} \cap A$; $\mathfrak{p}$ is a maximal ideal in A, and its residue field K' is a finite extension of K. Let $\bar{\mathfrak{n}}$ be the image of $\mathfrak{n}$ in $C \otimes_A K'$ and $\mathscr{I} \subset \mathscr{O}_{d(\operatorname{Spf} C_0)}$ the sheaf of ideals corresponding to I; the maximal ideal $\mathfrak{n}$ yields a point in $d(\operatorname{Spf} C_0)$, which we denote by $x(\mathfrak{n})$. We have an isomorphism on the $\mathfrak{n}$-adic completions ([12, §7.3.2, Prop.3]):

$$(7.2.44) \qquad (\mathscr{O}_{x(\mathfrak{n})})^\wedge \simeq (C_\mathfrak{n})^\wedge \qquad \text{and} \qquad (\mathscr{I}_{x(\mathfrak{n})})^\wedge \simeq (I_\mathfrak{n})^\wedge.$$

Moreover, there are natural maps :

$$\mathscr{I}_{x(\mathfrak{n})}/\mathscr{I}^2_{x(\mathfrak{n})} \to \bar{\mathfrak{n}}/\bar{\mathfrak{n}}^2 \to \Omega^{\mathrm{an}}_{C/A} \otimes_C C/\mathfrak{n}$$

and, by [14, Prop.2.5], there exists a set of generators $g_1, ..., g_k$ for $\mathscr{I}_{x(\mathfrak{n})}$ such that the images $dg_1, ..., dg_k$ in $\Omega^{\mathrm{an}}_{C/A} \otimes_C C/\mathfrak{n}$ are linearly independent; it follows that the images $\bar{g}_1, ..., \bar{g}_k$ in $\bar{\mathfrak{n}}/\bar{\mathfrak{n}}^2$ are also linearly independent. Due to (7.2.44), we can assume that $g_1, ..., g_k \in I_\mathfrak{n}$, and then it follows that $g_1 \otimes 1, ..., g_k \otimes 1$ are the first k elements of a regular system of parameters for the regular local ring $C_\mathfrak{n} \otimes_A K'$. From lemma 7.1.6 it follows that $C_\mathfrak{n}$ is a flat $A_\mathfrak{p}$-module; then by [28, Ch.0, Prop.15.1.16] we deduce that $g_1, ..., g_n$ is a regular sequence of elements of $C_\mathfrak{n}$, as required. $\Diamond$

Set $B := B_0 \otimes_{K^+} K$; it follows from claim 7.2.43 and [50, Ch.III, Prop.3.2.4] that $B_\mathfrak{n} \otimes_B \mathbb{L}_{B/C} \simeq \mathbb{L}_{B_\mathfrak{n}/C_\mathfrak{n}} \simeq I_\mathfrak{n}/I^2_\mathfrak{n}[1]$ for every maximal ideal $\mathfrak{n}$, hence $\mathbb{L}_{B/C} \simeq I/I^2[1]$. By proposition 7.1.31, we derive $\mathbb{L}^{\mathrm{an}}_{B_0/C_0} \otimes_{K^+} K \simeq I/I^2[1]$. Finally, by theorem 7.1.33(ii) and proposition 7.1.29 we deduce an isomorphism in $\mathsf{D}^-(B\text{-}\mathbf{Mod})$:

$$\mathbb{L}^{\mathrm{an}}_{B_0/A_0} \otimes_{K^+} K \simeq (0 \to I/I^2 \to B \otimes_{C_0} \Omega^{\mathrm{an}}_{C_0/A_0} \to 0)$$

and the latter complex is quasi-isomorphic to $\Omega^{\mathrm{an}}_{B/A}[0]$ by [14, Prop.2.5]. $\square$

Remark 7.2.45. In the situation of (7.2.30), let $(A_0, B_0) \in \mathscr{C}_f$; then the map $K \otimes_{K^+} \mathbb{L}^{an}_{B_0/A_0} \to \mathbb{L}^{an}_{B/A}$ is a quasi-isomorphism. Indeed, using transitivity triangles it suffices to show that if f is an isomorphism and (A_0, A_1) is any pair in $\mathscr{C}_f$, then $K \otimes_{K^+} \mathbb{L}^{an}_{A_1/A_0}$ is acyclic, which follows from theorem 7.2.42(iii).

Lemma 7.2.46. (i) *Let* $f : X \to Y$ *be a closed imbedding of adic spaces locally of finite type over* $\mathrm{Spa}(K, K^+)$. *Then the natural morphisms*

$$\mathbb{L}_{X/Y} \to \mathbb{L}^{an}_{X/Y} \qquad \mathbb{L}_{X^+/Y^+} \to \mathbb{L}^+_{X/Y}$$

are isomorphisms in $\mathsf{D}(\mathcal{O}_X\text{-}\mathbf{Mod})$. *(Notation of (7.2.33).)*

(ii) *Let* $A \to B$ *be a surjection of* K-*algebras of topologically finite type. then natural map* $\mathbb{L}_{B/A} \to \mathbb{L}^{an}_{B/A}$ *(defined as in (7.2.34)) is an isomorphism in* $\mathsf{D}^-(B\text{-}\mathbf{Mod})$.

Proof. (i): The question is local on X, so we can reduce to the case where $Y = \mathrm{Spa}(A, A^\circ)$ for some K-algebra of topologically finite type and $X = \mathrm{Spa}(B, B^\circ)$, where $B = A/I$ for some ideal $I \subset A$. Let $\mathscr{C}$ be the filtered family of all triples of the form (U, A_0, B_0) where $U \subset Y$ is an affinoid open neighborhood of X, $A_0 \subset \mathcal{O}_Y(U)^\circ$, $B_0 \subset B^\circ$ are two open K^+-subalgebras of topologically finite presentation with $\mathrm{Im}(A_0 \to B) \subset B_0$.

Claim 7.2.47. The family $\mathscr{C}' \subset \mathscr{C}_\pi$ of all (U, A_0, B_0) such that $\mathrm{Im}(A_0 \to B) = B_0$ is cofinal.

Proof of the claim. Let (U, A_0, B_0) be any triple in $\mathscr{C}$. We can find finitely many elements $\bar{b}_1, ..., \bar{b}_k \in B_0$ such that $A_0[\bar{b}_1, ..., \bar{b}_k]$ is dense in B_0. Choose elements $b_1, ..., b_k \in A$ that lift the $\bar{b}_i$; the subset $U' := \{x \in U \mid |b_i(x)| \le 1; \ i = 1, ..., k\}$ is an affinoid open neighborhood of X and the topological closure A'_0 of $A_0[b_1, ..., b_k]$ in $\mathcal{O}_X(U')^\circ$ is an open K^+-subalgebra of topologically finite presentation; moreover $\mathrm{Im}(A'_0 \to B) = B_0$, hence $(U', A'_0, B_0) \in \mathscr{C}$. $\diamond$

Finally, let $(U, A_0, B_0) \in \mathscr{C}'$; by proposition 7.1.31 we have $\mathbb{L}^{an}_{B_0/A_0} \simeq \mathbb{L}_{B_0/A_0}$. After taking colimits we deduce:

$$\underset{X \subset U \subset Y}{\mathrm{colim}} \ \mathbb{L}^{an}_{B/\mathcal{O}_Y(U)} \simeq \underset{X \subset U \subset Y}{\mathrm{colim}} \ \mathbb{L}_{B/\mathcal{O}_Y(U)}$$

where U runs over the cofiltered family of all open affinoid neighborhoods of X in Y One applies this with Y replaced by any affinoid subdomain in Y, and sheafifies to obtain the first stated isomorphism. Similarly one compares $\mathbb{L}_{X^+/Y^+}$ and $\mathbb{L}^+_{X/Y}$.

(ii) is analogous but simpler: one argues using proposition 7.1.31 and remark 7.2.45. $\square$

Proposition 7.2.48. *Let* $X \xrightarrow{f} Y \xrightarrow{g} Z$ *be morphisms of adic spaces locally of finite type over* $\mathrm{Spa}(K, K^+)$, *where* f *is a closed imbedding with defining ideal* $\mathscr{I} \subset \mathcal{O}_Y$, *and* g *is smooth. Then there is a natural isomorphism in* $\mathsf{D}(\mathcal{O}_X\text{-}\mathbf{Mod})$

$$(7.2.49) \qquad \tau_{[-1]} \mathbb{L}^{an}_{X/Z} \xrightarrow{\sim} (0 \to (\mathscr{I}/\mathscr{I}^2)_{|X} \xrightarrow{d} f^*\Omega^{an}_{Y/Z} \to 0)$$

where d *is the natural map.*

Proof. From lemma 7.2.46(i) and [50, Ch.III, Cor.1.2.8.1] we deduce a natural isomorphism

$$\tau_{[-1]}\mathbb{L}^{\mathrm{an}}_{X/Y} \simeq (\mathscr{I}/\mathscr{I}^2)_{|X}[1].$$

Taking into account proposition 7.2.39, theorem 7.2.42, one can repeat the proof of [50, Ch.III, Cor.1.2.9.1], which yields a distinguished triangle:

(7.2.50) $\qquad f^*H_0(\mathbb{L}_{Y/Z})[0] \to \tau_{[-1]}\mathbb{L}_{X/Z} \to H_1(\mathbb{L}_{X/Y})[1] \to$

such that the connecting morphism

$$(\mathscr{I}/\mathscr{I}^2)_{|X} \simeq H_1(\mathbb{L}_{X/Y}) \to f^*H_0(\mathbb{L}_{Y/Z}) \simeq f^*\Omega^{\mathrm{an}}_{Y/Z}$$

is naturally identified (up to sign) with the differential map $f \mapsto df$. There follows an isomorphism such as (7.2.49); however, the naturality of (7.2.49) is not explicitly verified in *loc.cit.* : in general, this kind of manipulations, when carried out in the derived category, do not lead to functorial identifications. The problem is that we can have morphisms of distinguished triangles:

$$
\begin{array}{ccccccc}
A^\bullet & \longrightarrow & B^\bullet & \longrightarrow & C^\bullet & \longrightarrow & A^\bullet[1] \\
\downarrow{\scriptstyle f} & & \downarrow{\scriptstyle g} & & \downarrow{\scriptstyle h} & & \downarrow{\scriptstyle f[1]} \\
A'^\bullet & \longrightarrow & B'^\bullet & \longrightarrow & C'^\bullet & \longrightarrow & A'^\bullet[1]
\end{array}
$$

such that both f and g are the zero maps, and yet h is not. The issue can be resolved if one remarks that (7.2.50) is deduced from a transitivity triangle via proposition 7.2.39, and thus it is actually well defined in the derived category of true triangles $\mathsf{T}(\mathscr{O}_X\text{-}\mathbf{Mod})$ (cp. remark 7.2.41 and [50, Ch.I, §3.2.4]). It suffices then to apply the following

Claim 7.2.51. Let $\mathscr{C}$ be any abelian category, $\underline{T} := (0 \to A^\bullet \to B^\bullet \to C^\bullet \to 0)$ a true triangle of $\mathsf{T}(\mathscr{C})$ and $n \in \mathbb{Z}$ an integer such that $H^{i+1}(A^\bullet) = 0 = H^i(C^\bullet)$ for every $i \neq n$. Then there is a natural isomorphism in $\mathsf{T}(\mathscr{C})$:

$$\underline{T} \simeq (0 \to H^{n+1}(A^\bullet)[n+1] \to \mathrm{Cone}(d_n) \to H^n(C^\bullet)[n] \to 0)$$

where $d_n : H^n(C^\bullet)[n] \to H^{n+1}(A^\bullet)[n+1]$ is the connecting morphism of the long exact homology sequence associated to $\underline{T}$.

Proof of the claim. Under the stated assumptions, the natural maps of complexes $\phi : \tau_{[n]}C^\bullet \to C$ and $\psi : A^\bullet \to \tau_{[n+1]}A^\bullet$ are quasi-isomorphisms (the conventions concerning truncation operators follow [50, I.1.4.7]); we set $\underline{T}' := \psi * \underline{T} * \phi$ (notation of (2.5.5)). Clearly the pullback and push out maps define a natural isomorphism $\underline{T}' \overset{\sim}{\to} \underline{T}$ in $\mathsf{T}(\mathscr{C})$. Say that $\underline{T}' = (0 \to A'^\bullet \to B'^\bullet \to C'^\bullet \to 0)$; since $H^{n+1}(C'^\bullet) = 0$, it follows that the complex

$$\underline{T}'' := \tau_{[n+1]}\underline{T}' := (0 \to \tau_{[n+1]}A'^\bullet \to \tau_{[n+1]}B'^\bullet \to \tau_{[n+1]}C'^\bullet \to 0)$$

is again a true triangle of $\mathsf{T}(\mathscr{C})$, naturally isomorphic to $\underline{T}'$. Likewise, $\underline{T}''$ is naturally isomorphic to the true triangle $\tau_{[n]}\underline{T}''$, and by inspection, one sees that the latter has the expected shape. $\qquad\square$

7.3. Deformations of formal schemes and adic spaces.

Lemma 7.3.1. *Let* $(\mathfrak{X}, \mathscr{O}_{\mathfrak{X}})$ *be a formal scheme locally of finite presentation over* $\operatorname{Spf} K^+$, *let* $\mathscr{I}$ *be a coherent* $\mathscr{O}_{\mathfrak{X}}$-*module and* $(0 \to \mathscr{I} \to \mathscr{O}_1 \to \mathscr{O}_{\mathfrak{X}} \to 0)$ *an extension of sheaves of* K^+-*algebras on* $\mathfrak{X}$. *Then* $(\mathfrak{X}, \mathscr{O}_1)$ *is a formal scheme locally of finite presentation over* $\operatorname{Spf} K^+$.

Proof. We may and do assume that $\mathfrak{X}$ is an affine formal scheme over $\operatorname{Spf} K^+$; then the assertion follows from claims 7.3.2 and 7.3.3 below.

Claim 7.3.2. $\mathscr{O}_1(\mathfrak{X})$ is a complete K^+-algebra of topologically finite presentation.

Proof of the claim. Since $\mathfrak{X}$ is affine, we derive a short exact sequence $0 \to \mathscr{I}(\mathfrak{X}) \to \mathscr{O}_1(\mathfrak{X}) \to \mathscr{O}_{\mathfrak{X}}(\mathfrak{X}) \to 0$. Since $\mathscr{I}$ is coherent, the $\mathscr{O}_{\mathfrak{X}}(\mathfrak{X})$-module $\mathscr{I}(\mathfrak{X})$ is finitely presented. By proposition 7.1.1(i), the K^+-torsion submodule T' of $\mathscr{O}_{\mathfrak{X}}(\mathfrak{X})$ is a finitely generated $\mathscr{O}_{\mathfrak{X}}(\mathfrak{X})$-module. Let $N := \operatorname{Ker}(\mathscr{O}_1(\mathfrak{X}) \to \mathscr{O}_{\mathfrak{X}}(\mathfrak{X})/T')$; in the usual way (cp. the proof of proposition 7.1.1(iii)) we derive that there exists $k_0 \in \mathbb{N}$ such that

$$a^k \mathscr{O}_1(\mathfrak{X}) \cap \mathscr{I}(\mathfrak{X}) = a^k N \cap \mathscr{I}(\mathfrak{X}) \subset a^{k-k_0} \mathscr{I}(\mathfrak{X}) \qquad \text{for every } k \geq k_0.$$

We deduce by [58, Th.8.1(ii)] a short exact sequence of complete K^+-algebras $0 \to \mathscr{I}(\mathfrak{X}) \to \mathscr{O}_1(\mathfrak{X})^\wedge \to \mathscr{O}_{\mathfrak{X}}(\mathfrak{X}) \to 0$, which shows that $\mathscr{O}_1(\mathfrak{X})$ is complete. Since $\mathscr{O}_{\mathfrak{X}}(\mathfrak{X})$ is topologically of finite type and the module $\mathscr{I}(\mathfrak{X})$ is finitely generated, we can then find a continuous surjection $A := K^+\langle T_1, ..., T_n \rangle \to \mathscr{O}_1(\mathfrak{X})$; by lemma 7.1.4 it follows that $\mathscr{O}_{\mathfrak{X}}(\mathfrak{X})$ is a finitely presented A-algebra, hence $\mathscr{I}(\mathfrak{X})$ is a finitely presented A-module, and then the same holds for $\mathscr{O}_1(\mathfrak{X})$, so the claim is proved. ◇

Claim 7.3.3. For $f \in \mathscr{O}_{\mathfrak{X}}(\mathfrak{X})$, let $\mathfrak{D}(f) \subset \mathfrak{X}$ be as in lemma 7.2.9. Then the natural map $(\mathscr{O}_1(\mathfrak{X})_f)^\wedge \to \mathscr{O}_1(\mathfrak{D}(f))$ is an isomorphism.

Proof of the claim. The existence of the said map is a consequence of claim 7.3.2; however, we have as well a natural short exact sequence $0 \to \mathscr{I}(\mathfrak{X})_f^\wedge \to \mathscr{O}_1(\mathfrak{D}(f)) \to \mathscr{O}_{\mathfrak{X}}(\mathfrak{X})_f^\wedge \to 0$, so the assertion is immediate. □

The meaning of lemma 7.3.1 is that the square zero deformations of $\mathfrak{X}$ by $\mathscr{I}$ in the category of ringed spaces are the same as those in the category of formal $\operatorname{Spf} K^+$-schemes locally of finite presentation. Especially, the latter are classified by the appropriate Ext-group of the (usual) cotangent complex of the map of ringed spaces $\mathfrak{X} \to \operatorname{Spf} K^+$; we aim to show that the same computation can be carried out with the analytic cotangent complex $\mathbb{L}^{an}_{\mathfrak{X}/K^+}$ introduced in definition 7.2.2.

7.3.4.　Let T be a topos, I a small category; the category T^I of all functors from I to T is a topos, and we define a functor $c : T \to T^I$ by assigning to an object X of T the constant functor $c_X : I \to T$ of value X (so $c_X(i) = X$ for every object i of I, and $c_X(\phi) = 1_X$ for every morphism ϕ of I). The functor c admits a right adjoint $\varprojlim_I : T^I \to T$, and the adjoint pair $(\varprojlim_I, c)$ defines a morphism of topoi

$\pi_T : T^I \to T$. If $\phi : S \to T$ is any morphism of topoi, we obtain a commutative diagram of topoi:

$$
\begin{array}{ccc}
S^I & \xrightarrow{\;\pi_S\;} & S \\
\phi^I \downarrow & & \downarrow \phi \\
T^I & \xrightarrow{\;\pi_T\;} & T
\end{array}
$$

whence two spectral sequences:

$$E_2^{pq} := R^p \phi_* \lim_{i \in I}{}^q \mathscr{F}_i \Rightarrow R^{p+q}(\phi \circ \pi_S)_*(\mathscr{F}_i \mid i \in I)$$
$$F_2^{pq} := \lim_{i \in I}{}^p R^q \phi_* \mathscr{F}_i \Rightarrow R^{p+q}(\phi \circ \pi_S)_*(\mathscr{F}_i \mid i \in I)$$

for every abelian sheaf $(\mathscr{F}_i \mid i \in I)$ on S^I.

In the following we shall denote by $\mathbb{N}$ the category associated to the reversed ordering on the set of natural numbers (so, for every pair $i, j \in \mathbb{N}$ with $i > j$ there is exactly one morphism $i \to j$ in the category $\mathbb{N}$).

Lemma 7.3.5. *Let $\mathfrak{X}$ be a formal scheme adic over* $\mathrm{Spf}\, K^+$*, $\mathscr{F}$ a quasi-coherent $\mathscr{O}_{\mathfrak{X}}$-module and set $\mathscr{F}_n := \mathscr{F}/a^n \mathscr{F}$ for every $n \in \mathbb{N}$. Then $(\mathscr{F}_n \mid n \in \mathbb{N})$ defines an abelian sheaf on $\mathfrak{X}^{\mathbb{N}}$, and we have $\lim_{n \in \mathbb{N}}{}^q \mathscr{F}_n = 0$ on $\mathfrak{X}$, for every $q > 0$.*

Proof. We apply the spectral sequences of (7.3.4) to the indexing category $\mathbb{N}$, and the morphism of topoi $\phi : \mathfrak{U} \to \mathrm{Spf}\, K^+$, where $\mathfrak{U} \subset \mathfrak{X}$ is any affine open subset. Hence $F_2^{pq} = 0$ whenever $q > 0$, therefore the abutment is isomorphic to $F_2^{p0} = \lim_{n \in \mathbb{N}}{}^p \mathscr{F}_n(\mathfrak{U})$; however, the inverse system $(\mathscr{F}_n(\mathfrak{U}) \mid n \in \mathbb{N})$ is surjective, hence $F_2^{p0} = 0$ for $p > 0$, and it follows that $E_2^{pq} = E_\infty^{pq} = 0$ whenever $p + q > 0$. Since $\mathfrak{U}$ is arbitrary, the claim follows by sheafification. $\qquad\square$

7.3.6. With the notation of (7.3.4), let $\mathscr{O}_\bullet := (\mathscr{O}_i \mid i \in I)$ be a ring object of the topos T^I (briefly: a T^I-ring), or which is the same, a functor $\mathscr{O}$ from I to the category of T-rings. Let also $\mathscr{O}_T$ be a T-ring, and $\pi_T^\sharp : \mathscr{O}_T \to \pi_{T*} \mathscr{O}_\bullet$ a morphism of T-rings. Then the pair $(\pi_T, \pi_T^\sharp)$ defines a morphism of ringed topoi $\pi_T : (T^I, \mathscr{O}_\bullet) \to (T, \mathscr{O}_T)$. The corresponding functor $\pi_{T*} = \lim_I : \mathscr{O}_\bullet\text{-}\mathbf{Mod} \to \mathscr{O}_T\text{-}\mathbf{Mod}$ admits a left adjoint π_T^*, defined by the rule:

$$\mathscr{F} \mapsto \pi_T^{-1}\mathscr{F} \otimes_{\pi_T^{-1}\mathscr{O}_T} \mathscr{O}_\bullet = (\mathscr{F} \otimes_{\mathscr{O}_T} \mathscr{O}_i \mid i \in I).$$

(Of course, π_T^{-1} is the same as the functor $c : T \to T^I$ of (7.3.4).) The adjoint pair (π_{T*}, π_T^*) extends to an adjoint pair of derived functors $(R\pi_{T*}, L\pi_T^*)$; more precisely, for every complex $K_1^\bullet \in \mathsf{D}^-(\mathscr{O}_T\text{-}\mathbf{Mod})$ and $K_2^\bullet \in \mathsf{D}^+(\mathscr{O}_\bullet\text{-}\mathbf{Mod})$ there is a natural isomorphism ("trivial duality")

$$\mathrm{Hom}_{\mathsf{D}(\mathscr{O}_\bullet\text{-}\mathbf{Mod})}(\pi_T^{-1}K_1^\bullet \overset{\mathbf{L}}{\otimes}_{\pi_T^{-1}\mathscr{O}_T} \mathscr{O}_\bullet, K_2^\bullet) \simeq \mathrm{Hom}_{\mathsf{D}(\mathscr{O}_T\text{-}\mathbf{Mod})}(K_1^\bullet, \underset{I}{\mathrm{Rlim}} K_2^\bullet)$$

(see [50, Ch.III, Prop.4.6]).

Proposition 7.3.7. *Let $\mathfrak{X}$ be formal scheme locally finitely presented over* Spf K^+ *and $\mathscr{F}$ any coherent $\mathscr{O}_{\mathfrak{X}}$-module. Then the natural morphism $\mathbb{L}_{\mathfrak{X}/K^+} \to \mathbb{L}^{an}_{\mathfrak{X}/K^+}$ induces isomorphisms*

$$\mathrm{Ext}^i_{\mathscr{O}_{\mathfrak{X}}}(\mathbb{L}^{an}_{\mathfrak{X}/K^+}, \mathscr{F}) \xrightarrow{\sim} \mathrm{Ext}^i_{\mathscr{O}_{\mathfrak{X}}}(\mathbb{L}_{\mathfrak{X}/K^+}, \mathscr{F}) \qquad \textit{for every } i \in \mathbb{N}.$$

Proof. We endow the topos $\mathfrak{X}^{\mathbb{N}}$ with the ring object $\mathscr{O}_\bullet := (\mathscr{O}_{\mathfrak{X}}/a^n \mathscr{O}_{\mathfrak{X}} \mid n \in \mathbb{N})$; the natural morphism $\mathscr{O}_{\mathfrak{X}} \to \lim_{\mathbb{N}} \mathscr{O}_\bullet$ determines a morphism of ringed topoi $\pi : (\mathfrak{X}^{\mathbb{N}}, \mathscr{O}_\bullet) \to (\mathfrak{X}, \mathscr{O}_{\mathfrak{X}})$ as in (7.3.6). By lemma 7.3.5 and proposition 7.1.1(iv), the natural map $\mathscr{F} \to R\lim_{\mathbb{N}} \pi^* \mathscr{F}$ is an isomorphism and by lemma 7.1.6, both $\mathbb{L}_{\mathfrak{X}/K^+}$ and $\mathbb{L}^{an}_{\mathfrak{X}/K^+}$ are complexes of flat $\mathscr{O}_{\mathfrak{X}}$-modules, hence the trivial duality isomorphism reads

$$\mathrm{Ext}^i_{\mathscr{O}_\bullet}(\pi^* \mathbb{L}^{an}_{\mathfrak{X}/K^+}, \pi^* \mathscr{F}) \simeq \mathrm{Ext}^i_{\mathscr{O}_{\mathfrak{X}}}(\mathbb{L}^{an}_{\mathfrak{X}/K^+}, \mathscr{F}) \qquad \text{for every } i \in \mathbb{N}$$

and likewise for $\mathbb{L}_{\mathfrak{X}/K^+}$. To conclude it remains only to remark that $\pi^* \mathbb{L}_{\mathfrak{X}/K^+} \simeq \pi^* \mathbb{L}^{an}_{\mathfrak{X}/K^+}$. $\qquad\square$

Lemma 7.3.1 and proposition 7.3.7 enable one to derive the standard results on nilpotent deformations, along the lines of section 3.2 : the statements are unchanged, except that the topos-theoretic cotangent complex is replaced by the (much more manageable) analytic one. The details will be left to the reader.

7.3.8. Let $f : X \to Y$ be a morphism of adic spaces locally of finite type over $\mathrm{Spa}(K, K^+)$. We would like to show that the infinitesimal deformation theory of f is captured by the analytic cotangent complex $\mathbb{L}^{an}_{X/Y}$, in analogy with the usual topos-theoretic situation, and with the treatment for formal schemes already presented. This turns out to be indeed the case, however the proofs are rather more delicate than those for formal schemes, due to the existence of square zero deformations of affinoid algebras that are not themselves affinoid. In homological terms, this reflects the fact that the natural map

(7.3.9) $$\mathrm{Ext}^1_{\mathscr{O}_X}(\mathbb{L}^{an}_{X/Y}, \mathscr{F}) \to \mathrm{Ext}^1_{\mathscr{O}_X}(\mathbb{L}_{X/Y}, \mathscr{F})$$

is not in general an isomorphism for arbitrary coherent $\mathscr{O}_X$-modules $\mathscr{F}$. Nevertheless, we have the following:

Proposition 7.3.10. *Let $f : X \to Y$ be as in (7.3.8). Then the map (7.3.9) is injective for every coherent $\mathscr{O}_X$-module $\mathscr{F}$.*

Proof. We start out with the following:

Claim 7.3.11. $\mathrm{Hom}_{\mathscr{O}_X}(\Omega^{na}_{X/Y}, \mathscr{F}) = 0$. (Notation of (7.2.35))

Proof of the claim. We reduce easily to the case where X and Y are affinoid, say $X = \mathrm{Spa}\,B$ and $Y = \mathrm{Spa}\,A$. The sheaf $\Omega^{na}_{X/Y}$ is the sheafification of the presheaf $U \mapsto \Omega^{na}_{\mathscr{O}_X(U)/A}$ on the site of affinoid subdomains U of X. The assertion is equivalent to its presheaf analogue, that holds by lemma 7.2.37(iii). $\qquad\diamond$

Next, let $\mathbb{C}_{X/Y} := \mathrm{Cone}(\mathbb{L}_{X/Y} \to \mathrm{L}^{\mathrm{an}}_{X/Y})$. Notice that $\mathbb{C}_{X/Y}$ is well defined as an actual complex, not just as an object in $\mathbf{D}^-(\mathcal{O}_{\mathfrak{X}}\text{-}\mathbf{Mod})$. Using the long exact Ext sequence, the proposition is reduced to the following:

Claim 7.3.12. $\mathrm{Hom}_{\mathcal{O}_X}(H_1(\mathbb{C}_{X/Y}), \mathscr{F}) = 0$.

Proof of the claim. We may assume that the spaces are affinoids, and then we can factor f as a closed immersion $g : X \to Z$ followed by a smooth map $h : Z \to Y$. Proposition 7.2.39 yields another natural transitivity triangle (for the naturality, cp. the remarks in the proof of proposition 7.2.48):

$$Lg^*\mathbb{C}_{Z/Y} \to \mathbb{C}_{X/Y} \to \mathbb{C}_{X/Z} \to Lg^*\mathbb{C}_{Z/Y}[1].$$

However, $\mathbb{C}_{X/Z} = 0$ by lemma 7.2.46(i). Moreover, $H_0(\mathbb{C}_{Z/Y}) = 0$ since $\Omega^{\mathrm{an}}_{Z/Y}$ is a quotient of $\Omega_{Z/Y}$. Therefore $H_1(\mathbb{C}_{X/Y}) \simeq H_1(Lg^*\mathbb{C}_{Z/Y}) \simeq g^*H_1(\mathbb{C}_{Z/Y})$. We can then replace f by the morphism g and $\mathscr{F}$ by $g_*\mathscr{F}$, thereby reducing the claim to the case where f is smooth. In this case, theorem 7.2.42(ii) implies that $H_1(\mathbb{C}_{X/Y}) = \Omega^{\mathrm{na}}_{X/Y}$. We conclude by claim 7.3.11. $\qquad\square$

7.3.13. Let $f : X \to Y$ be a morphism of adic spaces locally of finite type over $\mathrm{Spa}(K, K^+)$. An *analytic deformation* of X over Y is a datum of the form $(j, \mathscr{F}, \beta)$ consisting of:

(a) a closed imbedding $j : X \to X'$ of Y-adic spaces, such that the ideal $\mathscr{I} \subset \mathcal{O}_{X'}$ defining j satisfies $\mathscr{I}^2 = 0$, and

(b) a coherent $\mathcal{O}_X$-module $\mathscr{F}$ with an $\mathcal{O}_X$-linear isomorphism $\beta : j^*\mathscr{I} \xrightarrow{\sim} \mathscr{F}$.

One defines in the obvious way a morphism of analytic deformations, and we let $\mathrm{Exan}_Y(X, \mathscr{F})$ denote the set of isomorphism classes of analytic deformations of X by $\mathscr{F}$ over Y.

Proposition 7.3.14. *Let X be an affinoid adic space of finite type over $\mathrm{Spa}(K, K^+)$ and $(j : X \to X', \mathscr{F}, \beta)$ an analytic deformation. Then X' is affinoid of finite type over $\mathrm{Spa}(K, K^+)$.*

Proof. To start out, we notice that j is a homeomorphism on the topological spaces underlying X and X', hence $\mathcal{O}_{X'}(X') = \Gamma(X, j^*\mathcal{O}_{X'})$; we deduce a short exact sequence of continuous maps:

$$0 \to \mathscr{F}(X) \to \mathcal{O}_{X'}(X') \to \mathcal{O}_X(X) \to 0.$$

Claim 7.3.15. Let $\phi : B \to A$ be a surjective map of complete K-algebras of topologically finite type, such that $(\mathrm{Ker}\,\phi)^2 = 0$. Then $B^\circ = \phi^{-1}(A^\circ)$.

Proof of the claim. Quite generally, let C be any K-algebra of topologically finite type; according to [12, §6.2.3, Prop.1] one has $C^\circ = \{f \in C \mid |f(x)| \leq 1 \text{ for every } x \in \mathrm{Max}\, C\}$. In our situation, it is clear that $\mathrm{Max}\, A = \mathrm{Max}\, B$, whence the claim. $\qquad\diamond$

Claim 7.3.16. $\mathcal{O}_{X'}(X')$ is a complete K-algebra of topologically finite type.

Proof of the claim. Let $\bar{g}_1, ..., \bar{g}_m$ be a set of topological generators for $\mathscr{O}_X(X)$, and choose arbitrary liftings $g_1, ..., g_m \in \mathscr{O}_{X'}(X')$. Pick also a finite set of generators $g_{m+1}, ..., g_n$ for the $\mathscr{O}_X(X)$-module $\mathscr{F}(X)$. We define a map $\phi : K[T_1, ..., T_n] \to \mathscr{O}_{X'}(X')$ by the rule $T_i \mapsto g_i$ $(i = 1, ..., n)$. Let $X' = \bigcup_k U_k$ be a covering of X' by finitely many of its affinoid domains. We deduce a map $\psi : K[T_1, ..., T_n] \to \prod_k \mathscr{O}_{X'}(U_k)$. By viewing $0 \to j_*\mathscr{F} \to \mathscr{O}_{X'} \to j_*\mathscr{O}_X \to 0$ as a short exact sequence of coherent $\mathscr{O}_{X'}$-modules on X', we deduce short exact sequences

$$0 \to \mathscr{F}(U_k) \to \mathscr{O}_{X'}(U_k) \xrightarrow{\pi_k} \mathscr{O}_X(U_k) \to 0$$

for every k. By the open mapping theorem (see [12, §2.8.1]) one deduces easily that the topology of $\mathscr{O}_X(U_k)$ is the same as the quotient topology deduced from the surjection π_k, especially $\mathscr{O}_X(U_k)$ is a K-algebra of topologically finite type. Since the images of $\bar{g}_1, ..., \bar{g}_m$ in $\mathscr{O}_X(U_k)$ are power bounded for every k, it then follows from claim 7.3.15 that the images of $g_1, ..., g_m$ in $\mathscr{O}_{X'}(U_k)$ are power bounded for every k. Hence ψ extends to a map $\psi^\wedge : K\langle T_1, ..., T_n \rangle \to \prod_k \mathscr{O}_{X'}(U_k)$, and by construction, $\psi^\wedge$ factors through a continuous map $\phi^\wedge : K\langle T_1, ..., T_n \rangle \to \mathscr{O}_{X'}(X')$. It is easy to check that $\phi^\wedge$ is surjective, so the claim follows, again by the open mapping theorem. $\diamond$

Set $A := \mathscr{O}_{X'}(X')$; by claim 7.3.16 we have the affinoid adic space $X'' := \mathrm{Spa}(A, A^\circ)$ of finite type over $\mathrm{Spa}(K, K^+)$. By construction we get morphisms $X \xrightarrow{j} X' \xrightarrow{\alpha} X''$ of adic spaces inducing homeomorphisms on the underlying topologies. To conclude it suffices to show:

Claim 7.3.17. The morphism $\alpha^\sharp : \mathscr{O}_{X''} \to \alpha_*\mathscr{O}_{X'}$ is an isomorphism of sheaves of topological algebras.

Proof of the claim. Using claim 7.2.38 we see that it suffices to show that $\alpha^\sharp$ is an isomorphism of sheaves of K-algebras. However, for any affinoid open domain in X'' we have a commutative diagram with exact rows:

$$
\begin{array}{ccccccccc}
0 & \longrightarrow & \mathscr{F}(U) & \longrightarrow & \mathscr{O}_{X''}(U) & \longrightarrow & \mathscr{O}_X(U) & \longrightarrow & 0 \\
& & \| & & \downarrow & & \| & & \\
0 & \longrightarrow & \mathscr{F}(U) & \longrightarrow & \mathscr{O}_{X'}(U) & \longrightarrow & \mathscr{O}_X(U) & \longrightarrow & 0.
\end{array}
$$

Since the affinoid open subsets of X'' form a basis of the topology for both X'' and X', the claim follows. $\square$

7.3.18.   In the situation of (7.3.13), let $(j : X \to X', \mathscr{F}, \beta)$ be an analytic deformation of X over Y. Clearly j is a homeomorphism on the underlying topological spaces, hence we can view the given deformation as the datum of a map of $f^{-1}\mathscr{O}_Y$-algebras $\mathscr{O}_{X'} \to \mathscr{O}_X$, whence a transitivity distinguished triangle (proposition 7.2.39)

$$Lj^*\mathbb{L}^{\mathrm{an}}_{X'/S} \to \mathbb{L}^{\mathrm{an}}_{X/S} \to \mathbb{L}^{\mathrm{an}}_{X/X'} \to Lj^*\mathbb{L}^{\mathrm{an}}_{X'/S}[1]$$

which in turn yields a distinguished triangle:

$$RHom_{\mathcal{O}_X}(\mathbb{L}^{an}_{X/X'}, \mathscr{F}) \to RHom_{\mathcal{O}_X}(\mathbb{L}^{an}_{X/S}, \mathscr{F}) \to RHom_{\mathcal{O}_{X'}}(\mathbb{L}^{an}_{X'/S}, j_*\mathscr{F}).$$

Especially, we get a map

(7.3.19) $$\qquad Ext^1_{\mathcal{O}_X}(\mathbb{L}^{an}_{X/X'}, \mathscr{F}) \to Ext^1_{\mathcal{O}_X}(\mathbb{L}^{an}_{X/S}, \mathscr{F}).$$

Let $\mathscr{I} \subset \mathcal{O}_{X'}$ be the ideal that defines the imbedding j; by proposition 7.2.48 we have a natural isomorphism $\tau_{[-1]}\mathbb{L}^{an}_{X/X'} \xrightarrow{\sim} j^*\mathscr{I}[1]$, whence an isomorphism

(7.3.20) $$\qquad Ext^1_{\mathcal{O}_X}(\mathbb{L}^{an}_{X/X'}, \mathscr{F}) \xrightarrow{\sim} Hom_{\mathcal{O}_X}(j^*\mathscr{I}, \mathscr{F}).$$

Combining (7.3.19) and (7.3.20) we see that the given isomorphism $\beta : j^*\mathscr{I} \to \mathscr{F}$ determines a unique element $e^{an}(X', \beta) \in Ext^1_{\mathcal{O}_X}(\mathbb{L}^{an}_{X/S}, \mathscr{F})$. One verifies easily that $e^{an}(X', \beta)$ depends only on the isomorphism class of the analytic deformation $(j, \mathscr{F}, \beta)$, therefore it defines a map

$$e^{an} : Exan_Y(X, \mathscr{F}) \to Ext^1_{\mathcal{O}_X}(\mathbb{L}^{an}_{X/S}, \mathscr{F}).$$

By inspecting the construction, it is easy to check that e^{an} fits into a commutative diagram

(7.3.21)
$$\begin{array}{ccc} Exan_Y(X, \mathscr{F}) & \xrightarrow{\;e^{an}\;} & Ext^1_{\mathcal{O}_X}(\mathbb{L}^{an}_{X/S}, \mathscr{F}) \\ \downarrow & & \downarrow \\ Exal_{f^{-1}\mathcal{O}_Y}(\mathcal{O}_X, \mathscr{F}) & \xrightarrow{\;e\;} & Ext^1_{\mathcal{O}_X}(\mathbb{L}_{X/Y}, \mathscr{F}) \end{array}$$

where e is the isomorphism of [50, Ch.III, Th.1.2.3] and the right vertical arrow is (7.3.9).

Theorem 7.3.22. *For every coherent $\mathcal{O}_X$-module $\mathscr{F}$, the map e^{an} is a bijection.*

Proof. In view of (7.3.21), in order to show that e^{an} is injective, it suffices to verify that if $(j_i : X \to X_i, \beta_i)$ for $i = 1, 2$ are two analytic deformations of X by $\mathscr{F}$ over Y, and $(j_1, \beta_1) \xrightarrow{\sim} (j_2, \beta_2)$ is an isomorphism in the category of extensions of $f^{-1}\mathcal{O}_Y$-algebras, then the corresponding map $\mathcal{O}_{X_1} \to \mathcal{O}_{X_2}$ is *bicontinuous, i.e.* it is an isomorphism of sheaves of topological algebras (notice that in this case the map restricts to an isomorphism $\mathcal{O}^+_{X_1} \xrightarrow{\sim} \mathcal{O}^+_{X_2}$, since the latter sheaves are the preimages of $\mathcal{O}^+_X$). This can be checked locally, so we may assume that X is affinoid; in this case both X_1 and X_2 are affinoid as well, by proposition 7.3.14. Then the assertion is a straightforward consequence of claim 7.2.38.

The surjectivity is a local issue as well; indeed, any class in the target of e^{an} represents an extension of $f^{-1}\mathcal{O}_Y$-algebras $0 \to \mathscr{F} \to \mathscr{E} \to \mathcal{O}_X \to 0$, and the question amounts to showing that $\mathscr{E}$ represents an analytic deformation, which can be checked locally. Thus, suppose that $X = Spa(B, B^\circ)$, $Y = Spa(A, A^\circ)$. We can write $B = P/I$, where $P := A\langle T_1, ..., T_n \rangle$, and by proposition 7.2.48, the complex $\tau_{[-1]}\mathbb{L}^{an}_{X/Y}$ is naturally quasi-isomorphic to the complex of sheaves associated to the complex of B-modules $0 \to I/I^2 \to \Omega^{an}_{P/A} \otimes_P B \to 0$. Whence, a natural isomorphism

$$\mathbf{Ext}^1_{\mathscr{O}_X}(\mathbb{L}^{an}_{X/Y}, \mathscr{F}) \simeq \mathrm{Coker}(\mathrm{Hom}_{\mathscr{O}_X}(j^*\Omega^{an}_{Z/Y}, \mathscr{F}) \to \mathrm{Hom}_{\mathscr{O}_X}((I/I^2)^\sim, \mathscr{F}))$$
$$\simeq \mathrm{Coker}(\mathrm{Hom}_B(\Omega^{an}_{P/A} \otimes_P B, \mathscr{F}(X)) \to \mathrm{Hom}_B(I/I^2, \mathscr{F}(X)))$$

(where we have denoted by $j : X \to Z := \mathrm{Spa}(P, P^\circ)$ the induced closed imbedding). However, given $\alpha : I/I^2 \to \mathscr{F}(X)$, the corresponding $f^{-1}\mathscr{O}_Y$ algebra $\mathscr{E}$ is the sheaf of algebras associated to the A-algebra E defined as follows. Let $\beta : I/I^2 \to E_0 := (P/I^2) \oplus \mathscr{F}(X)$ be the map given by the rule: $x \mapsto (x, \alpha(x))$ for every $x \in I/I^2$; E_0 is endowed with an A-algebra structure given by the rule $(x, f) \cdot (y, g) = (xy, fy + gx)$ and one verifies easily that the image of β is an ideal in E_0; then set $E := E_0/\mathrm{Im}\,\beta$. It is clear that E_0 is an A-algebra of topologically finite type, and the projection $E \to B$ determines a closed imbedding $j : X \to X' := \mathrm{Spa}(E, E^\circ)$ that identifies $j^*\mathscr{O}_{X'}$ with $\mathscr{E}$. $\square$

With the foregoing results, one can derive the usual results on existence of deformations and thereof obstructions; again we leave the details to the industrious reader. To conclude this section we want to show that (7.3.9) fails to be surjective already in the simplest situations. To carry out this analysis requires the use of more refined commutative algebra : the following proposition 7.3.23 collects all the information that we shall be needing.

Proposition 7.3.23. *Let A be a complete K-algebra of topologically finite type, $n \in \mathbb{N}$ an integer, and set $P := A\langle T_1, ..., T_n\rangle$. Then:*

(i) *the natural morphism* $\mathrm{Spec}\,P \to \mathrm{Spec}\,A$ *is regular.*

(ii) $H_i(\mathbb{L}_{P/A}) = 0$ *for every $i > 0$ and $H_0(\mathbb{L}_{P/A})$ is a flat P-module.*

Proof. We begin with the following observation:

Claim 7.3.24. Let $R \to S$ be a faithfully flat morphism of noetherian local rings. If S is regular, then R is regular.

Proof of the claim. This is [28, Ch.0, Prop.17.3.3(i)]. $\Diamond$

Claim 7.3.25. Let $R \to S$ be a local morphism of local noetherian rings; let $\mathfrak{n} \subset S$ be the maximal ideal. Suppose that R is quasi-excellent and that S is formally smooth over R for its $\mathfrak{n}$-adic topology. Then the induced morphism $\mathrm{Spec}\,S \to \mathrm{Spec}\,R$ is regular.

Proof of the claim. This is [3, Th.]. $\Diamond$

In view of claim 7.3.25, the proof of (i) is reduced to the following:

Claim 7.3.26. Let $\mathfrak{n} \subset P$ be any maximal ideal and set $\mathfrak{q} := \mathfrak{n} \cap A$. Then:

(i) $P_\mathfrak{n}$ is formally smooth over $A_\mathfrak{q}$ for its $\mathfrak{n}$-adic topology.

(ii) $A_\mathfrak{q}$ is an excellent local ring.

Proof of the claim. (i): It is well known that $\mathfrak{q}$ is a maximal ideal and the residue field $K' := A/\mathfrak{q}$ is a finite extension of K. It follows easily from lemma 7.1.6(i) that $P_\mathfrak{n}$ is flat over A, hence by [28, Ch.0, Th.19.7.1] it suffices to show that $P_\mathfrak{n} \otimes_A K'$ is geometrically regular over K', which allows to reduce to the case where $A = K$. Then, in view of claim 7.3.24, it suffices to show that, for every finite field extension

$K \subset K'$ there exists a larger finite extension $K' \subset K''$ such that the semilocal ring $P_{\mathfrak{n}} \otimes_K K''$ is regular. However, the residue field $\kappa := P/\mathfrak{n}$ is a finite extension of K, and if $K \subset K'$ is any finite normal extension with $\kappa \subset K'$, every maximal ideal of $P \otimes_K K'$ containing $\mathfrak{n} \otimes_K K'$ is of the form $\mathfrak{p} := (T_1 - a_1, ..., T_n - a_n)$ for certain $a_1, ..., a_n \in K'$. We are therefore reduced to the case where $\mathfrak{n} = (T_1 - a_1, ..., T_n - a_n)$ for some $a_i \in K$, $i = 1, ..., n$; in this case the $\mathfrak{n}$-adic completion $P_{\mathfrak{n}}^{\wedge}$ of $P_{\mathfrak{n}}$ is isomorphic to $K[[X_1, ..., X_n]]$, which is regular. (ii) is a theorem due to Kiehl, whose complete proof is reproduced in [22, Th.1.1.3]. $\Diamond$

Finally, (ii) follows from (i) and from the following:

Claim 7.3.27. Let $R \to S$ be a map of noetherian rings. Then the following are equivalent:

(a) $H_i(\mathbb{L}_{S/R}) = 0$ for every $i > 0$ and $H_0(\mathbb{L}_{S/R})$ is a flat S-module.
(b) The induced morphism $\operatorname{Spec} S \to \operatorname{Spec} R$ is regular.

Proof of the claim. Taking into account [58, Th.28.7 and 28.9], this is seen to be a paraphrase of [2, Suppl.(c), Th.30, p.331]. □

7.3.28.   Let us now specialize to the case where $P := K\langle T \rangle$; it is well known that P is a principal ideal domain; since $\Omega_{P/K}$ is a free P-module, we can choose a splitting:

$$(7.3.29) \qquad \Omega_{P/K} \simeq \Omega_{P/K}^{\mathrm{an}} \oplus \Omega_{P/K}^{\mathrm{na}}$$

(notation of (7.2.35)). Our first aim is to show that P admits nilpotent extensions (by finitely generated P-modules) that are not affinoid algebras. In view of proposition 7.3.23(ii), the square zero extensions of P by a P-module F are classified by $\operatorname{Ext}_P^1(\Omega_{P/K}, F)$ (and then it is clear that any non-trivial element of this group cannot represent an analytic deformation, since P admits none). Hence, we come down to computing $\operatorname{Ext}_P^1(\Omega_{P/K}^{\mathrm{na}}, F)$.

Lemma 7.3.30. (i) $\Omega_{P/K}^{\mathrm{na}}$ *is a vector space over the fraction field E of P.*

(ii) *If either* $\operatorname{char}(K) = 0$ *or* $\operatorname{char}(K) = p > 0$ *and* $[K : K^p] = \infty$, *then* $\Omega_{P/K}^{\mathrm{na}} \neq 0$.

Proof. (i): By proposition 7.3.23(ii) we know that $\Omega_{P/K}^{\mathrm{na}}$ is a flat P-module, especially it is torsion-free. Let $f \in P$ be any irreducible element; then $\kappa := P/fP$ is a finite field extension of K, and we have $\operatorname{Hom}_P(\Omega_{P/K}, \kappa) = \operatorname{Hom}_P(\Omega_{P/K}^{\mathrm{an}}, \kappa)$ by lemma 7.2.37. In other words, $\Omega^{\mathrm{na}} \otimes_P \kappa = 0$, so multiplication by f is a bijection on $\Omega_{P/K}^{\mathrm{na}}$, and the claim follows.

(ii): Suppose first that $\operatorname{char}(K) = 0$; by (i) we have:

$$\Omega_{E/K} \simeq E \otimes_P \Omega_{P/K} \simeq \Omega_{P/K}^{\mathrm{na}} \oplus (E \otimes_P \Omega_{P/K}^{\mathrm{an}}).$$

Since $\Omega_{P/K}^{\mathrm{an}}$ is a free P-module of rank one with generator dT, and since E is a separable extension of K, it suffices to show that $\operatorname{tr.deg}(E/K(T)) > 0$. This can be checked explicitly; for instance, let $\pi \in K$ be a non-zero element such that $\log(1 + \pi T) \in P$; it is well known that this power series is transcendental over

$K(T)$. Finally, suppose that char$(K) = p > 0$ and $[K : K^p] = \infty$. We construct a splitting for the surjection $\Omega_{E/K} \to E \otimes_P \Omega_{P/K}^{an}$, as follows. The tower of field extensions $K \subset F := E^p \cdot K(T) \subset E$ yields an exact sequence

$$E \otimes_F \Omega_{F/K} \xrightarrow{\alpha} \Omega_{E/K} \xrightarrow{\beta} \Omega_{E/F} \to 0$$

and it is easy to check that α factors through a surjection $E \otimes_F \Omega_{F/K} \to E \otimes_P \Omega_{P/K}^{an}$ and the induced map $E \otimes_P \Omega_{P/K}^{an} \to \Omega_{E/K}$ is the sought splitting. Hence, it suffices to exhibit an element $f \in E$ such that $\beta(df) \neq 0$. To this aim, we will use the following general remark:

Claim 7.3.31. Let E be a field with char$(E) = p > 0$ and $F \subset E$ a subfield with $E^p \subset F$. Then: $\mathrm{Ker}(d : E \to \Omega_{E/F}) = F$.

Proof of the claim. Let $(x_i \mid i \in I)$ be a p-basis for the extension $F \subset E$, and for every $i \in I$ set $E_i := F[x_i \mid i \in I \setminus \{i\}]$. Clearly $E = E_i[x_i]$ and the minimal polynomial of x_i over E_i is $m_i(X) := X^p - x_i^p$. A simple calculation shows that $\mathrm{Ker}(d : E \to \Omega_{E/E_i}) = E_i$. Since $F = \bigcap_{i \in I} E_i$, it follows that

$$\mathrm{Ker}(d : E \to \Omega_{E/F}) = \bigcap_{i \in I} \mathrm{Ker}(d : E \to \Omega_{E/E_i}) = F.$$

$\diamond$

In view of claim 7.3.31, we need to exhibit an element $f \in E$ such that $f \notin E^p \cdot K(T)$. However, under the standing assumptions, we can find a countable sequence $(b_n \mid n \in \mathbb{N})$ of elements $b_n \in K^\circ$ that are linearly independent over K^p. Set $f := \sum_{n \in \mathbb{N}} a^{pn} b_n T^n$, so that $f \in P$, and suppose by way of contradiction, that $f \in E^p \cdot K(T)$; then we could find $g \in E \cdot K^{1/p}(T^{1/p})$ such that $g^p = f$. In turns, this means that $g \in K' \cdot E(T^{1/p})$ for some finite K-extension $K' \subset K^{1/p}$. However, $K' \cdot E(T^{1/p})$ is the field of fractions of $K'\langle T^{1/p} \rangle$, and since the latter ring is normal, g must lie in it. Then $g^p = \sum_{n \in \mathbb{N}} c_n^p T^n$, and consequently $c_n^p = a^{pn} b_n$ for every $n \in \mathbb{N}$, in other words, K' contains all the elements $b_n^{1/p}$ for every $n \in \mathbb{N}$, which is absurd. $\square$

7.3.32.   We shall complete our calculation for $F := P$. To this aim, recall that E is an injective P-module, therefore

$$\mathrm{Ext}_P^1(\Omega_{P/K}^{na}, P) \simeq \mathrm{Hom}_P(\Omega_{P/K}^{na}, E/P)/\mathrm{Hom}_P(\Omega_{P/K}^{na}, E).$$

Let A_E be the (finite) adele ring of E, *i.e.* the restricted product of the fields of fractions $E_{\mathfrak{p}}^\wedge$ of the completions $P_{\mathfrak{p}}^\wedge$:

$$\mathsf{A}_E := \prod_{\mathfrak{p} \in \mathrm{Max} P}' E_{\mathfrak{p}}^\wedge.$$

By the strong approximation theorem (cp. [20, Ch.II, §15, Th.]) the natural imbedding $E \subset \mathsf{A}_E$ has everywhere dense image, whence an isomorphism:

$$E/P \simeq \bigoplus_{\mathfrak{p} \in \mathrm{Max} P} E_{\mathfrak{p}}^\wedge / P_{\mathfrak{p}}^\wedge.$$

A simple calculation shows that the map $E_{\mathrm{p}}^\wedge \to \mathrm{Hom}_P(E, E_{\mathrm{p}}^\wedge/P_{\mathrm{p}}^\wedge)$ given by the rule: $f \mapsto (x \mapsto xf)$ is an isomorphism; whence a natural isomorphism:

$$A_E \xrightarrow{\sim} \mathrm{Hom}_P(E, E/P) \qquad a \mapsto (x \mapsto ax).$$

Finally, $\mathrm{Ext}^1_P(E, P) \simeq A_E/E$; since $\Omega^{\mathrm{na}}_{P/K}$ is a direct sum of copies of E, this achieves our aim of producing non-trivial square zero extensions of K-algebras: $0 \to P \to \mathcal{E} \to P \to 0$, whenever the hypotheses of lemma 7.3.30(ii) are fulfilled.

7.3.33. Now we want to carry out the local counterpart of the calculations of (7.3.32). Namely, let $X := \mathrm{Spa}(P, P^\circ)$; we will show that there exist non-trivial deformations of $(X, \mathcal{O}_X)$ by the $\mathcal{O}_X$-module $\mathcal{O}_X$, in the category of locally ringed spaces. To this aim, choose any K-rational point $p \in X$, and let $\mathcal{O}_X(\infty)$ denote the quasi-coherent $\mathcal{O}_X$-module whose local sections on any open subset $U \subset X$ are the meromorphic functions on U with poles (of arbitrary finite order) only at p. We deduce an exact sequence

$$\mathrm{Hom}_{\mathcal{O}_X}(\Omega^{\mathrm{na}}_{X/K}, \mathcal{O}_X(\infty)) \xrightarrow{f} \mathrm{Hom}_{\mathcal{O}_X}(\Omega^{\mathrm{na}}_{X/K}, \frac{\mathcal{O}_X(\infty)}{\mathcal{O}_X}) \xrightarrow{g} \mathrm{Ext}^1_{\mathcal{O}_X}(\Omega^{\mathrm{na}}_{X/K}, \mathcal{O}_X)$$

and we remark that $\Omega^{\mathrm{na}}_{X/K}$ is the sheaf obtained by sheafifying the presheaf

(7.3.34) $$U \mapsto \Omega^{\mathrm{na}}_{\mathcal{O}_X(U)/K}$$

on the site of affinoid subdomains of X. After choosing a global splitting (7.3.29), we can write $\Omega_{X/K} \simeq \Omega^{\mathrm{an}}_{X/K} \oplus \Omega^{\mathrm{na}}_{X/K}$.

Lemma 7.3.35. *With the notation of (7.3.33), $\mathrm{Hom}_{\mathcal{O}_X}(\Omega^{\mathrm{na}}_{X/K}, \mathcal{O}_X(\infty)) = 0$.*

Proof. Indeed, let $\phi : \Omega^{\mathrm{na}}_{X/K} \to \mathcal{O}_X(\infty)$ be any $\mathcal{O}_X$-linear map; we extend ϕ to a map $\phi' : \Omega_{X/K} \to \mathcal{O}_X(\infty)$ by prescribing that ϕ' restricts to the zero map on the direct factor $\Omega^{\mathrm{an}}_{X/K}$. Then ϕ' corresponds to a K-linear derivation $\partial : \mathcal{O}_X \to \mathcal{O}_X(\infty)$; the restriction of ∂ to $U := X \setminus \{p\}$ is a derivation of $\mathcal{O}_U$ with values in the coherent $\mathcal{O}_U$-module $\mathcal{O}_U$; by lemma 7.2.37(iii) the restriction of $\phi'_{|U}$ to $\Omega^{\mathrm{na}}_{U/K}$ vanishes identically. It follows that $\partial_{|U}$ vanishes identically. Let $j : U \to X$ be the imbedding; since the natural map $\mathcal{O}_X(\infty) \to j_*\mathcal{O}_U$ is injective, it follows that ∂ must vanish as well, whence $\phi = 0$. $\square$

Lemma 7.3.36. *With the notation of (7.3.33), let $x \in X$ be a K-rational point and suppose that the hypotheses of lemma 7.3.30(ii) hold for K. Then $\Omega^{\mathrm{na}}_{X/K,x} \neq 0$.*

Proof. By lemma 7.3.30(ii) we know already that the global sections of the presheaf (7.3.34) do not vanish. It suffices therefore to show that, for every affinoid subdomain $U := \mathrm{Spa}(A, A^\circ) \subset X$ containing x, the natural map $\Omega_{P/K} \to \Omega_{A/K}$ is injective; this can be factored as the composition of the map $\Omega_{P/K} \to A \otimes_P \Omega_{P/K}$ (which is injective, since $\Omega_{P/K}$ is flat) and the map $A \otimes_P \Omega_{P/K} \to \Omega_{A/K}$. However, the kernel of this latter map is a quotient of $H_1(\mathbb{L}_{A/P})$, hence it suffices to show the following

Claim 7.3.37. $H_i(\mathbb{L}_{A/P}) = 0$ for every $i > 0$.

Proof of the claim. In light of claim 7.3.27, it suffices to show that the map $P \to A$ is regular. Then, in view of claims 7.3.25 and 7.3.26(ii), it suffices to show that, for every maximal ideal $\mathfrak{n} \subset A$, the ring A is formally smooth over P for its $\mathfrak{n}$-adic topology. Let $\mathfrak{q} := \mathfrak{n} \cap P$; since A is flat over P, [28, Ch.0, Th.19.7.1] reduces to showing that the induced morphism $P/\mathfrak{q} \to A/\mathfrak{q}A$ is formally smooth, which is trivial, since the latter is an isomorphism. $\qquad\square$

7.3.38. It follows easily from proposition 7.3.23(ii) and lemma 7.2.9 that $\mathbb{L}_{X/K} \simeq \Omega_{X/K}[0]$, so the extensions $0 \to \mathscr{O}_X \to \mathscr{E} \to \mathscr{O}_X \to 0$ are classified by $\mathrm{Ext}^1_{\mathscr{O}_X}(\Omega^{na}_{X/K}, \mathscr{O}_X)$. In order to show that the latter is not trivial, it suffices, in view of lemma 7.3.35, to exhibit a nonzero map $\Omega^{na}_{X/K} \to Q := \mathscr{O}_X(\infty)/\mathscr{O}_X$. However, Q is a skyscaper sheaf sitting at the point p, with stalk equal to $F/\mathscr{O}_{X,p}$, where $F := \mathrm{Frac}(\mathscr{O}_{X,p})$. We are therefore reduced to showing the existence of a nonzero map $\Omega^{na}_{X/K,p} \to F$. However, we deduce easily from lemma 7.3.30 that $\Omega^{na}_{X/K,p}$ is an F-vector space, and by lemma 7.3.36 the latter does not vanish, provided that either $\mathrm{char}(K) = 0$ or K has characteristic $p > 0$ and $[K : K^p] = \infty$.

7.3.39. The foregoing results notwithstanding, there is at least one situation in which the abstract topos deformation theory of an adic space is the same as its analytic deformation theory. This is explained by the following final proposition.

Proposition 7.3.40. *Suppose that* $\mathrm{char}(K) = p > 0$ *and that* $[K : K^p] < \infty$. *Then, for every morphism* $X \to Y$ *of adic spaces locally of finite type over* $\mathrm{Spa}(K, K^+)$, *the natural map* $\mathbb{L}_{X/Y} \to \mathbb{L}^{an}_{X/Y}$ *is an isomorphism in* $\mathsf{D}(\mathscr{O}_X\text{-}\mathbf{Mod})$.

Proof. It suffices to show the corresponding statement for maps of affinoid rings $A \to B$, hence choose a presentation $B \simeq A\langle T_1, ..., T_n\rangle/I$. We have to verify that the induced maps $H_i(\mathbb{L}_{B/A}) \to H_i(\mathbb{L}^{an}_{B/A})$ are isomorphisms for every $i \in \mathbb{N}$. The sequence $A \to P := A\langle T_1, ..., T_n\rangle \to B$ yields by transitivity two distinguished triangles : one relative to the usual cotangent complex and one relative to the analytic cotangent complex. Hence, by the five lemma, we are reduced to showing that the maps $H_i(B \otimes_P \mathbb{L}_{P/A}) \to H_i(B \otimes_P \mathbb{L}^{an}_{P/A})$ and $H_i(\mathbb{L}_{B/P}) \to H_i(\mathbb{L}^{an}_{B/P})$ are isomorphisms for every $i \in \mathbb{N}$. For the latter we appeal to lemma 7.2.46(ii), and therefore it remains only to show that the natural map $\mathbb{L}_{P/A} \to \mathbb{L}^{an}_{P/A}$ is a quasi-isomorphism. Also, it follows from proposition 7.3.23 and theorem 7.2.42(iii) that both $H_i(\mathbb{L}_{P/A})$ and $H_i(\mathbb{L}^{an}_{P/A})$ vanish for $i > 0$, hence we are reduced to showing that the natural map $\Omega_{P/A} \to \Omega^{an}_{P/A}$ is an isomorphism. Let $C \subset P$ be the A-subalgebra generated by the image of the Frobenius endomorphism $\Phi : P \to P$; a standard calculation shows that $\Omega_{P/A} \simeq \Omega_{P/C}$. However, under the standing assumptions, P is finite over its subalgebra C, therefore $\Omega_{P/A}$ is a finite P-module, and the claim follows. $\qquad\square$

7.4. Analytic geometry over a deeply ramified base. In this section we assume throughout that $(K, |\cdot|)$ is a deeply ramified complete valued field, with valuation of rank one. Recall that $a \in K^\times$ denotes a topologically nilpotent element of K.

If $\mathfrak{X}$ is a formal scheme of finite presentation over $\mathrm{Spf}\, K^+$, we will sometimes write $\mathbb{L}^{\mathrm{an}}_{\mathfrak{X}/K^+}$ instead of $\mathbb{L}^{\mathrm{an}}_{\mathfrak{X}/\mathrm{Spf}\, K^+}$. Similarly we define $\mathbb{L}^+_{X/K}$ for an adic space X of finite type over $\mathrm{Spa}(K, K^+)$, and set $\Omega^+_{X/K} := H_0(\mathbb{L}^+_{X/K})$.

Theorem 7.4.1. *Let X be a smooth adic space over $\mathrm{Spa}(K, K^+)$. Then $\mathbb{L}^+_{X/K} \simeq \Omega^+_{X/K}[0]$ in $\mathsf{D}^-(\mathscr{O}^+_X\text{-}\mathbf{Mod})$, and $\Omega^+_{X/K}$ is a flat sheaf of $\mathscr{O}^+_X$-modules.*

Proof. Both assertions can be checked on the stalks, therefore let $x \in X$ be any point. The stalk $\mathscr{O}_{X,x}$ is a local ring and its residue field $\kappa(x)$ carries a natural valuation; the preimage in $\mathscr{O}_{X,x}$ of the corresponding valuation ring $\kappa(x)^+$ is the subring $\mathscr{O}^+_{X,x}$. Let $I := \bigcap_{n\in\mathbb{N}} a^n \mathscr{O}^+_{X,x}$; it follows from this description that $\kappa(x)^+ = \mathscr{O}^+_{X,x}/I$. Especially, we have

(7.4.2) $$\mathscr{O}^+_{X,x}/a\mathscr{O}^+_{X,x} \simeq \kappa(x)^+/a \cdot \kappa(x)^+.$$

Claim 7.4.3. Let M be an $\mathscr{O}^+_{X,x}$-module, and suppose that a is regular on M. Then M/IM is a flat $\kappa(x)^+$-module.

Proof of the claim. By snake lemma we derive $\mathrm{Ker}(M/IM \xrightarrow{\cdot a} M/IM) \subset \mathrm{Coker}(IM \xrightarrow{\cdot a} IM)$. However, it is clear that $I = aI$, so a is regular on M/IM, hence the latter is a torsion-free $\kappa(x)^+$-module and the claim follows. $\Diamond$

Let $\mathscr{U}$ be the cofiltered system of all affinoid open neighborhoods of x in X; for $U \in \mathscr{U}$ let F_U be the filtered system of all open K^+-subalgebras of $\mathscr{O}^+_X(U)$ of topologically finite presentation. We derive

$$(\mathbb{L}^+_{X/K})_x \overset{\mathbf{L}}{\otimes}_{K^+} K^+/aK^+ \simeq \underset{U\in\mathscr{U}}{\mathrm{colim}}\, \underset{A\in F_U}{\mathrm{colim}}\, \mathbb{L}^{\mathrm{an}}_{A/K^+} \overset{\mathbf{L}}{\otimes}_{K^+} K^+/aK^+$$

$$\simeq \underset{U\in\mathscr{U}}{\mathrm{colim}}\, \underset{A\in F_U}{\mathrm{colim}}\, \mathbb{L}_{A/K^+} \overset{\mathbf{L}}{\otimes}_{K^+} K^+/aK^+$$

$$\simeq \mathbb{L}_{\mathscr{O}^+_{X,x}/K^+} \overset{\mathbf{L}}{\otimes}_{K^+} K^+/aK^+$$

$$\simeq \mathbb{L}_{(\mathscr{O}^+_{X,x}/a\mathscr{O}^+_{X,x})/(K^+/aK^+)}$$

$$\simeq \mathbb{L}_{\kappa(x)^+/K^+} \overset{\mathbf{L}}{\otimes}_{K^+} K^+/aK^+.$$

Together with theorem 6.6.12, this implies already that scalar multiplication by a is an automorphism of $H_i(\mathbb{L}^+_{X/K})$, for every $i > 0$. However, according to theorem 7.2.42(ii), $H_i(\mathbb{L}^+_{X/K})$ is a K^+-torsion sheaf of $\mathscr{O}^+_X$-modules, for $i > 0$, whence the first assertion. It also follows that $(\Omega^+_{X/K})_x$ is a torsion-free, hence flat, K^+-module. To prove that $(\Omega^+_{X/K})_x$ is a flat $\mathscr{O}^+_{X,x}$-module, we remark first that $(\Omega^{\mathrm{an}}_{X/K})_x \simeq (\Omega^+_{X/K})_x \otimes_{K^+} K$, and the latter is a flat $\mathscr{O}_{X,x}$-module, since X is smooth over $\mathrm{Spa}(K, K^+)$. By lemma 5.2.1 it suffices therefore to show

Claim 7.4.4. $(\Omega^+_{X/K})_x \otimes_{K^+} K^+/aK^+$ is a flat $\mathscr{O}^+_{X,x} \otimes_{K^+} K^+/aK^+$-module.

Proof of the claim. By claim 7.4.3 we know that $(\Omega^+_{X/K})_x \otimes_{\mathscr{O}^+_{X,x}} \kappa(x)^+$ is flat over $\kappa(x)^+$. In view of (7.4.2), the claim follows after base change to $\kappa(x)^+/a \cdot \kappa(x)^+$. $\square$

Definition 7.4.5. Let $(\mathfrak{X}_\alpha \mid \alpha \in I)$ be a system of formal schemes of finite presentation over $\mathrm{Spf}\, K^+$, indexed by a small cofiltered category I.

(i) Let $\mathfrak{X}_\infty := \lim_{\alpha \in I} \mathfrak{X}_\alpha$, where the limit is taken in the category of locally ringed spaces. For every $\alpha \in I$, let $\pi_\alpha : \mathfrak{X}_\infty \to \mathfrak{X}_\alpha$ be the natural morphism of locally ringed spaces. We define

$$\Omega^{\mathrm{an}}_{\mathfrak{X}_\infty/K^+} := \operatorname*{colim}_{\alpha \in I^o} \pi_\alpha^*(\Omega^{\mathrm{an}}_{\mathfrak{X}_\alpha/K^+})$$

which is a sheaf of $\mathscr{O}_{\mathfrak{X}_\infty}$-modules. More generally, we let

$$\mathbb{L}^{\mathrm{an}}_{\mathfrak{X}_\infty/K^+} := \operatorname*{colim}_{\alpha \in I^o} \pi_\alpha^*(\mathbb{L}^{\mathrm{an}}_{\mathfrak{X}_\alpha/K^+}).$$

(ii) We say that the cofiltered system $(\mathfrak{X}_\alpha \mid \alpha \in I)$ is *deeply ramified* if the natural morphism $\Omega^{\mathrm{an}}_{\mathfrak{X}_\infty/K^+} \to \Omega^{\mathrm{an}}_{\mathfrak{X}_\infty/K^+} \otimes_{K^+} K$ is an epimorphism.

Lemma 7.4.6. *Let $(\mathfrak{X}_\alpha \mid \alpha \in I)$ be a cofiltered system as in definition 7.4.5. For any morphism $\beta \to \alpha$ of I, let $\mathfrak{f}_{\alpha\beta} : \mathfrak{X}_\beta \to \mathfrak{X}_\alpha$ be the corresponding morphism of formal $\mathrm{Spf}\, K^+$-schemes. Moreover, for every $\alpha \in I$, let $\Omega^{\mathrm{tf}}_{\mathfrak{X}_\alpha/K^+}$ be the image of the morphism $\Omega^{\mathrm{an}}_{\mathfrak{X}_\alpha/K^+} \to \Omega^{\mathrm{an}}_{\mathfrak{X}_\alpha/K^+} \otimes_{K^+} K$ ("tf" stands for torsion-free). The following two conditions are equivalent:*

(i) *The system $(\mathfrak{X}_\alpha \mid \alpha \in I)$ is deeply ramified.*

(ii) *For every $\alpha \in I$ there is a morphism $\beta \to \alpha$ of I, such that the image of the natural morphism $\mathfrak{f}_{\alpha\beta}^*(\Omega^{\mathrm{tf}}_{\mathfrak{X}_\alpha/K^+}) \to \Omega^{\mathrm{tf}}_{\mathfrak{X}_\beta/K^+}$ is contained in the subsheaf $a \cdot \Omega^{\mathrm{tf}}_{\mathfrak{X}_\beta/K^+}.$*

Proof. It is clear that (ii)⇒(i). We show that (i)⇒(ii). Under the above assumptions, every $\mathfrak{X}_\alpha$ is quasi-compact, hence we can cover it by finitely many affine formal schemes $\mathfrak{U}_i := \mathrm{Spf}\, A_i$ ($i = 1, ..., n$) of finite type over $\mathrm{Spf}\, K^+$. Then, for every $i = 1, ..., n$, the restriction of $\Omega^{\mathrm{an}}_{\mathfrak{X}_\alpha/K^+}$ to $\mathfrak{U}_i$ is the coherent sheaf $(\Omega^{\mathrm{an}}_{A_i/K^+})^\triangle$ (notation of [26, Ch.I, §10.10.1]). Hence, by proposition 7.1.1(i), $\Omega^{\mathrm{tf}}_{\mathfrak{X}_\alpha/K^+}$ is a coherent sheaf of $\mathscr{O}_{\mathfrak{X}_\alpha}$-modules. For every morphism $\beta \to \alpha$, let

$$U_{\alpha\beta} := \{x \in \mathfrak{X}_\beta \mid \mathrm{Im}(\mathfrak{f}_{\alpha\beta}^*(\Omega^{\mathrm{tf}}_{\mathfrak{X}_\alpha/K^+})_x \to (\Omega^{\mathrm{tf}}_{\mathfrak{X}_\beta/K^+})_x) \subset a \cdot (\Omega^{\mathrm{tf}}_{\mathfrak{X}_\beta/K^+})_x\}.$$

$U_{\alpha\beta}$ is therefore a constructible open subset of $\mathfrak{X}_\beta$, and we denote its complement by $Z_{\alpha\beta}$. By assumption (i) we know that

$$\lim_{\beta \to \alpha} Z_{\alpha\beta} = \bigcap_{\beta \to \alpha} \pi_\beta^{-1}(Z_{\alpha\beta}) = \varnothing.$$

If we retopologize the reduced schemes $Z_{\alpha\beta}$ by their constructible topologies, we get an inverse system of compact spaces, and we deduce that some $Z_{\alpha\beta}$ is empty by [18, Ch.I, §9, n.6, Prop.8]. ☐

Example 7.4.7. The prototype of deeply ramified systems is given by the tower of morphisms

$$(7.4.8) \qquad \cdots \to \mathbb{B}^d_{K^+}(0, \rho^{1/p^n}) \xrightarrow{\phi_n} \mathbb{B}^d_{K^+}(0, \rho^{1/p^{n-1}}) \xrightarrow{\phi_{n-1}} \cdots \xrightarrow{\phi_1} \mathbb{B}^d_{K^+}(0, \rho)$$

where $p > 0$ is the residue characteristic of K, and for any $r = (r_1, ..., r_d) \in (K^\times)^d$, we have denoted

$$\mathbb{B}^d_{K^+}(0, |r|) := \operatorname{Spf} K^+ \langle r_1^{-1} T_1, ..., r_d^{-1} T_d \rangle$$

(*i.e.*, the formal d-dimensional polydisc defined by the equations $|T_i| \leq |r_i|$, $i = 1, ..., d$). The morphisms ϕ_n are induced by the ring homomorphisms $T_i \mapsto T_i^p$ ($i = 1, ..., d$). Notice that the tower (7.4.8) is defined whenever $\rho_i \in \Gamma_K^{p^\infty} := \bigcap_{n \in \mathbb{N}} \Gamma_K^{p^n}$ for every $i = 1, ..., d$. But under the standing assumption that K is deeply ramified, we have $\Gamma_K = \Gamma_K^{p^\infty}$ (see the proof of proposition 6.6.6). We leave to the reader the verification that condition (ii) of lemma 7.4.6 is indeed satisfied.

Lemma 7.4.9. *Let $\mathfrak{X} := (\mathfrak{X}_\alpha \mid \alpha \in I)$ be a cofiltered system as in definition 7.4.5.*

(i) *If $\mathfrak{Y} := (\mathfrak{Y}_\alpha \mid \alpha \in I) \to (\mathfrak{X}_\alpha \mid \alpha \in I)$ is a morphism of cofiltered systems such that the induced morphisms of adic spaces $d(\mathfrak{Y}_\alpha) \to d(\mathfrak{X}_\alpha)$ are unramified for every $\alpha \in I$ (cp. (7.2.27)), then $\mathfrak{Y}$ is deeply ramified if $\mathfrak{X}$ is.*

(ii) *Let $\mathfrak{Z} := (\mathfrak{Z}_\beta \mid \beta \in J)$ be another such cofiltered system, and suppose that $\mathfrak{X}$ and $\mathfrak{Z}$ are isomorphic as pro-objects of the category of formal schemes. Then $\mathfrak{X}$ is deeply ramified if and only if $\mathfrak{Z}$ is.*

(iii) *If $\mathfrak{X}$ and $\mathfrak{Z} := (\mathfrak{Z}_\beta \mid \beta \in J)$ are two deeply ramified cofiltered systems, then the fibred product $\mathfrak{X} \times \mathfrak{Z} := (\mathfrak{X}_\alpha \times_{\operatorname{Spf}(K^+)} \mathfrak{Z}_\beta \mid (\alpha, \beta) \in I \times J)$ is deeply ramified.*

Proof. (i): By [14, Prop.2.2] the natural morphism

$$\Omega^{an}_{\mathfrak{X}_\infty / K^+} \otimes_{K^+} K \to \Omega^{an}_{\mathfrak{Y}_\infty / K^+} \otimes_{K^+} K$$

is an epimorphism; the claim follows easily. (ii) and (iii) are easy and shall be left to the reader. $\qquad\square$

The counterpart of the above definitions for adic spaces is the following:

Definition 7.4.10. Let $(X_\alpha \mid \alpha \in I)$ be a system of quasi-separated adic spaces of finite type over $\operatorname{Spa}(K, K^+)$, indexed by a small cofiltered category I.

(i) Let $(X_\infty, \mathscr{O}_{X_\infty}, \mathscr{O}^+_{X_\infty}) := \lim_{\alpha \in I} (X_\alpha, \mathscr{O}_{X_\alpha}, \mathscr{O}^+_{X_\alpha})$, where the limit is taken in the category of locally ringed spaces. For every $\alpha \in I$, denote by $\pi_\alpha : (X_\infty, \mathscr{O}^+_{X_\infty}) \to (X_\alpha, \mathscr{O}^+_{X_\alpha})$ the natural morphism of locally ringed spaces. We define

$$\Omega^+_{X_\infty / K} := \operatorname{colim}_{\alpha \in I^o} \pi_\alpha^*(\Omega^+_{X_\alpha / K})$$

which is a sheaf of $\mathscr{O}^+_{X_\infty}$-modules. More generally, we let

$$\mathbb{L}^+_{X_\infty / K} := \operatorname{colim}_{\alpha \in I^o} \pi_\alpha^*(\mathbb{L}^+_{X_\alpha / K}).$$

(ii) We say that the cofiltered system $(X_\alpha \mid \alpha \in I)$ is *deeply ramified* if the natural morphism $\Omega^+_{X_\infty / K} \to \Omega^+_{X_\infty / K} \otimes_{K^+} K$ is an epimorphism.

7.4.11.   Let $(X_\alpha \mid \alpha \in I)$ be as in definition 7.4.10. For every $x := (x_\alpha \mid \alpha \in I) \in X_\infty$, we let

$$\kappa(x)^+ := \operatorname*{colim}_{\alpha \in I^\circ} \kappa(x_\alpha)^+$$

Since every $\kappa(x_\alpha)^+$ is a valuation ring, the same holds for $\kappa(x)^+$. Moreover, the image of a in $\kappa(x)^+$ is topologically nilpotent for the valuation topology of $\kappa(x)^+$, hence $\kappa(x)^+ = \mathscr{O}^+_{X_\infty,x} / \bigcap_{n \in \mathbb{N}} a^n \mathscr{O}^+_{X_\infty,x}$, the a-adic separation of $\mathscr{O}^+_{X_\infty,x}$.

7.4.12.   Given a cofiltered system $\mathfrak{X} := (\mathfrak{X}_\alpha \mid \alpha \in I)$ of formal schemes, one obtains a cofiltered system of adic spaces $\underline{X} := (X_\alpha := d(\mathfrak{X}_\alpha) \mid \alpha \in I)$, and using (7.2.28) and lemma 7.4.9(i) one sees easily that $\underline{X}$ is deeply ramified whenever $\mathfrak{X}$ is. Together with example 7.4.7, this yields plenty of examples of deeply ramified systems of adic spaces.

Proposition 7.4.13. *Let $\underline{X} := (X_\alpha \mid \alpha \in I)$ be a deeply ramified cofiltered system of adic spaces. Then, for every point $x \in X_\infty$, the valuation ring $\kappa(x)^+$ is deeply ramified.*

Proof. For every K^+-module M, let us denote by $T_n(M)$ the annihilator of a^n in M. Furthermore, let $T(M) := \bigcup_{n \in \mathbb{N}} T_n(M)$. Since the cofiltered system $\underline{X}$ is deeply ramified, we have:

$$(\Omega^+_{X_\infty/K})_x = T(\Omega^+_{X_\infty/K})_x + a \cdot (\Omega^+_{X_\infty/K})_x.$$

To lighten notation, let $\mathscr{O}^+_x := \mathscr{O}^+_{X_\infty,x}$. From lemma 7.1.27 one deduces easily that the natural map $\Omega_{\mathscr{O}^+_x/K^+} \to (\Omega^+_{X_\infty/K})_x$ is surjective with a-divisible kernel. Hence, by the snake lemma, the induced map $T_n(\Omega_{\mathscr{O}^+_x/K^+}) \to T_n(\Omega^+_{X_\infty/K})_x$ is surjective for every n, and *a fortiori* the map $T(\Omega_{\mathscr{O}^+_x/K^+}) \to T(\Omega^+_{X_\infty/K})_x$ is onto. It follows easily that

$$\Omega_{\mathscr{O}^+_x/K^+} = T(\Omega_{\mathscr{O}^+_x/K^+}) + a \cdot \Omega_{\mathscr{O}^+_x/K^+}$$

and consequently:

$$\Omega_{\kappa(x)^+/K^+} = T(\Omega_{\kappa(x)^+/K^+}) + a \cdot \Omega_{\kappa(x)^+/K^+}.$$

However, it follows from theorem 6.6.12 that $T(\Omega_{\kappa(x)^+/K^+}) = 0$, so finally

$$\Omega_{\kappa(x)^+/K^+} = a \cdot \Omega_{\kappa(x)^+/K^+}$$

and

(7.4.14) $$\mathbb{L}_{\kappa(x)^+/K^+} \overset{\mathbf{L}}{\otimes}_{K^+} K^+/aK^+ \simeq 0.$$

On the other hand, if $(E, |\cdot|_E)$ is any valued field extension of $\kappa(x)^+$, we have

(7.4.15) $$H_i(\mathbb{L}_{E^+/K^+} \otimes_{K^+} K^+/aK^+) = 0 \quad \text{for all } i > 0$$

by theorem 6.6.12. From (7.4.14), (7.4.15) and transitivity for the tower $K^+ \subset \kappa(x)^+ \subset E^+$, we derive

$$H_i(\mathbb{L}_{E^+/\kappa(x)^+} \otimes_{K^+} K^+/aK^+) = 0 \qquad \text{for all } i > 0.$$

Again by theorem 6.6.12 and remark 6.6.15 we conclude. $\square$

7.4.16. Let $\underline{X}$ be a cofiltered system as in definition 7.4.10. Let $\mathscr{A}$ be a sheaf of $\mathscr{O}_{X_\infty}^+$-algebras. We say that $\mathscr{A}$ is a *weakly étale* $\mathscr{O}_{X_\infty}^+$-*algebra* if, for every $x \in X_\infty$, the stalk $\mathscr{A}_x^a$ is a weakly étale $\mathscr{O}_{X_\infty,x}^{+a}$-algebra. Here and throughout the rest of this section we consider the almost ring theory associated to the standard setup attached to K (cp. (6.1.15)). Notice that, since K is deeply ramified, $\mathfrak{m}$ is the maximal ideal of K^+.

Theorem 7.4.17. *Suppose that K is deeply ramified, and let $\underline{X} := (X_\alpha \mid \alpha \in I)$ be a deeply ramified cofiltered system. Let also $\underline{f} : \underline{Y} := (Y_\alpha \mid \alpha \in I) \to \underline{X}$ be a morphism of cofiltered systems, such that the morphisms $Y_\alpha \to X_\alpha$ are finite étale for every $\alpha \in I$. Then $f_{\infty*}\mathscr{O}_{Y_\infty}^+$ is a weakly étale $\mathscr{O}_{X_\infty}^+$-algebra.*

Proof. To lighten notation, let us write $\mathscr{O}^+$ (resp. $\mathscr{A}^+$) instead of $\mathscr{O}_{X_\infty}^+$ (resp. $f_{\infty*}\mathscr{O}_{Y_\infty}^+$). For every $\alpha \in I$ consider the cofiltered system $\underline{Z}(\alpha) := \underline{X} \times_{X_\alpha} Y_\alpha$ indexed by I/α (the category of morphisms $\beta \to \alpha$), which is defined by setting $\underline{Z}(\alpha)_{\beta \to \alpha} := X_\beta \times_{X_\alpha} Y_\alpha$. We have obvious morphisms of cofiltered systems $\underline{f}_{/\alpha} : \underline{Z}(\alpha) \to \underline{X}$. The limit space Y_∞ is identified with $\varinjlim_{\alpha \in I} Z(\alpha)_\infty$ and the sheaf $\mathscr{A}^+$ is the colimit of the sheaves $f_{/\alpha\infty*}\mathscr{O}_{Z(\alpha)_\infty}^+$ (see [5, Exp.VI, 8.5.9], where the result is stated in terms of inverse limits of topoi; under the standing assumptions one can check that the inverse limit of the system of topoi associated to the topological spaces X_α, is naturally identified with the topos associated to the topological space X_∞). Hence it suffices to prove the assertion for the latter sheaves, and therefore in order to show the theorem we can and do assume that there exists $\alpha \in I$ such that, for every $\beta \to \alpha$, the induced commutative diagram

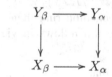

is cartesian. Let $x \in X_\infty$. Again by means of [5, Exp.VI, 8.5.9], one verifies that
$$\mathscr{A}_x^+ = \operatorname*{colim}_{\alpha \in I^\circ}(f_{\alpha*}\mathscr{O}_{Y_\alpha}^+)_{x_\alpha}$$

Claim 7.4.18. $\mathscr{A}_x^+$ is a flat $\mathscr{O}_x^+$-algebra.

Proof of the claim. On the one hand, by assumption $\mathscr{A}_x^+[1/a]$ is a flat $\mathscr{O}_x^+[1/a]$-algebra; on the other hand, $\mathscr{A}_x^+ \otimes_{\mathscr{O}_x^+} \kappa(x)^+$ is a flat $\kappa(x)^+$-module by the analogue of claim 7.4.3 for local rings of X_∞, so that $\mathscr{A}_x^+/a\mathscr{A}_x^+$ is a flat $\mathscr{O}_x^+/a\mathscr{O}_x^+$-module; thus the claim follows from lemma 5.2.1. ◇

Let $e \in C := \mathscr{A}_x^+ \otimes_{\mathscr{O}_x^+} \mathscr{A}_x^+[1/a]$ be the idempotent provided by lemma 3.1.4. In view of claim 7.4.18, we only have to show that $\varepsilon \cdot e \in C^+ := \mathscr{A}_x^+ \otimes_{\mathscr{O}_x^+} \mathscr{A}_x^+$ for every $\varepsilon \in \mathfrak{m}$. Let $\bar{e}$ be the image of e in $C \otimes_{\mathscr{O}_x} \kappa(x)$. Set $I := \bigcap_{n \geq 0} a^n \mathscr{O}_x^+$.

Claim 7.4.19. $\mathscr{A}_x^+/I\mathscr{A}_x^+$ is the integral closure of $\kappa(x)^+$ in $\mathscr{A}_x/I\mathscr{A}_x$.

Proof of the claim. First of all, since $I = aI$, we have $I\mathscr{A}_x = I\mathscr{A}_x^+$, and thus the natural map $\mathscr{A}_x^+/I\mathscr{A}_x^+ \to \mathscr{A}_x/I\mathscr{A}_x$ is injective. Next, we remark that $\mathscr{A}_x^+$ is the integral closure of $\mathscr{O}_x^+$ in $\mathscr{A}_x$; indeed, this follows from [12, §6.2.2, Lemma 3, Prop.4], after taking colimits. This already shows that $\mathscr{A}_x^+/I\mathscr{A}_x^+$ is a subalgebra of $\mathscr{A}_x/I\mathscr{A}_x$ integral over $\kappa(x)^+$. To conclude, suppose that $\overline{f} \in \mathscr{A}_x/I\mathscr{A}_x$ satisfies an integral equation:

$$\overline{f}^n + \overline{b}_1 \cdot \overline{f}^{n-1} + \dots + \overline{b}_n = 0$$

for certain $\overline{b}_1, \dots, \overline{b}_n \in \mathscr{A}_x^+/I\mathscr{A}_x^+$; pick arbitrary representatives $f \in \mathscr{A}_x$, $b_i \in \mathscr{A}_x^+$ of these elements. Then $f^n + b_1 \cdot f^{n-1} + \dots + b_n \in I\mathscr{A}_x$. Since $I\mathscr{A}_x = I\mathscr{A}_x^+$, we deduce that f is integral over $\mathscr{A}_x^+$, so $f \in \mathscr{A}_x^+$ and the claim follows. $\Diamond$

Claim 7.4.20. $(\mathscr{A}_x^+/I\mathscr{A}_x^+)^a$ is an étale $\kappa(x)^{+a}$-algebra.

Proof of the claim. In view of propositions 2.4.18(ii) and 2.4.19, it suffices to show that $(\mathscr{A}_x^+/I\mathscr{A}_x^+)^a$ is weakly étale over $\kappa(x)^{+a}$. Let $\mathfrak{p}$ be the height one prime ideal of $\kappa(x)^+$; then $\kappa(x)_{\mathfrak{p}}^+$ is a rank one valuation ring and the localization map induces isomorphisms

$$\kappa(x)^{+a} \xrightarrow{\sim} \kappa(x)_{\mathfrak{p}}^{+a} \qquad (\mathscr{A}_x^+/I\mathscr{A}_x^+)^a \xrightarrow{\sim} (\mathscr{A}_x^+/I\mathscr{A}_x^+)_{\mathfrak{p}}^a$$

(recall that the standing basic setup is the standard setup of K^+). Taking claim 7.4.19 into account, it suffices then to show that $(\mathscr{A}_x^+/I\mathscr{A}_x^+)_{\mathfrak{q}}^a$ is weakly étale over $\kappa(x)_{\mathfrak{p}}^{+a}$, for every prime ideal $\mathfrak{q} \subset \mathscr{A}_x^+/I\mathscr{A}_x^+$ of height one. By proposition 7.4.13, $\kappa(x)^+$ is deeply ramified, hence the same holds for $\kappa(x)_{\mathfrak{p}}^+$. Let $\kappa(x)^s$ be a separable closure of $\kappa(x)$, and $\kappa(x)^{s+}$ a valuation ring of $\kappa(x)^s$ dominating $\kappa(x)_{\mathfrak{p}}^+$; we can assume that $(\mathscr{A}_x^+/I\mathscr{A}_x^+)_{\mathfrak{q}} \subset \kappa(x)^{s+}$. From claim 7.4.19 we deduce that $(\mathscr{A}_x^+/I\mathscr{A}_x^+)_{\mathfrak{q}}$ is a valuation ring, hence $\kappa(x)^{s+}$ is a faithfully flat $(\mathscr{A}_x^+/I\mathscr{A}_x^+)_{\mathfrak{q}}$-algebra; then the claim follows in view of lemma 3.1.2(viii) and proposition 6.6.2. $\Diamond$

From claim 7.4.20 we deduce that $\varepsilon \cdot \overline{e} \in C^+ \otimes_{\mathscr{O}_x^+} \kappa(x)^+$ for every $\varepsilon \in \mathfrak{m}$. To conclude, it suffices therefore to remark:

Claim 7.4.21. The natural map: $C/C^+ \to (C/C^+) \otimes_{\mathscr{O}_x^+} \kappa(x)^+$ is an isomorphism.

Proof of the claim. Indeed, every element of C/C^+ is annihilated by some power of a, hence by $I = \mathrm{Ker}(\mathscr{O}_x^+ \to \kappa(x)^+)$ (see (7.4.11)). $\Box$

7.5. Semicontinuity of the discriminant.

Definition 7.5.1. Let $(V, \mathfrak{m})$ be a basic setup, A a V^a-algebra and P an almost projective A-module of constant rank $r \in \mathbb{N}$. Suppose moreover that P is endowed with a bilinear form $b : P \otimes_A P \to A$. We let $\beta : P \to P^*$ be the A-linear morphism defined by the rule: $\beta(x)(y) := b(x \otimes y)$ for every $x, y \in P_*$. The *discriminant* of the pair (P, b) is the ideal

$$\mathfrak{d}_A(P, b) := \mathrm{Ann}_A \, \mathrm{Coker}(\Lambda_A^r \beta : \Lambda_A^r P \to \Lambda_A^r P^*).$$

7.5.2.   As a special case, we can consider the pair $(B, t_{B/A})$ consisting of an A-algebra B which is almost projective of constant rank r over A, and its trace form $t_{B/A}$. In this situation, we let

$$\mathfrak{d}_{B/A} := \mathfrak{d}_A(B, t_{B/A})$$

and we call this ideal the *discriminant of the A-algebra B*.

Lemma 7.5.3. *Let B be an almost projective A-algebra of constant rank r as an A-module. Then B is étale over A if and only if $\mathfrak{d}_{B/A} = A$.*

Proof. By theorem 4.1.14, it is clear that $\mathfrak{d}_{B/A} = A$ when B is étale over A. Suppose therefore that $\mathfrak{d}_{B/A} = A$; it follows that $\Lambda_A^r \tau_{B/A}$ is an epimorphism. However, by proposition 4.3.27, $\Lambda_A^r B$ and $\Lambda_A^r B^*$ are invertible A-modules. It then follows by lemma 4.1.5(iv) that $\Lambda_A^r \tau_{B/A}$ is an isomorphism, hence $\tau_{B/A}$ is an isomorphism, by virtue of proposition 4.4.28. One concludes again by theorem 4.1.14.   □

Lemma 7.5.4. *Let $(V, \mathfrak{m})$ be the standard setup associated to a rank one valued field $(K, |\cdot|)$ (cp. (6.1.15), especially, $V := K^+$). Let $P' \subset P$ be two almost projective V^a-modules of constant rank equal to r. Let $b : P \otimes_{V^a} P \to V^a$ be a bilinear form, such that $b \otimes_{V^a} 1_{K^a}$ is a perfect pairing, and denote by b' the restriction of b to $P' \otimes_{V^a} P'$. Then we have:*

$$\mathfrak{d}_{V^a}(P', b') = F_0(P/P')^2 \cdot \mathfrak{d}_{V^a}(P, b).$$

Proof. Let $j : P' \to P$ be the imbedding, $\beta : P \to P^*$ (resp. $\beta' : P' \to P'^*$) the V^a-linear morphism associated to b (resp. to b'). The assumptions imply that $\Lambda_{V^a}^r j$, $\Lambda_{V^a}^r j^*$, $\Lambda_{V^a}^r \beta$ and $\Lambda_{V^a}^r \beta'$ are all injective, and clearly we have

$$\Lambda_{V^a}^r j^* \circ \Lambda_{V^a}^r \beta \circ \Lambda_{V^a}^r j = \Lambda_{V^a}^r \beta'.$$

There follow short exact sequences:

$$0 \to \text{Coker } \Lambda_{V^a}^r \beta \to \text{Coker } \Lambda_{V^a}^r (j^* \circ \beta) \to \text{Coker } \Lambda_{V^a}^r j^* \to 0$$
$$0 \to \text{Coker } \Lambda_{V^a}^r j \to \text{Coker } \Lambda_{V^a}^r \beta' \to \text{Coker } \Lambda_{V^a}^r (j^* \circ \beta) \to 0.$$

Using lemma 6.3.1 and remark 6.3.5(ii), we deduce:

$$F_0(\text{Coker } \Lambda_{V^a}^r \beta') = F_0(\text{Coker } \Lambda_{V^a}^r j) \cdot F_0(\text{Coker } \Lambda_{V^a}^r \beta) \cdot F_0(\text{Coker } \Lambda_{V^a}^r j^*).$$

Claim 7.5.5. Let Q be an almost finitely generated projective A-module of constant rank r, which is also uniformly almost finitely generated. Let $Q' \subset Q$ be any submodule. Then:

$$F_0(Q/Q') = \text{Ann}_A(\text{Coker}(\Lambda_A^r Q' \to \Lambda_A^r Q)).$$

Proof of the claim. By lemma 2.4.6 and proposition 2.3.25(iii) we can check the identity after a faithfully flat base change, hence we can suppose that $Q \simeq A^r$, in which case the identity holds by definition of the Fitting ideal F_0.   ◇

Notice that $\Lambda_{V^a}^r P$ and $\Lambda_{V^a}^r P'$ are invertible V^a-modules by virtue of proposition 4.3.27, consequently claim 7.5.5 applies and yields: $F_0(\text{Coker } \Lambda_{V^a}^r \beta') = \mathfrak{d}_{V^a}(P', b')$, $F_0(\text{Coker } \Lambda_{V^a}^r \beta) = \mathfrak{d}_{V^a}(P, b)$ and $F_0(P/P') = F_0(\text{Coker } \Lambda_{V^a}^r j) = F_0(\text{Coker } \Lambda_{V^a}^r j^*)$, by lemma 4.1.23(ii.c).   □

7.5.6.   Keep the notation of lemma 7.5.4 and let R be a flat V-algebra such that $R_K := K \otimes_V R$ is a finite étale K-algebra, say of dimension n as a K-vector space; we assume also that $B := R^a$ is almost finitely generated. It follows from proposition 6.3.6(i),(ii) and its proof, that B is almost projective and uniformly almost finitely generated as a V^a-module, with uniform bound n. Let $e \in R_K \otimes_K R_K$ be the idempotent characterized by proposition 3.1.4, and set $I := \{a \in V \mid a \cdot e \in R \otimes_V R\}$.

Lemma 7.5.7. *With the notation of* (7.5.6) *:*

(i) $I^a = \mathrm{Ann}_{V^a}(\mathrm{Coker}\, \tau_{B/V^a})$.

(ii) $(I^a)^n \subset \mathfrak{d}_{B/V^a} \subset I^a$.

Proof. In view of lemma 2.4.29(i), the morphism

$$\phi : B \otimes_{V^a} B \to \mathrm{Hom}_{V^a}(B^*, B) \qquad a \otimes b \mapsto (f \mapsto a \cdot f(b))$$

is an isomorphism. By remark 4.1.17 we derive that $\phi \otimes_{V^a} 1_{K^a}(e) = \tau_{R_K/K}^{-1}$ (notation of (4.1.12)), from which (i) follows easily. Moreover, by claim 7.5.5 we have $\mathfrak{d}_{B/V^a} = F_0(\mathrm{Coker}\, \tau_{B/V^a})$, so (ii) holds by [54, Ch.XIX, Prop.2.5] (which extends to uniformly almost finitely generated modules). $\square$

7.5.8.   Consider an étale K-algebra L, say of dimension n as a K-vector space; we denote by W_L the integral closure of K^+ in L. L is the product of finitely many separable field extensions of K, therefore W_L^a is an almost projective K^{+a}-module of constant rank n, by proposition 6.3.8. Hence, the discriminant of W_L^a over K^{+a} is defined, and to lighten notation, we will denote it by $\mathfrak{d}_{L/K}^+$. Furthermore, since L is étale over K, it is clear that $\mathfrak{d}_{L/K}^+$ is a fractional ideal of K^{+a} (cp. (6.1.16)). Let $|\cdot| : \mathrm{Div}(K^{+a}) \to \Gamma_K^\wedge$ be the isomorphism provided by lemma 6.1.19. We obtain an element $|\mathfrak{d}_{L/K}^+| \in \Gamma_K^\wedge$; then, after choosing (cp. example (6.1.21)(vi)) an order preserving isomorphism

$$(7.5.9) \qquad ((\Gamma_K \otimes_{\mathbb{Z}} \mathbb{Q})^\wedge, \leq) \xrightarrow{\sim} (\mathbb{R}_{>0}, \leq)$$

on the multiplicative group of positive real numbers, we can view $|\mathfrak{d}_{L/K}^+| \in (0, 1]$.

Lemma 7.5.10. *Let K, L be as in* (7.5.8) *and denote by $K^\wedge$ the completion of K for the valuation topology. Set $L^\wedge := K^\wedge \otimes_K L$. Then $K^{\wedge+} \otimes_{K^+} \mathfrak{d}_{L/K}^+ = \mathfrak{d}_{L^\wedge/K^\wedge}^+$.*

Proof. Since the base change $K^+ \to K^{\wedge+}$ is faithfully flat, everything is clear from the definitions, once we have established that $W_{L^\wedge} \simeq K^{\wedge+} \otimes_{K^+} W_L$. However, both rings can be identified with the a-adic completion $(W_L)^\wedge$ of W_L, so the assertion follows. $\square$

7.5.11.   After these generalities, we return to the standard setup $(K^+, \mathfrak{m})$ of this chapter, associated to a complete valued field $(K, |\cdot|)$ of rank one (cp. (6.1.15)). Let X be an adic space locally of finite type over $\mathrm{Spa}(K, K^+)$. X is a locally spectral space, and every point $x \in X$ admits a unique maximal generization $r(x) \in X$. The valuation ring $\kappa(r(x))^+$ has rank one, and admits a natural imbedding $K^+ \subset \kappa(r(x))^+$, continuous for the valuation topologies; especially, the image of the topologically nilpotent element a is topologically nilpotent in $\kappa(r(x))^+$. This imbedding induces therefore a natural isomorphism of completed value groups

$$(\Gamma_K \otimes_{\mathbb{Z}} \mathbb{Q})^\wedge \xrightarrow{\sim} (\Gamma_{\kappa(r(x))} \otimes_{\mathbb{Z}} \mathbb{Q})^\wedge.$$

In particular, our original choice of isomorphism (7.5.9) fixes univocally a similar isomorphism for every point $r(x)$. We denote by $\mathcal{M}(X)$ the set $r(X)$ endowed with the quotient topology induced by the mapping $X \to r(X) : x \mapsto r(x)$. This topology is coarser than the subspace topology induced by the imbedding into X. The mapping $x \mapsto r(x)$ is a retraction of X onto the subset $\mathcal{M}(X)$ of its maximal points. If X is a quasi-separated quasi-compact adic space, $\mathcal{M}(X)$ is a compact Hausdorff topological space ([49, 8.1.8]).

7.5.12. Let X be as in (7.5.11), and let $f : Y \to X$ be a finite étale morphism of adic spaces. For every point $x \in X$, the fibre $E(x) := (f_* \mathcal{O}_Y)_x \otimes_{\mathcal{O}_{X,x}} \kappa(x)$ is a finite étale $\kappa(x)$-algebra. If now $x \in \mathcal{M}(X)$, we can consider the discriminant $\mathfrak{d}^+_{E(x)/\kappa(x)}$ defined as in (7.5.8) (warning: notice that the definition makes sense when we choose the standard setup associated to the valuation ring $\kappa(x)^+$; since it may happen that the valuation of K is discrete and that of $\kappa(x)$ is not discrete, the setups relative to K and to $\kappa(x)$ may not agree in general). Upon passing to absolute values, we finally obtain a real valued function:

$$\mathfrak{d}^+_{Y/X} : \mathcal{M}(X) \to (0, 1] \qquad x \mapsto |\mathfrak{d}^+_{E(x)/\kappa(x)}|.$$

The study of the function $\mathfrak{d}^+_{Y/X}$ is reduced easily to the case where X (hence Y) are affinoid. In such case, one can state the main result in a more general form, as follows.

Definition 7.5.13. Let A be any (commutative unitary) ring.

(i) We denote by $\mathcal{N}(A)$ the set consisting of all multiplicative ultrametric seminorms $|\cdot| : A \to \mathbb{R}$. For every $x \in \mathcal{N}(A)$ and $f \in A$ we write usually $|f(x)|$ in place of $x(f)$. $\mathcal{N}(A)$ is endowed with the coarsest topology such that, for every $f \in A$, the real-valued map $|f| : \mathcal{N}(A) \to \mathbb{R}$ given by the rule: $x \mapsto |f(x)|$, is continuous.

(ii) For every $x \in \mathcal{N}(A)$, we let $\mathrm{Supp}(x) := \{f \in A \mid |f(x)| = 0\}$. Then $\mathrm{Supp}(x)$ is a prime ideal and we set $\kappa(x) := \mathrm{Frac}(A/\mathrm{Supp}(x))$. The seminorm x induces a valuation on the residue field $\kappa(x)$, and as usual we denote by $\kappa(x)^+$ its valuation ring.

(iii) Let $A \to B$ be a finite étale morphism. For every $x \in \mathcal{N}(A)$, we let $E(x) := B \otimes_A \kappa(x)$. Then $E(x)$ is an étale $\kappa(x)$-algebra, so we can define

$$\mathfrak{d}^+_{B/A}(x) := \mathfrak{d}^+_{E(x)/\kappa(x)}$$

(cp. the warning in (7.5.12)). By setting $x \mapsto |\mathfrak{d}^+_{B/A}(x)|$ we obtain a well-defined function

$$|\mathfrak{d}^+_{B/A}| : \mathcal{N}(A) \to (0, 1].$$

7.5.14. If $X = \mathrm{Spa}(A, A^+)$, with A a complete K-algebra of topologically finite type, then $\mathcal{M}(X)$ is naturally homeomorphic to the subspace $\mathcal{M}(A)$ of $\mathcal{N}(A)$ consisting of the continuous seminorms that extend the absolute value of K given by (7.5.9). It is shown in [10, §1.2] that $\mathcal{M}(A)$ is a compact Hausdorff space, for every Banach K-algebra A.

Proposition 7.5.15. *Let A be a ring, B a finite étale A-algebra. Then the function $|\partial^+_{B/A}|$ is lower semi-continuous (i.e. it is continuous for the topology of $(0, 1]$ whose open subsets are of the form $(c, 1]$, $c \in [0, 1]$).*

Proof. Let $f \in A$ be any element; notice that $\mathcal{N}(A[1/f])$ is naturally homeomorphic to the open subset $U(f) := \{x \in \mathcal{N}(A) \mid |f(x)| \neq 0\}$. Hence, after replacing A by some localization, we can assume that B is a free A-module, say of rank n. For every $b \in B$, let $\chi(b, T) := T^n + s_1(b) \cdot T^{n-1} + ... + s_n(b)$ be the characteristic polynomial of the A-linear endomorphism $B \to B$ given by the rule $b' \mapsto b' \cdot b$.

Claim 7.5.16. For every $x \in \mathcal{N}(A)$ and every $b \in B$, the following are equivalent:

    (a) $b \otimes 1 \in W_{E(x)}$. (Notation of (7.5.8).)
    (b) $|s_i(b)(x)| \leq 1$ for $i = 1, ..., n$.

Proof of the claim. Indeed, if (b) holds, then the image of $\chi(b, T)$ in $\kappa(x)[T]$ is a monic polynomial with coefficients in $\kappa(x)^+$ and $b \otimes 1$ is one of its roots (Cayley-Hamilton), hence $b \otimes 1$ is integral over $\kappa(x)^+$, which is (a). Conversely, if (a) holds, let L be a finite Galois extension of $\kappa(x)$ such that there exists an isomorphism $\zeta : E(x) \otimes_{\kappa(x)} L \to L^n$ of L-algebras; for each $j := 1, ..., n$, the projection $\pi_j : L^n \to L$ onto the j-th factor is an L-algebra homomorphism, and we set $\zeta_j := \pi_j \circ \zeta$; let also $e_j : L \to L^n$ be the injection of the j-th direct summand. Denote by $\mu_b : L^n \to L^n$ (resp. by $s_i(b)(x)$) the L-linear homomorphism induced by b (resp. the image of $s_i(b)$ in $\kappa(x)$); it follows that

$$\mu_b = \sum_{j=1}^{n} e_j \circ (\zeta_j(b \otimes 1) \cdot \pi_j).$$

Hence, the image of $s_i(b)(x)$ in L is the element $(-1)^i \cdot \sigma_i(\zeta_1(b \otimes 1), ..., \zeta_n(b \otimes 1))$, the i-th elementary symmetric polynomial in $\zeta_1(b \otimes 1), ..., \zeta_n(b \otimes 1)$, so it is integral over $\kappa(x)^+$, therefore it actually lies in $\kappa(x)^+$, which is (b). $\diamond$

Let $t_{B/A}$ be the trace form of the A-algebra B, and let $x \in \mathcal{N}(A)$. Then the trace morphism $t_x := t_{E(x)/\kappa(x)}$ equals $t_{B/A} \otimes_A 1_{\kappa(x)}$.

Claim 7.5.17. For every real number $\varepsilon > 0$ we can find a free $\kappa(x)^+$-submodule W_ε of $W_{E(x)}$, such that $W_\varepsilon \otimes_{\kappa(x)^+} \kappa(x) = E(x)$ and

$$(7.5.18) \qquad |\partial_{\kappa(x)^+}(W_\varepsilon, t_x)| + \varepsilon > |\partial^+_{B/A}(x)|.$$

Proof of the claim. From claim 6.3.10 we derive that for every positive real number $\delta < 1$ there exists a free finitely generated $\kappa(x)^+$-submodule $W_\delta \subset W_{E(x)}$ such that $|F_0(W^a_{E(x)}/W^a_\delta)| > \delta$; in view of lemma 7.5.4, the claim follows easily. $\diamond$

Let $w_1, ..., w_n$ be a basis of W_ε; up to replacing A by a localization $A[1/g]$ for some $g \in A$, we can write $w_i = b_i \otimes 1$, for some $b_i \in B$ $(i = 1, ..., n)$. Consequently:

$$(7.5.19) \qquad |\partial_{\kappa(x)^+}(W_\varepsilon, t_x)| = |\det(t_{B/A}(b_i \otimes b_j))(x)|.$$

By claim 7.5.16 we have $|s_j(b_i)(x)| \le 1$ for $i, j = 1, ..., n$. Let $1 > \delta > 0$ be a real number; for every $i, j \le n$, we define an open neighborhood U_{ij} of x in $\mathcal{N}(A)$ as follows. Suppose first that $|s_j(b_i)(x)| < 1$; since the real-valued function $y \mapsto |s_j(b_i)(y)|$ is continuous on $\mathcal{N}(A)$ (for the standard topology of $\mathbb{R}$), we can find U_{ij} such that $|s_j(b_i)(y)| \le 1$ for all $y \in U_{ij}$.

Suppose next that $|s_j(b_i)(x)| = 1$; then, up to replacing A by $A[1/s_j(b_i)]$, we can assume that $s_j(b_i)$ is invertible in A. We pick U_{ij} such that

$$(7.5.20) \qquad |s_j(b_i)(y)| \le 1 + \delta \quad \text{for every } y \in U_{ij}.$$

We set $U := \bigcap_{1 \le i, j \le n} U_{ij}$. Next, we define, for every $y \in U$, an element $c_y \in A$, as follows. Choose $\alpha, \beta \le n$ such that

$$|s_\beta(b_\alpha)(y)| = \max_{1 \le i, j \le n} |s_j(b_i)(y)|.$$

If $|s_\beta(b_\alpha)(y)| \le 1$, then set $c_y := 1$; if $|s_\beta(b_\alpha)(y)| > 1$, set $c_y := s_\beta(b_\alpha)^{-1}$. Then $|s_j(c_y \cdot b_i)(y)| \le 1$ for every $i, j = 1, ..., n$ and every $y \in U$. Let $\overline{W}_y$ be the $\kappa(y)^+$-submodule of $B \otimes_A \kappa(y)$ spanned by the images of $c_y \cdot b_1, ..., c_y \cdot b_n$. It follows that $\overline{W}_y \subset W_{E(y)}$ for every $y \in U$. We compute:

$$|\partial^+_{B/A}(y)| \ge |\partial_{\kappa(y)^+}(\overline{W}_y, t_y)| = |\det(t_{B/A}(c_y \cdot b_i \otimes c_y \cdot b_j))(y)|$$
$$= |c_y(y)|^{2n} \cdot |\det(t_{B/A}(b_i \otimes b_j))(y)|.$$

However, the real-valued function $y \mapsto |\det(t_{B/A}(b_i \otimes b_j))(y)|$ is continuous on $\mathcal{N}(A)$, therefore, combining (7.5.18) and (7.5.19), we see that, up to shrinking further the open neighborhood U, we can assume that

$$|\det(t_{B/A}(b_i \otimes b_j))(y)| + \varepsilon > |\partial^+_{B/A}(x)| \quad \text{for every } y \in U$$

so finally:

$$|\partial^+_{B/A}(y)| \ge (1 - \delta)^{2n} \cdot (|\partial^+_{B/A}(x)| - \varepsilon) \quad \text{for every } y \in U$$

which implies the claim. $\qquad \square$

Theorem 7.5.21. *Let* $Y \to X$ *be as in* (7.5.12). *Then the map* $\partial^+_{Y/X}$ *is lower semi-continuous.*

Proof. The claim is equivalent to saying that the map $\partial^+_{Y/X} \circ r : X \to \mathbb{R}$ is lower semi-continuous (notation of (7.5.11)). The latter assertion is easily reduced to the affinoid case, hence we can assume that $X = \mathrm{Spa}(A, A^+)$, where A is a complete K-algebra of topologically finite type, and therefore $Y = \mathrm{Spa}(B, B^+)$, for a finite étale A-algebra B. Then there is a natural homeomorphism $\omega : \mathcal{M}(X) \overset{\sim}{\to} \mathcal{M}(A)$, so the theorem follows from proposition 7.5.15 and :

Claim 7.5.22. $\partial^+_{Y/X} = \partial^+_{B/A} \circ \omega.$

Proof of the claim. Let $x \in \mathcal{M}(X)$; x corresponds to a rank one valuation of A, whose value group we identify with (a subgroup of) $\mathbb{R}_{>0}$ according to (7.5.11). The resulting multiplicative seminorm is $\omega(x)$. We derive easily a natural imbedding $\iota : \kappa(\omega(x)) \subset \kappa(x)$, compatible with the identifications of value groups. One knows moreover that ι induces an isomorphism on completions $\iota^{\wedge} : \kappa(\omega(x))^{\wedge} \xrightarrow{\sim} \kappa(x)^{\wedge}$, so the claim follows from lemma 7.5.10. $\square$

Lemma 7.5.23. *Let $(K_{\alpha}, | \cdot |_{\alpha} \mid \alpha \in I)$ be a system of valued field extensions of $(K, | \cdot |)$, indexed by a filtered small category I and such that K_{α}^{+} is a valuation ring of rank one for every $\alpha \in I$. Let moreover L be a finite étale K_{β}-algebra, for some $\beta \in I$. Set $L_{\alpha} := L \otimes_{K_{\beta}} K_{\alpha}$ for every morphism $\beta \to \alpha$ in I. Then :*

(i) *$(K_{\infty}, | \cdot |_{\infty}) := \operatorname*{colim}_{\alpha \in I}(K_{\alpha}, | \cdot |_{\alpha})$ is a valued field extension of $(K, | \cdot |)$, with valuation ring of rank one.*

(ii) *Set $L_{\infty} := L \otimes_{K_{\beta}} K_{\infty}$. Then, for every sequence of morphisms $\beta \to \gamma \to \alpha$ in I, we have $|\partial_{L_{\alpha}/K_{\alpha}}^{+}| \geq |\partial_{L_{\gamma}/K_{\gamma}}^{+}|$, and moreover:*

$$\lim_{(\beta \to \alpha) \in I} |\partial_{L_{\alpha}/K_{\alpha}}^{+}| = |\partial_{L_{\infty}/K_{\infty}}^{+}|.$$

Proof. (i) is obvious. The proof of the first assertion in (ii) is easy and shall be left to the reader. For the second assertion in (ii) we remark that, due to claim 6.3.10, for every $\varepsilon < 1$ there exists a free K_{∞}^{+}-submodule $W_{\varepsilon} \subset W_{L_{\infty}}$ of finite type, such that $|F_0(W_{L_{\infty}}^{a}/W_{\varepsilon}^{a})| > \varepsilon$. We can find $\alpha \in I$ such that $W_{\varepsilon} = W_0 \otimes_{K_{\alpha}^{+}} K_{\infty}^{+}$ for some free K_{α}^{+} submodule $W_0 \subset L_{\alpha}^{+}$. It then follows from lemma 7.5.4 that

$$|\partial_{L_{\beta}/K_{\beta}}^{+}| \geq |\partial_{K_{\alpha}^{+}}(W_0, t_{L_{\alpha}/K_{\alpha}})| = |\partial_{K_{\infty}^{+}}(W_{\varepsilon}, t_{L_{\infty}/K_{\infty}})| > \varepsilon^2 \cdot |\partial_{L_{\infty}/K_{\infty}}^{+}|$$

for every morphism $\alpha \to \beta$ in I. $\square$

7.5.24. Suppose now that $(K, | \cdot |)$ is deeply ramified and let $X := (X_{\alpha} \mid \alpha \in I)$ be a deeply ramified cofiltered system of adic spaces of finite type over $\mathrm{Spa}(K, K^{+})$. Suppose furthermore that it is given, for some $\beta \in I$, a finite étale morphism $Y_{\beta} \to X_{\beta}$ of adic $\mathrm{Spa}(K, K^{+})$-spaces of finite type. For every morphism $\alpha \to \beta$ of I we set $Y_{\alpha} := Y_{\beta} \times_{X_{\beta}} X_{\alpha}$ and denote by $f_{\alpha} : Y_{\alpha} \to X_{\alpha}$ the induced morphism of adic spaces.

Theorem 7.5.25. *In the situation of (7.5.24), for every positive real number $\varepsilon < 1$ there exists a morphism $\alpha \to \beta$ in I such that, for every morphism $\gamma \to \alpha$ we have*

$$|\partial_{Y_{\gamma}/X_{\gamma}}^{+}(x)| > \varepsilon \quad \text{for every } x \in \mathcal{M}(X_{\gamma}).$$

Proof. Notice that $\underline{Y} := (Y_{\alpha} \mid \alpha \to \beta)$ is a cofiltered system, hence we can define X_{∞} and Y_{∞} as in definition 7.4.10, and we obtain a morphism of locally ringed spaces $f_{\infty} : (Y_{\infty}, \mathscr{O}_{Y_{\infty}}^{+}) \to (X_{\infty}, \mathscr{O}_{X_{\infty}}^{+})$. For every $\alpha \in I$, let $\pi_{\alpha} : X_{\infty} \to X_{\alpha}$ be the natural morphism. Moreover, let $\mathcal{M}(X_{\infty}) := \lim_{\alpha \in I} \mathcal{M}(X_{\alpha})$; as a topological space, it is compact, by Tychonoff's theorem and the fact that $\mathcal{M}(X_{\alpha})$ is compact (see (7.5.11)); as a set, it admits an injective (usually non-continuous) map $\mathcal{M}(X_{\infty}) \to X_{\infty}$, so we can identify it as a subset of the latter.

Let $x \in \mathcal{M}(X_\infty)$; by proposition 7.4.13, the valuation ring $\kappa(x)^+$ is deeply ramified. Set $\kappa(x) := \kappa(x)^+ \otimes_{K^+} K$; it is clear that the morphism

$$\kappa(x) \rightarrow E(x) := (f_{\infty *}\mathcal{O}_{Y_\infty}^+)_x \otimes_{\mathcal{O}_{X_\infty,x}^+} \kappa(x)$$

is finite and étale. Let W_x be the integral closure of $\kappa(x)^+$ in $E(x)$. By proposition 6.6.2 and lemma 3.1.2(viii) we deduce easily that the induced morphism of K^{+a}-algebras $\kappa(x)^{+a} \rightarrow W_x^a$ is weakly étale, hence étale by proposition 6.3.8 and lemma 3.1.2(vii). Consequently $|\mathfrak{d}_{E(x)/\kappa(x)}^+| = 1$, in light of lemma 7.5.3. For every $\alpha \rightarrow \beta$, let $x_\alpha := \pi_\alpha(x)$. Then $\kappa(x)$ is the colimit of the filtered system $(\kappa(x_\alpha) \mid \alpha \rightarrow \beta)$, and similarly $E(x)$ is the colimit of the finite étale $\kappa(x_\alpha)$-algebras $(f_{\beta *}\mathcal{O}_{Y_\beta})_{x_\beta} \otimes_{\mathcal{O}_{X_\beta,x_\beta}^+} \kappa(x_\alpha)$ (for all $\alpha \rightarrow \beta$). In this situation, lemma 7.5.23 applies and shows that, for every $\varepsilon < 1$ there exists $\alpha(\varepsilon, x)$ such that

$$(7.5.26) \qquad |\mathfrak{d}_{Y_\alpha/X_\alpha}^+(x_\alpha)| > \varepsilon \quad \text{for every } \alpha \rightarrow \alpha(\varepsilon, x).$$

In light of theorem 7.5.21, for every $\alpha \rightarrow \beta$, the subset

$$X_\alpha(\varepsilon) := \{y \in \mathcal{M}(X_\alpha) \mid |\mathfrak{d}_{Y_\alpha/X_\alpha}^+(y)| \le \varepsilon\}$$

is closed in $\mathcal{M}(X_\alpha)$, hence compact. From (7.5.26) we see that $\lim_{\alpha \rightarrow \beta} X_\alpha(\varepsilon) = \varnothing$, therefore one of the $X_\alpha(\varepsilon)$ must be empty ([18, Ch.I, §9, n.6, Prop.8]), and the claim follows. $\qquad \square$

7.5.27. Let us choose an imbedding $\rho : (\Gamma_K, \cdot, \le) \hookrightarrow (\mathbb{R}, +, \le)$ as in example 6.1.21(vi). For every $r \in \mathbb{R}$ let $\mathbb{B}_K(r) := \{x \in K \mid \rho(|\varepsilon|) \le -r\}$. Let $f : Y \rightarrow X$ be a finite étale morphism of adic spaces of finite type over $\mathrm{Spa}(K, K^+)$. For every $x \in X$, set

$$A(x) := (f_*\mathcal{O}_Y^+)_x \otimes_{\mathcal{O}_{X,x}^+} (f_*\mathcal{O}_Y^+)_x.$$

We denote by $e_x \in A(x) \otimes_{K^+} K$ the unique idempotent characterized by the conditions of proposition 3.1.4. We define the *defect* of the morphism f as the real number

$$\mathrm{def}(f) := \inf\{r \in \mathbb{R}_{\ge 0} \mid \varepsilon \cdot e_x \in A(x) \text{ for every } x \in X \text{ and every } \varepsilon \in \mathbb{B}_K(r)\}.$$

Clearly $\mathrm{def}(f) \ge 0$ and $\mathrm{def}(f) = 0$ if and only if $(f_*\mathcal{O}_Y^+)_x^a$ is an étale $\mathcal{O}_{X,x}^{+a}$-algebra for every $x \in X$. Furthermore we remark that, by proposition 2.4.19, the map $\mathcal{O}_{X,x}^{+a} \rightarrow (f_*\mathcal{O}_Y^+)_x^a$ is weakly étale if and only if it is étale.

Corollary 7.5.28. *In the situation of* (7.5.24), *for every real number* $r > 0$ *there exists* $\alpha \in I$ *such that, for every morphism* $\gamma \rightarrow \alpha$, *we have:* $\mathrm{def}(f_\gamma) < r$.

Proof. Let $r > 0$; according to theorem 7.5.25, claim 7.4.21 and lemma 7.5.7(ii), there exists $\alpha \in I$ such that $\varepsilon \cdot e_x \in A(x)$ for every $\gamma \rightarrow \alpha$, every $x \in \mathcal{M}(X_\gamma)$ and every $\varepsilon \in K^+$ with $\rho(|\varepsilon|) \le -r$. If now $y \in X_\gamma$ is any point, there is a unique generization x of y in $\mathcal{M}(X_\gamma)$. Let $\phi : A(y) \rightarrow A(x)$ be the induced specialization map, and set $\phi_K := \phi \otimes_{K^+} 1_K$. One verifies easily that $\phi_K^{-1}(\mathfrak{m} \cdot A(x)) \subset A(y)$. Since $\phi_K(e_y) = e_x$, the claim follows easily. $\qquad \square$

7.5.29. To conclude, we want to explain briefly what kind of Galois cohomology calculations are enabled by the results of this section. Let $f : Y \to X$ be a finite étale *Galois* morphism of connected $\mathrm{Spa}(K, K^+)$-adic spaces of finite type, and let G denote the group of X-automorphisms of Y. Denote by $f_* \mathscr{O}_Y^+[G]$-**Mod** the category of $f_* \mathscr{O}_Y^+$-modules on X, endowed with a semilinear action of G. Let

$$\underline{\Gamma}^G : f_* \mathscr{O}_Y^+[G]\text{-}\mathbf{Mod} \to \mathscr{O}_X^+\text{-}\mathbf{Mod}$$

be the functor that associates to an $f_* \mathscr{O}_Y^+[G]$-module the sheaf of its G-invariant local sections. A standard argument shows that, for every $f_* \mathscr{O}_Y^+[G]$-module $\mathscr{F}$, the cone of the natural morphism in $\mathrm{D}(f_* \mathscr{O}_Y^+\text{-}\mathbf{Mod})$

$$(7.5.30) \qquad\qquad f_* \mathscr{O}_Y^+ \otimes_{\mathscr{O}_X^+} R\underline{\Gamma}^G \mathscr{F} \to \mathscr{F}$$

is annihilated by all $\varepsilon \in \mathfrak{m}$ such that $\rho(\varepsilon) < -\mathrm{def}(f)$. However, for applications one is rather more interested in understanding the Galois cohomology groups $H^i := H^i(G, H^0(X, \mathscr{F}))$. One can try to study H^i via (7.5.30); indeed, a bridge between these two objects is provided by the higher derived functors of the related functor

$$\Gamma^G : f_* \mathscr{O}_Y^+[G]\text{-}\mathbf{Mod} \to \Gamma(X, \mathscr{O}^+)\text{-}\mathbf{Mod} \quad : \quad \mathscr{F} \mapsto \Gamma(X, \underline{\Gamma}^G \mathscr{F}) = \Gamma(X, \mathscr{F})^G.$$

We have two spectral sequences converging to $R\Gamma^G \mathscr{F}$, namely

$$E_2^{pq} : H^p(X, R^q \underline{\Gamma}^G \mathscr{F}) \Rightarrow R^{p+q}\Gamma^G \mathscr{F}$$
$$F_2^{pq} : H^p(G, H^q(X, \mathscr{F})) \Rightarrow R^{p+q}\Gamma^G \mathscr{F}.$$

Using (7.5.30) one deduces that E_2^{pq} degenerates up to some torsion, which can be estimated precisely in terms of the defect of the morphism f. However, the spectral sequence F_2^{pq} contains the terms $H^q(X, \mathscr{F})$, about which not much is currently known. In this direction, the only results that we could find in the literature concern the calculation of $H^i(Y, \mathscr{O}_Y^+)$, for an affinoid space, under some very restrictive assumptions : in [7] these groups are shown to be almost zero modules for $i > 0$, in case Y admits a smooth formal model over K^+; in [76] the case of generalized polydiscs is taken up, and the same kind of almost vanishing is proven.

8. APPENDIX

In this appendix we have gathered a few miscellaneous results that were found in the course of our investigation, and which may be useful for other applications.

8.1. Simplicial almost algebras. We need some preliminaries on simplicial objects : first of all, a *simplicial almost algebra* is just an object in the category $s.(V^a\text{-}\mathbf{Alg})$. Then for a given simplicial almost algebra A we have the category $A\text{-}\mathbf{Mod}$ of A-modules : it consists of all simplicial almost V-modules M such that $M[n]$ is an $A[n]$-module and such that the face and degeneracy morphisms $d_i : M[n] \to M[n-1]$ and $s_i : M[n] \to M[n+1]$ $(i = 0, 1, ..., n)$ are $A[n]$-linear.

8.1.1. We will need also the derived category of A-modules, defined below. A bit more generally, let $\mathscr{C}$ be any abelian category. For an object X of $s.\mathscr{C}$ let $N(X)$ be the normalized chain complex (defined as in [50, I.1.3]). By the theorem of Dold-Kan ([77, Th. 8.4.1]) $X \mapsto N(X)$ induces an equivalence $N : s.\mathscr{C} \to \mathbf{C}_\bullet(\mathscr{C})$ with the category $\mathbf{C}_\bullet(\mathscr{C})$ of complexes of objects of $\mathscr{C}$ that vanish in positive degrees. Now we say that a morphism $X \to Y$ in $s.\mathscr{C}$ is a *quasi-isomorphism* if the induced morphism $N(X) \to N(Y)$ is a quasi-isomorphism of chain complexes.

8.1.2. In the following we fix a simplicial almost algebra A.

Definition 8.1.3. We say that A is *exact* if the almost algebras $A[n]$ are exact for all $n \in \mathbb{N}$. A morphism $\phi : M \to N$ of A-modules (or A-algebras) is a *quasi-isomorphism* if the morphism ϕ of underlying simplicial almost V-modules is a quasi-isomorphism. We define the category $\mathbf{D}(A\text{-}\mathbf{Mod})$ (resp. the category $\mathbf{D}(A\text{-}\mathbf{Alg})$) as the localization of the category $A\text{-}\mathbf{Mod}$ (resp. $A\text{-}\mathbf{Alg}$) with respect to the class of quasi-isomorphisms.

8.1.4. As usual, the morphisms in $\mathbf{D}(A\text{-}\mathbf{Mod})$ can be computed via a left and right calculus of fractions on the category $\mathrm{Hot}(A)$ of simplicial modules up to homotopy. Moreover, if A_1 and A_2 are two simplicial almost algebras, then the "extension of scalars" functors define equivalences of categories

$$\mathbf{D}(A_1 \times A_2\text{-}\mathbf{Mod}) \xrightarrow{\sim} \mathbf{D}(A_1\text{-}\mathbf{Mod}) \times \mathbf{D}(A_2\text{-}\mathbf{Mod})$$
$$\mathbf{D}(A_1 \times A_2\text{-}\mathbf{Alg}) \xrightarrow{\sim} \mathbf{D}(A_1\text{-}\mathbf{Alg}) \times \mathbf{D}(A_2\text{-}\mathbf{Alg}).$$

Proposition 8.1.5. *Let A be a simplicial V^a-algebra.*

(i) *The functor on A-algebras given by $B \mapsto (s.V^a \times B)_{!!}$ preserves quasi-isomorphisms and therefore induces a functor*

$$\mathbf{D}(A\text{-}\mathbf{Alg}) \to \mathbf{D}((s.V^a \times A)_{!!}\text{-}\mathbf{Alg}).$$

(ii) *The localization functor $R \mapsto R^a$ followed by "extension of scalars" via $s.V^a \times A \to A$ induces a functor*

$$\mathbf{D}((s.V^a \times A)_{!!}\text{-}\mathbf{Alg}) \to \mathbf{D}(A\text{-}\mathbf{Alg})$$

and the composition of this and the above functor is naturally isomorphic to the identity functor on $\mathbf{D}(A\text{-}\mathbf{Alg})$.

Proof. (i): Let $B \to C$ be a quasi-isomorphism of A-algebras. Clearly the induced morphism $s.V^a \times B \to s.V^a \times C$ is still a quasi-isomorphism of V-algebras. But by remark 2.2.28, $s.V^a \times B$ and $s.V^a \times C$ are exact simplicial almost V-algebras; moreover, it follows from corollary 2.2.24 that $(s.V^a \times B)_! \to (s.V^a \times C)_!$ is a quasi-isomorphism of V-modules. Then the claim follows easily from the exactness of the sequence (2.2.26). Now (ii) is clear. □

Remark 8.1.6. In case $\mathfrak{m}$ is flat, then all A-algebras are exact, and the same argument shows that the functor $B \mapsto B_{!!}$ induces a functor $\mathsf{D}(A\text{-}\mathbf{Alg}) \to \mathsf{D}(A_{!!}\text{-}\mathbf{Alg})$. In this case, composition with localization is naturally isomorphic to the identity functor on $\mathsf{D}(A\text{-}\mathbf{Alg})$.

Proposition 8.1.7. *Let* $f : R \to S$ *be a map of V-algebras. We have:*

(i) *If* $f^a : R^a \to S^a$ *is an isomorphism, then* $\mathbb{L}_{S/R}^a \simeq 0$ *in* $\mathsf{D}(s.S^a\text{-}\mathbf{Mod})$.

(ii) *The natural map* $\mathbb{L}_{S^a/R^a}^a \to \mathbb{L}_{S/R}^a$ *is an isomorphism in* $\mathsf{D}(s.S^a\text{-}\mathbf{Mod})$.

Proof. (ii) is an easy consequence of (i), of transitivity and of localization. To show (i) we prove by induction on q that

$$\mathbf{VAN}(q; S/R) \qquad\qquad H_q(\mathbb{L}_{S/R}^a) = 0.$$

For $q = 0$ the claim follows immediately from [50, II.1.2.4.2]. Therefore suppose that $q > 0$ and that $\mathbf{VAN}(j; D/C)$ is known for all almost isomorphisms of V-algebras $C \to D$ and all $j < q$. Let $\overline{R} := f(R)$. Then by transitivity ([50, II.2.1.2]) we have a distinguished triangle in $\mathsf{D}(s.S^a\text{-}\mathbf{Mod})$

$$(S \otimes_{\overline{R}} \mathbb{L}_{\overline{R}/R})^a \xrightarrow{u} \mathbb{L}_{S/R}^a \xrightarrow{v} \mathbb{L}_{S/\overline{R}}^a \to \sigma(S \otimes_{\overline{R}} \mathbb{L}_{\overline{R}/R})^a.$$

We deduce that $\mathbf{VAN}(q; \overline{R}/R)$ and $\mathbf{VAN}(q; S/\overline{R})$ imply $\mathbf{VAN}(q; S/R)$, thus we can assume that f is either injective or surjective. Let $S_\bullet \to S$ be the simplicial V-algebra augmented over S defined by $S_\bullet := P_V(S)$. It is a simplicial resolution of S by free V-algebras, in particular the augmentation is a quasi-isomorphism of simplicial V-algebras. Set $R_\bullet := S_\bullet \times_S R$. This is a simplicial V-algebra augmented over R via a quasi-isomorphism. Moreover, the induced morphisms $R[n]^a \to S[n]^a$ are isomorphisms. By [50, II.1.2.6.2] there is a quasi-isomorphism $\mathbb{L}_{S/R} \simeq \mathbb{L}_{S_\bullet/R_\bullet}^\Delta$. On the other hand we have a spectral sequence

$$E_{ij}^1 := H_j(\mathbb{L}_{S[i]/R[i]}) \Rightarrow H_{i+j}(\mathbb{L}_{S_\bullet/R_\bullet}^\Delta).$$

It follows easily that $\mathbf{VAN}(j; S[i]/R[i])$ for all $i \geq 0, j \leq q$ implies $\mathbf{VAN}(q; S/R)$. Therefore we are reduced to the case where S is a free V-algebra and f is either injective or surjective. We examine separately these two cases. If $f : R \to V[T]$ is surjective, then we can find a right inverse $s : V[T] \to R$ for f. By applying transitivity to the sequence $V[T] \to R \to V[T]$ we get a distinguished triangle

$$(V[T] \otimes_R \mathbb{L}_{R/V[T]})^a \xrightarrow{u} \mathbb{L}_{V[T]/V[T]}^a \xrightarrow{v} \mathbb{L}_{V[T]/R}^a \to \sigma(V[T] \otimes_R \mathbb{L}_{R/V[T]})^a.$$

Since $\mathbb{L}_{V[T]/V[T]}^a \simeq 0$, there follows an isomorphism:

$$H_q(\mathbb{L}_{V[T]/R})^a \simeq H_{q-1}(V[T] \otimes_R \mathbb{L}_{R/V[T]})^a.$$

Furthermore, since f^a is an isomorphism, s^a is an isomorphism as well, hence by induction (and by a spectral sequence of the type [50, I.3.3.3.2]) $H_{q-1}(V[T] \otimes_R \mathbb{L}_{R/V[T]})^a \simeq 0$. The claim follows in this case.

Finally suppose that $f : R \to V[T]$ is injective. Write $V[T] = \mathrm{Sym}_V^\bullet(F)$, for a free V-module F and set $\widetilde{F} = \widetilde{\mathfrak{m}} \otimes_V F$; since f^a is an isomorphism, $\mathrm{Im}(\mathrm{Sym}_V^\bullet(\widetilde{F}) \to \mathrm{Sym}_V^\bullet(F)) \subset R$. We apply transitivity to the sequence

$$\mathrm{Sym}_V^\bullet(\widetilde{F}) \to R \to V[T].$$

By arguing as above we are reduced to showing that $\mathbb{L}^a_{V[T]/\mathrm{Sym}_V^\bullet(\widetilde{F})} \simeq 0$. We know already that $H_0(\mathbb{L}^a_{V[T]/\mathrm{Sym}_V^\bullet(\widetilde{F})}) \simeq 0$, hence it remains only to show:

Claim 8.1.8. $H_j(\mathbb{L}^a_{V[T]/\mathrm{Sym}_V^\bullet(\widetilde{F})}) \simeq 0$ for all $j > 0$.

Proof of the claim. We apply transitivity to the sequence $V \to \mathrm{Sym}_V^\bullet(\widetilde{F}) \to V[T]$. As F and $\widetilde{F}$ are flat V-modules, [50, II.1.2.4.4] yields

$$H_j(\mathbb{L}_{V[T]/V}) \simeq H_j(\mathbb{L}_{\mathrm{Sym}_V^\bullet(\widetilde{F})/V}) \simeq 0 \qquad \text{for all } j > 0$$

and $H_0(\mathbb{L}_{\mathrm{Sym}_V^\bullet(\widetilde{F})/V})$ is a flat $\mathrm{Sym}_V^\bullet(\widetilde{F})$-module. In particular

$$H_j(V[T] \otimes_{\mathrm{Sym}_V^\bullet(\widetilde{F})} \mathbb{L}_{\mathrm{Sym}_V^\bullet(\widetilde{F})/V}) \simeq 0 \qquad \text{for all } j > 0.$$

Consequently $H_{j+1}(\mathbb{L}_{V[T]/\mathrm{Sym}_V^\bullet(\widetilde{F})}) \simeq 0$ for all $j > 0$ and

$$H_1(\mathbb{L}_{V[T]/\mathrm{Sym}_V^\bullet(\widetilde{F})}) \simeq \mathrm{Ker}\,(V[T] \otimes_{\mathrm{Sym}_V^\bullet(\widetilde{F})} \Omega_{\mathrm{Sym}_V^\bullet(\widetilde{F})/V} \to \Omega_{V[T]/V}).$$

The latter module is easily seen to be almost zero. □

Theorem 8.1.9. *Let $\phi : R \to S$ be a map of simplicial V-algebras inducing an isomorphism $R^a \xrightarrow{\sim} S^a$ in $\mathsf{D}(R^a\text{-}\mathbf{Mod})$. Then $(\mathbb{L}^\Delta_{S/R})^a \simeq 0$ in $\mathsf{D}(S^a\text{-}\mathbf{Mod})$.*

Proof. Apply the base change theorem [50, II.2.2.1] to the (flat) projections of $s.V \times R$ onto R and respectively $s.V$ to deduce that the natural map $\mathbb{L}^\Delta_{s.V \times S/s.V \times R} \to \mathbb{L}^\Delta_{S/R} \oplus \mathbb{L}^\Delta_{s.V/s.V} \to \mathbb{L}^\Delta_{S/R}$ is a quasi-isomorphism in $\mathsf{D}(s.V \times S\text{-}\mathbf{Mod})$. By proposition 8.1.5 the induced morphism $(s.V \times R)_{!!}^a \to (s.V \times S)_{!!}^a$ is still a quasi-isomorphism. There are spectral sequences

$$E_{ij}^1 := H_j(\mathbb{L}_{(V \times R[i])/(V \times R[i])_{!!}^a}) \Rightarrow H_{i+j}(\mathbb{L}^\Delta_{(s.V \times R)/(s.V \times R)_{!!}^a})$$
$$F_{ij}^1 := H_j(\mathbb{L}_{(V \times S[i])/(V \times S[i])_{!!}^a}) \Rightarrow H_{i+j}(\mathbb{L}^\Delta_{(s.V \times S)/(s.V \times S)_{!!}^a}).$$

On the other hand, by proposition 8.1.7(i) we have $\mathbb{L}^a_{(V \times R[i])/(V \times R[i])_{!!}^a} \simeq 0 \simeq \mathbb{L}^a_{(V \times S[i])/(V \times S[i])_{!!}^a}$ for all $i \in \mathbb{N}$. Then the theorem follows directly from [50, II.1.2.6.2(b)] and transitivity. □

8.1.10. In view of proposition 8.1.7(i) we have $\mathbb{L}^a_{V \times A_{!!}/(V^a \times A)_{!!}} \simeq 0$ in $\mathsf{D}(V^a \times A\text{-}\mathbf{Mod})$. From this, transitivity and localization ([50, II.2.3.1.1]) we derive that $\mathbb{L}^a_{B/A} \to \mathbb{L}^a_{B_{!!}/A_{!!}}$ is a quasi-isomorphism for all A-algebras B. If A and B are exact (*e.g.* if $\mathfrak{m}$ is flat), we conclude from proposition 2.5.43 that the natural map $\mathbb{L}_{B/A} \to \mathbb{L}_{B_{!!}/A_{!!}}$ is a quasi-isomorphism.

8.1.11. Finally we want to discuss left derived functors of (the almost version of) some notable non-additive functors that play a role in deformation theory. Let R be a simplicial V-algebra. Then we have an obvious functor $G : \mathrm{D}(R\text{-}\mathbf{Mod}) \to \mathrm{D}(R^a\text{-}\mathbf{Mod})$ obtained by applying dimension-wise the localization functor. Let Σ be the multiplicative set of morphisms of $\mathrm{D}(R\text{-}\mathbf{Mod})$ that induce almost isomorphisms on the cohomology modules. An argument as in section 2.4 shows that G induces an equivalence of categories $\Sigma^{-1}\mathrm{D}(R\text{-}\mathbf{Mod}) \to \mathrm{D}(R^a\text{-}\mathbf{Mod})$.

8.1.12. Now let R be a V-algebra and $\mathscr{F}_p$ one of the functors $\otimes^p$, Λ^p, Sym^p, Γ^p defined in [50, I.4.2.2.6].

Lemma 8.1.13. *Let* $\phi : M \to N$ *be an almost isomorphism of R-modules. Then* $\mathscr{F}_p(\phi) : \mathscr{F}_p(M) \to \mathscr{F}_p(N)$ *is an almost isomorphism.*

Proof. Let $\psi : \widetilde{\mathfrak{m}} \otimes_V N \to M$ be the map corresponding to $(\phi^a)^{-1}$ under the bijection (2.2.4). By inspection, the compositions $\phi \circ \psi : \widetilde{\mathfrak{m}} \otimes_V N \to N$ and $\psi \circ (1_{\widetilde{\mathfrak{m}}} \otimes \phi) : \widetilde{\mathfrak{m}} \otimes_V M \to M$ are induced by scalar multiplication. Pick any $s \in \mathfrak{m}$ and lift it to an element $\widetilde{s} \in \widetilde{\mathfrak{m}}$; define $\psi_s : N \to M$ by $n \mapsto \psi(\widetilde{s} \otimes n)$ for all $n \in N$. Then $\phi \circ \psi_s = s \cdot 1_N$ and $\psi_s \circ \phi = s \cdot 1_M$. This easily implies that s^p annihilates $\mathrm{Ker}\, \mathscr{F}_p(\phi)$ and $\mathrm{Coker}\, \mathscr{F}_p(\phi)$. In light of proposition 2.1.7(ii), the claim follows. $\qquad\qquad\square$

8.1.14. Let B be an almost V-algebra. We define a functor $\mathscr{F}_p^a$ on $B\text{-}\mathbf{Mod}$ by letting $M \mapsto (\mathscr{F}_p(M_!))^a$, where $M_!$ is viewed as a $B_{!!}$-module or a B_*-module (to show that these choices define the same functor it suffices to observe that $B_* \otimes_{B_{!!}} N \simeq N$ for all B_*-modules N such that $N = \mathfrak{m} \cdot N$). For all $p > 0$ we have diagrams :

$$
\begin{array}{ccc}
R\text{-}\mathbf{Mod} & \xrightarrow{\ \mathscr{F}_p\ } & R\text{-}\mathbf{Mod} \\
\updownarrow & & \updownarrow \\
R^a\text{-}\mathbf{Mod} & \xrightarrow{\ \mathscr{F}_p^a\ } & R^a\text{-}\mathbf{Mod}
\end{array}
$$

(8.1.15)

where the downward arrows are localization and the upward arrows are the functors $M \mapsto M_!$. Lemma 8.1.13 implies that the downward arrows in the diagram commute (up to a natural isomorphism) with the horizontal ones. It will follow from the following proposition 8.1.17 that the diagram commutes also going upward.

8.1.16. For any V-module N we have an exact sequence $\Gamma^2 N \to \otimes^2 N \to \Lambda^2 N \to 0$. As observed in the proof of proposition 2.1.7, the symmetric group S_2 acts trivially on $\widetilde{\mathfrak{m}}^{\otimes 2}$ and $\Gamma^2 \widetilde{\mathfrak{m}} \simeq \widetilde{\mathfrak{m}}^{\otimes 2}$, so $\Lambda^2 \widetilde{\mathfrak{m}} = 0$. Also we have natural isomorphisms $\Gamma^p \widetilde{\mathfrak{m}} \simeq \widetilde{\mathfrak{m}}$ for all $p > 0$.

Proposition 8.1.17. *Let R be a commutative ring, L and N two R-modules. Then:*

(i) *For every $p \geq 0$ we have a natural map*

(8.1.18) $\Gamma_R^p(L) \otimes_R \mathscr{F}_p(N) \to \mathscr{F}_p(L \otimes_R N)$.

(ii) *If L is a flat R-module with $\Lambda_R^2 L = 0$, then (8.1.18) is an isomorphism.*

Proof. Fix an element $x \in \mathscr{F}_p(N)$. For each R-algebra R' and each element $l \in R' \otimes_R L$ we get a map $\phi_l : R' \otimes_R N \to R' \otimes_R L \otimes_R N$ by $y \mapsto l \otimes y$, hence a map $\mathscr{F}_p(\phi_l) : R' \otimes_R \mathscr{F}_p(N) \simeq \mathscr{F}_p(R' \otimes_R N) \to \mathscr{F}_p(R' \otimes_R L \otimes_R N) \simeq R' \otimes_R \mathscr{F}_p(L \otimes_R N)$. For varying l we obtain a map of sets

$$\psi_{R',x} : R' \otimes_R L \to R' \otimes_R \mathscr{F}_p(L \otimes_R N) \quad : \quad l \mapsto \mathscr{F}_p(\phi_l)(1 \otimes x).$$

According to the terminology of [67], the system of maps $\psi_{R',x}$ for R' ranging over all R-algebras forms a homogeneous polynomial law of degree p from L to $\mathscr{F}_p(L \otimes_R N)$, so it factors through the universal homogeneous degree p polynomial law $\gamma_p : L \to \Gamma_R^p(L)$. The resulting R-linear map $\overline{\psi}_x : \Gamma_R^p(L) \to \mathscr{F}_p(L \otimes_R N)$ depends R-linearly on x, hence we derive an R-linear map $\psi : \Gamma_R^p(L) \otimes_R \mathscr{F}_p(N) \to \mathscr{F}_p(L \otimes_R N)$ as claimed in (i). Next notice that under the assumptions of (ii), S_2 acts trivially on $L^{\otimes 2}$ so S_p acts trivially on $L^{\otimes p}$ and we get an isomorphism $\beta : \Gamma_R^p(L) \xrightarrow{\sim} L^{\otimes p}$. We deduce a natural map $(L^{\otimes p}) \otimes_R \mathscr{F}_p(N) \to \mathscr{F}_p(L \otimes_R N)$. Now, in order to prove the proposition for the case $\mathscr{F}_p = \otimes^p$, it suffices to show that this last map is just the natural isomorphism that "reorders the factors". Indeed, let $x_1, ..., x_n \in L$ and $q := (q_1, ..., q_n) \in \mathbb{N}^n$ such that $|q| := \sum_i q_i := p$; then β sends the generator $x_1^{[q_1]} \cdot ... \cdot x_n^{[q_n]}$ to $\binom{p}{q_1,...,q_n} \cdot x_1^{\otimes q_1} \otimes ... \otimes x_n^{\otimes q_n}$. On the other hand, pick any $y \in N^{\otimes p}$ and let $R[T] := R[T_1, ..., T_n]$ be the polynomial R-algebra in n variables; write

$$(T_1 \otimes x_1 + ... + T_n \otimes x_n)^{\otimes p} \otimes y = \psi_{R[T],y}(T_1 \otimes x_1 + ... + T_n \otimes x_n) = \sum_{r \in \mathbb{N}^n} T^r \otimes w_r$$

with $w_r \in (L \otimes_R N)^{\otimes p}$. Then $\psi((x_1^{[q_1]} \cdot ... \cdot x_n^{[q_n]}) \otimes y) = w_q$ (see [67, pp.266-267]) and the claim follows easily. Next notice that $\Gamma_R^p(L)$ is flat, so that tensoring with $\Gamma_R^p(L)$ commutes with taking coinvariants (resp. invariants) under the action of the symmetric group; this implies the assertion for $\mathscr{F}_p := \mathrm{Sym}^p$ (resp. $\mathscr{F}_p := TS^p$). To deal with $\mathscr{F}_p := \Lambda^p$ recall that for any V-module M and $p > 0$ we have the antisymmetrization operator $a_M := \sum_{\sigma \in S_p} \mathrm{sgn}(\sigma) \cdot \sigma : M^{\otimes p} \to M^{\otimes p}$ and a surjection $\Lambda^p(M) \to \mathrm{Im}(a_M)$ which is an isomorphism for M free, hence for M flat. The result for $\mathscr{F}_p = \otimes^p$ (and again the flatness of $\Gamma_R^p(L)$) then gives $\Gamma_R^p(L) \otimes_R \mathrm{Im}(a_N) \simeq \mathrm{Im}(a_{L \otimes_R N})$, hence the assertion for $\mathscr{F}_p = \Lambda^p$ and N flat. For general N let $F_1 \xrightarrow{\partial} F_0 \xrightarrow{\varepsilon} N \to 0$ be a presentation with F_i free. Define $j_0, j_1 : F_0 \oplus F_1 \to F_0$ by $j_0(x,y) := x + \partial(y)$ and $j_1(x,y) := x$. By functoriality we derive an exact sequence

$$\Lambda^p(F_0 \oplus F_1) \rightrightarrows \Lambda^p(F_0) \longrightarrow \Lambda^p(N) \longrightarrow 0$$

which reduces the assertion to the flat case. For $\mathscr{F}_p := \Gamma_R^p$ the same reduction argument works as well (cp. [67, p.284]) and for flat modules the assertion for Γ_R^p follows from the corresponding assertion for TS^p. $\qquad\square$

Lemma 8.1.19. *Let A be a simplicial almost algebra, L, E and F three A-modules, $f : E \to F$ a quasi-isomorphism. If L is flat or E, F are flat, then $L \otimes_A f : L \otimes_A E \to L \otimes_A F$ is a quasi-isomorphism.*

Proof. It is deduced directly from [50, I.3.3.2.1] by applying $M \mapsto M_!$. $\square$

8.1.20.   As usual, this allows one to show that

$$\otimes : \mathrm{Hot}(A) \times \mathrm{Hot}(A) \to \mathrm{Hot}(A)$$

admits a left derived functor $\overset{L}{\otimes} : \mathsf{D}(A\text{-}\mathbf{Mod}) \times \mathsf{D}(A\text{-}\mathbf{Mod}) \to \mathsf{D}(A\text{-}\mathbf{Mod})$. If R is a simplicial V-algebra, then we have essentially commutative diagrams

$$
\begin{array}{ccc}
\mathsf{D}(R\text{-}\mathbf{Mod}) \times \mathsf{D}(R\text{-}\mathbf{Mod}) & \xrightarrow{\ \overset{L}{\otimes}\ } & \mathsf{D}(R\text{-}\mathbf{Mod}) \\
\updownarrow & & \updownarrow \\
\mathsf{D}(R^a\text{-}\mathbf{Mod}) \times \mathsf{D}(R^a\text{-}\mathbf{Mod}) & \xrightarrow{\ \overset{L}{\otimes}\ } & \mathsf{D}(R^a\text{-}\mathbf{Mod})
\end{array}
$$

where again the downward (resp. upward) functors are induced by localization (resp. by $M \mapsto M_!$).

8.1.21.   We mention the derived functors of the non-additive functor $\mathscr{F}_p$ defined above in the simplest case of modules over a constant simplicial ring. Let A be a (commutative) V^a-algebra.

Lemma 8.1.22. *If* $u : X \to Y$ *is a quasi-isomorphism of flat s.A-modules then* $\mathscr{F}_p^a(u) : \mathscr{F}_p^a(X) \to \mathscr{F}_p^a(Y)$ *is a quasi-isomorphism.*

Proof. This is deduced from [50, I.4.2.2.1] applied to $N(X_!) \to N(Y_!)$ which is a quasi-isomorphism of chain complexes of flat $A_{!!}$-modules. We note that *loc. cit.* deals with a more general mixed simplicial construction of $\mathscr{F}_p$ which applies to bounded above complexes, but one can check that it reduces to the simplicial definition for complexes in $\mathbf{C}_{\bullet}(A_{!!})$. $\square$

8.1.23.   Using the lemma one can construct

$$L\mathscr{F}_p^a : \mathsf{D}(s.A\text{-}\mathbf{Mod}) \to \mathsf{D}(s.A\text{-}\mathbf{Mod}).$$

If R is a V-algebra we have the derived category version of the essentially commutative squares (8.1.15), relating $L\mathscr{F}_p : \mathsf{D}(s.R\text{-}\mathbf{Mod}) \to \mathsf{D}(s.R\text{-}\mathbf{Mod})$ and $L\mathscr{F}_p^a : \mathsf{D}(s.R^a\text{-}\mathbf{Mod}) \to \mathsf{D}(s.R^a\text{-}\mathbf{Mod})$.

8.2. Fundamental group of an almost algebra.

We will need some generalities from [44, Exp.V] and [5, Exp.VI]. In the following we fix a universe U and suppose that all our categories are U-categories and all our topoi are U-topoi in the sense of [4, Exp.I, Déf.1.1 and Exp.IV, Déf.1.1]. Especially, **Set** denotes the category of sets in U. Recall that a category is said to be *essentially small* if it is equivalent to a U-small category.

8.2.1.   Let $\mathscr{C}$ be a site. Recall ([5, Exp.VI, Déf.1.1]) that an object X of $\mathscr{C}$ is called *quasi-compact* if, for every covering family $(X_i \to X \mid i \in I)$ there is a finite subset $J \subset I$ such that the subfamily $(X_j \to X \mid j \in J)$ is still covering.

8.2.2. Let E be a topos; in the following we always endow E with its canonical topology ([4, Exp.II, Déf 2.5]), so E is a site in a natural way and the terminology of (8.2.1) applies to the objects of E. Moreover, if $\mathscr{C}$ is any site and $\varepsilon : \mathscr{C} \to \mathscr{C}^{\sim}$ the natural functor to the category $\mathscr{C}^{\sim}$ of sheaves on $\mathscr{C}$, then an object X of $\mathscr{C}$ is quasi-compact in $\mathscr{C}$ if and only if $\varepsilon(X)$ is quasi-compact in $\mathscr{C}^{\sim}$ ([5, Exp.VI, Prop.1.2]).

Furthermore, since in E all finite limits are representable, we can make the following further definitions ([5, Exp.VI, Déf.1.7]). A morphism $f : X \to Y$ in E is called *quasi-compact* if, for every morphism $Y' \to Y$ in E with quasi-compact Y', the object $X \times_Y Y'$ is quasi-compact. We say that f is *quasi-separated* if the diagonal morphism $X \to X \times_Y X$ is quasi-compact. We say that f is *coherent* if it is quasi-compact and quasi-separated.

8.2.3. Let X be an object of a topos E. We say that X is *quasi-separated* if, for every quasi-compact object S of E, every morphism $S \to X$ is quasi-compact. We say that X is *coherent* if it is quasi-compact and quasi-separated ([5, Exp.VI, Déf.1.13]). We denote by E_{coh} the full subcategory of E consisting of all the coherent objects.

Suppose that the object Y of E is coherent and let $f : X \to Y$ be a coherent morphism; by [5, Exp.VI, Prop.1.14(ii)], it follows that X is coherent.

Definition 8.2.4. (cp. [5, Exp.VI, Déf.2.3]) We say that a topos E is *coherent* if it satisfies the following conditions:

  (i) E admits a full generating subcategory $\mathscr{C}$ consisting of coherent objects.
  (ii) Every object X of $\mathscr{C}$ is quasi-separated over the final object of E, *i.e.* the diagonal morphism $X \to X \times X$ is quasi-compact.
  (iii) The final object of E is coherent.

8.2.5. If E is a coherent topos, then E_{coh} is stable under arbitrary finite limits (of E) ([5, Exp.VI, 2.2.4]). Moreover, a topos E is coherent if and only if it is equivalent to a topos of the form $\mathscr{C}^{\sim}$, where $\mathscr{C}$ is a small site whose objects are quasi-compact and whose finite limits are representable.

It is possible to characterize the categories of the form E_{coh} arising from a coherent topos : this leads to the following definition.

Definition 8.2.6. A *pretopos* is an essentially small category $\mathscr{C}$ satisfying the following conditions ([5, Exp.VI, Exerc.3.11]).

  (PT1) All finite limits are representable in $\mathscr{C}$.
  (PT2) All finite sums are representable in $\mathscr{C}$ and they are universal and disjoint.
  (PT3) Every equivalence relation in $\mathscr{C}$ is effective, and every epimorphism is universally effective.

8.2.7. As in [5, Exp.VI, Exerc.3.11], we leave to the reader the verification that, for every coherent topos E, the subcategory E_{coh} is a pretopos, and E induces on E_{coh} the *precanonical* topology, *i.e.* the topology whose covering families $(X_i \to X \mid i \in I)$ are those admitting a finite subfamily which is covering for the canonical topology of E_{coh}. One deduces that E is equivalent to $(E_{\text{coh}})^\sim$, the topos of sheaves on the precanonical topology of E_{coh}.

Conversely, if $\mathscr{C}$ is a pretopos, let $E := \mathscr{C}^\sim$ be the topos of sheaves on the precanonical topology of $\mathscr{C}$; then E is a coherent topos and the natural functor $\varepsilon : \mathscr{C} \to E$ induces an equivalence of $\mathscr{C}$ with E_{coh}.

8.2.8.   Furthermore, if $\mathscr{C}$ is a pretopos (endowed with the precanonical topology), the natural functor $\mathscr{C} \to \mathscr{C}^\sim$ commutes with finite sums, with quotients under equivalence relations, and it is left exact (*i.e.* commutes with finite limits) ([5, Exp.VI, Exerc.3.11]).

8.2.9.   Recall ([4, Exp.IV, Déf.6.1]) that a *point* of a topos E is a morphism of topoi

$$p : \mathbf{Set} \to E$$

(where one views the category **Set** as the topos of sheaves on the one-point topological space). By [4, Exp.IV, Cor.1.5], the assignment $p \mapsto p^*$ defines an equivalence from the category of points of E to the opposite of the category of all functors $F : E \to \mathbf{Set}$ that commute with all colimits and are left exact. A functor $F : E \to \mathbf{Set}$ with these properties is called a *fibre functor* for E. By [4, Exp.IV, Cor.1.7], a functor $E \to \mathbf{Set}$ is a fibre functor if and only if it is left exact and it takes covering families to surjective families.

8.2.10.   By a theorem of Deligne ([5, Exp.VI, Prop.9.0]) every coherent non-empty topos admits at least one fibre functor. (Actually Deligne's result is both more precise and more general, but for our purposes, the foregoing statement will suffice.)

In several contexts, it is useful to attach fibre functors to categories that are not quite topoi. These situations are axiomatized in the following definition.

Definition 8.2.11. A *Galois category* ([44, Exp.V, §5]) is the datum of an essentially small category $\mathscr{C}$ and a functor F from $\mathscr{C}$ to the category **f.Set** of finite sets, satisfying the following conditions:

(G1) all finite limits exist in $\mathscr{C}$ (in particular $\mathscr{C}$ has a final object).

(G2) Finite sums exist in $\mathscr{C}$ (in particular $\mathscr{C}$ has an initial object). Also, for every object X of $\mathscr{C}$ and every finite group G of automorphisms of X, the quotient X/G exists in $\mathscr{C}$.

(G3) Every morphism $u : X \to Y$ in $\mathscr{C}$ factors as a composition $X \xrightarrow{u'} Y' \xrightarrow{u''} Y$, where u' is a strict epimorphism and u'' is both a monomorphism and an isomorphism onto a direct summand of Y.

(G4) The functor F is left exact.

(G5) F commutes with finite direct sums and with quotients under actions of finite groups of automorphisms. Moreover F takes strict epimorphisms to epimorphisms.

(G6) Let $u : X \to Y$ be a morphism of $\mathscr{C}$. Then u is an isomorphism if and only if $F(u)$ is.

8.2.12. Given a Galois category $(\mathscr{C}, F)$, one says that F is a *fibre functor* of $\mathscr{C}$. It is shown in [44, Exp.V, §4] that for any Galois category $(\mathscr{C}, F)$ the functor F is pro-representable and its automorphism group is a profinite group π in a natural way. Furthermore, F factors naturally through the forgetful functor π-f.Set $\rightarrow$ f.Set from the category π-f.Set of (discrete) finite sets with continuous π-action, and the resulting functor $\mathscr{C} \rightarrow \pi$-f.Set is an equivalence.

8.2.13. The category of étale coverings $\mathbf{Cov}(A)$ of an almost algebra A (to be defined in (8.2.22)) is not directly presented as a Galois category, since it does not afford an a priori choice of fibre functor; rather, the existence of a fibre functor is deduced from Deligne's theorem. The argument only appeals to some general properties of the category $\mathbf{Cov}(A)$, which are abstracted in the following definition 8.2.14 and lemma 8.2.15.

Definition 8.2.14. A *pregalois category* is a category $\mathscr{C}$ satisfying the following conditions.

(PG1) Every monomorphism $X \rightarrow Y$ in $\mathscr{C}$ induces an isomorphism of X onto a direct summand of Y.

(PG2) $\mathscr{C}$ admits a final object e which is connected and non-empty (that is, e is not an initial object).

(PG3) For every object X of $\mathscr{C}$, there exists $n \in \mathbb{N}$ such that, for every non-empty object Y of $\mathscr{C}$, the product $X \times Y$ exists in $\mathscr{C}$ and is not Y-isomorphic to an object of the form $(\overset{n}{\underset{}{\amalg}} Y) \amalg Z$ (where Z is an object of $\mathscr{C}/Y$).

Lemma 8.2.15. *Let $\mathscr{C}$ be a pregalois pretopos. Then there exists a functor $F : \mathscr{C} \rightarrow$ f.Set such that $(\mathscr{C}, F)$ is a Galois category.*

Proof. (G1) holds because it is the same as (PT1). (G2) follows easily from (PT1), (PT2) and (PT3). In order to show (G3) we will need the following:

Claim 8.2.16. A morphism $u : X \rightarrow Y$ of $\mathscr{C}$ is an isomorphism if and only if it is both a monomorphism and an epimorphism.

Proof of the claim. It follows from (PT3) and the general categorical fact that a morphism is an epimorphism if and only if it is both a monomorphism and a strict epimorphism. ◇

Now, let $u : X \rightarrow Y$ be a morphism in $\mathscr{C}$; the induced morphisms

$$\mathrm{pr}_1, \mathrm{pr}_2 : X \times_Y X \rightrightarrows X$$

define an equivalence relation; by (PT3) there is a corresponding quotient morphism $u' : X \rightarrow Y'$ and moreover u' is a strict epimorphism. Clearly u factors via a morphism $u'' : Y' \rightarrow Y$. We need to show that u'' is a monomorphism, or equivalently, that the induced diagonal morphism $\delta : Y' \rightarrow Y' \times_Y Y'$ is an isomorphism. However, there is a natural commutative diagram

$$X \times_{Y'} X \xrightarrow{\ \alpha\ } X \times_Y X$$

$$\downarrow \qquad\qquad \downarrow$$

$$Y' \xrightarrow{\ \delta\ } Y' \times_Y Y'$$

where α is an isomorphism by construction and by (PT3) both vertical arrows are epimorphisms. It follows that δ is an epimorphism; since it is obviously a monomorphism as well, we deduce (G3) in view of (PG1) and claim 8.2.16. Let $\mathscr{C}^\sim$ be the topos of sheaves on the precanonical topology of $\mathscr{C}$; by (8.2.7), (PG2) and Deligne's theorem (8.2.10), there exists a fibre functor $\mathscr{C}^\sim \to \mathbf{Set}$. Composing with the natural functor $\mathscr{C} \to \mathscr{C}^\sim$ we obtain a functor $F : \mathscr{C} \to \mathbf{Set}$ which is left exact, commutes with finite sums and quotients under equivalence relations, in view of (8.2.8), so (G4) and (G5) hold for F.

Claim 8.2.17. Let X be a non-empty object of $\mathscr{C}$. Then $F(X) \neq \varnothing$.

Proof of the claim. Note that by (PT2) there is no morphism from X to the initial object. Since we know already that (G3) holds in $\mathscr{C}$, we deduce using (PG2) that the unique morphism $X \to e$ is an epimorphism. Then (PT3) says that e can be written as the quotient of X under the induced equivalence relation $\mathrm{pr}_1, \mathrm{pr}_2$: $X \times_e X \rightrightarrows X$. Since F commutes with quotients under such equivalence relations, the claim follows after remarking that $F(e) \neq \varnothing$. $\qquad\qquad \Diamond$

Claim 8.2.18. Let $u : X \to Y$ be a morphism in $\mathscr{C}$ such that $F(u)$ is surjective. Then u is an epimorphism.

Proof of the claim. We use (G3) to factor u as an epimorphism $u' : X \to Y'$ followed by a monomorphism of the form $Y' \to Y' \amalg Z$. We need to show that $Z = \varnothing$ or equivalently, in view of claim 8.2.17, that $F(Z) = \varnothing$. However, the assumption implies that $F(Y')$ maps onto $F(Y' \amalg Z)$; on the other hand, F commutes with finite sums, so the claim holds. $\qquad\qquad \Diamond$

Claim 8.2.19. Let $u : X \to Y$ be a morphism in $\mathscr{C}$ such that $F(u)$ is injective. Then u is a monomorphism.

Proof of the claim. The assumption means that the induced diagonal map $F(X) \to F(X) \times_{F(Y)} F(X)$ is bijective. Then claim 8.2.18 implies that the diagonal morphism $X \to X \times_Y X$ is an epimorphism. The latter is also obviously a monomorphism, hence an isomorphism, in view of claim 8.2.16; but this means that u is a monomorphism. $\qquad\qquad \Diamond$

Now, taking into account claims 8.2.16, 8.2.18 and 8.2.19 we deduce that (G6) holds for F. It remains only to show that F takes values in finite sets. So, suppose by contradiction that $F(X)$ is an infinite set for some object X of $\mathscr{C}$. We define inductively a sequence of objects $(Y_i \mid i \in \mathbb{N})$, with morphisms $\phi_{i+1} : Y_{i+1} \to Y_i$ for every $i \in \mathbb{N}$, as follows. Let $Y_0 := e$, $Y_1 := X$ and ϕ_1 the unique morphism. Let then $i > 0$ and suppose that Y_i and ϕ_i have already been given. Using the diagonal morphism, we can write $Y_i \times_{Y_{i-1}} Y_i \simeq Y_i \amalg Z$ for some object Z; we set

$Y_{i+1} := Z$ and let ϕ_{i+1} be the restriction of the projection $\text{pr}_1 : Y_i \times_{Y_{i-1}} Y_i \to Y_i$. Notice that $F(Y_i) \neq \varnothing$ for every $i \in \mathbb{N}$ (indeed all the fibers of the induced map $F(Y_i) \to F(Y_{i-1})$ are infinite whenever $i > 0$); in particular $Y_i \neq \varnothing$ for every $i \in \mathbb{N}$. On the other hand, for every $n > 0$, $X \times Y_n$ admits a decomposition of the form $(\overset{n}{\amalg} Y_n) \amalg Y_{n+1}$, which is against (PG3). The contradiction concludes the proof of the lemma. □

8.2.20. Let E be a topos. Recall that an object X of E is said to be *constant* if it is a direct sum of copies of the final object e of E ([6, Exp.IX, §2.0]). The object X is *locally constant* if there exists a covering $(Y_i \to e \mid i \in I)$ of e, such that, for every $i \in I$, the restriction $X \times Y_i$ is constant on the induced topos $E_{/Y_i}$ ([4, Exp.IV, §5.1]). If additionally there exists an integer n such that each $X \times Y_i$ is a direct sum of at most n copies of Y_i, then we say that X is *bounded*.

Lemma 8.2.21. *Let E be a topos. Denote by E_{lcb} the full subcategory of all locally constant bounded objects of E. Then:*

(i) *E_{lcb} is a pretopos.*
(ii) *If the final object of E is connected and non-empty, E_{lcb} is pregalois.*

Proof. Let X be an object of E_{lcb}, and $(Y_i \to e \mid i \in I)$ a covering of the final object of E by non-empty objects, such that $X \times Y_i$ is constant on $E_{/Y_i}$ for every $i \in I$, say $X \times Y_i \simeq Y_i \times S_i$, where S_i is some set. Since X is bounded, there exists $n(X) \geq 0$ such that the cardinality of every S_i is no greater than $n(X)$. Since all finite limits are representable in E, in order to check that axiom (PT1) holds, it suffices to show that the limit of any finite inverse system $\underline{Z} := (Z_j \mid j \in J)$ of objects of E_{lcb} is in E_{lcb}. Set $n(\underline{Z}) := \max(n(Z_j) \mid j \in J)$. The fact that $\lim_J \underline{Z}$ is locally constant can be checked locally on E, so we can reduce to the case of an inverse system $\underline{W} := (W_j \mid j \in J)$ of constant bounded objects $W_j := e \times S_j$, where the cardinality of every S_j does not exceed $n(\underline{Z})$. Furthermore, thanks to [6, Exp.IX, Lemme 2.1(i)] we can assume that the inverse system is induced from an inverse system $\underline{S} := (S_j \mid j \in J)$ of maps of sets, in which case it is easy to check that $\lim_J \underline{W} \simeq e \times \lim_J \underline{S}$; moreover, it is then clear that the cardinality of $\lim_J \underline{S}$ is at most the product of the cardinalities of the S_j, which implies (PT1).

(PT2) is immediate. Notice that the foregoing argument also shows that E_{lcb} is closed under finite colimits. However, for every category $\mathscr{C}$ in which finite colimits exist, one can characterize the epimorphisms of $\mathscr{C}$ as those morphisms $u : X \to Y$ such that the two natural morphisms $Y \to \text{colim}\,(Y \overset{u}{\leftarrow} X \overset{u}{\rightarrow} Y)$ coincide. Clearly this condition holds in E_{lcb} if and only if it holds in E. In other words, the (categorical) epimorphisms of E_{lcb} are epimorphisms of E, and we deduce easily that all epimorphisms in E_{lcb} are universally effective.

Similarly, if R is a locally constant bounded equivalence relation on an object X of E_{lcb}, then X/R is again in E_{lcb}; indeed, since equivalence relations in E are universally effective, this can be checked locally on a covering $(Y_i \to e \mid i \in I)$. Then again, by [6, Exp.IX, Lemme 2.1(i)] we can reduce to the case of an equivalence relation on sets, where everything is obvious. This shows that (PT3) holds, and proves (i). Suppose next that e is connected and non-empty; since e is in E_{lcb}, it follows that (PG2) holds in E_{lcb}. To show (PG1), consider a morphism $u : X \to Y$ in E_{lcb}. As in the foregoing, we can find a covering $(Z_i \to e \mid i \in I)$ such that $X \times Z_i \simeq Z_i \times S_i$, $Y \times Z_i \simeq Z_i \times T_i$ for some sets S_i, T_i, and $u_{|Z_i}$ is induced by a map $u_i : S_i \to T_i$. Especially, one deduces easily that $u(X)$ is in E_{lcb}. Let $S'_i := T_i \setminus u_i(S_i)$. Clearly we have an isomorphism $Y \times Z_i \simeq (u(X) \times Z_i) \amalg (Z_i \times S'_i)$ for every $i \in I$. Since the induced decompositions agree on $Z_i \times Z_j$ for every $i, j \in I$, the constant objects $Z_i \times S'_i$ glue to a locally constant object X' of E, and u induces an isomorphism $Y \simeq u(X) \amalg X'$. Finally, if X is in E_{lcb}, find sets $(S_i \mid i \in I)$ and a covering $(Y_i \to e \mid i \in I)$ by non-empty objects such that $X \times Y_i \simeq S_i \times Y_i$, and let m be the maximum of the cardinalities of the sets S_i. Clearly (PG3) holds for X, if one chooses $n := m + 1$. $\square$

8.2.22.   We work with universes $U \in U'$. Let A be an almost algebra in U, and consider the U'-site $\mathscr{S}_A := (A\text{-}\mathbf{Alg})^o_{\text{fpqc}}$ of affine A-schemes in U with the fpqc topology (see (4.4.23)), as well as the corresponding U'-topos $\mathscr{S}_A^{\sim}$. Moreover, the category of *étale coverings* of Spec A is defined as the full subcategory $\mathbf{Cov}(\text{Spec } A)$ of the category of affine A-schemes in U consisting of all étale A-schemes of finite rank. (It is essentially small and essentially independent of U.)

Proposition 8.2.23. *The natural functor* $\varepsilon : \mathscr{S}_A \to \mathscr{S}_A^{\sim}$ *induces an equivalence of the category* $\mathbf{Cov}(\text{Spec } A)$ *to the category of locally constant bounded sheaves on* $\mathscr{S}_A$.

Proof. Let B be an almost finite projective and étale A-algebra of finite rank $\leq r$. By proposition 4.3.27 there is an isomorphism of almost algebras $A \simeq \prod_{i=0}^{r} A_i$ such that $B_i := B \otimes_A A_i$ is an A_i-algebra of constant rank i for every $i = 0, ..., r$. In particular, $B_0 = 0$ and B_i is a faithfully flat étale A_i-algebra for every $i > 0$. We use the diagonal morphism to obtain a decomposition $B_i \otimes_{A_i} B_i \simeq B_i \times C_i$, where C_i is again an étale B_i-algebra of constant rank $i - 1$. Iterating this procedure we find faithfully flat A_i-algebras D_i such that $B \otimes_A D_i$ is D_i-isomorphic to a direct product of i copies of D_i, for every $i > 0$. Setting $D_0 := A_0$, we obtain a covering $(\varepsilon(\text{Spec } D_i) \to \text{Spec } A) \mid i = 0, ..., r)$ of $\varepsilon(\text{Spec } A)$ in $\mathscr{S}_A^{\sim}$ such that the restriction of $\varepsilon(\text{Spec } B)$ to each $\varepsilon(\text{Spec } D_i)$ is a bounded constant sheaf. This shows that the restriction of ε to $\mathbf{Cov}(\text{Spec } A)$ lands in the category of locally constant bounded sheaves. Since the fpqc topology is coarser than the canonical topology, ε is fully faithful. To show that ε is essentially surjective amounts to an exercise in faithfully flat descent : clearly every constant sheaf is represented by an étale A-scheme; then one uses [6, Exp.IX, Lemme 2.1(i)] to show that any descent datum of bounded constant sheaves is induced by a cocycle system of morphisms for the corresponding representing algebras, and one can descend the latter. We leave the details to the reader. $\square$

8.2.24.   Let us say that an affine almost scheme S is *connected* if $\varepsilon(S)$ is a connected object of the category $\mathscr{S}_{\mathscr{O}_S}^{\sim}$, which simply means that $\mathscr{O}_S$ in non-zero and the only non-zero idempotent of $\mathscr{O}_{S*}$ is the identity. In this case, proposition 8.2.23

and lemma 8.2.21 show that $\mathbf{Cov}(S)$ is a pregalois pretopos, hence it admits a fibre functor $F : \mathbf{Cov}(S) \to \mathbf{f.Set}$ by lemma 8.2.15.

Definition 8.2.25. Suppose that the affine almost scheme S is connected. The *fundamental group* of S is the group $\pi_1(S)$ defined as the automorphism group of any fibre functor $F : \mathbf{Cov}(S) \to \mathbf{f.Set}$, endowed with its natural profinite topology (see (8.2.12)).

8.2.26. It results from the general theory ([44, Exp.V, §5]) that $\pi_1(S)$ is independent (up to isomorphism) on the choice of fibre functor. We refer to *loc.cit.* for a general study of fundamental groups of Galois categories. In essence, several of the standard results for schemes admit adequate almost counterpart. We conclude this section with a sample of such statements.

Definition 8.2.27. Let $A \subset B$ be a pair of V^a-algebras; the *integral closure* of A in B is the subalgebra $\mathrm{i.c.}(A, B) := \mathrm{i.c.}(A_*, B_*)^a$, where for a pair of rings $R \subset S$ we let $\mathrm{i.c.}(R, S)$ be the integral closure of R in S.

Lemma 8.2.28. *If $R \subset S$ is a pair of V-algebras, then* $\mathrm{i.c.}(R^a, S^a) = \mathrm{i.c.}(R, S)^a$.

Proof. It suffices to show the following:

Claim 8.2.29. Given a commutative diagram of V-algebras

$$
\begin{array}{ccc}
R & \longrightarrow & S \\
\downarrow & & \downarrow \\
R_1 & \longrightarrow & S_1
\end{array}
$$

whose vertical arrows are almost isomorphism, the induced map $\mathrm{i.c.}(R, S) \to \mathrm{i.c.}(R_1, S_1)$ is an almost isomorphism.

Proof of the claim. Clearly the kernel of the map is almost zero; it remains to show that for every $b \in \mathrm{i.c.}(R_1, S_1)$ and $\varepsilon \in \mathfrak{m}$, the element εb lifts to $\mathrm{i.c.}(R, S)$. By assumption we have a relation $b^N + \sum_{i=1}^N a_i b^{N-i} = 0$, with $a_i \in R_1$, so $(\varepsilon b)^N + \sum_{i=1}^N \varepsilon^i a_i (\varepsilon b)^{N-i} = 0$. By lifting $\varepsilon^i a_i$ to some $\bar{a}_i \in R$, we get a monic polynomial $P(T)$ over R such that $P(\varepsilon b) = 0$, so if $\bar{b}$ is a lifting of εb, we have $\mathfrak{m} \cdot P(\bar{b}) = 0$. Since the restriction $\mathfrak{m}S \to \mathfrak{m}S_1$ is surjective, we can choose $\bar{b} \in \mathfrak{m}S$, so $\bar{b} \cdot P(\bar{b}) = 0$. $\square$

Remark 8.2.30. (i) If $A \subset B$ are V^a-algebras, then A is integrally closed in B if and only if A_* is integrally closed in B_*. Indeed, the "if" part holds by definition; for the "only if" part, we know that the integral closure of A_* in B_* is almost equal to A_* and any such V-subalgebra must be contained in A_*.

(ii) If $(A_i \subset B_i \mid i \in I)$ is a (possibly infinite) family of pairs of V^a-algebras, then $\prod_{i \in I} A_i$ is integrally closed in $\prod_{i \in I} B_i$ if and only if A_i is integrally closed in B_i for every $i \in I$.

The following proposition is an analogue of [30, Prop.18.12.15].

Proposition 8.2.31. *Let $A \subset B$ be a pair of V^a-algebras such that $A = \text{i.c.}(A, B)$.*

   (i) *For any étale almost finite projective A-algebra A_1 of almost finite rank we have $A_1 = \text{i.c.}(A_1, A_1 \otimes_A B)$.*

  (ii) *Suppose that A and B are connected, and choose a fibre functor F for the category $\mathbf{Cov}(B)$. Then the functor*

$$\mathbf{Cov}(A) \to \mathbf{f.Set} \quad : \quad C \mapsto F(C \otimes_A B)$$

is a fibre functor, and the induced group homomorphism:

$$\pi_1(\text{Spec } B) \to \pi_1(\text{Spec } A)$$

   is surjective.

Proof. (i): Using remark 8.2.30(ii) we reduce to the case where A_1 has constant finite rank over A. Set $B_1 := A_1 \otimes_A B$, and suppose that $x \in B_{1*}$ is integral over A_{1*}, or equivalently, over A_*. Consider the element $e \in (A_1 \otimes_A A_1)_*$ provided by proposition 3.1.4; for given $\varepsilon \in \mathfrak{m}$ write $\varepsilon \cdot e = \sum_{i=1}^{k} c_i \otimes d_i$ for some $c_i, d_i \in A_{1*}$. According to remark 4.1.17 we have: $\sum_{i=1}^{k} c_i \cdot \text{Tr}_{B_1/B}(x d_i) = \varepsilon \cdot x$ for every $x \in B_{1*}$. Corollary 4.4.31 implies that $\text{Tr}_{B_1/B}(x d_i)$ is integral over A_* for every $i \leq k$, hence it lies in A_*, by remark 8.2.30(i). Hence $\varepsilon \cdot x \in A_{1*}$, as claimed.

   (ii) is an immediate consequence of (i) and of the general theory of [44, Exp.V, §5].   □

8.2.32. As a special case, let R be a V-algebra whose spectrum is connected; we suppose that there exists an element $\varepsilon \in \mathfrak{m}$ which is regular in R. Suppose moreover that R is integrally closed in $R[\varepsilon^{-1}]$, consequently $\text{Spec } R[\varepsilon^{-1}]$ is connected and $\pi_1(\text{Spec } R[\varepsilon^{-1}])$ is well defined. It follows as well that $\text{Spec } R^a$ is connected. Indeed, if $\text{Spec } R^a$ were not connected, neither would be $\text{Spec } R^a_*$; but since $R^a_* \subset R[\varepsilon^{-1}]$, this is absurd. Then $\pi_1(\text{Spec } R^a)$ is also well defined and, after a fibre functor for $\mathbf{Cov}(R[\varepsilon^{-1}])$ is chosen, the functors $\mathbf{Cov}(\text{Spec } R) \to \mathbf{Cov}(\text{Spec } R^a) \to \mathbf{Cov}(\text{Spec } R[\varepsilon^{-1}]) : B \mapsto B^a \mapsto B^a_*[\varepsilon^{-1}]$ induce continuous group homomorphisms ([44, Exp.V, Cor.6.2])

$$(8.2.33) \qquad \pi_1(\text{Spec } R[\varepsilon^{-1}]) \to \pi_1(\text{Spec } R^a) \to \pi_1(\text{Spec } R).$$

Proposition 8.2.34. *Under the assumptions of (8.2.32), we have:*

   (i) *R^a_* is integrally closed in $R[\varepsilon^{-1}]$.*

  (ii) *The maps (8.2.33) are surjective.*

Proof. (i): By assumption and lemma 8.2.28, R^a is integrally closed in $R[\varepsilon^{-1}]^a$, so the assertion follows from remark 8.2.30(i).

   (ii): It suffices to show the assertion for the leftmost map and for the composition of the two maps. However, the composition of the two maps is actually a special case of the leftmost map (for the classical limit $V = \mathfrak{m}$), so we need only consider the leftmost map. Then the assertion follows from (i) and proposition 8.2.31(ii).   □

REFERENCES

[1]   S.ANANTHARAMAN, Schémas en groupes, espaces homogènes et espaces algébriques sur une base de dimension 1. *Bull. Soc. Math. France* 33 (1973) pp.5–79.

[2]   M.ANDRÉ, Homologie des algèbres commutatives. *Springer Grundl. Math. Wiss.* 206 (1974).

[3]   M.ANDRÉ, Localisation de la lissité formelle. *Manuscr. Math.* 13 (1974) pp.297–307.

[4]   M.ARTIN ET AL., Théorie des topos et cohomologie étale des schémas – tome 1. *Springer Lect. Notes Math.* 269 (1972).

[5]   M.ARTIN ET AL., Théorie des topos et cohomologie étale des schémas – tome 2. *Springer Lect. Notes Math.* 270 (1972).

[6]   M.ARTIN ET AL., Théorie des topos et cohomologie étale des schémas – tome 3. *Springer Lect. Notes Math.* 305 (1973).

[7]   W.BARTENWERFER, Die höheren metrischen Kohomologiegruppen affinoider Räume. *Math. Ann.* 241 (1979) pp.11–34.

[8]   H.BASS, Algebraic K-theory. *W.A. Benjamin* (1968).

[9]   A.BEAUVILLE, Y.LASZLO, Un lemme de descente. *C.R. Acad. Sc. Paris* 320 (1995) pp.335–340.

[10] V.BERKOVICH, Spectral theory and analytic geometry over non-archimedean fields. *AMS Math. Surveys and Monographs* 33 (1990).

[11] P.BERTHELOT ET AL., Théorie des Intersection et Théorèmes de Riemann-Roch. *Springer Lect. Notes Math.* 225 (1971).

[12] S.BOSCH, U.GÜNTZER, R.REMMERT, Non-Archimedean analysis. *Springer Grundl. Math. Wiss.* 261 (1984).

[13] S.BOSCH, W.LÜTKEBOHMERT, Formal and rigid geometry I. Rigid spaces. *Math. Ann.* 295 (1993) pp.291–317.

[14] S.BOSCH, W.LÜTKEBOHMERT, M.RAYNAUD, Formal and rigid geometry III. The relative maximum principle. *Math. Ann.* 302 (1995) pp.1–29.

[15] N.BOURBAKI Algèbre. *Hermann* (1970).

[16] N.BOURBAKI, Algèbre Commutative. *Hermann* (1961).

[17] N.BOURBAKI, Algèbre Homologique. *Masson* (1980).

[18] N.BOURBAKI, Topologie Générale. *Hermann* (1971).

[19] F.BRUHAT, J.TITS, Groupes réductifs sur un corps local – Chapitre II. Schémas en groupes. Existence d'une donnée radicielle valuée. *Publ. Math. IHES* 60 (1984) pp.1–194.

[20] J.W.S.CASSELS, A.FRÖHLICH, Algebraic number theory. *Academic Press* (1967).

[21] J.COATES, R.GREENBERG, Kummer theory for abelian varieties over local fields. *Invent. Math.* 124 (1996) pp.129–174.

[22] B.CONRAD, Irreducible components of rigid spaces. *Ann. Inst. Fourier* 49 (1999) pp.473–541.

[23] P.DELIGNE, Catégories tannakiennes. *Grothendieck Festschrift vol.II, Birkhauser Progress in Math.* 87 (1990) pp.111–195.

[24] P.DELIGNE, J.MILNE, Tannakian categories. *Springer Lect. Notes Math.* 900 (1982) pp.101–228.

[25] M.DEMAZURE, A.GROTHENDIECK ET AL., Schémas en Groupes I. *Springer Lect. Notes Math.* 151 (1970).

[26] J.DIEUDONNÉ, A.GROTHENDIECK, Éléments de Géométrie Algébrique – Chapitre I. *Publ. Math. IHES* 4 (1960).

[27] J.DIEUDONNÉ, A.GROTHENDIECK, Éléments de Géométrie Algébrique – Chapitre II. *Publ. Math. IHES* 8 (1961).

[28] J.DIEUDONNÉ, A.GROTHENDIECK, Éléments de Géométrie Algébrique – Chapitre IV, partie 1. *Publ. Math. IHES* 20 (1964).

[29] J.DIEUDONNÉ, A.GROTHENDIECK, Éléments de Géométrie Algébrique – Chapitre IV, partie 3. *Publ. Math. IHES* 28 (1966).

[30] J.DIEUDONNÉ, A.GROTHENDIECK, Éléments de Géométrie Algébrique – Chapitre IV, partie 4. *Publ. Math. IHES* 32 (1967).

[31] R.ELKIK, Solutions d'équations à coefficients dans un anneau hensélien. *Ann. Sci. E.N.S.* 6 (1973) pp.553–604.

[32] G.ELLIOTT, On totally ordered groups and K_0. *Springer Lect. Notes Math.* 734 (1979) pp.1–49.

[33] G.FALTINGS, p-adic Hodge theory. *J. Amer. Math. Soc.* 1 (1988) pp.255–299.

[34] G.FALTINGS, Almost étale extensions. *Astérisque* 279 (2002) pp.185–270.

[35] D.FERRAND, Descente de la platitude par un homomorphisme fini. *C.R. Acad. Sc. Paris* 269 (1969) pp.946–949.

[36] D.FERRAND, M.RAYNAUD, Fibres formelles d'un anneau local noethérien. *Ann. Sci. E.N.S.* 3 (1970) pp.295–311.

[37] J.-M.FONTAINE, Sur certains types de représentations p-adiques du groupe de Galois d'un corps local; construction d'un anneau de Barsotti-Tate. *Ann. of Math.* 115 (1982) pp.529–577.

[38] J.FRESNEL, M.MATIGNON, Produit tensoriel topologique de corps valués. *Canadian J. Math.* 35 (1983) pp.218–273.

[39] W.FULTON, Young tableaux. *Cambridge Univ. Press* (1997).

[40] P.GABRIEL, Des catégories abéliennes. *Bull. Soc. Math. France* 90 (1962) pp.323–449.

[41] J.GIRAUD, Cohomologie non abélienne. *Springer Grundl. Math. Wiss.* 179 (1971).

[42] A.GROTHENDIECK, On the de Rham cohomology of algebraic varieties. *Publ. Math. IHES* 29 (1966) pp.95–103.

[43] A.GROTHENDIECK, Groupes de Barsotti-Tate et cristaux de Dieudonné. *Presses de l'Univ. de Montréal – Séminaire de Math. Supérieures* 45 (1974).

[44] A.GROTHENDIECK ET AL., Revêtements Étales et Groupe Fondamental. *Springer Lect. Notes Math.* 224 (1971).

[45] L.GRUSON, Dimension homologique des modules plats sur un anneau commutatif noetherien. *Symposia Mathematica* Vol. XI; Academic Press, London (1973) pp. 243–254.

[46] L.GRUSON, M.RAYNAUD, Critères de platitude et de projectivité. *Invent. Math.* 13 (1971) pp.1–89.

[47] M.HAKIM, Topos annelés et schémas relatifs. *Springer Ergebnisse Math. Grenz.* 64 (1972).

[48] R.HUBER, Bewertungsspektrum und rigide Geometrie. *Regensburger Math. Schriften* 23 (1993).

[49] R.HUBER, Étale cohomology of rigid analytic varieties and adic spaces. *Vieweg Aspects of Math.* 30 (1996).

[50] L.ILLUSIE, Complexe cotangent et déformations I. *Springer Lect. Notes Math.* 239 (1971).

[51] L.ILLUSIE, Complexe cotangent et déformations II. *Springer Lect. Notes Math.* 283 (1972).

[52] K.KATO, Logarithmic structures of Fontaine-Illusie. *Algebraic analysis, geometry, and number theory – Johns Hopkins Univ. Press* (1989) pp.191–224.

[53] G.KEMPF ET AL., Toroidal embeddings I. *Springer Lect. Notes Math.* 339 (1973).

[54] S.LANG, Algebra – Third edition. *Addison-Wesley* (1993).

[55] D.LAZARD, Autour de la platitude. *Bull. Soc. Math. France* 97 (1969) pp.81–128.

[56] M.LAZARD, Commutative formal groups. *Springer Lect. Notes Math.* 443 (1975).

[57] S.MAC LANE, Categories for the working mathematician. *Springer Grad. Text Math.* 5 (1971).

[58] H.MATSUMURA, Commutative ring theory. *Cambridge Univ. Press* (1986).

[59] B.MITCHELL, Rings with several objects. *Advances in Math.* 8 (1972) pp.1–161.

[60] L.MORET-BAILLY, Un problème de descente. *Bull. Soc. Math. France* 124 (1996) pp.559–585.

[61] D.MUMFORD, Abelian varieties. *Oxford U.Press* (1970).

[62] A.NEEMAN, A counterexample to a 1961 "theorem" in homological algebra. *Invent. Math.* 148 (2002) pp.397–420.

[63] W.NIZIOL, Crystalline conjecture via K-theory. *Ann. Sci. E.N.S.* 31 (1998) pp.659–681.

[64] J.-P.OLIVIER, Descente par morphismes purs. *C.R. Acad. Sc. Paris* 271 (1970) pp.821–823.

[65] M.RAYNAUD, Faisceaux amples sur les schémas en groupes et les espaces homogènes. *Springer Lect. Notes Math.* 119 (1970).

[66] M.RAYNAUD, Anneaux locaux henséliens. *Springer Lect. Notes Math.* 169 (1970).

[67] N.ROBY, Lois polynômes et lois formelles en théorie des modules. *Ann. Sci. E.N.S.* 80 (1963) pp.213–348.

[68] J.-E.ROOS, Caractérisation des catégories qui sont quotients des catégories de modules par des sous-catégories bilocalisantes. *C.R. Acad. Sc. Paris* 261 (1965) pp.4954–4957.

[69] J.-E.ROOS, Sur les foncteurs dérivés des produits infinis dans les catégories de Grothendieck. Exemples et contre-exemples. *C.R. Acad. Sc. Paris* 263 (1966) pp.A895–A898.

[70] A.ROSENBERG, Noncommutative algebraic geometry and representations of quantized algebras. *Kluwer Math. and its Applications* 330 (1995).

[71] P.SAMUEL, O.ZARISKI, Commutative algebra vol.II. *Springer Grad. Text Math.* 29 (1975).

[72] J.-P.SERRE, Groupes de Grothendieck des schémas en groupes réductifs déployes. *Publ. Math. IHES* 34 (1968) pp.37–52.

[73] H.SUMIHIRO, Equivariant completion II. *J. Math. Kyoto*, 15 (1975) pp.573–605.

[74] J.TATE, p-divisible groups. *Proc. conf. local fields, Driebergen* (1967) pp.158–183.

[75] T.TSUJI, p-adic étale cohomology and crystalline cohomology in the semi-stable reduction case. *Invent. Math.* 137 (1999) pp.233–411.

[76] M.VAN DER PUT, Cohomology on affinoid spaces. *Compositio. Math.*, 45 (1982) pp.165–198.

[77] C.WEIBEL, An introduction to homological algebra. *Cambridge Univ. Press* (1994).

[78] O.ZARISKI, Reduction of the singularities of algebraic 3-dimensional varieties. *Ann. of Math.* (1944) pp.472–542.

Printing and Binding: Strauss GmbH, Mörlenbach

Lecture Notes in Mathematics

For information about Vols. 1–1639
please contact your bookseller or Springer-Verlag

Vol. 1788: A. Vasil'ev, Moduli of Families of Curves for Conformal and Quasiconformal Mappings.IX, 211 pages. 2002.

Vol. 1789: Y. Sommerhäuser, Yetter-Drinfel'd Hopf algebras over groups of prime order. V, 157 pages. 2002.

Vol. 1790: X. Zhan, Matrix Inequalities. VII, 116 pages. 2002.

Vol. 1791: M. Knebusch, D. Zhang, Manis Valuations and Prüfer Extensions I: A new Chapter in Commutative Algebra. VI, 267 pages. 2002.

Vol. 1792: D. D. Ang, R. Gorenflo, V. K. Le, D. D. Trong, Moment Theory and Some Inverse Problems in Potential Theory and Heat Conduction. VIII, 183 pages. 2002.

Vol. 1793: J. Cortés Monforte, Geometric, Control and Numerical Aspects of Nonholonomic Systems. XV, 219 pages. 2002.

Vol. 1794: N. Pytheas Fogg, Substitution in Dynamics, Arithmetics and Combinatorics. Editors: V. Berthé, S. Ferenczi, C. Mauduit, A. Siegel. XVII, 402 pages. 2002.

Vol. 1795: H. Li, Filtered-Graded Transfer in Using Noncommutative Gröbner Bases. IX, 197 pages. 2002.

Vol. 1796: J.M. Melenk, hp-Finite Element Methods for Singular Perturbations. XIV, 318 pages. 2002.

Vol. 1797: B. Schmidt, Characters and Cyclotomic Fields in Finite Geometry. VIII, 100 pages. 2002.

Vol. 1798: W.M. Oliva, Geometric Mechanics. XI, 270 pages. 2002.

Vol. 1799: H. Pajot, Analytic Capacity, Rectifiability, Menger Curvature and the Cauchy Integral. XII,119 pages. 2002.

Vol. 1800: O. Gabber, L. Ramero, Almost Ring Theory. VI, 307 pages. 2003.

Vol. 1801: J. Azéma, M. Émery, M. Ledoux, M. Yor, Séminaire de Probabilités XXXVI. VIII, 499 pages. 2003.

Vol. 1802: V. Capasso, E. Merzbach, B.G. Ivanoff, M. Dozzi, R. Dalang, T. Mountford, Topics in Spatial Stochastic Processes. Martina Franca, Italy 2001. Editor: E. Merzbach. VIII, 253 pages. 2003.

Vol. 1803: G. Dolzmann, Variational Methods for Crystalline Microstructure - Analysis and Computation. VIII, 212 pages. 2003.

Vol. 1804: I. Cherednik, Ya. Markov, R. Howe, G. Lusztig, Iwahori-Hecke Algebras and their Representation Theory. Martina Franca, Italy 1999. Editors: V. Baldoni, D. Barbasch. X, 103 pages. 2003.

Vol. 1805: F. Cao, Geometric Curve Evolution and Image Processing. X, 187 pages. 2003.

Vol. 1806: H. Broer, I. Hoveijn. G. Lunther, G. Vegter, Bifurcations in Hamiltonian Systems. Computing Singularities by Gröbner Bases. XIV, 169 pages. 2003.

Vol. 1807: V. D. Milman, G. Schechtman, Geometric Aspects of Functional Analysis. Israel Seminar 2000-2002. VIII, 429 pages. 2003.

Vol. 1808: W. Schindler, Measures with Symmetry Properties.IX, 167 pages. 2003.

Vol. 1809: O. Steinbach, Stability Estimates for Hybrid Coupled Domain Decomposition Methods. VI, 120 pages. 2003.

Vol. 1810: J. Wengenroth, Derived Functors in Functional Analysis. VIII, 134 pages. 2003.

Vol. 1811: J. Stevens, Deformations of Singularities. VII, 157 pages. 2003.

Vol. 1812: L. Ambrosio, K. Deckelnick, G. Dziuk, M. Mimura, V. A. Solonnikov, H. M. Soner, Mathematical Aspects of Evolving Interfaces. Madeira, Funchal, Portugal 2000. Editors: P. Colli, J. F. Rodrigues. X, 237 pages. 2003.

Vol. 1813: L. Ambrosio, L. A. Caffarelli, Y. Brenier, G. Buttazzo, C. Villani, Optimal Transportation and its Applications. Martina Franca, Italy 2001. Editors: L. A. Caffarelli, S. Salsa. X, 164 pages. 2003.

Vol. 1814: P. Bank, F. Baudoin, H. Föllmer, L.C.G. Rogers, M. Soner, N. Touzi, Paris-Princeton Lectures on Mathematical Finance. X,172 pages. 2003.

Vol. 1815: A. M. Vershik (Ed.), Asymptotic Combinatorics with Applications to Mathematical Physics. St. Petersburg, Russia 2001. IX, 246 pages. 2003.

Vol. 1816: S. Albeverio, W. Schachermayer, M. Talagrand, Lectures on Probability Theory and Statistics. Ecole d'Eté de Probabilités de Saint-Flour XXX-2000. Editor: P. Bernard. VIII, 296 pages. 2003.

Vol. 1817: E. Koelink (Ed.), Orthogonal Polynomials and Special Functions. Leuven 2002. X, 249 pages. 2003.

Vol. 1818: M. Bildhauer, Convex Variational Problems with Linear, nearly Linear and/or Anisotropic Growth Conditions. X, 217 pages. 2003.

Vol. 1819: D. Masser, Yu. V. Nesterenko, H. P. Schlickewei, W. M. Schmidt, M. Waldschmidt, Diophantine Approximation. Cetraro, Italy 2000. Editors: F. Amoroso, U. Zannier. XI,353 pages. 2003.

Vol. 1820: F. Hiai, H. Kosaki, Means of Hilbert Space Operators. VIII, 148 pages. 2003.

Vol. 1821: S. Teufel, Adiabatic Perturbation Theory in Quantum Dynamics.VI, 236 pages. 2003.

Vol. 1822: S.-N. Chow, R. Conti, R. Johnson, J. Mallet-Paret, R. Nussbaum, Dynamical Systems. Cetraro, Italy 2000. Editors: J. W. Macki, P. Zecca. XII, 345 pages. 2003.

Vol. 1823: A. M. Anile, W. Allegretto, C. Ringhofer, Mathematical Problems in Semiconductor Physics. Cetraro, Italy 1998. Editor: A. M. Anile. X, 135 pages. 2003.

Vol. 1824: J. A. Navarro González, J. B. Sancho de Salas, C^∞ - Differentiable Spaces. XIII, 188 pages. 2003.

Vol. 1825: J. H. Bramble, A. Cohen, W. Dahmen, Multiscale Problems and Methods in Numerical Simulations, Martina Franca, Italy 2001. Editor: C. Canuto. XII, 155 pages. 2003.

Recent Reprints and New Editions

Vol. 1200: V. D. Milman, G. Schechtman, Asymptotic Theory of Finite Dimensional Normed Spaces. 1986. – Corrected Second Printing. X, 156 pages. 2001.

Vol. 1618: G. Pisier, Similarity Problems and Completely Bounded Maps. 1995 – Second, Expanded Edition VII, 198 pages. 2001.

Vol. 1629: J. D. Moore, Lectures on Seiberg-Witten Invariants. 1997 – Second Edition. VIII, 121 pages. 2001.

Vol. 1638: P. Vanhaecke, Integrable Systems in the realm of Algebraic Geometry. 1996 – Second Edition. X, 256 pages. 2001.

Vol. 1702: J. Ma, J. Yong, Forward-Backward Stochastic Differential Equations and Their Applications. 1999. – Corrected Second Printing. XIII, 270 pages. 2000.